Lecture Notes in Computer Science 12868

More information about this subseries at https://link.springer.com/bookseries/7409

Helen C. Purchase · Ignaz Rutter (Eds.)

Graph Drawing and Network Visualization

29th International Symposium, GD 2021
Tübingen, Germany, September 14–17, 2021
Revised Selected Papers

 Springer

Editors
Helen C. Purchase 🆔
University of Glasgow
Glasgow, UK

Ignaz Rutter 🆔
University of Passau
Passau, Germany

ISSN 0302-9743   ISSN 1611-3349   (electronic)
Lecture Notes in Computer Science
ISBN 978-3-030-92930-5   ISBN 978-3-030-92931-2   (eBook)
https://doi.org/10.1007/978-3-030-92931-2

LNCS Sublibrary: SL3 – Information Systems and Applications, incl. Internet/Web, and HCI

This Springer imprint is published by the registered company Springer Nature Switzerland AG
The registered company address is: Gewerbestrasse 11, 6330 Cham, Switzerland

Preface

This volume contains the papers presented at GD 2021, the 29th International Symposium on Graph Drawing and Network Visualization, held during September 14–17, 2021, in Tübingen, Germany. Graph drawing is concerned with the geometric representation of graphs and constitutes the algorithmic core of network visualization. Graph drawing and network visualization are motivated by applications where it is crucial to visually analyze and interact with relational datasets. Information about the conference series and past symposia is maintained at http://www.graphdrawing.org.

This 2021 conference was held under extraordinary circumstances. After the GD 2020 conference, which sadly had to be held wholly online due to the COVID-19 pandemic, we were delighted to be able to meet our colleagues face-to-face again at a hybrid GD 2021 conference at Universität Tübingen. The credit for this remarkable achievement in such uncertain times goes wholly to Michael A. Bekos and Michael Kaufmann as co-chairs of the Organizing Committee, whose early optimism and determination ensured that the conference went ahead in the form that we all so desired: in-person, with live presentations, poster sessions, and social events. We are much indebted to them.

In effect, the Organizing Committee arranged two conferences, since they were committed to also offering remote attendance, thus making the event accessible to researchers unable to travel to Germany. Remote participation was facilitated through Gather, Zoom, and a live video stream from the conference room in Tübingen. Hybrid conferences may indeed be the way forward in the future – and this Organizing Committee is well aware of the extensive additional effort this model requires.

A total of 70 participants attended the conference in person, with a further 100 registered participants online.

With regards to the program itself, regular papers could be submitted to one of two distinct tracks: Track 1 for papers on combinatorial and algorithmic aspects of graph drawing and Track 2 for papers on experimental, applied, and network visualization aspects. Short papers were given a separate category, which welcomed both theoretical and applied contributions. An additional track was devoted to poster submissions. All the tracks were handled by a single Program Committee. In response to the call for papers, the Program Committee received a total of 74 submissions, consisting of 59 papers (38 in Track 1, 12 in Track 2, and nine in the short paper category) and 15 posters. More than 220 single-blind reviews were provided, with almost a third contributed by external sub-reviewers. After extensive electronic discussions via Easy-Chair, the Program Committee selected 28 papers and 13 posters for inclusion in the scientific program of GD 2021. This resulted in an overall paper acceptance rate of 47% (50% in Track 1, 33% in Track 2, and 56% in the short paper category). Nine of the 30 oral presentations (including two invited talks) were delivered remotely using Zoom; the remaining 21 were delivered on-site. Posters were displayed on-site as well as in Gather.

Authors published an electronic version of their accepted papers on the arXiv e-print repository; a conference index with links to these contributions was made available before the conference.

There were two invited lectures at GD 2021. Kim Marriott from Monash University (Australia) asked "Node Link Diagrams: Are they (actually) useful?", while Meirav Zehavi from Ben-Gurion University (Israel) discussed "Parameterized Complexity in Graph Drawing." Abstracts of both invited lectures are included in these proceedings.

The conference gave out best paper awards in Track 1 and Track 2, as well as a best presentation award and a best poster award. The award for the best paper in Track 1 was given to "Edge-Minimum Saturated k-Planar Drawings" by Steven Chaplick, Fabian Klute, Irene Parada, Jonathan Rollin, and Torsten Ueckerdt, and the award for the best paper in Track 2 was assigned to "A User Study on Hybrid Graph Visualizations" by Emilio Di Giacomo, Walter Didimo, Fabrizio Montecchiani, and Alessandra Tappini. Based on a majority vote of conference participants, the best presentation award was given to Henry Förster for his presentation of the paper "Recognizing and Embedding Simple Optimal 2-Planar Graphs" and the best poster award was given to "The Universe Beyond Planarity" by Michael A. Bekos, Paul Goehring, Michael Kaufmann, and Axel Kuckuk. Many thanks to Springer whose sponsorship funded the prize money for these awards.

A PhD School was held on the two days prior to the conference, attended by 23 participants from five different countries. Four half-day sessions led by five invited lecturers covered both theoretical and practical topics in graph drawing and network visualization, each including a hands-on activity.

As is traditional, the 29th Annual Graph Drawing Contest was held during the conference. The contest was divided into two parts, creative topics and the live challenge. The creative topics task featured two graphs, a Movie Remake graph (modeling remakes of movies by different directors), and an Argumentation Network (a logical reconstruction of an historical scientific debate). The live challenge focused on minimizing the planar polyline edge-length ratio on a fixed grid, with planar undirected inputs. There were two categories: manual and automatic. We thank the Contest Committee, chaired by Philipp Kindermann, for preparing interesting and challenging contest problems. A report about the contest is included in these proceedings.

Many people and organizations contributed to the success of GD 2021. We would like to thank all members of the Program Committee and the external reviewers for carefully reviewing and discussing the submitted papers and posters; this was crucial for putting together a strong and interesting program. Thanks too to all authors who chose GD 2021 as the publication venue for their research.

We are grateful for the support of our "Premium" sponsor yWorks, our "Diamond" sponsor DFG, our "Gold" sponsors Tom Sawyer Software and Universität Tübingen, and our "Bronze" sponsor Springer.

Our special thanks go to all the members of the Organizing Committee based at Universität Tübingen: Michael A. Bekos, Henry Förster, Renate Hallmayer, Michael Kaufmann, Axel Kuckuk, Maximilian Pfister, and Lena Schlipf – they performed a miracle that we did not dare to hope for in early 2021.

The 30th International Symposium on Graph Drawing and Network Visualization (GD 2022) will take place during September 13–16, 2022, in Tokyo, Japan. Reinhard von Hanxleden and Patrizio Angelini will co-chair the Program Committee, and Takayuki Itoh will chair the Organizing Committee.

October 2021   Helen Purchase
                                                      Ignaz Rutter

Organization

Steering Committee

Patrizio Angelini	John Cabot University, Italy
David Auber	LaBRI and Université de Bordeaux, France
Giuseppe Di Battista	Roma Tre University, Italy
Emilio Di Giacomo	University of Perugia, Italy
Reinhard von Hanxleden	University of Kiel, Germany
Stephen G. Kobourov (Chair)	University of Arizona, USA
Anna Lubiw	University of Waterloo, Canada
Helen Purchase	University of Glasgow, UK
Ignaz Rutter	University of Passau, Germany
Roberto Tamassia	Brown University, USA
Ioannis G. Tollis	ICS-FORTH and University of Crete, Greece
Pavel Valtr	Charles University, Czech Republic
Alexander Wolff	University of Würzburg, Germany

Program Committee

Md. Jawaherul Alam	Amazon Inc., USA
Michael Bekos	University of Tübingen, Germany
Carla Binucci	University of Perugia, Italy
Romain Bourqui	Université de Bordeaux, France
Vida Dujmovic	McGill University, Canada
Cody Dunne	Northeastern University, USA
Seok-Hee Hong	University of Sydney, Australia
Takayuki Itoh	Ochanomizu University, Japan
Radu Jianu	City University of London, UK
Michael Kaufmann	University of Tübingen, Germany
Andreas Kerren	Linköping University, Sweden
Linda Kleist	TU Braunschweig, Germany
Giuseppe Liotta	University of Perugia, Italy
Anna Lubiw	University of Waterloo, Canada
Guy Melançon	Université de Bordeaux, France
Yoshio Okamoto	University of Electro-Communications, Japan
Maurizio Patrignani	Roma Tre University, Italy
Helen Purchase (Co-chair)	University of Glasgow, UK
Chrysanthi Raftopoulou	NTU Athens, Greece
Ignaz Rutter (Co-chair)	University of Passau, Germany
Marcus Schaefer	DePaul University, USA
Geza Toth	Renyi Institute, Hungary

Arthur van Goethem	TU Eindhoven, The Netherlands
Tatiana von Landesberger	University of Cologne, Germany
Hsiang-Yun Wu	TU Wien, Austria
Kai Xu	Middlesex University, UK

Organizing Committee

Michael A. Bekos (Co-chair)	University of Tübingen, Germany
Henry Förster	University of Tübingen, Germany
Renate Hallmayer	University of Tübingen, Germany
Michael Kaufmann (Co-chair)	University of Tübingen, Germany
Axel Kuckuk	University of Tübingen, Germany
Maximilian Pfister	University of Tübingen, Germany
Lena Schlipf	University of Tübingen, Germany

Contest Committee

Philipp Kindermann (Chair)	University of Trier, Germany
Tamara Mchedlidze	Utrecht University, The Netherlands
Wouter Meulemans	TU Eindhoven, The Netherlands

External Reviewers

Ahmed, Abu Reyan
Aichholzer, Oswin
Alegría, Carlos
Angelini, Patrizio
Archambault, Daniel
Bläsius, Thomas
Chen, Kun-Ting
Crnovrsanin, Tarik
Di Bartolomeo, Sara
Di Giacomo, Emilio
Didimo, Walter
Durocher, Stephane
Eppstein, David
Felsner, Stefan
Fink, Simon D.
Frati, Fabrizio
Förster, Henry
Giovannangeli, Loann
Grilli, Luca
Gronemann, Martin

Grosso, Fabrizio
Hosobe, Hiroshi
Karim, Md. Rezaul
Keszegh, Balázs
Kindermann, Philipp
Kryven, Myroslav
Kuckuk, Axel
Lahiri, Abhiruk
Li, Guangping
Löffler, Maarten
Martins, Rafael M.
Meulemans, Wouter
Miltzow, Till
Montecchiani, Fabrizio
Morin, Pat
Moy, Cameron
Nishat, Rahnuma Islam
Nöllenburg, Martin
Ortali, Giacomo
Parada, Irene

Pfister, Maximilian
Pupyrev, Sergey
Saffo, David
Schlipf, Lena
Schnider, Patrick
Schröder, Felix
Sonke, Willem

Sorge, Manuel
Spence, Richard
Stumpf, Peter
T. P., Sandhya
Tappini, Alessandra
Verbeek, Kevin
Wood, David R.

Sponsors

Premium Sponsor

Diamond Sponsor

Gold Sponsors

Bronze Sponsor

Abstracts of Invited Talks

Parameterized Complexity in Graph Drawing

Meirav Zehavi

Ben-Gurion University of the Negev, Beer-Sheva, Israel
meiravze@bgu.ac.il

Abstract. Research at the intersection of graph drawing and parameterized complexity—in particular, parameterized algorithms—is in its infancy. Most early efforts have been directed at variants of the classic Crossing Minimization problem, introduced by Turán in 1940, parameterized by the number of crossings. However, in the past few years, there is an increasing interest in the analysis of a variety of other problems in graph drawing from the perspective of parameterized complexity.

In this talk, I will first give an overview of the field of parameterized complexity. Then, I will briefly discuss some results at the intersection of parameterized complexity and graph drawing, with emphasis on crossing minimization. Lastly, I will discuss in more detail a joint work with Agrawal, Guspiel, Madathil and Saurabh, which analyzes a class of crossing minimization problems from the perspective of parameterized complexity.

Supported by Israel Science Foundation (ISF) grant no. 1176/18, and United States - Israel Binational Science Foundation (BSF) grant no. 2018302.

Node Link Diagrams: Are They (Actually) Useful?

Kim Marriott

Department of Human-Centred Computing, Monash University, Australia
Kim.Marriott@monash.edu

Abstract

Over the last thirty years computer scientists and mathematicians in the graph drawing community have devoted considerable effort to developing algorithms and software for the automated layout of node-link diagrams. The underlying assumption is that node-link representations of graphs are useful. But is this assumption warranted?

Few user studies have compared node-link diagrams with other possible graph representations [1]. However a number of studies have compared node-link diagrams with an adjacency matrix, e.g. [3, 6]. These have found that adjacency matrices are better for most tasks except those that are path related. And even for path related tasks node-link diagrams are only useful for quite small graphs [9].

What does this mean? Should we only focus on the layout of small graphs and strive to obtain layouts that are more like those created manually [4]? Should we look at modifications of node-link diagrams that scale to larger graphs? For instance, by laying the diagram out on the surface of a torus [2]. Or should we explore completely new representations for larger graphs, e.g. [8]? But perhaps task effectiveness isn't the only thing we should be considering and instead we should be investigating other possible benefits of node-link diagrams. For example, perhaps they are more intuitive and natural than adjacency matrices?

There is some evidence that node-link diagrams are, in fact, more intuitive and natural than adjacency matrices. This comes from a user study comparing the effectiveness of different tactile representations of graphs for blind readers [7]. The study compared adjacency list, adjacency matrix and node-link representations of social networks. None of the blind participants had seen graphical representations of networks before They rated the node-link diagrams as

- More understandable
- More imaginable (I imagined the social network in my head)
- More intuitive/natural representation

than the other representations. As one participant said:

Love these [node-link] graphics, they make it so easy. This is how I would show someone what a social network is.

The obvious question to ask is why might this be true and what does it mean to be a "more intuitive/natural representation"? I am not sure of the answer but feel that conceptual metaphor theory [5] may provide least part of the answer. This theory argues that metaphor is at the heart of human cognition: that we leverage from our knowledge of the concrete world to reason about abstract ideas. This is revealed through language. For example the UNDERSTANDING-IS-SEEING metaphor is revealed through phrases such as "Now I *see* what you are getting at."

Conceptual metaphors can also be revealed through graphics. Thus, the CATEGORY-IS-A-CONTAINER metaphor underlies both the phrase "The class of mammals *contains* dogs and cats" and the use of Venn and Euler diagrams. Similarly, the SOCIAL RELATIONSHIP-IS-A-PHYSICAL CONNECTION conceptual metaphor underlies how we reason about social relationships including kinship. It is revealed through the phrase "I feel *connected* to you." I believe it also underpins our use of family trees and sociograms. This is why a node-link diagram feels like a natural way of representing a social network: it corresponds to a conceptual metaphor that we already have in our head.

References

1. Burch, M., Huang, W., Wakefield, M., Purchase, H.C., Weiskopf, D., Hua, J.: The state of the art in empirical user evaluation of graph visualizations. IEEE Access **9**, 4173–4198 (2020)
2. Chen, K.T., Dwyer, T., Bach, B., Marriott, K.: It'sa wrap: toroidal wrapping of network visualisations supports cluster understanding tasks. In: Proceedings of the 2021 CHI Conference on Human Factors in Computing Systems, pp. 1–12 (2021)
3. Ghoniem, M., Fekete, J.D., Castagliola, P.: A comparison of the readability of graphs using node-link and matrix-based representations. In: IEEE Symposium on Information Visualization, pp. 17–24. IEEE (2004)
4. Kieffer, S., Dwyer, T., Marriott, K., Wybrow, M.: HOLA: Human-like orthogonal network layout. IEEE Trans. Visual. Comput. Graphics **22**(1), 349–358 (2015)
5. Lakoff, G., Johnson, M.: Metaphors We Live By. University of Chicago Press, Chicago (2003)
6. Okoe, M., Jianu, R., Kobourov, S.: Node-link or adjacency matrices: Old question, new insights. IEEE Trans. Visual. Comput. Graphics **25**(10), 2940–2952 (2018)
7. Yang, Y., Marriott, K., Butler, M., Goncu, C., Holloway, L.: Tactile presentation of network data: text, matrix or diagram? In: Proceedings of the 2020 CHI Conference on Human Factors in Computing Systems, pp. 1–12 (2020)
8. Yoghourdjian, V., Dwyer, T., Klein, K., Marriott, K., Wybrow, M.: Graph thumbnails: identifying and comparing multiple graphs at a glance. IEEE Trans. Visual. Comput. Graphics **24**(12), 3081–3095 (2018)
9. Yoghourdjian, V., Yang, Y., Dwyer, T., Lawrence, L., Wybrow, M., Marriott, K.: Scalability of network visualisation from a cognitive load perspective. IEEE Trans. Visual. Comput. Graphics **27**(2), 1677–1687 (2020)

Contents

xxii   Contents

Best Paper (Track 1: Combinatorial and Algorithmic Aspects)

Edge-Minimum Saturated k-Planar Drawings

Steven Chaplick[1], Fabian Klute[2]([✉]), Irene Parada[3], Jonathan Rollin[4], and Torsten Ueckerdt[5]

[1] Department of Data Science and Knowledge Engineering, Maastricht University, Maastricht, The Netherlands
s.chaplick@maastrichtuniversity.nl

[2] Utrecht University, Utrecht, The Netherlands
f.m.klute@uu.nl

[3] TU Eindhoven, Eindhoven, The Netherlands
i.m.de.parada.munoz@tue.nl

[4] Department of Mathematics and Computer Science, FernUniversität in Hagen, Hagen, Germany
jonathan.rollin@fernuni-hagen.de

[5] Institute of Theoretical Informatics, Karlsruhe Institute of Technology, Karlsruhe, Germany
torsten.ueckerdt@kit.edu

Abstract. For a class $\mathcal{D}$ of drawings of loopless (multi-)graphs in the plane, a drawing $D \in \mathcal{D}$ is *saturated* when the addition of any edge to D results in $D' \notin \mathcal{D}$—this is analogous to saturated graphs in a graph class as introduced by Turán (1941) and Erdős, Hajnal, and Moon (1964). We focus on k-planar drawings, that is, graphs drawn in the plane where each edge is crossed at most k times, and the classes $\mathcal{D}$ of all k-planar drawings obeying a number of restrictions, such as having no crossing incident edges, no pair of edges crossing more than once, or no edge crossing itself.

While saturated k-planar drawings are the focus of several prior works, tight bounds on how sparse these can be are not well understood. We establish a generic framework to determine the minimum number of edges among all n-vertex saturated k-planar drawings in many natural classes. For example, when incident crossings, multicrossings and selfcrossings are all allowed, the sparsest n-vertex saturated k-planar drawings have $\frac{2}{k-(k \bmod 2)}(n-1)$ edges for any $k \geq 4$, while if all that is forbidden, the sparsest such drawings have $\frac{2(k+1)}{k(k-1)}(n-1)$ edges for any $k \geq 6$.

1 Introduction

Graph *saturation problems* concern the study of edge-extremal n-vertex graphs under various restrictions. They originate in the works of Turán [20] and Erdős,

Supported by the Netherlands Organisation for Scientific Research (NWO) under project no. 612.001.651.

H. C. Purchase and I. Rutter (Eds.): GD 2021, LNCS 12868, pp. 3–17, 2021.
https://doi.org/10.1007/978-3-030-92931-2_1

Fig. 1. Saturated 4-planar drawing of the 8-cycle (left), 3-planar drawing of the 8-clique (middle), and saturated 6-planar drawing of the 7-matching (right).

Hajnal, and Moon [9]. For a family $\mathcal{F}$ of graphs, a graph G without loops or parallel edges is called $\mathcal{F}$-*saturated* when no subgraph of G belongs to $\mathcal{F}$ and for every $u, v \in V(G)$, where $uv \notin E(G)$, some subgraph of the graph $G + uv$ belongs to $\mathcal{F}$. Turán [20] described, for each t, the n-vertex graphs that are $\{K_t\}$-saturated and have the maximum number of edges—this led to the introduction of the Turán Numbers where the setting moves from graphs to hypergraphs, see for example the surveys [14,19]. Analogously, Erdős, Hajnal, and Moon [9] studied the n-vertex graphs G that are $\{K_t\}$-saturated and have the minimum number of edges. This sparsest saturation view has also received much subsequent study [10], and our work fits into this latter direction but concerns "drawings of (multi-)graphs", also called *topological (multi-)graphs*.

There has been increasing interest in saturation problems on drawings of (multi-)graphs in addition to the abstract graphs above. A *drawing* is a graph together with a cyclic order of edges around each vertex and the sequence of crossings along each edge so that it can be realized in the plane (or on another specified surface). The saturation conditions usually concern the crossings (which can be thought of as avoiding certain topological subgraphs). The majority of work has been on *Turán-type* results regarding the maximum number of edges which can occur in an n-vertex drawing (without loops and homotopic parallel edges) of a particular drawing style, e.g., n-vertex planar drawings are well known to have at most $3n - 6$ edges for any $n \geq 3$. In the case of planar drawings (i.e., crossing-free in the plane), the sparsest saturation version (as in Erdős, Hajnal, and Moon [9]) is also equal to the Turán version: Every saturated planar drawing has $3n - 6$ edges.

However, for drawing styles that allow crossings in a limited way, these two measures become non-trivial to compare and can indeed be quite different. This interesting phenomenon happens for example for k-*planar* drawings where at most k crossings on each edge are allowed; and which are the focus of the present paper. The left of Fig. 1 depicts a drawing of the 8-cycle C_8 in which each edge is crossed exactly four times and one cannot add a ninth (non-loop) edge to the drawing while maintaining 4-planarity, i.e., this is a saturated 4-planar drawing of C_8. On the other hand, note that even the complete graph K_8 in fact admits 3-planar drawings as shown in the middle of Fig. 1.

In this sense, we call a drawing that attains the Turán-type maximum number of edges a *max-saturated*[1] drawing, while a sparsest saturated drawing is called *min-saturated*. The target of this paper is to determine the number of edges in min-saturated k-planar drawings of loopless (multi-)graphs, i.e., the *smallest* number of edges among all saturated k-planar drawings with n vertices. The answer will always be of the form $\alpha_k \cdot (n-1)$. However, it turns out that the precise value of α_k depends on numerous subtleties of what precisely we allow in the considered k-planar drawings. Such subtleties are formalized by drawing styles Γ, each one with its own constant α_Γ. As we always require k-planarity, we omit k from the notation α_Γ.

For example, restricting to connected graphs, we immediately have at least $n-1$ edges on n vertices, i.e., $\alpha_\Gamma \geq 1$. And in fact we also have $\alpha_\Gamma \leq 1$ for all $k \geq 4$ as testified by entangled drawings of cycles like in the left of Fig. 1. Allowing disconnected graphs but restricting to contiguous drawings, we immediately have $\alpha_\Gamma \geq 1/2$ since we have minimum degree at least 1 in that case. And again we also have $\alpha_\Gamma \leq 1/2$ for all $k \geq 6$ as one can find saturated k-planar drawings of matchings like in the right of Fig. 1. Other subtleties occur when we distinguish whether selfcrossing edges, repeatedly crossing edges, crossing incident edges, etc., are allowed or forbidden. We enable a concise investigation of all possible combinations by first deriving lower bounds on α_Γ for any drawing style that satisfies only some mild assumptions. We can then consider each drawing style Γ and swiftly determine the exact value of α_Γ, thus determining the smallest number of edges among all k-planar drawings of that style on n vertices. Our results for multigraphs are summarized in Table 1.

1.1 Related Work

For k-planar graphs the Turán-type question, the edge count in max-saturated drawings, is well studied. Any k-planar *simple*[2] drawing on n vertices contains at most $3.81\sqrt{k}n$ edges [1], and better (and tight) bounds are known for small k [1,16,17]. Specifically 1-planar drawings contain at most $4n - 8$ edges which is tight [17]. For $k \leq 3$, any k-planar drawing with the fewest crossings (among all k-planar drawings of the abstract graph) is necessarily simple [16]. Therefore the tight bounds for $k \leq 3$ also hold for drawings that are not necessarily simple. However, already for $k = 4$, Schaefer [18, p. 58] has constructed k-planar graphs having no k-planar simple drawings, and these easily generalize to all $k > 4$. Pach et al. [16] conjectured that for every k there is a max-saturated k-planar graph with a simple k-planar drawing. For $k = 2, 3$, the max-saturated k-planar homotopy-free multigraphs have been characterized [5].

In the sparsest saturation setting not only min-saturated k-planar drawings are of interest but also min-saturated k-planar (abstract) graphs: sparse k-planar graphs that are no longer k-planar after adding any edge [2,4,6,8]. The questions we address in this work have also been explicitly asked [13, Section 3.2].

---

[1] Sometimes these drawings are called *optimal* in the literature [5].

[2] A drawing is simple if any two edges share at most one point. In particular there are no parallel edges.

Table 1. Overview of results (see also Theorem 1): The minimum number of edges of saturated k-planar drawings on n vertices of a drawing style defined by a set of restrictions.

k	restrictions	minimum number of edges of saturated k-planar drawings on n vertices	tight example[a]
$k \geq 4$	no restriction	$\frac{2}{k-(k \bmod 2)} \cdot (n-1)$	Figure 2
	I no incident crossings		
$k \geq 4$	**S** no selfcrossings	$\frac{2}{k-1} \cdot (n-1)$	[7]
	S no self- and **I** no incident crossings		
$k \geq 4$	**M** no multicrossings	$\frac{2(k-1)}{(k-1)(k-2)+2} \cdot (n-1)$	[7]
$k \geq 4$	**S** no self- and **M** no multicrossings	$\frac{2(k+1)}{k(k-1)} \cdot (n-1)$	Figure 3
$k = 4$	**I** no incident and **M** no multicrossings	$\frac{4}{5} \cdot (n-1)$	[7]
$k \geq 5$		$\frac{2(k-1)}{(k-1)(k-2)+2} \cdot (n-1)$	
$k \geq 6$	**S** no self-, **M** no multi-, and **I** no incident crossings	$\frac{2(k+1)}{k(k-1)} \cdot (n-1)$	Figure 1 (right), Fig. 4
	S no self-, **M** no multi-, **I** no incident crossings, and **H** homotopy-free		

[a] To attain the stated bound via these constructions, insert an isolated vertex in each empty cell.

Recently, the case of saturation problems for simple drawings has come into focus. The Turán-type question is trivial here as all complete graphs have simple drawings. However, there are constructions of saturated simple drawings (and generalizations thereof) with only $O(n)$ edges [12,15].

1.2 Drawings, Crossing Restrictions, and Drawing Types

Throughout the paper, we consider topological drawings in the plane, that is, vertices are represented by distinct points in $\mathbb{R}^2$ and edges are represented by continuous curves connecting their respective endpoints. We allow parallel edges but forbid loops. As usual, edges do not pass through vertices, any two edges have only finitely many interior points in common, each of which is a proper crossing, and no three edges cross in a common point. An edge may cross itself but it uses any crossing point at most twice. Also, each of these selfcrossings are counted twice when considering the number of times that edge is crossed.

The *planarization* of a drawing D is the planar drawing obtained from D by making each crossing into a new vertex, thereby subdividing the edges involved in the crossing. Although we forbid loops in D its planarization might have loops due to selfcrossing edges. In a drawing, an edge involved in at least one crossing is a *crossed* edge, while those involved in no crossing are the *planar* or *uncrossed* edges. The *cells* of a drawing are the connected components of the plane after the removal of every vertex and edge in D. In other words, the cells of D are the faces of its planarization. A vertex v is incident to a cell c if v is contained in the closure of c, i.e., one could at least start drawing an uncrossed edge from v into cell c.

Two distinct parallel edges e and f in a drawing D are called *homotopic*, if there is a homotopy of the sphere between e and f, that is, the curves of e and f can be continuously deformed into each other along the surface of the sphere while all vertices of D are treated as holes.

In what follows, we investigate drawings that satisfy a specific set of restrictions, where we focus on those with frequent appearance in the literature:

- k-*planar*: Each edge is crossed at most k times.
- **H** *homotopy-free*: No two distinct parallel edges are homotopic.
- **M** *single-crossing*: Any pair of edges crosses at most once and any edge crosses itself at most once (edges with $t \in \{0, 1, 2\}$ common endpoints have at most $t + 1$ common points).
- **I** *locally starlike*[3]: Incident edges do not cross (selfcrossing edges are allowed).
- **S** *selfcrossing-free*: No edge crosses itself.
- *branching*: The drawing is **M** single-crossing, **I** locally starlike, **S** selfcrossing-free, and **H** homotopy-free.

A *drawing style* is just a class Γ of drawings, i.e., a predicate whether any given drawing D is in Γ or not. A drawing style Γ is *monotone* if removing any edge or vertex from any drawing $D \in \Gamma$ results again in a drawing $D' \in \Gamma$, i.e., Γ is closed under edge/vertex removal.

We consider drawing styles given by all k-planar drawings of finite, loopless multigraphs obeying a subset X of the restrictions above. Such a drawing style is denoted by Γ_X. We focus on the restrictions **M** forbidding multicrossings, **S** forbidding selfcrossings, **I** forbidding incident crossings, and **H** forbidding homotopic edges. Note that the k-planar drawing style is monotone, and so is Γ_X for each $X \subseteq \{\mathbf{S}, \mathbf{I}, \mathbf{M}\}$. However, the style of all homotopy-free drawings is not monotone, as removing a vertex may render two edges homotopic.

We are interested in k-planar drawings in Γ_X to which no further edge can be added without either violating k-planarity or any of the restrictions in X, and particularly in how sparse these drawings can be; namely, the sparsest saturated such drawings.

Definition 1. *A drawing D is Γ-saturated for drawing style Γ if $D \in \Gamma$ and the addition of any new edge to D results in a drawing $D' \notin \Gamma$.*

1.3 Our Results

In order to determine the sparsest k-planar Γ_X-saturated drawings for restrictions in X, we introduce in Sect. 2 the concept of *filled* drawings in general monotone drawing styles and give lower bounds on the number of edges in these. Using the lower bounds for filled drawings and constructing particularly sparse Γ_X-saturated drawings, we then give in Sect. 3 the precise answer for all $X \subseteq \{\mathbf{S}, \mathbf{I}, \mathbf{M}\}$ and for the branching style, i.e., $X = \{\mathbf{S}, \mathbf{I}, \mathbf{M}, \mathbf{H}\}$, leaving open

---

[3] In other papers this is also called *star simple* or *semi simple* [3,11] and may not allow selfcrossing edges.

only a few cases for $k \in \{4, 5, 6\}$. Our results for multigraphs are summarized in Table 1 and formalized in Theorem 1. In Sect. 4 we discuss saturated drawings of simple graphs instead of multigraphs. Finally, in Sect. 5 we discuss further extensions.

Proofs of statements marked with (⋆) can be found in the full version [7].

2 Lower Bounds and Filled Drawings

Throughout this section, let Γ be an arbitrary monotone drawing style; not necessarily k-planar or defined by any of the restrictions in Sect. 1.2. Recall that Γ is monotone if it is closed under the removal of vertices and/or edges.

Definition 2. *A drawing D is* filled *if any two distinct vertices that are incident to the same cell c of D are connected by an uncrossed edge that lies completely in the boundary of c.*

For example, the filled crossing-free homotopy-free drawings are exactly the planar drawings of loopless multigraphs with every face bounded by three edges. Using Euler's formula, such drawings on $n \geq 3$ vertices have exactly $m = 3n - 6$ edges. In this section we derive lower bounds on the number of edges in n-vertex filled drawings in drawing style Γ. Another important example of filled drawings are those in which every cell has at most one incident vertex. Note that every cell in a filled drawing has at most three incident vertices. Generally, for a drawing D we use the following notation:

$$n_D = \text{\# vertices} \quad c_i(D) = \text{\# cells with exactly } i \text{ incident vertices, } i \geq 0$$
$$m_D = \text{\# edges} \quad c_2'(D) = \text{\# cells with 2 uncrossed edges in their boundary}$$

For a drawing D, let G be its graph and P be its planarization. A *component* of D is a connected component of P. A *cut-vertex* of D is a cut-vertex of G that is also a cut-vertex of P. And finally, D is *essentially 2-connected* if one component has at least one edge, all other components are isolated vertices and along the boundary of each cell each vertex appears at most once (that is, D has no cut-vertex). This means that for each simple closed curve that intersects D in exactly one vertex or not at all, either the interior or the exterior contains no edges from D.

Lemma 1 (⋆). *For every monotone drawing style Γ and every filled drawing $D \in \Gamma$ we have $m_D \geq \alpha_\Gamma \cdot (n_D + c_0(D) - 1)$ where*

$$\alpha_\Gamma = \min \left\{ \frac{m_{D'}}{n_{D'} + c_0(D') - 1} : D' \in \Gamma \text{ is filled and essentially 2-connected} \right\}.$$

As suggested by Lemma 1, we shall now focus on filled drawings that are essentially 2-connected. Our goal is to determine the parameter α_Γ. First, we give an exact formula for the number of edges in any filled essentially 2-connected drawing. The parameter k in the following lemma will later be the k for the k-planar drawings in Sect. 3. However, we do not require any drawing to be k-planar here.

Lemma 2 ($\star$). *For any $k > 2$, if D is a filled, essentially 2-connected drawing with $n_D \geq 3$ vertices, then $m_D = \frac{2}{k-2}(n_D + c_0(D) - 2 + \varepsilon(D))$, where*

$$\varepsilon(D) = (\tfrac{k}{2}m_x - \mathrm{cr}) + \tfrac{k-4}{4}m_p + c_2' + c_3, \qquad such \ that$$
$$m_p = \#planar\ edges, \qquad \mathrm{cr} = \#crossings, and \qquad m_x = \#crossed\ edges.$$

Lemmas 1 and 2 together imply that for any filled drawing $D \in \Gamma$ we have

$$\frac{m_D}{n_D - 1} \geq \frac{m_D}{n_D + c_0(D) - 1} \geq \min_{D'} \frac{m_{D'}}{n_{D'} + c_0(D') - 1}$$
$$= \min_{D'} \frac{2}{k-2} \cdot \frac{n_{D'} + c_0(D') - 2 + \varepsilon(D')}{n_{D'} + c_0(D') - 1},$$

where both minima are taken over all filled, essentially 2-connected drawings $D' \in \Gamma$ and $\varepsilon(D')$ can be thought of as an error term for the drawing D', which we seek to minimize. Indeed, if D' is k-planar, i.e., each edge is crossed at most k times, then $2\mathrm{cr} \leq km_x$. Thus for $k \geq 4$ we have $\varepsilon(D') \geq 0$. In the next section we shall see that (in many cases) the minimum is indeed attained by drawings D' with $\varepsilon(D') = 0$.

3   Exact Bounds and Saturated Drawings

Recall that we seek to find the sparsest k-planar, Γ_X-saturated drawings in a drawing style Γ_X that is given by a set $X \subseteq \{\mathbf{S}, \mathbf{I}, \mathbf{M}, \mathbf{H}\}$ of additional restrictions. These Γ_X-saturated drawings are related to the filled drawings from Sect. 2.

Lemma 3 ($\star$). *For any $k \geq 0$ and any $X \subseteq \{\mathbf{S}, \mathbf{I}, \mathbf{M}\}$, as well as for $X = \{\mathbf{S}, \mathbf{I}, \mathbf{M}, \mathbf{H}\}$, every k-planar, Γ_X-saturated drawing is filled.*

In order to determine the exact edge-counts for min-saturated drawings, we shall find for each drawing style some essentially 2-connected, Γ_X-saturated drawings that attain the minimum in Lemma 1. Motivated by the error term $\varepsilon(D) = (\tfrac{k}{2}m_x - \mathrm{cr}) + \tfrac{k-4}{4}m_p + c_2' + c_3$ in Lemma 2, we define *tight drawings* as those k-planar drawings in which **1)** every edge is crossed exactly k times (so $\tfrac{k}{2}m_x = \mathrm{cr}$) and **2)** every cell contains exactly one vertex (so $m_p = c_0 = c_2' = c_3 = 0$). Observe that tight drawings are indeed Γ_X-saturated and filled and exist only in case $k \geq 4$. Note that, to aid readability, isolated vertices are omitted from the drawings in the figures. Namely, the actual drawings have one isolated vertex in each cell shown empty in the figures. This is also mentioned in the figure captions.

Lemma 4 ($\star$). *For every $k \geq 4$ and every monotone drawing style Γ of k-planar drawings, if $D \in \Gamma$ is a tight drawing, then $\alpha_\Gamma \leq \frac{2}{k-2} \cdot \frac{n_D-2}{n_D-1} < 1$.*

Theorem 1 (See also Table 1). *Let $k \geq 4$, $X \subseteq \{S, I, M, H\}$ be a set of restrictions, and $\Gamma = \Gamma_X$ be the corresponding drawing style of k-planar drawings.*

For infinitely many values of n, the minimum number of edges in any n-vertex Γ-saturated drawing is

$$\frac{2}{k - (k \bmod 2)}(n-1) \quad for\, X = \{I\}\, and\, X = \emptyset.$$

$$\frac{2}{k-1}(n-1) \quad for\, X = \{S\}\, and\, X = \{S, I\}.$$

$$\frac{2(k-1)}{(k-1)(k-2)+2}(n-1) \quad for\, X = \{M\}.$$

$$\frac{2(k+1)}{k(k-1)}(n-1) \quad for\, X = \{S, M\}.$$

$$\frac{4}{5}(n-1) \quad for\, X = \{I, M\}\, and\, k = 4.$$

$$\frac{2(k-1)}{(k-1)(k-2)+2}(n-1) \quad for\, X = \{I, M\}\, and\, k \geq 5.$$

$$\frac{2(k+1)}{k(k-1)}(n-1) \quad for\, X = \{S, I, M\}\, and\, k \geq 6.$$

$$\frac{2(k+1)}{k(k-1)}(n-1) \quad for\, X = \{S, I, M, H\}\, and\, k \geq 6.$$

Proof. For space requirements we present four out of seven cases in detail, the other cases can be found in the full version [7]. We start with the cases when $X \subseteq \{S, I, M\}$. Here the drawing style Γ_X is monotone and every Γ_X-saturated drawing is filled by Lemma 3. Thus, by Lemma 4, we have $\alpha_\Gamma \leq \frac{2}{k-2} \cdot \frac{n_{D_0}-2}{n_{D_0}-1}$ for every tight drawing D_0. This gives the smallest bound when n_{D_0} is minimized. In this case D_0 is essentially 2-connected and $m_{D_0} = \frac{2}{k-2}(n_{D_0} - 2)$ by Lemma 2, since $n_{D_0} \geq 3$ for tight drawings. So it suffices to consider a tight drawing D_0 with the smallest possible number m_{D_0} of edges.

Next, we shall go through the possible subsets X of $\{S, I, M\}$ and determine exactly the value α_Γ for $\Gamma = \Gamma_X$ in two steps.

- First, we present a tight (hence filled) drawing D_0 with the smallest possible number m_{D_0} of edges, which gives by Lemma 4 the upper bound

$$\alpha_\Gamma \leq \frac{2}{k-2} \cdot \frac{n_{D_0} - 2}{n_{D_0} - 1}.$$

- Second, we argue that for every filled (hence also every Γ_X-saturated), essentially 2-connected drawing $D' \in \Gamma_X$ we have

$$\frac{n_{D'} + c_0(D') - 2 + \varepsilon(D')}{n_{D'} + c_0(D') - 1} \geq \frac{n_{D_0} - 2}{n_{D_0} - 1}, \tag{1}$$

Fig. 2. Smallest tight drawings for even $k \geq 4$ (left) and odd $k \geq 4$ (right) in case $X = \emptyset$ and $X = \{I\}$, i.e. nothing, resp. incident crossings, are forbidden. (Isolated vertices in empty cells are omitted.)

which by Lemmas 1 and 2 then proves the matching lower bound:

$$\alpha_\Gamma = \min_{D'} \frac{m_{D'}}{n_{D'} + c_0(D') - 1} = \min_{D'} \frac{2}{k-2} \cdot \frac{n_{D'} + c_0(D') - 2 + \varepsilon(D')}{n_{D'} + c_0(D') - 1}$$

$$\overset{(1)}{\geq} \frac{2}{k-2} \cdot \frac{n_{D_0} - 2}{n_{D_0} - 1}$$

In order to verify (1), observe that if $\varepsilon(D') \geq 1$, then the lefthand side is at least 1, while the righthand side is less than 1. Thus it is enough to verify (1) when $\varepsilon(D') < 1$. In particular we may assume $c_2' = c_3 = 0$ and $2\mathrm{cr} \geq km_\mathrm{x} - 1$ for D'. Similarly, as $\varepsilon(D') \geq 0$, we may assume that $n_{D'} + c_0(D') \leq n_{D_0} - 1$. Altogether this implies that (1) is fulfilled unless

$$m_{D'} = \frac{2}{k-2}(n_{D'} + c_0(D') - 2 + \varepsilon(D')) < \frac{2}{k-2}(n_{D_0} - 1 - 2 + 1) = m_{D_0}.$$

In summary, for each X we shall give a tight drawing D_0 with as few edges as possible, and argue that every filled, essentially 2-connected drawing D' with fewer edges satisfies the inequality (1). Note that $m_{D'} \geq 1$ as essentially 2-connected drawings have at least one edge. In fact, we may assume that D' contains at least one crossed edge. Otherwise D' is filled, planar and hence connected. Thus $m_{D'} \geq n_{D'} - 1$ and $c_0(D') = 0$ which verifies (1) as follows:

$$\frac{n_{D'} - c_0(D') - 2 + \varepsilon(D')}{n_{D'} - c_0(D') - 1} = \frac{k-2}{2} \cdot \frac{m_{D'}}{n_{D'} - 1} \geq 1 > \frac{n_{D_0} - 2}{n_{D_0} - 1}$$

Case 1. $X = \{I\}$ and $X = \emptyset$

Figure 2 shows drawings D_0 with $m_{D_0} = 1$ edge when k is even, and $m_{D_0} = 2$ edges when k is odd, which are tight for $\Gamma = \Gamma_X$ for both $X = \{I\}$ and $X = \emptyset$, as incident edges do not cross. Thus $m_{D_0} = 1 + (k \bmod 2)$ and $n_{D_0} = \frac{k+2}{2}$ for k even, respectively $n_{D_0} = k$ for k odd. Together this gives
$$\alpha_\Gamma \leq \frac{2}{k-2} \cdot \frac{n_{D_0} - 2}{n_{D_0} - 1} = \frac{2}{k-(k \bmod 2)}.$$
On the other hand, let $D' \in \Gamma_X$ be any filled, essentially 2-connected drawing. As argued above, we may assume that $1 \leq m_\mathrm{x} \leq m_{D'} < m_{D_0}$. For even k, there is nothing to show as $m_{D'} \geq 1 = m_{D_0}$. For odd k, we may assume that D' consists of exactly one edge, which has exactly $(k-1)/2$ selfcrossings (since $2\mathrm{cr} \geq km_\mathrm{x} - 1$), and some of the resulting cells may contain an isolated vertex. In particular, $\varepsilon(D') \geq \frac{k}{2}m_\mathrm{x} - \mathrm{cr} = 1/2$. Applying Euler's formula to

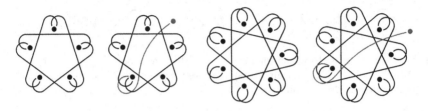

Fig. 3. Smallest tight drawings for $k \geq 4$ in case $X = \{\mathbf{S}, \mathbf{M}\}$, i.e. selfcrossings and multicrossings are forbidden. (Isolated vertices in empty cells are omitted.)

the planarization of D' we get $n_{D'} + c_0(D') = (k+1)/2$, which verifies (1) as follows:

$$\frac{n_{D'} + c_0(D') - 2 + \varepsilon(D')}{n_{D'} + c_0(D') - 1} \geq \frac{(k+1)/2 - 2 + 1/2}{(k+1)/2 - 1} = \frac{k-2}{k-1} = \frac{n_{D_0} - 2}{n_{D_0} - 1}.$$

Case 2. $X = \{\mathbf{S}\}$ and $X = \{\mathbf{S}, \mathbf{I}\}$ $(\star)$
Case 3. $X = \{\mathbf{M}\}$ $(\star)$
Case 4. $X = \{\mathbf{S}, \mathbf{M}\}$

Figure 3 shows tight drawings D_0 with $m_{D_0} = k + 1$ edges. Thus $n_{D_0} = \frac{k-2}{2} m_{D_0} + 2 = \binom{k}{2} + 1$, which gives

$$\alpha_\Gamma \leq \frac{2}{k-2} \cdot \frac{n_{D_0} - 2}{n_{D_0} - 1} = \frac{2}{k-2} \cdot \frac{\binom{k}{2} - 1}{\binom{k}{2}} = \frac{2(k+1)}{k(k-1)}.$$

On the other hand, let D' be any drawing in Γ_X. Again (1) holds, unless $k m_{\mathsf{x}} - 1 \leq 2\mathrm{cr}$ and $1 \leq m_{\mathsf{x}} \leq m_{D'} < m_{D_0} = k + 1$. As there are no multicrossings and no selfcrossings, we have $\mathrm{cr} \leq \binom{m_{\mathsf{x}}}{2}$. However, this would imply $k m_{\mathsf{x}} - 1 \leq 2\mathrm{cr} \leq m_{\mathsf{x}}(m_{\mathsf{x}} - 1) \leq k(m_{\mathsf{x}} - 1) = k m_{\mathsf{x}} - k \leq k m_{\mathsf{x}} - 4$, which is a contradiction.

Case 5. $X = \{\mathbf{I}, \mathbf{M}\}$ $(\star)$
Case 6. $X = \{\mathbf{S}, \mathbf{I}, \mathbf{M}\}$

The right of Fig. 1 (with isolated vertices added to both empty cells) and Fig. 4 show tight drawings D_0 with $m_{D_0} = k + 1$ edges for $k \geq 6$. Analogous to Case 4 $n_{D_0} = \binom{k}{2} + 1$, which gives

$$\alpha_\Gamma \leq \frac{2}{k-2} \cdot \frac{n_{D_0} - 2}{n_{D_0} - 1} = \frac{2}{k-2} \cdot \frac{\binom{k}{2} - 1}{\binom{k}{2}} = \frac{2(k+1)}{k(k-1)}.$$

On the other hand, any drawing $D' \in \Gamma_X$ is also a drawing in $\Gamma_{\{\mathbf{S}, \mathbf{M}\}}$ for $\{\mathbf{S}, \mathbf{M}\} \subset X = \{\mathbf{S}, \mathbf{I}, \mathbf{M}\}$. However, we already argued in Case 4 that there is no drawing $D' \in \Gamma_{\{\mathbf{S}, \mathbf{M}\}}$ with $k m_{\mathsf{x}} - 1 \leq \mathrm{cr}$ and $m_{\mathsf{x}} < m_{D_0} = k + 1$.

Case 7. $X = \{\mathbf{S}, \mathbf{I}, \mathbf{M}, \mathbf{H}\}$

We can not proceed with $X = \{\mathbf{S}, \mathbf{I}, \mathbf{M}, \mathbf{H}\}$ as before, since Γ_X is not monotone in that case. However, we see that the tight drawings D_0 in Fig. 1 (right)

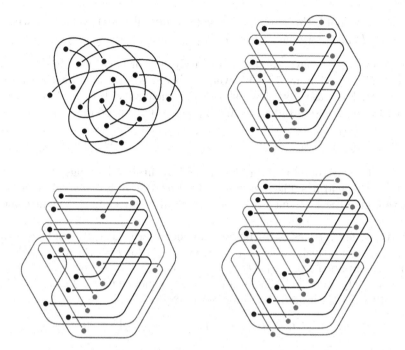

Fig. 4. Smallest tight drawings for $k \geq 7$ (for $k = 6$, see Fig. 1 (right)) in case $X = \{\mathbf{S}, \mathbf{I}, \mathbf{M}\}$, i.e. selfcrossings, incident crossings, and multicrossings are forbidden. Top-Left: The 8-matching for $k = 7$. Top-Right: The 9-matching for $k = 8$. Bottom-Left: The 10-matching for $k = 9$. Bottom-Right: The 11-matching for $k = 10$. (Isolated vertices in empty cells are omitted.)

and Fig. 4 for drawing style $\Gamma_{\{\mathbf{S},\mathbf{I},\mathbf{M}\}}$ are also in Γ_X as there are no parallel edges and hence no homotopic edges. Thus

$$\frac{m_{D_0}}{n_{D_0} - 1} \geq \min\left\{\frac{m_D}{n_D - 1} : D \in \Gamma_X \text{ is } \Gamma_X\text{-saturated}\right\}$$

$$\geq \min\left\{\frac{m_D}{n_D - 1} : D \in \Gamma_X \text{ is filled}\right\}$$

$$\geq \min\left\{\frac{m_D}{n_D - 1} : D \in \Gamma_{\{\mathbf{S},\mathbf{I},\mathbf{M}\}} \text{ is filled}\right\}$$

$$= \min\left\{\frac{m_D}{n_D + c_0(D) - 1} : D \in \Gamma_{\{\mathbf{S},\mathbf{I},\mathbf{M}\}}\right.$$

$$\left. \text{is filled and essentially 2-connected}\right\}$$

$$= \alpha_{\Gamma_{\{\mathbf{S},\mathbf{I},\mathbf{M}\}}} = \frac{2}{k-2} \cdot \frac{n_{D_0} - 2}{n_{D_0} - 1} = \frac{m_{D_0}}{n_{D_0} - 1}$$

and equality holds throughout. Hence, for every filled and every Γ_X-saturated drawing D in Γ_X we have $m_D \geq \alpha_{\Gamma_{\{S,I,M\}}} \cdot (n_D - 1) = \frac{2(k+1)}{k(k-1)} \cdot (n - 1)$.

In Cases 1–6 we have determined exactly α_Γ for each considered drawing style $\Gamma = \Gamma_X$. By Lemmas 1 and 3 every Γ-saturated drawing D satisfies $m_D \geq \alpha_\Gamma(n_D - 1)$. For Case 7 we have shown this inequality directly. Moreover, we presented in each case a tight drawing D_0 attaining this bound:

$$m_{D_0} = \frac{2}{k-2}(n_{D_0} - 2) = \frac{2}{k-2} \cdot \frac{n_{D_0} - 2}{n_{D_0} - 1} \cdot (n_{D_0} - 1) = \alpha_\Gamma(n_{D_0} - 1)$$

It remains to construct an infinite family of Γ-saturated drawings attaining this bound. To this end it suffices to take tight drawings with $\alpha_\Gamma(n - 1)$ edges and iteratively glue these at single vertices, which always results in a tight drawing again.

Formally, for vertices v_1, v_2 in two (not necessarily distinct) tight drawings D_1 and D_2, respectively, with $m_{D_i} = \alpha_\Gamma(n_{D_i} - 1)$ for $i = 1, 2$, we consider the drawing D obtained from D_1, D_2 by identifying v_1 and v_2 into a single vertex and putting D_2 completely inside a cell of D_1 incident to v_1. Then D is again tight and thus Γ-saturated. Moreover we have $n_D = n_{D_1} + n_{D_2} - 1$ and

$$m_D = m_{D_1} + m_{D_2} = \alpha_\Gamma(n_{D_1} - 1) + \alpha_\Gamma(n_{D_2} - 1) = \alpha_\Gamma(n_D - 1).$$

$\square$

4   Bounds for Simple Graphs

We define a *simple filled* drawing D of a simple graph G as a drawing in which any two vertices that are incident to the same cell c of D are connected. In contrast to filled drawings (according to Definition 2) the connecting edge may (partially or completely) lie outside of the boundary of c. With this definition in mind, Lemmas 1 and 3 directly translate to the simple graph setting (note that $\mathbf{H} \not\subseteq X$ for any drawing style Γ_X in this setting). Lemma 2 though does not translate and consequently neither does the bound in Lemma 4. We obtain the following bound on m_D.

Lemma 5 ($\star$). *For any k-planar simple filled and essentially 2-connected drawing D it holds that $m_D \geq \frac{2}{k+2}(n_D - 1)$.*

Consequently we get for any simple filled drawing $D \in \Gamma$ (and hence for every saturated k-planar drawing of a simple graph) that

$$\frac{m_D}{n_D - 1} \geq \min_{D'} \frac{m_{D'}}{n_{D'} - 1} \geq \min_{D'} \frac{2}{k+2} \cdot \frac{n_{D'} - 1}{n_{D'} - 1} = \frac{2}{k+2},$$

where both minima are taken over all k-planar simple filled, essentially 2-connected drawings $D' \in \Gamma$.

Considering upper bounds on the minimum number of edges in any Γ_X-saturated k-planar drawing of a simple graph, we show in the following theorem that for any drawing style $X \subseteq \{\mathbf{S}, \mathbf{I}, \mathbf{M}\}$ there exist sparser drawings than for multigraphs. Moreover, for $X = \emptyset$ and $X = \{\mathbf{I}\}$ the resulting bound is tight.

Fig. 5. Modifications of the constructions used in Theorem 1 for $X = \emptyset$ and $X = \{I\}$ on the left and $X = \{S\}$ and $X = \{S, I\}$ on the right.

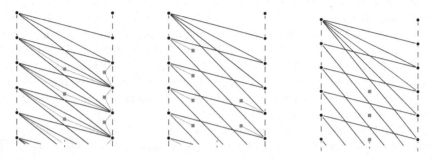

Fig. 6. Construction for saturated simple k-plane drawings. The dashed left and right sides of the drawings are identified.

Theorem 2 ($\star$). *Let $X \subseteq \{S, I, M\}$ be a set of restrictions, and $\Gamma = \Gamma_X$ be the corresponding drawing style of k-planar drawings of simple graphs. For infinitely many values of n, the minimum number of edges in any n-vertex Γ-saturated drawing is upper bounded by (Figs. 5 and 6)*

$$\frac{2}{k+2}(n-1) \quad for\, X = \{I\}\, and\, X = \emptyset\, and\, k \geq 2.$$

$$\frac{2}{k + ((k+1) \mod 2)}(n-1) \quad for\, X = \{S\}\, and\, X = \{S, I\}\, and\, k \geq 4.$$

$$\frac{2}{k-1}(n-1) \quad for\, M \in X\, and\, X \subseteq \{S, I\}\, and\, k \geq 1.$$

5   Concluding Remarks

With respect to multicrossings, in this work we either disallowed their existence (**M**) or did not restrict their number. It is possible to make a more fine-grained analysis and consider the maximum number of times that a pair of edges (or an edge with itself) is allowed to cross as a parameter μ. Modifications of our constructions, for example retracing a side of each edge in the construction in Fig. 4 from both endpoints, yield tight bounds for arbitrarily many values of k and μ.

Our drawings typically contain a large amount of isolated vertices. We discussed the case that isolated vertices are not desired already in the introduction: in this case the sparsest graphs possible are matchings and saturated k-planar drawings of matchings indeed exist for $k \geq 6$ (see Figs. 1 and 4). These drawings are simple and hence contained in all specific drawing styles that we consider. Disjoint unions of these drawings also yield arbitrarily large saturated drawings of matchings for any fixed $k \geq 6$. For $k \leq 5$ saturated k-planar drawings of matchings do not exist provided homotopic parallel edges are allowed. For other drawing styles and for simple graphs saturated drawings of matchings may exist also in case $k \leq 5$. For connected graphs, our best answers are the saturated drawings of cycles depicted in Figs. 1 and 3 and (non-simple) drawings of trees; in the full version see [7, Fig. 8] for an illustration. It is an interesting question to characterize those trees that admit saturated drawings for some fixed k (with respect to the drawing styles discussed here).

For simple graphs, it is a relevant open question to determine the minimum number of edges in a saturated k-planar simple drawing. Finally, our techniques only work for fixed drawings. It remains open to determine the min-saturated k-planar (abstract) graphs and the sizes of their edge sets.

References

1. Ackerman, E.: On topological graphs with at most four crossings per edge. Comput. Geom. **85**, 101574 (2019). https://doi.org/10.1016/j.comgeo.2019.101574
2. Auer, C., Brandenburg, F.J., Gleißner, A., Hanauer, K.: On sparse maximal 2-planar graphs. In: Didimo, W., Patrignani, M. (eds.) GD 2012. LNCS, vol. 7704, pp. 555–556. Springer, Heidelberg (2013). https://doi.org/10.1007/978-3-642-36763-2_50
3. Balko, M., Fulek, R., Kynčl, J.: Crossing numbers and combinatorial characterization of monotone drawings of K_n. Discrete Comput. Geom. **53**(1), 107–143 (2014). https://doi.org/10.1007/s00454-014-9644-z
4. Barát, J., Tóth, G.: Improvements on the density of maximal 1-planar graphs. J. Graph Theor. **88**(1), 101–109 (2018). https://doi.org/10.1002/jgt.22187
5. Bekos, M.A., Kaufmann, M., Raftopoulou, C.N.: On optimal 2- and 3-planar graphs. In: Proceedings 33rd International Symposium on Computational Geometry (SoCG), pp. 16:1–16:16. LIPIcs 77, Schloss Dagstuhl-Leibniz-Zentrum fuer Informatik (2017). https://doi.org/10.4230/LIPIcs.SoCG.2017.16
6. Brandenburg, F.J., Eppstein, D., Gleißner, A., Goodrich, M.T., Hanauer, K., Reislhuber, J.: On the density of maximal 1-planar graphs. In: Didimo, W., Patrignani, M. (eds.) GD 2012. LNCS, vol. 7704, pp. 327–338. Springer, Heidelberg (2013). https://doi.org/10.1007/978-3-642-36763-2_29
7. Chaplick, S., Klute, F., Parada, I., Rollin, J., Ueckerdt, T.: Edge-minimum saturated k-planar drawings. ArXiv e-Prints abs/2012.08631v3 (2020)
8. Eades, P., Hong, S., Katoh, N., Liotta, G., Schweitzer, P., Suzuki, Y.: A linear time algorithm for testing maximal 1-planarity of graphs with a rotation system. Theor. Comput. Sci. **513**, 65–76 (2013). https://doi.org/10.1016/j.tcs.2013.09.029
9. Erdős, P., Hajnal, A., Moon, J.W.: A problem in graph theory. Am. Math. Mon. **71**(10), 1107–1110 (1964). https://doi.org/10.2307/2311408

10. Currie, B.L., Faudree, J.R., Faudree, R.J., Schmitt, J.R.: A survey of minimum saturated graphs. Electron. J. Comb., DS19 (2011). https://doi.org/10.37236/41
11. Felsner, S., Hoffmann, M., Knorr, K., Parada, I.: On the maximum number of crossings in star-simple drawings of K_n with no empty lens. In: GD 2020. LNCS, vol. 12590, pp. 382–389. Springer, Cham (2020). https://doi.org/10.1007/978-3-030-68766-3_30
12. Hajnal, P., Igamberdiev, A., Rote, G., Schulz, A.: Saturated simple and 2-simple topological graphs with few edges. J. Graph Algorithms Appl. **22**(1), 117–138 (2018). https://doi.org/10.7155/jgaa.00460
13. Hong, S.H., Kaufmann, M., Kobourov, S.G., Pach, J.: Beyond-planar graphs: algorithmics and combinatorics (Dagstuhl seminar 16452). Dagstuhl Rep. **6**(11), 35–62 (2017). https://doi.org/10.4230/DagRep.6.11.35
14. Keevash, P.: Hypergraph Turán problems. Surv. Comb. **392**, 83–140 (2011). https://doi.org/10.1017/CBO9781139004114.004
15. Kynčl, J., Pach, J., Radoičić, R., Tóth, G.: Saturated simple and k-simple topological graphs. Comput. Geom. **48**(4), 295–310 (2015)
16. Pach, J., Radoičić, R., Tardos, G., Tóth, G.: Improving the Crossing Lemma by finding more crossings in sparse graphs. Discrete Comput. Geom. **36**(4), 527–552 (2006). https://doi.org/10.1007/s00454-006-1264-9
17. Pach, J., Tóth, G.: Graphs drawn with few crossings per edge. Combinatorica **17**(3), 427–439 (1997). https://doi.org/10.1007/BF01215922
18. Schaefer, M.: The graph crossing number and its variants: a survey. Electron. J. Comb. DS21 (2013). https://doi.org/10.37236/2713, version 6, May 21, 2021
19. Sidorenko, A.: What we know and what we do not know about Turán numbers. Graphs Comb. **11**(2), 179–199 (1995). https://doi.org/10.1007/BF01929486
20. Turán, P.: Eine Extremalaufgabe aus der Graphentheorie. Mat. Fiz. Lapok **48**, 436–452 (1941). in Hungarian, German summary

Best Paper (Track 2: Experimental, Applied, and Network Visualization Aspects)

A User Study on Hybrid Graph Visualizations

Emilio Di Giacomo⬤, Walter Didimo⬤, Fabrizio Montecchiani⬤,
and Alessandra Tappini[✉]⬤

Department of Engineering, University of Perugia, Perugia, Italy
{emilio.digiacomo,walter.didimo,fabrizio.montecchiani,
alessandra.tappini}@unipg.it

Abstract. Hybrid visualizations mix different metaphors in a single layout of a network. In particular, the popular NODETRIX model, introduced
by Henry, Fekete, and McGuffin in 2007, combines node-link diagrams
and matrix-based representations to support the analysis of real-world
networks that are globally sparse but locally dense. That idea inspired
a series of works, proposing variants or alternatives to NODETRIX. We
present a user study that compares the classical node-link model and
three hybrid visualization models designed to work on the same types of
networks. The results of our study provide interesting indications about
advantages/drawbacks of the considered models on performing classical
tasks of analysis. At the same time, our experiment has some limitations
and opens up to further research on the subject.

1 Introduction

Many real-world networks, in a variety of application domains, exhibit a het-
erogeneous structure with a double nature: they are globally sparse but locally
dense, i.e., they contain *clusters* of highly connected nodes (also called *communities* in social network analysis) that are loosely connected to each other (see,
e.g., [28,30,48]). Examples include social and financial networks [14,25,47,56],
as well as biological and information networks [27,42].

The visualization of such networks through classical node-link diagrams is
often unsatisfactory, due to the visual clutter caused by the high number of edges
in the dense portions of the network (Fig. 1(a)). To overcome this limit, *hybrid
visualizations* have been proposed. A hybrid visualization combines different
graph visualization models in a unique drawing, with the aim of conveying the
high-level cluster structure of the network and, at the same time, facilitating in
the analysis of its communities. One of the seminal ideas in this regard is the

This work is partially supported by: (*i*) MIUR, grant 20174LF3T8 "AHeAD: efficient
Algorithms for HArnessing networked Data", (*ii*) Dipartimento di Ingegneria - Università degli Studi di Perugia, grants RICBA19FM: "Modelli, algoritmi e sistemi per
la visualizzazione di grafi e reti" and RICBA20EDG: "Algoritmi e modelli per la rappresentazione visuale di reti".

H. C. Purchase and I. Rutter (Eds.): GD 2021, LNCS 12868, pp. 21–38, 2021.
https://doi.org/10.1007/978-3-030-92931-2_2

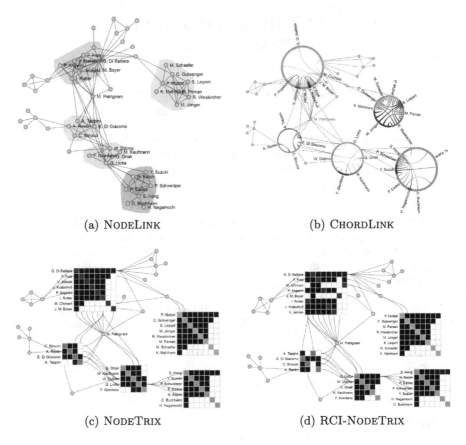

(a) NodeLink

(b) ChordLink

(c) NodeTrix

(d) RCI-NodeTrix

Fig. 1. The same clustered network with our four visualization models (please refer to the online version for colors).

NodeTrix model, introduced by Henry, Fekete, and McGuffin [34], which adopts a node-link diagram to represent the (sparse) global structure of the network, and a matrix representation for denser subgraphs identified and selected by the user (Fig. 1(c)). After the introduction of NodeTrix, hybrid visualizations have become an emerging topic in graph drawing and network visualization, and inspired an array of both theoretical and application results (see, e.g., [7–9,11,12,20,22,23,32,38,40,57]).

Contribution. Motivated by the growing interest in hybrid visualizations, this paper focuses on network layouts with a given set of clusters, and addresses two broad research questions: RQ1 – "Are hybrid visualizations more effective than node-link diagrams for the visual analysis of clustered networks?"; RQ2 – "When considering specific tasks of analysis, are there differences in terms of response time or accuracy among different hybrid visualization models?"

To investigate these questions, we designed a user study that compares three hybrid visualization models and the classical node-link model. Namely, we considered two hybrid models that are designed to work on similar types of networks: the aforementioned NODETRIX model [34] and the CHORDLINK model [9], which represents clusters as chord diagrams instead of adjacency matrices (Fig. 1(b)). Additionally, we considered the RCI-NODETRIX model [40], a variant of NODE-TRIX that adopts independent orderings for the matrix rows and columns so to reduce crossings between inter-cluster edges (Fig. 1(d)).

To the best of our knowledge, our study is the first that addresses research question RQ1, and that considers RQ2 for hybrid visualizations that adopt different styles to represent clusters. Our work is also motivated by open questions from [9,40], namely: [9] suggests to perform a user study to compare CHORDLINK and other hybrid visualizations; [40] asks what is the impact of reducing crossings between inter-cluster edges at the expenses of independent row/column orderings in NODETRIX. The results of our study provide some hints about the usefulness of hybrid visualizations in the execution of topology-based tasks with respect to node-link diagrams. At the same time, our experiment has some limitations and opens up to new research to further investigate the subject.

The paper is structured as follows. Section 2 briefly surveys the scientific literature related to our work. Section 3 explains in detail the design of our user study and describes the rationale behind each of our choices. Section 4 discusses both the quantitative and qualitative results of our experiment, as well as its limitations. Section 5 lists some future research directions. All the experimental data are available at http://mozart.diei.unipg.it/tappini/hybridUserStudy/.

2   Related Work

The focus of our work is on hybrid graph representations that mix different visual metaphors to visually convey both the global structure of a sparse network and its locally dense subgraphs. In this direction, Henry et al. [34] introduce the NODETRIX model for social network analysis in one of the most cited papers of the InfoVis conference [1]; the model is implemented in a system where the user can select (dense) portions of a node-link diagram to be represented as adjacency matrices. NODETRIX visualizations have also been exploited to analyze other real-world graphs, such as ontology graphs [11] and brain networks [57].

Angori et al. [9] introduce the CHORDLINK model. Similarly to NODETRIX, this model is designed to work in a system where the user can visually identify and select clusters on a node-link diagram; differently from NODETRIX, the selected cluster regions are represented as chord diagrams. CHORDLINK aims to represent all edges as geometric links and to preserve the drawing outside clusters by possibly duplicating nodes within a cluster (but each node can only appear in at most one cluster, as for NODETRIX).

Our study focuses on comparing NODETRIX and CHORDLINK, as they are conceived to work on networks with similar structure and within systems with similar characteristics. Our study also considers the RCI-NODETRIX model [40], a variant of NODETRIX that allows independent orderings of the rows and columns in a matrix, to possibly reduce crossings between inter-cluster edges.

For social network analysis, the NODETRIX model has also been proposed with a variant that considers "overlapping clusters", i.e., where a node can occur in multiple clusters at the same time [32]. This kind of node duplication may help in the execution of community-related tasks, but sometimes interferes with other graph readability tasks.

Batagelj et al. [12] propose a system where the user can choose to represent each cluster according to a desired drawing style. Differently from NODETRIX and CHORDLINK, this system is designed to automatically compute a set of clusters that guarantees desired properties (e.g., planarity) for the graph of clusters and adopts an orthogonal drawing convention (instead of a straight-line node-link diagram) to represent the outside of the clusters. Hybrid visualizations have also been exploited in the context of dynamic network analysis (see, e.g., [13,31,53]). We finally mention several theoretical results on hybrid visualizations that concentrate on the complexity of minimizing the number of inter-cluster edge crossings (see, e.g., [6–8,15,20,22,23,38,40]).

Our work falls into the research line devoted to the design of user experiments in graph drawing and network visualization. We recall here the contributions that are mainly related to our study; refer to [17] for a comprehensive survey on the subject. There is a series of works that compare node-link diagrams with matrix-based representations (see, e.g., [3,4,18,24,29,35,45,46]). An insight that seems to emerge from these studies is that node-link diagrams have usually better performance on topology and connectivity tasks when graphs are not too large and dense, while matrices perform better on group tasks. Our study does not aim to further compare node-link and matrix representations, but rather to investigate hybrid visualizations that mix these two, or others, drawing conventions.

In the context of hybrid graph visualizations, Henry and Fekete [33] conduct a user study on MatLink, a model that combines adjacency matrices overlaid with node-link diagrams using curvature for the links. They find that MatLink outperforms the two individual metaphors (node-link diagrams and adjacency matrices) for most of the considered tasks, including path-related tasks, where matrices are usually worse than node-link. However, differently from our study, [33] does not focus on the visualization of clustered networks. Henry et al. [32] present a user study aimed to understand whether node duplication for non-disjoint clusters improves the performance of NODETRIX for some types of tasks. Since the majority of hybrid visualizations are designed to deal with disjoint clusters, our study focuses on this setting; moreover, we consider tasks that are mostly different from those addressed in [32].

3   Study Design

This section describes in detail the design of our user study. Our target population are researchers and analysts (including practitioners, academics, and students) that make use of network visualization to accomplish tasks of analysis on real-world networks. In the following we discuss the visualization models, the tasks and the hypothesis, the stimuli, and the experimental procedure.

3.1 Visualization Models

The conditions compared in our study are four different models for the visualization of undirected clustered networks, where subsets of nodes are grouped into clusters (see Fig. 1 for an illustration). We consider networks with neither self-loops nor multiple edges. An edge connecting two nodes in the same cluster is an *intra-cluster edge*; every other edge is an *inter-cluster edge*. The models are:

– NODELINK (NL). This is the classical node-link model, where nodes are represented as small disks and edges are straight-line segments connecting their end-nodes. In this model, we visually highlight each cluster through a colored convex region that includes all the nodes in the cluster.

– CHORDLINK (CL). This model has been introduced in [9,10]. Nodes outside clusters and inter-cluster edges are drawn as in the NODELINK model. Clusters are represented as chord diagrams. A node in a cluster may have multiple copies, each represented as a colored circular arc along the circumference of the chord diagram; all copies of the same node have the same color. An intra-cluster edge is drawn as a "ribbon" connecting two of the copies representing its end-nodes.

– NODETRIX (NT). This is the model introduced in [34]; each cluster C of size n is represented by a (symmetric) $n \times n$ adjacency matrix. Nodes outside clusters, and edges between them, are drawn as in NODELINK. An inter-cluster edge having an end-node v in a cluster C is drawn as a curve incident to the row or to the column associated with v, on one of the sides of the matrix representing C.

– RCI-NODETRIX (RC). This is a variant of the NODETRIX model, introduced in [39,40]. The difference with the NODETRIX model is that in each adjacency matrix, the row and the column associated with the same node may have different indices, in order to save some crossings between inter-cluster edges. As a consequence the matrices may not be symmetric.

Rationale. Among the various types of hybrid visualizations described in the literature, we selected NT and CL as they are designed to work similarly within visualization systems for the analysis of real-world networks. In particular, we exploited the system in [9], which implements both these models in a unique interface, where the implementation of NT reflects that given in [2] by the authors of [34]. The system in [9] allows direct support for clustered drawings in the NL model and makes it possible to create drawings in all the supported models by defining the same set of clusters on the same node-link diagram. For the purposes of our experiment, we enriched the system with the RC model.

3.2 Tasks

We defined six different tasks, listed in Table 1. We classify each task according to the taxonomy by Lee et al. [36], which we refer to as LeeTax. Moreover, following

Table 1. Tasks used in our study.

Task	LeeTax	AmarTax
T1. Is there an edge that links the two highlighted nodes?	Topology-based (adjacency)	Retrieve value
T2. Which of the two highlighted nodes has higher degree?	Topology-based (adjacency)	Retrieve value; sort
T3. Is there a path of length at most k that connects the two highlighted nodes?	Topology-based (connectivity)	Retrieve value; compute derived value; filter
T4. Which of the following three node labels appear in the highlighted portion of the network?	Attribute-based (on the nodes)	Retrieve value; filter
T5. What is the denser* cluster between the two highlighted?	Overview	Filter; compute derived value; sort
T6. How many edges directly connect the two highlighted parts of the drawing?	Overview	Filter; compute derived value

*The cluster density is the ratio between the number of edges and the number nodes in a cluster

the taxonomy by Amar et al. [5], which we refer to as AmarTax, we indicate the low-level visual analytics operations needed to execute each task.

Rationale. We designed the user study with a set of tasks that requires to explore the drawing locally and globally. Moreover, each task is easy to explain, it can be executed in a reasonably short time, and it can be easily measured. Concentrating on representative tasks is a common approach for this kind of experiments (e.g., [50]), which supports generalizability to more complex tasks that include these representatives as subroutines. Most of our tasks have already been used in previous graph visualization user studies (e.g., [46,49,51,54]) and they cover all task categories in the taxonomy of Lee et al. [36], with the exception of the browsing category. We excluded the latter because it requires the users to interact with the visualization and we decided to avoid interaction to keep the test execution as simple as possible and avoid possible confounding factors. According to the top-level task classification by Burch et al. [17], all our tasks are *interpretation tasks*, as our goal is to evaluate the differences of the considered visualization models in terms of readability, understandability, and effectiveness. About task T5, we point out that there are two commonly used definitions for the density of a graph with n nodes and m edges: $d_1 = \frac{m}{n}$ and $d_2 = \frac{2m}{n(n-1)}$. We adopted definition d_1 for two reasons: (a) it is simpler to explain to a user; (b) according to previous research work [44], d_1 is a better descriptor of the complexity of real-world networks. Indeed, the visual perception of the density of a cluster region is affected by the number of nodes in the cluster; if a drawing contains two clusters with different sizes, the largest one may be perceived as a denser portion of the drawing, even if it has lower density according to d_2.

3.3   Hypotheses

Similarly to previous works (e.g., [32,46]), we define our hypotheses based on tasks, structuring them according to the task categories of LeeTax.

H1: On topology-based tasks (T1, T2, T3), we expect that NODELINK outperforms hybrid visualizations in terms of response time. On the other hand, we expect hybrid visualizations to have a lower error rate than NODELINK, and CHORDLINK to behave better than NODETRIX and RCI-NODETRIX.

H2: On attribute-based tasks (T4), we expect NODETRIX and RCI-NODETRIX to outperform the other two models in terms of response time and error rate.

H3: On overview tasks (T5, T6), we expect hybrid visualizations to perform better than NODELINK in terms of both response time and error rate. Among the hybrid visualizations, we expect NODETRIX and RCI-NODETRIX to be better than CHORDLINK, especially when one needs to estimate cluster density.

Rationale. About **H1**, our expectations in terms of response time are motivated by the fact that NODELINK is quite intuitive and widely used. Moreover, hybrid visualizations intrinsically require to switch from a visualization metaphor to another during the visual exploration, which may represent a cognitive effort. Concerning the error rate, we think that, by reducing the visual clutter, hybrid visualizations may be able to avoid visual ambiguities (such as edges that are almost collinear) and therefore may better support topology-based tasks. Also, since topology-based tasks are known to be harder when dealing with matrices, we expect CHORDLINK to have better performance than NODETRIX and RCI-NODETRIX in terms of error rate. About **H2**, we believe that placing labels on a matrix side is more effective than placing them around chord diagrams or near nodes in a node-link diagram. In chord diagrams labels may be harder to read due to their rotation, while in node-link diagrams they may be hidden by edges. About **H3**, we expect hybrid visualizations to behave better than NODELINK due to their capability to represent clusters more clearly. For tasks that require to estimate cluster density, both NODETRIX and RCI-NODETRIX have the advantage that the proportion between black (edges) and white (non-edges) cells immediately conveys the density of a cluster; the same estimation in CHORDLINK is more difficult due to node duplication, which may give the impression that a cluster is sparser than it actually is.

3.4   Stimuli

Our experimental objects are three real-word networks of small/medium size. The first one, weavers, is an animal social network with 64 nodes and 177 edges, describing the interactions of a colony of weavers in the usage of nests [26,52]. The second one, e.coli, is a biological network with 97 nodes and 212 edges that describes transcriptional interactions in the Escherichia coli bacterium [43]. The third one, dblp, is a co-authorship network obtained from the DBLP repository [37] by searching for the keyword "network visualization" and considering only the largest connected component, which has 118 nodes and 322 edges.

For each of the four visualization models described in Sect. 3.1, we produced a diagram of the three networks above. The diagrams for the NODELINK model are computed through the force-directed algorithm available in the D3 library [16]. Starting from these drawings, we defined some geometric clusters with the technique based on the K-means algorithm [41] described in [9]. As explained in Sect. 3.1, the system presented in [9] is then used to compute the diagrams in the CHORDLINK, NODETRIX, and RCI-NODETRIX models with the same sets of clusters. Further details about the stimuli creation can be found in [21].

Each of the 12 stimuli obtained by applying each of the 4 conditions (models) to the 3 experimental objects (networks) is used in all of the 6 tasks described in Sect. 3.2, for a total of $4 \times 3 \times 6 = 72$ trials. For T1, T2, and T3, we highlighted the node labels with a yellow background; to help the user to locate the nodes, we also put a red cross close to the clusters containing them. For T4 and T6, we highlighted the regions of interest by enclosing them inside a colored polygonal area. Finally, for T5 we indicated the two clusters of interest with large red labels. The trials for the network weavers can be found in [21].

Rationale. The visualization models that we compare are suitable for networks with up to few thousand nodes and edges, while for significantly larger networks ad-hoc techniques are required that typically reduce the amount of displayed information. The choice of using networks with few hundred elements avoids an excessive burden for the participants. Namely, we wanted that each trial could be executed in a reasonable amount of time without an excessive fatigue and that the whole test could be completed in about 30 min. Further, since we decided to show static images (without zoom), the whole picture of the network should be displayed with a level of zoom that keeps the labels readable. Since hybrid visualizations are mainly used to visualize networks that are globally sparse but locally dense, we selected three networks that exhibit this structure. Moreover, we designed the specific trials so that the user was required to explore both the sparse parts of the network, represented by the node-link metaphor, and the dense parts, represented in different ways depending on the model.

3.5 Experimental Setting and Procedure

We designed a between-subject experiment where each participant was exposed to one of the four conditions and hence to 18 trials. The users executed the test fully on-line. The questionnaire was prepared using the LimeSurvey tool (https://www.limesurvey.org/) and is structured as follows. First, some information about the user are collected, namely: gender, age, educational level, expertise in graph visualization, screen size, and possible color vision deficiency. Then, the visualization model to be assigned to the user is decided in a round robin fashion. Based on this assignment, a video tutorial is presented, followed by a training phase in which the user has to answer a trial for each task with an explanatory feedback in case of wrong answer. Next, the 18 trials are presented in random order. Finally, the user is asked for some qualitative feedback: two Likert scale questions about the aesthetic quality of the drawings and about the easiness of the questions, plus an optional free comment. While no time limit was given to complete the test, the participants were asked to answer each question as fast as they could but, at the same time, trying to be accurate. For each user, we collected the answers and the time spent on each question. We recruited the participants with announcements to the gdnet, ieee_vis, infovis mailing lists and to the computer engineering students of the universities of Perugia and Roma Tre. The actual experiment was preceded by a pilot study (see [21]).

Rationale. As previously explained, exposing the users to all four conditions would imply each user solving 72 trials. We believe that keeping the same level of attention in such a long experiment is difficult, and may cause many participants prematurely quitting the test. Besides such undesired fatigue effect, a within-subject design would also imply that each user sees the same experimental object 24 times, which makes it difficult to avoid the learning effect. Hence, we adopted a between-subject design, where each participant is exposed to only one condition. This choice limited the number of trials per user to 18, thus mitigating both the fatigue and the learning effect, which is further counteracted by presenting the trials in a random order. Finally, since the test also includes a video tutorial and a training phase to make the user familiar with the given visualization model, an additional advantage of the between-subject design is that these phases can be focused on one model only. About the execution of the experiment, we opted for a fully on-line test for two reasons: (i) the difficulties to perform a controlled in-person experiment due to the COVID-19 pandemic; (ii) the possibility of recruiting a larger number of participants that better represent our target population, through announcements on the aforementioned mailing lists.

4   Study Results

Participants. We collected questionnaires from 89 participants. We discarded seven tests for various reasons, reported in [21]. Of the remaining 82 tests, 19 were for CHORDLINK and 21 for each of the other models. Regarding the participants, 66 (80.49%) were males, 15 (18.29%) were females, and 1 (1.22%) preferred not to answer. The majority of them (82.72%) were aged below 40. 85.37% of the participants has at least a Bachelor's degree, with 34.15% of them having a doctoral degree. 62.2% of the participants declared a medium or high familiarity with graph visualization and 68.29% used a screen of size at least 15". Refer to [21] for more details.

Quantitative Results. We compared the performance of the four models in terms of error rate and response time. For T1–T5, the error rate of a user is the ratio between the number of wrong answers and the total number of questions. Recall that there are three questions per task and that in T4 the user has to find three labels for each question. About T6, the error on a question is computed as $1 - \frac{1}{1+|u-r|}$, where u is the value given by the user and r is the correct value; the error rate for T6 is the average of the errors on the three questions of the task.

Table 2. Results for error rate (top) and response time (bottom) for each task.

Task	Models ranked by average error rate	Kruskal-Wallis H(3)	p-value	NL-CL	NL-NT	NL-RC	CL-NT	CL-RC	NT-RC
				Pairwise comparisons (p-value)					
T1	CL (0.175), NL (0.191), RC (0.254), NT (0.350)	**8.471**	**0.037**	1.000	0.179	1.000	**0.036**	1.000	0.549
T2	CL (0.140), NL (0.143), RC (0.191), NT (0.238)	2.419	0.490	–	–	–	–	–	–
T3	CL (0.140), RC (0.143), NT (0.270), NL (0.333)	**10.882**	**0.012**	**0.035**	0.788	**0.024**	1.000	1.000	1.000
T4	CL (0.105), NT (0.143), RC (0.206), NL (0.302)	6.703	0.082	–	–	–	–	–	–
T5	NT (0.270), NL (0.302), CL (0.386), RC (0.400)	2.407	0.492	–	–	–	–	–	–
T6	RC (0.350), CL (0.386), NT (0.418), NL (0.429)	0.478	0.924	–	–	–	–	–	–

Task	Models ranked by average response time	Kruskal-Wallis H(3)	p-value	NL-CL	NL-NT	NL-RC	CL-NT	CL-RC	NT-RC
				Pairwise comparisons (p-value)					
T1	NL (16.64), NT (19.91), RC (23.47), CL (26.12)	**20.084**	**<0.001**	**0.000**	1.000	**0.031**	**0.016**	0.832	0.720
T2	NL (25.23), CL (36.35), NT (37.63), RC (39.95)	**12.632**	**0.006**	0.058	0.061	**0.006**	1.000	1.000	1.000
T3	NL (24.58), RC (39.56), NT (42.68), CL (47.69)	**19.533**	**<0.001**	**0.000**	**0.005**	**0.011**	1.000	1.000	1.000
T4	NT (35.88), RC (37.04), NL (42.36), CL (50.60)	**9.793**	**0.020**	1.000	0.579	1.000	**0.018**	0.143	1.000
T5	NL (29.71), NT (37.77), CL (42.30), RC (44.24)	3.657	0.301	–	–	–	–	–	–
T6	RC (31.56), NL (34.07), CL (35.83), NT (38.27)	1.265	0.737	–	–	–	–	–	–

By performing the Shapiro-Wilk test with significance level $\alpha = 0.05$, we found that data were not normally distributed. Hence, we performed the non-parametric Kruskal-Wallis test with significance level $\alpha = 0.05$, which is suitable for comparing multiple independent samples. We finally performed post-hoc pairwise comparisons by using Bonferroni corrections. (See also [19,55].)

Table 2 summarizes the results of our analysis both for the error rate (top) and for the response time (bottom). For each task, we list the models sorted by increasing values of the average error rate or response time. These values are shown in parentheses together with the model. Table 2 reports the statistic (column $H(3)$) and the p-value of the Kruskal-Wallis test. Finally, we report the adjusted (after Bonferroni corrections) significance for each pairwise comparison. Values that are statistically significant are highlighted in bold. Figures 2 and 3 depict the box-plots of the error rate and response time for all the tasks.

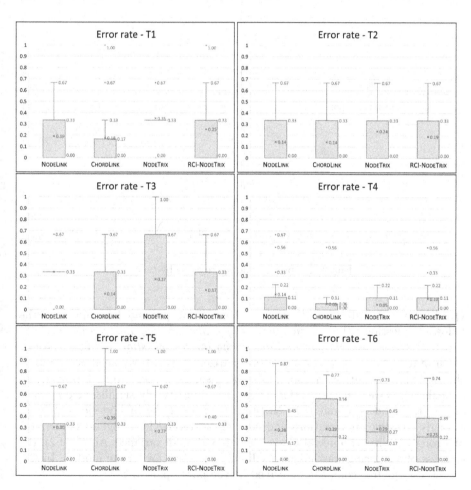

Fig. 2. Error rate aggregated by task.

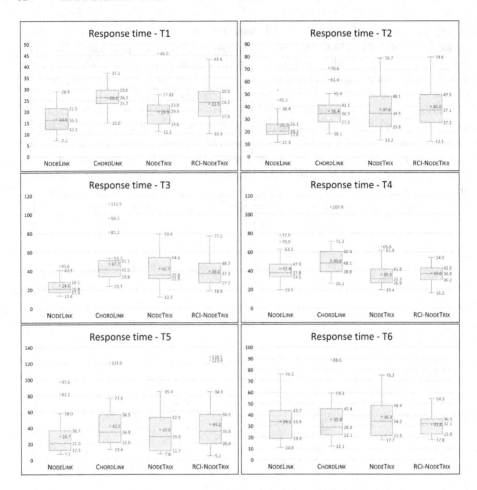

Fig. 3. Response time aggregated by task.

Qualitative Results. At the end of the test, we presented to the users the following questions: (F1) *"How much do you like the diagrams you have seen?"* and (F2) *"How easy did you find answering the test questions?"*. The answers, given in a 5-point Likert scale, are summarized in Fig. 4, where we also report the answer distributions as box-plots; we assigned a score from 1 (lowest) to 5 (highest) to each answer. While there is no statistically significant difference among the models, about (F1) NODELINK received the highest percentage of strongly negative appreciations and NODETRIX received the highest percentage of strongly positive appreciations, although with high variance. About (F2), the easiness of answering was judged medium on average for all the models. In [21] we discuss the free comments received by the participants.

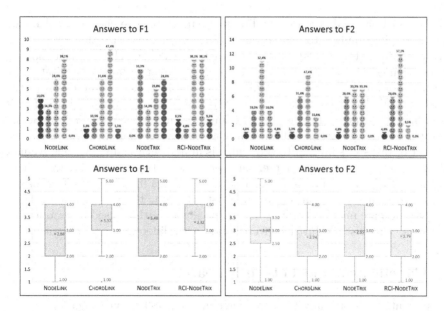

Fig. 4. Qualitative results (please refer to the online version for colors).

Discussion and Limitations. The following highlights summarize our results:

– Hypothesis **H1** is largely supported by the results in terms of response time. More precisely, NODELINK behaves better (with statistical significance) than: CHORDLINK and RCI-NODETRIX for task T1; RCI-NODETRIX for task T2; and for all hybrid models for T3. In particular, the slower performance of RCI-NODETRIX with respect to NODELINK for all the topology-based tasks seems to confirm the difficulty pointed out by some participants about dealing with non-symmetric matrices. In terms of error rate, **H1** is partially supported. Indeed, while there is no statistically significant difference for task T2, we observe that for task T3 both CHORDLINK and RCI-NODETRIX yield better accuracy than NODELINK, and for task T1 CHORDLINK behaves better than NODETRIX. One may wonder why the same behavior is not observable between CHORDLINK and RCI-NODETRIX on task T1; our interpretation is that this might depend on the smaller number of crossings that RCI-NODETRIX usually causes between edges that are incident to the matrices with respect to NODETRIX.

– Hypothesis **H2** is also partially supported by the results (see task T4). In terms of response time, the two models based on matrices seem to lead to better performance than the other two models, with statistical significance when comparing NODETRIX and CHORDLINK. In terms of error rate, we do not observe any statistically significant difference; in particular, the high accuracy achieved with all models seems to reveal that this task was generally easy to perform.

– Hypothesis **H3** is not supported by our results, as we do not observe any statistically significant difference among all the four models.

We conclude by discussing the limits of our study. The choice of not allowing interaction implied to use networks of small/medium size that fit into the screen window; also, it required to have a set of predefined clusters that the user cannot change. On the other hand, a non-interactive environment facilitated the execution of an on-line test; we believe that enabling visual interaction for the considered models would require a different study design, preferably based on a controlled experiment. Further, interactions may introduce confounding factors and it is difficult to design interaction features that are fair to all models. The number of tasks was limited to six, which is in line with many previous studies. Although some works use a larger number of tasks (see, e.g., [46]), we believe that more tasks may cause long execution times and a high fatigue effect for the users, which may result in less reliable data. Finally, the visualization models that we compare may be sensitive to the specific algorithms used to produce the drawings. This justifies further investigation with different layout algorithms.

5 Conclusions and Future Research

We presented a user study that compares different hybrid visualization models and the popular node-link model. As a preliminary answer to RQ1, the results suggest that hybrid visualizations may help to overcome some limits of node-link diagrams in accurately executing topology-based tasks on globally sparse but locally dense networks, at the expenses of the execution time. About RQ2, we could not conclude that any of the considered hybrid models is superior; however, for some topology-based tasks, we observed better accuracy with CHORDLINK and faster execution with NODETRIX. Our study has some limitations and cannot be generalized to settings significantly different from ours. This motivates further experiments with larger networks, interaction features, and additional tasks. Enlarging the set of participants is also an interesting future objective. In this direction, we tried to collect an additional data set by using the Amazon's Mechanical Turk crowdsourcing service. To keep the population sample homogeneous, we required users with background in computer science. However, we report a negative response on this side: the participation was low and the collected data was strongly unreliable (the error rates were very high and the response times too short for a reasoned answer). For these reasons, we decided to discard this additional data set.

Acknowledgements. We thank Giuseppe Liotta for useful discussions, Lorenzo Angori for his help in the implementation, and all the participants to the study.

References

1. CiteVis: Visualizing citations among InfoVis conference papers. http://www.cc.gatech.edu/gvu/ii/citevis
2. NodeTrix Javascript. https://github.com/IRT-SystemX/nodetrix

3. Abuthawabeh, A., Beck, F., Zeckzer, D., Diehl, S.: Finding structures in multi-type code couplings with node-link and matrix visualizations. In: VISSOFT, pp. 1–10. IEEE Computer Society (2013)
4. Alper, B., Bach, B., Riche, N.H., Isenberg, T., Fekete, J.: Weighted graph comparison techniques for brain connectivity analysis. In: CHI, pp. 483–492. ACM (2013)
5. Amar, R.A., Eagan, J., Stasko, J.T.: Low-level components of analytic activity in information visualization. In: IEEE Symposium on Information Visualization (InfoVis 2005), 23–25 October 2005, Minneapolis, MN, USA, pp. 111–117 (2005). https://doi.org/10.1109/INFVIS.2005.1532136
6. Angelini, P., Da Lozzo, G.: Beyond clustered planar graphs. In: Hong, S.-H., Tokuyama, T. (eds.) Beyond Planar Graphs, pp. 211–235. Springer, Singapore (2020). https://doi.org/10.1007/978-981-15-6533-5_12
7. Angelini, P., Da Lozzo, G., Di Battista, G., Frati, F., Patrignani, M., Rutter, I.: Intersection-link representations of graphs. J. Graph Algorithms Appl. 21(4), 731–755 (2017). https://doi.org/10.7155/jgaa.00437
8. Angelini, P., et al.: Graph planarity by replacing cliques with paths. Algorithms 13(8), 194 (2020). https://doi.org/10.3390/a13080194
9. Angori, L., Didimo, W., Montecchiani, F., Pagliuca, D., Tappini, A.: Hybrid graph visualizations with ChordLink: algorithms, experiments, and applications. IEEE Trans. Vis. Comput. Graph. (2020). https://doi.org/10.1109/TVCG.2020.3016055
10. Angori, L., Didimo, W., Montecchiani, F., Pagliuca, D., Tappini, A.: CHORDLINK: a new hybrid visualization model. In: Archambault, D., Tóth, C.D. (eds.) GD 2019. LNCS, vol. 11904, pp. 276–290. Springer, Cham (2019). https://doi.org/10.1007/978-3-030-35802-0_22
11. Bach, B., Pietriga, E., Liccardi, I.: Visualizing populated ontologies with OntoTrix. Int. J. Semant.Web Inf. Syst. 9(4), 17–40 (2013). https://doi.org/10.4018/ijswis.2013100102
12. Batagelj, V., Brandenburg, F., Didimo, W., Liotta, G., Palladino, P., Patrignani, M.: Visual analysis of large graphs using (X, Y)-clustering and hybrid visualizations. IEEE Trans. Vis. Comput. Graph. 17(11), 1587–1598 (2011). https://doi.org/10.1109/TVCG.2010.265
13. Beck, F., Burch, M., Diehl, S., Weiskopf, D.: A taxonomy and survey of dynamic graph visualization. Comput. Graph. Forum 36(1), 133–159 (2017)
14. Bedi, P., Sharma, C.: Community detection in social networks. Wiley Interdisc. Rev. Data Min. Knowl. Discov. 6(3), 115–135 (2016). https://doi.org/10.1002/widm.1178
15. Besa Vial, J.J., Da Lozzo, G., Goodrich, M.T.: Computing k-modal embeddings of planar digraphs. In: ESA. LIPIcs, vol. 144, pp. 19:1–19:16. Schloss Dagstuhl - Leibniz-Zentrum für Informatik (2019)
16. Bostock, M., Ogievetsky, V., Heer, J.: D³ data-driven documents. IEEE Trans. Vis. Comput. Graph. 17(12), 2301–2309 (2011). https://doi.org/10.1109/TVCG.2011.185
17. Burch, M., Huang, W., Wakefield, M., Purchase, H.C., Weiskopf, D., Hua, J.: The state of the art in empirical user evaluation of graph visualizations. IEEE Access 9, 4173–4198 (2021). https://doi.org/10.1109/ACCESS.2020.3047616
18. Christensen, J., Bae, J.H., Watson, B., Rappa, M.: Understanding which graph depictions are best for viewers. In: Christie, M., Li, T.-Y. (eds.) SG 2014. LNCS, vol. 8698, pp. 174–177. Springer, Cham (2014). https://doi.org/10.1007/978-3-319-11650-1_17

19. Conover, W.J.: Practical Nonparametric Statistics. Wiley Series in Probability and Mathematical Statistics, Wiley, Chichester (1980)
20. Da Lozzo, G., Di Battista, G., Frati, F., Patrignani, M.: Computing NodeTrix representations of clustered graphs. J. Graph Algorithms Appl. **22**(2), 139–176 (2018). https://doi.org/10.7155/jgaa.00461
21. Di Giacomo, E., Didimo, W., Montecchiani, F., Tappini, A.: A user study on hybrid graph visualizations. CoRR abs/2108.10270 (2021). http://arxiv.org/abs/2108.10270
22. Di Giacomo, E., Lenhart, W.J., Liotta, G., Randolph, T.W., Tappini, A.: (k, p)-planarity: a relaxation of hybrid planarity. In: WALCOM: Algorithms and Computation - 13th International Conference, WALCOM 2019, Guwahati, India, 27 February – 2 March 2019, Proceedings, pp. 148–159 (2019). https://doi.org/10.1007/978-3-030-10564-8_12
23. Di Giacomo, E., Liotta, G., Patrignani, M., Rutter, I., Tappini, A.: NodeTrix planarity testing with small clusters. Algorithmica **81**(9), 3464–3493 (2019). https://doi.org/10.1007/s00453-019-00585-6
24. Didimo, W., Kornaropoulos, E.M., Montecchiani, F., Tollis, I.G.: A visualization framework and user studies for overloaded orthogonal drawings. Comput. Graph. Forum **37**(1), 288–300 (2018)
25. Didimo, W., Liotta, G., Montecchiani, F.: Network visualization for financial crime detection. J. Vis. Lang. Comput. **25**(4), 433–451 (2014). https://doi.org/10.1016/j.jvlc.2014.01.002
26. van Dijk, R.E., Kaden, J.C., Argüelles-Ticó, A., Dawson, D.A., Burke, T., Hatchwell, B.J.: Cooperative investment in public goods is kin directed in communal nests of social birds. Ecol. Lett. **17**(9), 1141–1148 (2014). https://doi.org/10.1111/ele.12320
27. Flake, G.W., Lawrence, S., Giles, C.L., Coetzee, F.: Self-organization and identification of web communities. IEEE Comput. **35**(3), 66–71 (2002). https://doi.org/10.1109/2.989932
28. Fortunato, S.: Community detection in graphs. Phys. Rep. **486**(3–5), 75–174 (2010). https://doi.org/10.1016/j.physrep.2009.11.002
29. Ghoniem, M., Fekete, J., Castagliola, P.: On the readability of graphs using node-link and matrix-based representations: a controlled experiment and statistical analysis. Inf. Vis. **4**(2), 114–135 (2005). https://doi.org/10.1057/palgrave.ivs.9500092
30. Girvan, M., Newman, M.E.J.: Community structure in social and biological networks. Proc. Natl. Acad. Sci. USA **99**(12), 7821–7826 (2002). https://doi.org/10.1073/pnas.122653799
31. Hadlak, S., Schulz, H., Schumann, H.: In situ exploration of large dynamic networks. IEEE Trans. Vis. Comput. Graph. **17**(12), 2334–2343 (2011)
32. Henry, N., Bezerianos, A., Fekete, J.: Improving the readability of clustered social networks using node duplication. IEEE Trans. Vis. Comput. Graph. **14**(6), 1317–1324 (2008). https://doi.org/10.1109/TVCG.2008.141
33. Henry, N., Fekete, J.-D.: MatLink: enhanced matrix visualization for analyzing social networks. In: Baranauskas, C., Palanque, P., Abascal, J., Barbosa, S.D.J. (eds.) INTERACT 2007, Part II. LNCS, vol. 4663, pp. 288–302. Springer, Heidelberg (2007). https://doi.org/10.1007/978-3-540-74800-7_24
34. Henry, N., Fekete, J., McGuffin, M.J.: NodeTrix: a hybrid visualization of social networks. IEEE Trans. Vis. Comput. Graph. **13**(6), 1302–1309 (2007). https://doi.org/10.1109/TVCG.2007.70582

35. Keller, R., Eckert, C.M., Clarkson, P.J.: Matrices or node-link diagrams: which visual representation is better for visualising connectivity models? Inf. Vis. **5**(1), 62–76 (2006). https://doi.org/10.1057/palgrave.ivs.9500116

36. Lee, B., Plaisant, C., Parr, C.S., Fekete, J., Henry, N.: Task taxonomy for graph visualization. In: Proceedings of the 2006 AVI Workshop on BEyond time and errors: novel evaluation methods for information visualization, BELIV 2006, Venice, Italy, 23 May 2006, pp. 1–5 (2006). https://doi.org/10.1145/1168149.1168168

37. Ley, M.: The DBLP computer science bibliography: evolution, research issues, perspectives. In: Laender, A.H.F., Oliveira, A.L. (eds.) SPIRE 2002. LNCS, vol. 2476, pp. 1–10. Springer, Heidelberg (2002). https://doi.org/10.1007/3-540-45735-6_1. https://dblp.uni-trier.de

38. Liotta, G., Rutter, I., Tappini, A.: Graph planarity testing with hierarchical embedding constraints. CoRR abs/1904.12596 (2019). http://arxiv.org/abs/1904.12596

39. Liotta, G., Rutter, I., Tappini, A.: Simultaneous FPQ-ordering and hybrid planarity testing. In: Chatzigeorgiou, A., et al. (eds.) SOFSEM 2020. LNCS, vol. 12011, pp. 617–626. Springer, Cham (2020). https://doi.org/10.1007/978-3-030-38919-2_51

40. Liotta, G., Rutter, I., Tappini, A.: Simultaneous FPQ-ordering and hybrid planarity testing. Theoret. Comput. Sci. **874**, 59–79 (2021). https://doi.org/10.1016/j.tcs.2021.05.012

41. Lloyd, S.P.: Least square quantization in PCM. IEEE Trans. Inf. Theory **28**(2), 129–137 (1982)

42. Mahmoud, H., Masulli, F., Rovetta, S., Russo, G.: Community detection in protein-protein interaction networks using spectral and graph approaches. In: Formenti, E., Tagliaferri, R., Wit, E. (eds.) CIBB 2013 2013. LNCS, vol. 8452, pp. 62–75. Springer, Cham (2014). https://doi.org/10.1007/978-3-319-09042-9_5

43. Mangan, S., Alon, U.: Structure and function of the feed-forward loop network motif. Proc. Natl. Acad. Sci. **100**(21), 11980–11985 (2003). https://doi.org/10.1073/pnas.2133841100

44. Melancon, G.: Just how dense are dense graphs in the real world? A methodological note. In: Proceedings of the 2006 AVI Workshop on BEyond Time and Errors: Novel Evaluation Methods for Information Visualization, BELIV 2006, pp. 1–7. Association for Computing Machinery, New York (2006). https://doi.org/10.1145/1168149.1168167

45. Okoe, M., Jianu, R.: GraphUnit: evaluating interactive graph visualizations using crowdsourcing. Comput. Graph. Forum **34**(3), 451–460 (2015)

46. Okoe, M., Jianu, R., Kobourov, S.G.: Node-link or adjacency matrices: Old question, new insights. IEEE Trans. Vis. Comput. Graph. **25**(10), 2940–2952 (2019). https://doi.org/10.1109/TVCG.2018.2865940

47. Onnela, J.-P., Kaski, K., Kertész, J.: Clustering and information in correlation based financial networks. Eur. Phys. J. B **38**(2), 353–362 (2004). https://doi.org/10.1140/epjb/e2004-00128-7

48. Porter, M.A., Onnela, J.P., Mucha, P.J.: Communities in networks. Not. Am. Math. Soc. **56**(1082–1097), 1164–1166 (2009)

49. Purchase, H.C.: Performance of layout algorithms: comprehension, not computation. J. Vis. Lang. Comput. **9**(6), 647–657 (1998). https://doi.org/10.1006/jvlc.1998.0093

50. Purchase, H.C.: Experimental Human-Computer Interaction - A Practical Guide with Visual Examples. Cambridge University Press, Cambridge (2012)

51. Purchase, H.C., Hamer, J., Nöllenburg, M., Kobourov, S.G.: On the usability of Lombardi graph drawings. In: Didimo, W., Patrignani, M. (eds.) GD 2012. LNCS, vol. 7704, pp. 451–462. Springer, Heidelberg (2013). https://doi.org/10.1007/978-3-642-36763-2_40

52. Rossi, R.A., Ahmed, N.K.: The network data repository with interactive graph analytics and visualization. In: AAAI (2015). http://networkrepository.com

53. Rufiange, S., McGuffin, M.J.: DiffAni: visualizing dynamic graphs with a hybrid of difference maps and animation. IEEE Trans. Vis. Comput. Graph. 19(12), 2556–2565 (2013)

54. Saket, B., Simonetto, P., Kobourov, S.G., Börner, K.: Node, node-link, and node-link-group diagrams: an evaluation. IEEE Trans. Vis. Comput. Graph. 20(12), 2231–2240 (2014). https://doi.org/10.1109/TVCG.2014.2346422

55. Thode, H.C.: Testing for Normality. Marcel Dekker, New York (2002)

56. Wu, H., He, J., Pei, Y., Long, X.: Finding research community in collaboration network with expertise profiling. In: Huang, D.-S., Zhao, Z., Bevilacqua, V., Figueroa, J.C. (eds.) ICIC 2010. LNCS, vol. 6215, pp. 337–344. Springer, Heidelberg (2010). https://doi.org/10.1007/978-3-642-14922-1_42

57. Yang, X., Shi, L., Daianu, M., Tong, H., Liu, Q., Thompson, P.M.: Blockwise human brain network visual comparison using NodeTrix representation. IEEE Trans. Vis. Comput. Graph. 23(1), 181–190 (2017). https://doi.org/10.1109/TVCG.2016.2598472

Crossing Minimization
and Beyond-Planarity

Star-Struck by Fixed Embeddings: Modern Crossing Number Heuristics

Markus Chimani⬤, Max Ilsen(✉)⬤, and Tilo Wiedera⬤

Theoretical Computer Science, Osnabrück University, Osnabrück, Germany
{markus.chimani,max.ilsen,tilo.wiedera}@uos.de

Abstract. We present a thorough experimental evaluation of several crossing minimization heuristics that are based on the construction and iterative improvement of a planarization, i.e., a planar representation of a graph with crossings replaced by dummy vertices. The evaluated heuristics include variations and combinations of the well-known planarization method, the recently implemented star reinsertion method, and a new approach proposed herein: the mixed insertion method. Our experiments reveal the importance of several implementation details such as the detection of non-simple crossings (i.e., crossings between adjacent edges or multiple crossings between the same two edges). The most notable finding, however, is that the insertion of stars in a fixed embedding setting is not only significantly faster than the insertion of edges in a variable embedding setting, but also leads to solutions of higher quality.

Keywords: Crossing number · Experimental evaluation · Algorithm engineering

1 Introduction

Given a graph G, the *crossing number* problem asks for the minimum number of edge crossings in any drawing of G, denoted by $cr(G)$. This problem is NP-complete [20], even when G is restricted to cubic graphs [24] or graphs that become planar after removing a single edge [7]. While the currently known integer linear programming approaches to the problem [6,16,17] solve sparse instances within a reasonable time frame [12], dense instances require the use of heuristics.

One such heuristic is the well-known *planarization method* [1,22], which constructs a *planarization*, i.e., a planar representation of G with crossings replaced by dummy vertices of degree 4. The heuristic first computes a spanning planar subgraph of G and then iteratively inserts the remaining edges. Several variants of the planarization method have been thoroughly evaluated, including different edge insertion algorithms and postprocessing strategies; see [10] for the latest study. In a recent paper [18], Clancy et al. present an alternative heuristic—the *star reinsertion method*—, which differs in two key aspects from the planarization

Supported by the German Research Foundation (DFG) grant CH 897/2-2.

H. C. Purchase and I. Rutter (Eds.): GD 2021, LNCS 12868, pp. 41–56, 2021.
https://doi.org/10.1007/978-3-030-92931-2_3

method: It (i) starts with a full planarization (instead of a planar subgraph) that is iteratively improved by reinserting elements, and (ii) the reinserted elements are stars (vertices with their incident edges) rather than individual edges. These star insertions are performed using a straight-forward but never tried algorithm from literature [13]. Clancy et al. were faced with the problem that the implementations of the aforementioned heuristics were written in different languages, leading to incomparable running times. In their evaluation, they thus focus on variants of the star reinsertion method; their comparison with the planarization method only gives averages over (a quite limited number of) full instance sets and relies on old data from previous experiments.

Herein, we present a comprehensive experimental evaluation of a wide array of crossing minimization heuristics based on edge and star insertion encompassing all known strong candidates. This includes not only variants of the planarization and star reinsertion methods but also *combined* approaches. In addition, we present and evaluate a new heuristic that builds up a planarization from a planar subgraph using *both* star and edge insertions. All of these algorithms are implemented as part of the same framework, enabling us to accurately compare their running times. Furthermore, we suggest ways of simplifying the implementation of the heuristics, increasing their speed in practice, and improving their results—e.g., by properly handling crossings between adjacent edges and multiple crossings between the same two edges.

2   Preliminaries

In the following, we consider a connected undirected graph G (that is usually simple, i.e., does not contain parallel edges or self-loops) with n vertices and m edges, denoted by $V(G)$ and $E(G)$ respectively. Let Δ be the maximum degree of any vertex in $V(G)$ and $N(v) := \{w \mid (v, w) \in E\}$ the neighborhood of a vertex v. Then, v along with a subset of its incident edges $F \subseteq \{(v, w) \in E\}$ is collectively called a *star*, denoted by (v, F). Furthermore, a (combinatorial) *embedding* of a planar graph G corresponds to a cyclic ordering of the edges around each vertex in $V(G)$ such that the resulting drawing can be realized without any edge crossings. This induces a set of cycles that bound the *faces* of the embedding. Based on a combinatorial embedding of the *primal graph* G, we can define the *dual graph* G^*, whose vertices correspond to the faces of G, and vice versa. For each primal edge $e \in E(G)$, there exists a dual edge $e^* \in E(G^*)$ between the dual vertices corresponding to the e-incident primal faces. Note that G^* may be a multi-graph with self-loops even if G is simple.

For the purpose of this paper, it is of particular concern how to insert an edge (v_1, v_2) into a planarization. First, it is necessary to find a corresponding *insertion path*, i.e., a sequence of faces $f_1, \ldots, f_k$ such that v_1 is incident to f_1, v_2 incident to f_k, and f_i adjacent to f_{i+1} for $i \in \{1, \ldots, k-1\}$. An edge between v_1 and v_2 can then be inserted into a planarization by subdividing a common edge for each face pair (f_i, f_{i+1}) and routing the new edge as a sequence of edges from v_1 along the subdivision vertices to v_2. By extension, the *insertion spider* of a star (v, F) is a set of insertion paths, one for each edge in F. These insertion paths necessarily share a common face into which v can be inserted.

3 Algorithms

3.1 Solving Insertion Problems

Insertion problems, and their efficient solutions, form the cornerstone of all known strong crossing minimization heuristics.

Definition 1 (EIF, SIF). *Given a planar graph G, an embedding Π of G, and an edge (or star) not yet in G, insert this edge (star) into Π such that the number of crossings in Π is minimized. We refer to these problems as the edge (star) insertion problem with fixed embedding EIF (SIF, resp.).*

Given a primal vertex v, let $\hat{v}$ be the vertex that is created by contracting the dual vertices that correspond to v-incident faces. Then, the EIF for any given edge (v_1, v_2) can be solved optimally in $\mathcal{O}(n)$ time by computing the shortest path from $\hat{v_1}$ to $\hat{v_2}$ in the dual graph G^* via breadth-first search (BFS) [1]. By extension, the SIF for a star (v, F) can be solved in $\mathcal{O}(|F| \cdot n)$ time as follows [13]: For each edge $(v, w) \in F$, solve the single-source shortest path problem in G^* with $\hat{w}$ as the source (via BFS). For each face f, the sum over all of the resulting distance values at this f then represents the number of crossings that would be created if v was to be inserted into f. Hence, the face with the minimum distance sum is the optimal face to insert v into, and the computed shortest paths to this face collectively form the insertion spider. To avoid crossings between these shortest paths (due to them not being necessarily unique), we can construct the insertion spider using a final BFS starting at the optimal face.

Definition 2 (EIV, MEIV, SIV). *Given a planar graph G and an edge (a set of k edges, or a star) not yet in G, find an embedding Π among all possible embeddings of G such that optimally inserting the edge (set of k edges, star) into this Π results in the minimum number of crossings. We refer to these problems as the edge (multiple edge, star) insertion problem with variable embedding EIV (MEIV, SIV, resp.).*

The EIV can be solved in $\mathcal{O}(n)$ time using an algorithm by Gutwenger et al. [23], which finds a suitable embedding (with the help of SPR-trees) and then executes the EIF-algorithm described above. Now consider the MEIV: Solving it for general k is NP-hard [28], however there exists an $\mathcal{O}(kn + k^2)$-time approximation algorithm with an additive guarantee of $\Delta k \log k + \binom{k}{2}$ [14] that performs well in practice [10]. Put briefly, the EIV-algorithm is run for each of the k edges independently, and a single final embedding is identified by combining the individual (potentially conflicting) solutions via voting. Then, the EIF-algorithm can be executed once for each edge. Note that the SIV can be solved optimally in polynomial time by using dynamic programming techniques [13]. However, for graphs that are not series-parallel, the resulting running times are exorbitant and there is no known implementation of this algorithm. In fact, our results herein suggest that in the context of crossing minimization heuristics, the solution power of the SIV-algorithm is fortunately not necessary in practice.

Each problem discussed above has a *weighted* version which can be solved in the same manner if each c_e-weighted edge e is replaced by c_e parallel 1-weighted edges beforehand. In practice it is worthwhile to compute the shortest paths during the EIF/SIV-algorithm on the weighted instance directly. However, this does not allow for the same theoretical upper bounds of the running times since the weights may be arbitrarily large.

3.2 Crossing Minimization Heuristics

We start with reviewing several crossing minimization heuristics that iteratively build up a planarization, starting with a planar subgraph:

The planarization method (plm) is the longest studied and best-known approach considered, achieving strong results in previous evaluations [1,10,22]. First, we compute a spanning planar subgraph $G' = (V, E') \subseteq G$, usually by employing a maximum planar subgraph heuristic and extending the result such that it becomes (inclusion-wise) maximal. Then, the remaining edges $F := E \setminus E'$ are either inserted one after another—by solving the respective EIF (*fix*) or EIV (*var*)—or simultaneously using the MEIV-approximation algorithm (*multi*). Gutwenger and Mutzel [22] describe a postprocessing strategy for *plm* based on edge insertion: Each edge is deleted from the planarization and reinserted one after another (*all*). To incrementally improve the planarization, *all* can also be executed once after each individual edge insertion (*inc*) [10]. In the following, we represent the use of these postprocessing strategies by appending the respective shorthand to the algorithm's abbreviation, e.g. *fix-all*. When neither *all* nor *inc* is employed, we use the specifier *none* instead.

The chordless cycle method (ccm) realizes the idea of extending a *vertex-induced* planar subgraph to a full planarization via star insertion [13]. It corresponds to the best-performing scheme for the star insertion algorithm as examined by Clancy et al. [18]: Search for a chordless cycle in G, e.g., via breadth-first search. Let G' denote the subgraph of G that is already embedded and initialize it with this chordless cycle. Iteratively (until the whole graph is embedded) select a vertex $v \notin V(G')$ such that there exists at least one edge (v, w) that connects v with the already embedded subgraph G'; insert v into G' by solving the SIF for the star $(v, \{(v, w) \in E \mid w \in V(G')\})$.

The mixed insertion method (mim) is a novel approach that we propose as an alternative to the planarization schemes above. It proceeds in a fashion that is similar to *plm* but relies on star insertion instead of edge insertion in as many cases as possible. Accordingly, let G' denote the subgraph of G that is already embedded and initialize it with a spanning planar subgraph $(V, E') \subseteq G$. Then, (attempt to) insert the remaining edges $F := E \setminus E'$ by reinserting at least one endpoint of each edge $e \in F$ via star insertion. Since removing and then reinserting a *cut vertex* of the planar subgraph G' would temporarily disconnect it, the cut vertices of the planar subgraph are computed (cf. [25]) and each edge $e \in F$ is processed as follows: If both endpoints of e are cut vertices of G', insert the edge via edge insertion (we choose to do so in a variable embedding setting as such edge insertions happen rarely). If only one endpoint of the edge is a cut vertex, reinsert the other one. If neither endpoint of the edge is a cut vertex, the endpoint to be reinserted can be chosen freely—globally, this corresponds to finding a vertex cover on the graph induced by F that has to include all vertices neighboring a cut vertex in G'. Finding an optimal vertex cover is NP-hard [26]; therefore we compare several heuristics: For each edge e, choose one of the endpoints randomly (*random*), choose the one with the higher or lower degree in G ($high_G$, low_G), choose the one with the higher or lower degree in the graph induced by all edges in F not incident to a cut vertex in G' ($high_F$, low_F), or choose both endpoints (*both*). Each of the chosen vertices is then deleted from the planar subgraph and reinserted together with all of its edges in the original graph by solving the corresponding SIF.

Herein, we evaluate the aforementioned heuristics not only on their own but also in combination with the *star reinsertion method* (*srm*) by Clancy et al. [18], a postprocessing strategy based on star insertion. It starts with an already existing planarization, which may be constructed using any of the methods outlined above (or even more trivial ones, such as extracting a planarization from a circular layout of the vertices, which, however, is known to perform worse [18]). To represent that the result of an algorithm is improved via *srm*, we append "*srm*" to its abbreviation, e.g. *fix-none-srm*. The given planarization is thereby processed as follows: Iteratively choose a vertex v, delete v from G, and reinsert it again by solving the SIF for the star $(v, v \times N(v))$. Continue the loop until there is no more vertex whose reinsertion improves the solution (in which case the latter is said to be *locally optimal*). Clancy et al. propose different methods for choosing v; here, we consider the scheme they report to be the best compromise between solution quality and running time: In each iteration, try to reinsert every vertex once and continue with the next iteration as soon as a vertex is found whose reinsertion improves the number of crossings in the planarization.

The original algorithm only updates a planarization once an actual improvement is found and resets it to its original state otherwise. We propose to never reset it. This approach is permissible as the SIF is solved optimally and the number of crossings hence never increases after the reinsertion of a star. Not resetting the planarization has the potential to save time in practice as it allows for a simpler implementation without any need to copy the dual graph.

4   A Note on Non-simple Crossings

It is well-known that any crossing-optimal drawing can be assumed to be *simple*: No edge self-intersects and each pair of edges intersects at most once (either in a crossing or an endpoint). In particular, a simple drawing may not contain

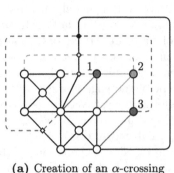 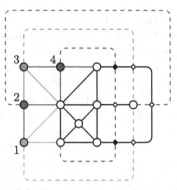

(a) Creation of an α-crossing   **(b)** Creation of a β-crossing

Fig. 1. A non-simple crossing on the red dashed edge as the result of incrementally solving the same kind of insertion problem. When starting with the black planar subgraph, this may happen by solving the SIV using the described algorithm for the colored vertices in the order of their label numbers. Alternatively, if all solid edges constitute the initial planar subgraph, solving the EIV for the dashed edges in the order of their label numbers can have the same result. The examples apply both in the fixed and the variable embedding setting. Dummy vertices for (non-simple) crossings are represented by small (black) diamonds. (Color figure online)

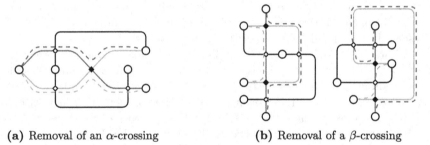

(a) Removal of an α-crossing   **(b)** Removal of a β-crossing

Fig. 2. Non-simple crossings between the red and green edges. After their removal (new edge paths drawn as dashed), the red edge is involved in a new non-simple crossing of the same type and the green edge in a new non-simple crossing of the opposite type. Thus, the removal procedure may have to be iterated. (Color figure online)

crossings between adjacent edges (α-crossings) or multiple crossings between the same two edges (β-crossings). We may hence call any such undesired crossings *non-simple*. Surprisingly, earlier implementations of the planarization method did not consider the emergence and removal of any non-simple crossings [10] while the implementation of the star reinsertion method by Clancy et al. only considers β- but not α-crossings [18]. However, we show in Fig. 1 that incrementally solving the same kind of insertion problem may result in a planarization with α- or β-crossings, even when starting with a planar subgraph. Non-simple crossings can be removed by reassigning edges in the planarization to different edges in the original graph and then deleting the respective dummy vertices (see Fig. 2). Doing so leads to better results overall, see [15, Appendix C].

5 Experiments

Setup: All algorithms are implemented in C++ as part of the Open Graph Drawing Framework (OGDF, www.ogdf.net, based on the release "2020.02 Catalpa") [11], and compiled with GCC 8.3.0. Each computation is performed on a single physical processor of a Xeon Gold 6134 CPU (3.2 GHz), with a memory limit of 4 GB but no time limit. All instances and results are available for download at http://tcs.uos.de/research/cr.

Instances: Table 1 lists the instance sets used for our evaluation (see [15, Appx. A] for further statistical analysis). To enable a proper comparison of the tested algorithms (and potentially in the future, their competitors), we consider multiple well-known benchmark sets as well as constructed, random, and real-world instances with varying characteristics. These are preprocessed by computing the *non-planar core* (NPC) [9] for each non-planar biconnected component. We consider only those instances that have at least 25 vertices after the NPC reduction unless the instance is part of the Complete, Complete-Bip., or KnownCR instance sets. Moreover, we precompute a planar subgraph and chordless cycle for each instance such that different runs of *plm*, *mim* and *ccm* can be started with the same initialization. The planar subgraph is computed by using Chalermsook and Schmid's diamond algorithm [8] and extending the result to a maximal planar subgraph. On average, this computation took only 0.77% of the time needed to execute the fastest evaluated heuristic *fix-none*—a comparatively negligible amount of time that is not further taken into consideration during the evaluation.

The precomputed chordless cycle almost always consists of 3–6 vertices, containing 7–11 vertices for only 15 instances overall. How many edges are deleted to create the planar subgraph, on the other hand, varies greatly depending on the size and density of the graph. Of particular interest is the number of deleted edges that are incident to one or two cut vertices of the planar subgraph: During *mim*, the former ones have a fixed endpoint that must be reinserted via star insertion (disallowing a choice of the reinserted endpoint) while the latter ones must be inserted via edge insertion. Clearly, more dense instances such as the complete (bipartite) ones and the expanders require more edges to be deleted to form a planar subgraph. At the same time, due to their high connectivity, these instances also have less deleted edges that are connected to cut vertices in the planar subgraph. In particular, the complete (bipartite) instances do not have a single such edge. However, even on the sparser instances, *mim* inserts almost all edges via star insertion and one can usually choose the endpoint to be reinserted (see the *mim*-variants described in Subsect. 3.2).

Table 1. Considered instance sets. "#" denotes the number of graphs and $|V(G)|$ the (range of the) numbers of nodes—both values refer to the instance sets *after* preprocessing. Further, let δ denote the node degree, $\square$ the Cartesian product of two graphs, C_i the cycle with i edges, P_j the path with j edges, and G_k the 21 non-isomorphic connected graphs on 5 vertices indexed by k.

Name	#	$|V(G)|$	Description
Rome	3668	25–58	Well-known benchmark set [3], sparse
North	106	25–64	Well-known benchmark set collected by S. North [2]
Webcompute	75	25–112	Instances sent to our online tool [17] for the exact computation of crossing numbers, crossings.uos.de
Expanders	240	30–100	20 random regular graphs [27] (*expander graphs* with high probability) for each parameterization $(|V(G)|, \delta) \in \{30, 50, 100\} \times \{4, 6, 10, 20\}$
Circuit-Based *ISCAS-85* [5] *ISCAS-89* [4] *ITC-99* [19]	45 9 24 12	26–3045 180–3045 60–584 26–980	Hypergraphs from real world electrical networks, transformed into traditional graphs by replacing each hyperedge h by a new hypervertex connected to all vertices contained in h
KnownCR $C \square C$ $G \square P$ $G \square C$ $P(\_,\_)$	1946 251 893 624 178	9–250 9–250 15–245 15–250 10–250	Benchmark set with cr known through proofs [21]: $\rightarrow C_i \square C_j$ with $3 \le i \le 7$, $j \ge i$ such that $i \cdot j \le 250$ $\rightarrow$ Subset of $G_i \square P_j$ with $1 \le i \le 21$, $3 \le j \le 49$ $\rightarrow$ Subset of $G_i \square C_j$ with $1 \le i \le 21$, $3 \le j \le 50$ $\rightarrow$ Generalized Petersen graphs $P(2k+1, 2)$ with $2 \le k \le 62$ and $P(m, 3)$ with $9 \le m \le 125$
Complete	46	5–50	Complete graphs K_n for $5 \le n \le 50$
Complete-Bip.	666	10–80	Complete bipartite graphs K_{n_1,n_2} for $5 \le n_1, n_2 \le 40$

5.1 Fast Heuristics: Mixed Insertion Method, Chordless Cycle Method And Fixed Embedding Edge Insertion

The *mim*-variants, *ccm*, and *fix-none* (all without *srm*-postprocessing) are very fast but yield a comparably high number of crossings. Figure 3 displays some representative results on the expanders, contrasting them with the *BEST* solution found by 50 random permutations of any heuristic tested herein (cf. Subsection 5.4). Among the *mim*-variants, there are only little differences in computation speed and resulting number of crossings. However, reinserting *both* endpoints whenever a choice between two endpoints can be made clearly provides the best results across all instances while only taking an insignificant amount of additional time. The variant leads to the highest amount of reinserted stars and hence also to more chances for an improvement of the number of crossings. In contrast, *high$_F$* needs the lowest amount of star insertions and is thus the fastest variant (but provides results of mixed quality).

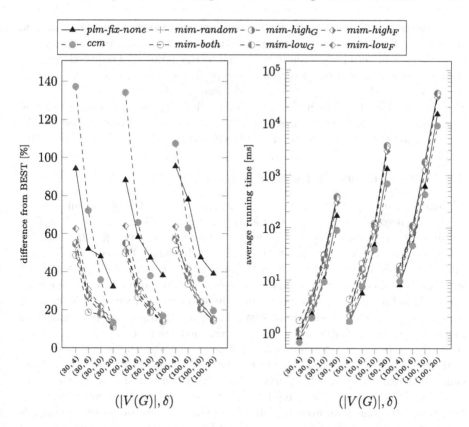

Fig. 3. Comparison of the *mim*-variants, *ccm* and *fix-none* on the expanders.

Compared with *fix-none* and *ccm*, *mim* (from now on always referring to the *both*-variant) provides better results on almost all instances. The fastest of the algorithms, on the other hand, is *fix-none*. The last of the three, *ccm*, should only be considered when examining particularly dense instances: On sparse instance sets such as Rome or KnownCR, it is slower and yields far worse results than *fix-none* (which in turn yields worse results than *mim*), but the solution and speed disparity between the algorithms becomes smaller on instances with a higher density—see, e.g., Fig. 3. On complete (bipartite) instances, *ccm* even surpasses *mim* both in terms of solution quality and speed.

5.2 Planarization Method

The different edge insertion algorithms and postprocessing strategies for the planarization method allow to greatly improve the final planarizations at the cost of additional running time. A detailed experimental comparison of these *plm*-variants was already carried out in 2012 [10]. We are able to replicate the results of that study and corroborate its claims with findings on additional instances:

In terms of solution quality, *none* provides much worse results than *all* and *inc* across all instance sets. However, postprocessing and *inc* in particular has the drawback of very high running times and a large amount of required memory. Among the edge insertion algorithms, *var* performs better (but is also slower) than *multi*, which in turn performs better than *fix*. Overall, *fix-all* is the fastest *plm*-variant that still benefits from the quality improvements of postprocessing. The best compromise between solution quality and speed is provided by the *multi*-variants while the best results are achieved by *var-inc* (cf. [15, Appx. B]).

5.3   Improvements via the Star Reinsertion Method

We tested *srm* as a postprocessing method for the eight most promising and interesting algorithms that construct an initial planarization: The three fast algorithms *mim*, *ccm*, and *fix-none*, as well as the more involved *fix-all*, *multi-all*, *multi-inc*, *var-all*, and *var-inc*. In the case of the latter five, a form of postprocessing is already used, and the additional application of *srm* only leads to a small increase in running time, comparatively speaking. In the case of the former three, the additional postprocessing via *srm* significantly increases the running times (*fix-none-srm* becomes even slower than *fix-all-srm*), but the algorithms are still surprisingly fast: On sparse instances, the running times are comparable to *multi-inc* (without *srm*); on dense instances, the algorithms are even faster than *fix-all*. This is especially interesting as all *srm*-enhanced algorithms typically outperform even the best previously known heuristic variant *var-inc* (see Figs. 4 and 5). In spite of its simplicity, star insertion in a fixed embedding setting is able to greatly improve intermediate planarizations by inserting multiple edges at once. It provides better results and is faster than edge insertion in a variable embedding setting even if the latter uses incremental postprocessing.

When observing the solution quality of the *srm*-algorithms, the same hierarchy as for the algorithms without *srm* emerges: *fix-none-srm* performs worse than the other *plm*-based *srm*-variants, with *var-inc-srm* providing the best results overall. However, *var-inc-srm* is rarely worth the additional running time since the three significantly faster *mim-srm*, *ccm-srm* and *fix-none-srm* perform similarly well or even surpass it on many instances such as several circuit-based ones and the expanders. In comparison to *mim-srm* for example, *var-inc-srm*'s solution quality difference to BEST is only 1.7% smaller but its median running time is eight times higher (when averaged over all instances). The running times of the faster algorithms seem to coincide with the quality of the planarization delivered by the base algorithm: While *fix-none-srm* is generally faster than *ccm-srm* on sparse instances, the opposite is true on denser ones. On complete (bipartite) instances, *ccm-srm* becomes even faster than *mim-srm*. However, *mim-srm* is the otherwise fastest among these algorithms, and thus we recommend to use it.

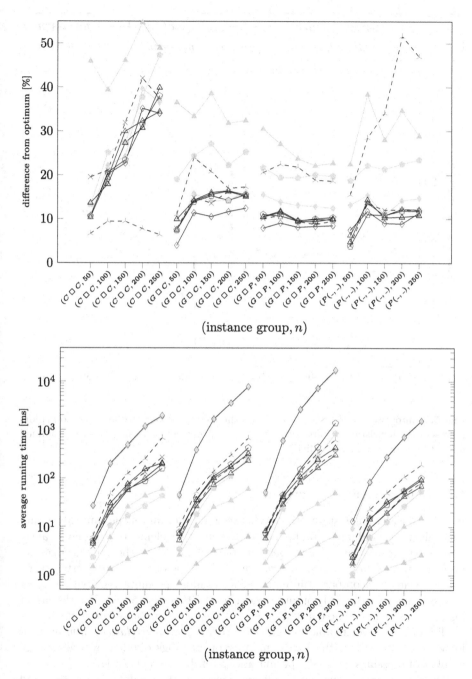

Fig. 4. Comparison of the *srm*-variants on the KnownCR instances. The legend of Fig. 5 applies. Instance sizes are rounded up to the nearest multiple of fifty. Note that the results of *ccm-srm* heavily depend on the structure of the instance; they also vary a lot across other instance sets (with middling results on average).

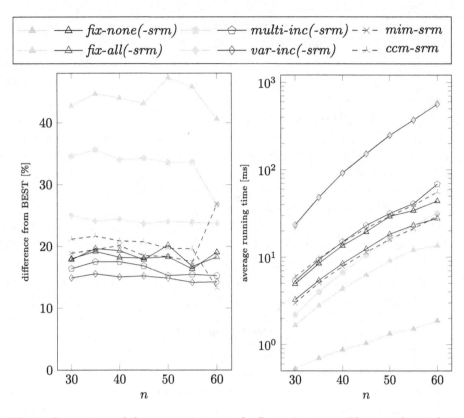

Fig. 5. Comparison of the *srm*-variants on the Rome instances. The grayed out plots represent the heuristic variants without *srm*-postprocessing. Instance sizes are rounded up to the nearest multiple of five.

5.4   Improvements via Permutations

We will consider one last question: Whether multiple runs of the same algorithm with different random permutations of the inserted elements can significantly improve the results. For *plm*, we permute the order in which the deleted edges are inserted, and for *mim*, *ccm* and *srm*, we permute the order of (re)inserted stars. Our experiments compare the effect of 50 random permutations with respect to the Rome, North, Webcompute and KnownCR instance sets. For the larger instances and more time-consuming algorithms, this number of permutations is the limit of what we are able to compute. We focus on the *(relative) improvement* for each instance, i.e., the lowest number of crossings divided by the average number of crossings across 50 permutations (cf. [15, Appendix D]).

Overall, permutations can significantly improve the results of *mim*, *ccm*, and *plm-none* at the cost of little additional time. However, when more time is available, *plm* with postprocessing is clearly preferable. Multiple permutations of *all* and *inc* can be of use if one tries to marginally improve already good solutions.

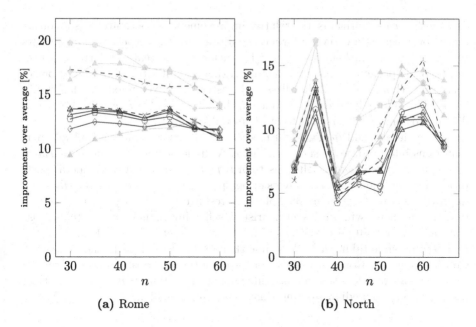

Fig. 6. Comparison of relative improvements for 50 permutations over their average on the Rome and North instances. The legend of Fig. 5 applies.

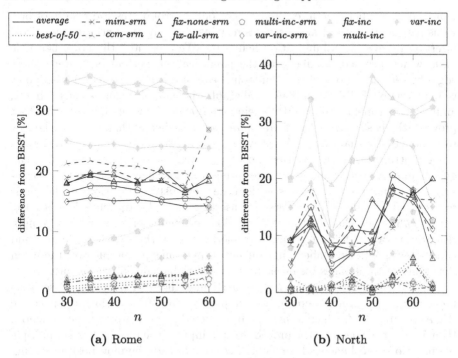

Fig. 7. Comparison of high-solution-quality heuristics (with a single or 50 permutations) on the Rome and North instances.

Among the *srm*-algorithms, the relative improvement via permutations is consistently low with little variance; for a comparison with the respective *plm*-variants see Fig. 6. The one outlier is *ccm-srm*, which achieves the greatest relative improvements for 50 permutations. Note, however, that we initialize all permutations of *ccm-srm* with a fixed small chordless cycle instead of a fixed maximal planar subgraph. This allows for greater variance in the solutions of *ccm-srm* and makes it difficult to compare the results to other *srm*-algorithms.

The general trend of high-solution-quality algorithms, taking multiple permutations into account, is shown in Fig. 7: A single permutation of *mim-srm* or *ccm-srm* will yield better solutions than a *plm*-variant with incremental postprocessing (but no *srm*). Two layers of postprocessing, i.e., *-all-srm* or *-inc-srm*, improve the results even more. Solutions resulting from 50 permutations are in a tier of their own, with *srm*-heuristics achieving higher quality than those without. Overall, 50 permutations of *mim-srm* or *ccm-srm* provide some of the best results while taking a lot less time than other algorithms in their category. Consider, e.g., the Rome instances in a 50-permutations setting; *var-inc-srm* can reduce the average solution quality difference to BEST by only 1.2% more than *mim-srm*, but its median running time is ten times as high.

6 Conclusion

Our in-depth experimental evaluation not only corroborates the results of previous papers [10, 18] but also provides new insights into the performance of star insertion in crossing minimization heuristics. We presented the novel heuristic *mim*, which proceeds similarly to the planarization method but inserts most edges by reinserting one of their endpoints as a star. Whenever neither endpoint is a cut vertex of the initial planar subgraph, the endpoint can be chosen freely, and our experiments indicate that reinserting *both* endpoints one after another provides the best results. In general, *mim* performs better than the basic heuristics from [10, 18] that have a similarly low running time (i.e., *ccm* and *fix-none*).

A central observation is that postprocessing via star insertion (*srm*) can greatly improve the planarizations resulting from fast heuristics: *mim-srm*, *ccm-srm*, and *fix-none-srm* are all faster than the previously best-performing heuristic *var-inc* and provide better results. By inserting multiple adjacent edges at once, star (re-)insertion changes the planarization and its underlying graph decomposition in a way that is sufficient to properly explore the search space and find good solutions. Fixed embedding star insertion is thus preferable over the much slower insertion of edges (or even stars) in a variable embedding setting.

We note that many heuristics—in particular those without edge-wise postprocessing—are prone to create non-simple crossings (due to lack of space see [15, Appendix C]). Such crossings can be detected and it is worthwhile to remove them in order to speed up the procedure and improve the results. Lastly, multiple permutations are beneficial for heuristics that already employ postprocessing. In particular, their application to *mim-srm* and *ccm-srm* provides very high solution quality at moderate running times.

References

1. Batini, C., Talamo, M., Tamassia, R.: Computer aided layout of entity relationship diagrams. J. Syst. Softw. **4**(2–3), 163–173 (1984). https://doi.org/10.1016/0164-1212(84)90006-2
2. Battista, G.D., et al.: Drawing directed acyclic graphs: an experimental study. Int. J. Comput. Geom. Appl. **10**(6), 623–648 (2000). https://doi.org/10.1142/S0218195900000358
3. Battista, G.D., Garg, A., Liotta, G., Tamassia, R., Tassinari, E., Vargiu, F.: An experimental comparison of four graph drawing algorithms. Comput. Geom. **7**, 303–325 (1997). https://doi.org/10.1016/S0925-7721(96)00005-3
4. Brglez, F., Bryan, D., Kozminski, K.: Notes on the ISCAS 1989 benchmark circuits. North-Carolina State University, Technical report, October 1989
5. Brglez, F., Fujiwara, H.: A neutral netlist of 10 combinational circuits and a targeted translator in FORTRAN. In: Proceedings of the ISCAS; Special Session on ATPG and Fault Simulation, pp. 151–158, June 1985
6. Buchheim, C., et al.: A branch-and-cut approach to the crossing number problem. Discrete Optim. **5**(2), 373–388 (2008). https://doi.org/10.1016/j.disopt.2007.05.006
7. Cabello, S., Mohar, B.: Crossing number and weighted crossing number of near-planar graphs. Algorithmica **60**(3), 484–504 (2011). https://doi.org/10.1007/s00453-009-9357-5
8. Chalermsook, P., Schmid, A.: Finding triangles for maximum planar subgraphs. In: Poon, S.-H., Rahman, M.S., Yen, H.-C. (eds.) WALCOM 2017. LNCS, vol. 10167, pp. 373–384. Springer, Cham (2017). https://doi.org/10.1007/978-3-319-53925-6_29
9. Chimani, M., Gutwenger, C.: Non-planar core reduction of graphs. Discrete Math. **309**(7), 1838–1855 (2009). https://doi.org/10.1016/j.disc.2007.12.078
10. Chimani, M., Gutwenger, C.: Advances in the planarization method: effective multiple edge insertions. J. Graph Algorithms Appl. **16**(3), 729–757 (2012). https://doi.org/10.7155/jgaa.00264
11. Chimani, M., Gutwenger, C., Jünger, M., Klau, G.W., Klein, K., Mutzel, P.: The Open Graph Drawing Framework (OGDF). In: Handbook on Graph Drawing and Visualization, pp. 543–569. Chapman and Hall/CRC (2013)
12. Chimani, M., Gutwenger, C., Mutzel, P.: Experiments on exact crossing minimization using column generation. ACM J. Exp. Algorithmics **14**, 3–4 (2009). https://doi.org/10.1145/1498698.1564504
13. Chimani, M., Gutwenger, C., Mutzel, P., Wolf, C.: Inserting a vertex into a planar graph. In: Proceedings of the SODA 2009, pp. 375–383. SIAM (2009). http://dl.acm.org/citation.cfm?id=1496770.1496812
14. Chimani, M., Hliněný, P.: A tighter insertion-based approximation of the crossing number. J. Comb. Optim. **33**(4), 1183–1225 (2016). https://doi.org/10.1007/s10878-016-0030-z
15. Chimani, M., Ilsen, M., Wiedera, T.: Star-struck by fixed embeddings: Modern crossing number heuristics (2021). https://arxiv.org/abs/2108.11443, extended version of this paper including appendix: arXiv:2108.11443 [cs.DM]
16. Chimani, M., Mutzel, P., Bomze, I.: A new approach to exact crossing minimization. In: Halperin, D., Mehlhorn, K. (eds.) ESA 2008. LNCS, vol. 5193, pp. 284–296. Springer, Heidelberg (2008). https://doi.org/10.1007/978-3-540-87744-8_24

17. Chimani, M., Wiedera, T.: An ILP-based proof system for the crossing number problem. In: Proceedings of the ESA 2016. LIPIcs, vol. 57, pp. 29:1–29:13 (2016), https://doi.org/10.4230/LIPIcs.ESA.2016.29
18. Clancy, K., Haythorpe, M., Newcombe, A.: An effective crossing minimisation heuristic based on star insertion. J. Graph Algorithms Appl. **23**(2), 135–166 (2019). https://doi.org/10.7155/jgaa.00487
19. Corno, F., Reorda, M.S., Squillero, G.: RT-level ITC'99 benchmarks and first ATPG results. IEEE Des. Test Comput. **17**(3), 44–53 (2000). https://doi.org/10.1109/54.867894
20. Garey, M.R., Johnson, D.S.: Crossing number is NP-complete. SIAM J. Algebraic Discrete Methods **4**(3), 312–316 (1983)
21. Gutwenger, C.: Application of SPQR-Trees in the Planarization Approach for Drawing Graphs. Ph.D. thesis, TU Dortmund, Dortmund, Germany (2010). http://hdl.handle.net/2003/27430
22. Gutwenger, C., Mutzel, P.: An experimental study of crossing minimization heuristics. In: Liotta, G. (ed.) GD 2003. LNCS, vol. 2912, pp. 13–24. Springer, Heidelberg (2004). https://doi.org/10.1007/978-3-540-24595-7_2
23. Gutwenger, C., Mutzel, P., Weiskircher, R.: Inserting an edge into a planar graph. Algorithmica **41**(4), 289–308 (2005). https://doi.org/10.1007/s00453-004-1128-8
24. Hlinený, P.: Crossing number is hard for cubic graphs. J. Comb. Theory Ser. B **96**(4), 455–471 (2006). https://doi.org/10.1016/j.jctb.2005.09.009
25. Hopcroft, J.E., Tarjan, R.E.: Efficient algorithms for graph manipulation [H] (algorithm 447). Commun. ACM **16**(6), 372–378 (1973). https://doi.org/10.1145/362248.362272
26. Karp, R.M.: Reducibility among combinatorial problems. In: Complexity of Computer Computations. The IBM Research Symposia Series, pp. 85–103. Plenum Press, New York (1972). https://doi.org/10.1007/978-1-4684-2001-2_9
27. Steger, A., Wormald, N.C.: Generating random regular graphs quickly. Comb. Probab. Comput. **8**(4), 377–396 (1999). https://doi.org/10.1017/S0963548399003867
28. Ziegler, T.: Crossing minimization in automatic graph drawing. Ph.D. thesis, Saarland University, Saarbrücken, Germany (2001). http://d-nb.info/961610808

Simplifying Non-simple Fan-Planar Drawings

Boris Klemz[1]📧, Kristin Knorr[2]📧, Meghana M. Reddy[3(✉)]📧,
and Felix Schröder[4]📧

[1] Institut für Informatik, Universität Würzburg, Würzburg, Germany
boris.klemz@uni-wuerzburg.de
[2] Institut für Informatik, Freie Universität Berlin, Berlin, Germany
knorrkri@inf.fu-berlin.de
[3] Department of Computer Science, ETH, Zürich, Switzerland
meghana.mreddy@inf.ethz.ch
[4] Institut für Mathematik, Technische Universität Berlin, Berlin, Germany
fschroed@math.tu-berlin.de

Abstract. A drawing of a graph is fan-planar if the edges intersecting a common edge a share a vertex A on the same side of a. More precisely, orienting e arbitrarily and the other edges towards A results in a consistent orientation of the crossings. So far, fan-planar drawings have only been considered in the context of simple drawings, where any two edges share at most one point, including endpoints. We show that every non-simple fan-planar drawing can be redrawn as a simple fan-planar drawing of the same graph while not introducing additional crossings. Combined with previous results on fan-planar drawings, this yields that n-vertex-graphs having such a drawing can have at most $6.5n$ edges and that the recognition of such graphs is NP-hard. We thereby answer an open problem posed by Kaufmann and Ueckerdt in 2014.

Keywords: Simple topological graphs · Fan-planar graphs · Beyond-planar graphs · Graph drawing

1 Introduction

In a *fan-planar* drawing of a graph, each edge a is either not involved in any crossing or its crossing edges $c_1, \ldots, c_k$ have a common endpoint A that is on a

---

This work was initiated at the 5th DACH Workshop on Arrangements and Drawings, which was conducted online, via gathertown, in March 2021. The authors thank the organizers of the workshop for inviting us and providing a productive working atmosphere. B. K. was partially supported by DFG project WO 758/11-1. M. M. R. is supported by the Swiss National Science Foundation within the collaborative DACH project *Arrangements and Drawings* as SNSF Project 200021E-171681. K. K. is supported by DFG Project MU 3501/3-1 and within the Research Training Group GRK 2434 *Facets of Complexity*. F. S. is supported by the German Research Foundation DFG Project FE 340/12-1.

H. C. Purchase and I. Rutter (Eds.): GD 2021, LNCS 12868, pp. 57–71, 2021.
https://doi.org/10.1007/978-3-030-92931-2_4

common side of a, i.e., orienting a arbitrarily and the edges $c_1, \dots, c_k$ towards A results in a consistent orientation of the crossings on a (either a crosses each c_i from left to right at each crossing, or it crosses each c_i from right to left at each crossing); for illustrations refer to Fig. 1. We call A the *special* vertex of a. All *graphs* in this paper are simple, that is, we do not allow parallel edges or self-loops. Hence, the vertex A is uniquely defined if $k \geq 2$. If $k = 1$, then A is an arbitrary endpoint of c_1.

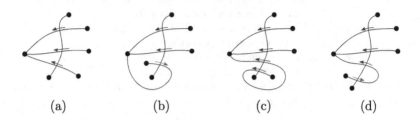

(a)   (b)   (c)   (d)

Fig. 1. Drawings that are (a) simple and fan-planar, (b) simple and not fan-planar, (c) non-simple and fan-planar, and (d) non-simple and not fan-planar.

Previous literature is exclusively concerned with fan-planar drawings that are also *simple*, meaning that each pair of edges intersects in at most one point, which can be either an endpoint or a proper crossing. Simple drawings can be characterized in terms of two forbidden crossing configurations[1] (see Fig. 2):

S1 Two adjacent edges cross.
S2 Two edges cross at least twice.

Simple drawings that are fan-planar can be characterized in terms of two additional forbidden crossing configurations [17] (see Fig. 2):

SF1 Two independent edges cross a common third edge.
SF2 Two adjacent edges cross a third edge a such that their common endpoint A
    is not on a common side of a.

In this paper, we study non-simple fan-planar drawings and how to turn them into simple fan-planar drawings.

Previous and Related Work. A drawing is k-*planar* if each edge is crossed at most k times and a graph is k-*planar*, if it admits such a drawing [20]. A k-*quasiplanar* graph can be drawn such that no k edges mutually cross – such a drawing is called k-*quasiplanar* [2]. Kaufmann and Ueckerdt [17] introduced the notion of fan-planarity in 2014. They describe the class of graphs representable by simple

---

[1] In the literature, usually more obstructions are mentioned, which we exclude for *all* drawings (simple or not), see Sect. 2.

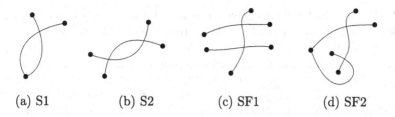

(a) S1   (b) S2   (c) SF1   (d) SF2

Fig. 2. Forbidden configurations in simple fan-planar drawings.

fan-planar drawings[2] as somewhere between 1-planar graphs and 3-quasiplanar graphs. Indeed, every 1-planar graph admits a simple 1-planar drawing. Since such a drawing cannot contain configuration SF1 or SF2, it is fan-planar. Moreover, a simple fan-planar drawing cannot contain three mutually crossing edges and, therefore, it is 3-quasiplanar. Binucci et al. [9] have shown that for each $k \geq 2$ the class of graphs admitting simple k-planar drawings and the class of graphs admitting simple fan-planar drawings are incomparable. In contrast, every so-called optimal 2-planar graph admits a simple fan-planar drawing [7]. This follows from the fact that these graphs can be characterized as the graphs obtained by drawing a pentagram in the interior of each face of a pentagulation [7], which yields a fan-planar drawing. Angelini et al. [3] introduced a drawing style that combines fan-planarity with a visualization technique called edge bundling [14,15,21]. Each of their so-called 1-sided 1-fan-bundle-planar drawings represents a graph that is also realizable as a simple fan-planar drawing, but the converse is not true [3]. Brandenburg [10] examines *fan-crossing* drawings, where all edges crossing a common edge share a common endpoint (in particular, this implies that SF1 is forbidden), as well as *adjacency-crossing* drawings, where SF1 is the only obstruction. Simple fan-planar drawings are somewhat opposite to simple *k-fan-crossing-free* [11] drawings, where no $k \geq 2$ adjacent edges cross another common edge.

The maximum number of edges in a simple fan-planar drawing on n vertices is upperbounded by $6.5n - 20$ [17], which follows from the known density bounds for 3-quasiplanar graphs [1]. A better upper bound of $5n - 10$ edges was claimed in a preprint [17]. However, the corresponding proof appears to be flawed. We spoke with the authors and they confirmed that the current version of their proof is not correct and that they do not see a simple way to fix it[3]. Kaufmann and Ueckerdt [17]

---

[2] In [17], these graphs are called *fan-planar*. We do not use this terminology to avoid mix-ups with the class of graphs admitting (not necessarily simple) fan-planar drawings.

[3] More specifically, the statement and proof of [17, Lemma 1] are incorrect. A counterexample can be obtained by removing the edge g from the construction illustrated in Fig. 8 (vertices R, B correspond to the vertices u, w in [17, Lemma 1]); for a formal description of the construction see Lemma 3.

    After our submission to GD'21, the authors of [17] have uploaded a new version [18] of their preprint in which they state a different definition of fan-planarity with an additional forbidden crossing configuration; also see [18, last paragraph of Sect. 1].

described an infinite family of simple fan-planar drawings with $5n - 10$ edges. The same lower bound also follows from the aforementioned connection to optimal 2-planar graphs [7].

The recognition of graphs realizable as simple fan-planar drawings is NP-hard [9]. The same statement also holds in the fixed rotation system setting [5], where the cyclic order of edges incident to each vertex is prescribed as part of the input. Consequently, efficient algorithms have only been discovered for special graph classes [5] and for restricted drawing styles [5,8].

For a more comprehensive overview of previous work related to fan-planarity, we refer to a very recent survey article dedicated to fan-planarity due to Bekos and Grilli [6]. The study of fan-planarity also falls in line with the recent trend of studying so-called beyond-planar graph classes, whose corresponding drawing styles permit crossings in restricted ways only. Apart from k-planar [20], k-quasiplanar [2], k-fan-crossing-free [11], fan-bundle-planar [3], fan-crossing [10], adjacency-crossing [17], and fan-planar [17] drawings, which have already been mentioned above, several other classes of beyond-planar graphs and their corresponding drawing styles have been studied, e.g.: k-gap-planar drawings [4] (each crossing is assigned to one of the involved edges such that each edge is assigned at most k crossings), RAC-drawings [12] (straight-line drawings with right angle crossings), and many more. We refer to [13,16] for recent surveys on beyond-planar graphs.

Contribution. A fan-planar drawing that is not simple may contain configuration S1. Configuration S2 is allowed in a partial sense: two edges may cross any number of times, but only if orienting them arbitrarily results in a consistent orientation of their crossings; cf. Figs. 1(c) and (d). Recall that every simple fan-planar drawing is 3-quasiplanar. In contrast, Fig. 3(a) depicts a non-simple fan-planar drawing that is not 3-quasiplanar, which suggests that graphs admitting non-simple fan-planar drawings are not necessarily 3-quasiplanar. Consequently, the density bound of $6.5n - 20$ [1] for 3-quasiplanar graphs does not directly carry over. However, the depicted graph is just a K_3, which can obviously be redrawn as a simple (fan-)planar drawing. This raises two very natural questions:

1. Is the largest number of edges in a n-vertex non-simple fan-planar drawing larger than the number of edges in any n-vertex simple fan-planar drawing?
2. Which non-simple fan-planar drawings can be redrawn as simple fan-planar drawings of the same graph?

Question 1 is also mentioned as an open problem by Kaufmann and Ueckerdt [17]. Regarding question 2, we remark that the standard method for simplifying the configurations S1 and S2 does not necessarily maintain fan-planarity, see Figs. 3(b) and (c). As our main result, we answer both questions, thereby solving the open problem by Kaufmann and Ueckerdt:

Theorem 1. *Every non-simple fan-planar drawing can be redrawn as a simple fan-planar drawing of the same graph without introducing additional crossings.*

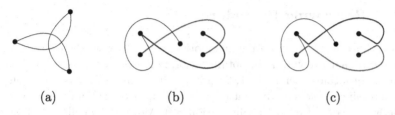

(a)   (b)   (c)

Fig. 3. (a) A non-3-quasiplanar non-simple fan-planar drawing. (b) A non-simple fan-planar drawing. Applying the standard procedure for simplifying configuration S2 yields the drawing in (c), which is not fan-planar since the unmodified (black) edge crosses two independent edges.

The proof of Theorem 1 is constructive and gives rise to an efficient algorithm for simplifying a given fan-planar drawing. Combined with the aforementioned previous results regarding the density [1,17] and the recognition complexity [9] of graphs realizable as simple fan-planar drawings, we obtain:

Corollary 1. *Every (not necessarily simple) fan-planar drawing realizes a 3-quasiplanar graph.*

Corollary 2. *Every (not necessarily simple) fan-planar drawing on n vertices has at most $6.5n - 20$ edges.*

Corollary 3. *Recognizing graphs that admit (not necessarily simple) fan-planar drawings is NP-hard.*

We start with some basic terminology and conventions in Sect. 2. The algorithm for simplifying non-simple fan-planar drawings is described in Sect. 3.

2   Terminology

In all drawings in this paper, edges are represented by simple curves. We assume no two edges touch, that is, meet tangentially. Further, we assume that no three edges share a common crossing and that edges do not contain vertices except their endpoints. Let Γ be a drawing of a graph G. A *redrawing* of Γ is a drawing of G. *Redrawing* an edge e in Γ refers to the process of obtaining a redrawing Γ' of Γ such that $(\Gamma - e) = (\Gamma' - e)$.

In the beginning of Sect. 1, we introduced the notion of special vertices for crossed edges. To streamline the arguments, we also assign an arbitrarily chosen *special* vertex to each uncrossed edge. Let e and f be edges that cross and let E be the special vertex of e. We define the ith crossing of f with e as the ith crossing between f and e encountered when traversing f from endpoint E. For example, in Fig. 5(a), the first crossing of g with b is x and the second crossing is y.

3   The Redrawing Procedure

We prove Theorem 1 by providing an algorithm that redraws the edges of a non-simple fan-planar drawing Γ to obtain a simple fan-planar drawing. It is based on three subroutines (Lemmata 1, 2 and 4), which can be iteratively applied to remove crossings between adjacent edges (configuration S1) and multiple crossings between pairs of edges (configuration S2). More specifically, the first procedure (Lemma 1) eliminates a particular type of adjacent crossings, namely, those that involve an edge that is incident to its special vertex. The second procedure (Lemma 2) removes multiple crossings between edge pairs. Both procedures reduce the overall number of crossings. Hence, they can be exhaustively applied to obtain a redrawing Γ' of Γ that does not contain multiple crossings between edge pairs and where adjacent crossings only involve edges that are not incident to their special vertices (Corollary 4). The procedure (Lemma 4) for removing these remaining crossings is quite involved and based on a structural analysis (Lemma 3) of the drawing Γ'.

The first procedure, for getting rid of some of the adjacent crossings, is very simple; the proof is deferred to the full version of this paper [19], but illustrated in Fig. 4.

Lemma 1. *Let Γ be a non-simple fan-planar drawing. Let $b = (B, R)$ be an edge in Γ that is incident to its special vertex B. If b has at least one crossing, then one of the edges in the drawing can be redrawn such that the total number of crossings in the drawing decreases. Moreover, the redrawing is fan-planar.*

We continue by describing the second procedure, which eliminates crossings between pairs of edges (independent or adjacent) that cross more than once.

Lemma 2. *Let Γ be a non-simple fan-planar drawing. Let $b = (G, R)$ be an edge in Γ whose special vertex B is not incident to b. If edge b has multiple crossings with at least one other edge, then an edge that crosses b multiple times, say $g = (B, W)$ (where W could also be incident to b), can be redrawn such that at least one crossing between b and g is eliminated and the total number of crossings in the drawing decreases. Moreover, the redrawing is fan-planar.*

Proof. We start by describing a procedure to pick the edge that will be redrawn. We traverse b from G to R, until the second crossing of an edge $g = (B, W)$ with b is encountered such that the first crossing of g with b appeared before its second crossing, i.e., the second crossing y with b is closer to R than its first crossing x with b, see Fig. 5(a). If no such edge exists, we exchange the roles of R and G and repeat the procedure. We are guaranteed to find an edge g with the desired properties, since there is an edge crossing b multiple times.

So without loss of generality, assume that the edge g has its second crossing y with b closer to R than its first crossing x. We then walk from y towards G along b until we encounter a crossing z between an edge p and b. The edge p must also be incident to B, the special vertex of b.

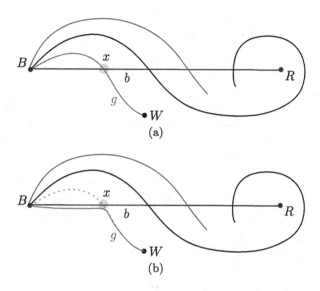

Fig. 4. Illustration of Lemma 1. If b is incident to its special vertex B, then all crossings on b are adjacent crossings. We redraw the edge g whose crossing x with b is closest to B along b. Redrawing the part of g between x and B along b cannot introduce any new crossings.

We can now describe the redrawing procedure. The edge g is redrawn to follow its previous drawing from W to y, cross b at y, follow the drawing of b from y to z, and, finally, closely follow p from z to B; for an illustration see Fig. 5. The following statements are proved in the full version [19].

Proposition 1. *The crossing z is the first crossing of p with b.*

Proposition 2. *Redrawing g maintains fan-planarity. Moreover, there is an injective mapping that assigns each crossing on the redrawn part of g to a crossing on the replaced part of g that involves the same edges.*

The described redrawing of g eliminates the crossing between g and b at x. Moreover, by Proposition 1, the redrawn version of g does not contain any new crossings between b and g. Combined with Proposition 2, it follows that the total number of crossings decreases. Moreover, fan-planarity is maintained. □

Equipped with Lemmata 1 and 2, we can apply the following normalization to the drawing (the proof can be found in the full version [19]):

Corollary 4. *Let Γ be a non-simple fan-planar drawing. There is a fan-planar redrawing Γ' of Γ such that*

- *no two edges cross more than once in Γ';*
- *no edge is incident to its special vertex; and*
- *Γ' does not have more crossings than Γ.*

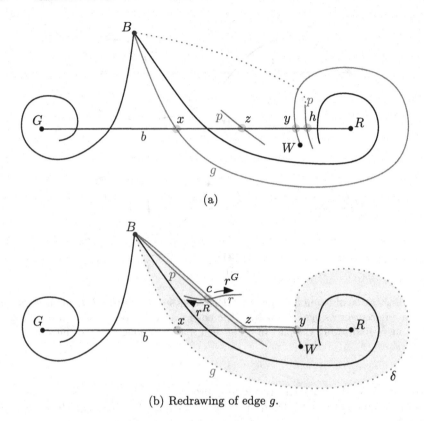

(a)

(b) Redrawing of edge g.

Fig. 5. Edges b and g cross multiple times and the special vertex B of b is not incident to b. The figures do not reflect the case when W is incident to b. Any new crossing with the redrawn version of g involves an edge r crossing p between z and B, which has to cross the replaced part of g since it is incident to G or R.

Adjacent crossings between edges that are not incident to their special vertices may lead to configurations where the previous edge-rerouting strategies would incur additional crossings. In the following lemma, we deal with some unproblematic cases and characterize the remaining, more challenging, configurations in terms of a sequence of conflicting edges.

Lemma 3. *Let Γ be a non-simple fan-planar drawing in which no two edges cross more than once and such that no edge is incident to its special vertex. Let $b = (G, R)$ and $g = (R, B)$ be (adjacent) edges which cross each other at x.*

We can redraw g such that the total number of crossings decreases and fan-planarity is maintained; or, alternatively, we can determine a sequence of edges $r_0, b_1, r_2, b_3, r_4, \ldots, r_k$ such that the edges $b, g, r_0, b_1, r_2, b_3, r_4, \ldots, r_k$ are pairwise distinct and the following properties are satisfied (we call the edges r_i "red" and the edges b_i "black"; for an illustration, see Fig. 6, as well as Fig. 8, which also depicts r_k):

I1 B *is the special vertex of the black edges and incident to the red edges.*

I2 R *is the special vertex of the red edges and incident to the black edges.*

I3 *For any odd* i, *the first crossing* x_{i+1} *of* b_i *starting from* R *is with* r_{i+1}. *For any even* $i < k$, *the first crossing* x_{i+1} *of* r_i *starting from* B *is with* b_{i+1}.

I4 r_0 *crosses* b_1 *but no other black edge.* b *crosses* r_0 *and* r_k *but no other red edges.*

I5 *For the purposes of the final two invariants, we define* $q_{-1} = b$. *For* $0 \leq i < k$, *let* α_i *be the closed curve defined by* g, *the arc of* q_i *and the arc of* q_{i-1}, *where* $q \in \{r, b\}$, *that connect* R, B *and* x_i. *For* $0 \leq i < k$, *let* Γ_i *be the drawing induced by the edges* b, g, r_0, b_1, r_2, \ldots, q_i.

For $0 \leq i < k$, *the curve* α_i *is simple and bounds a region* f_i *that contains only* G, *an arc of* b *that connects* G *to* $x \in \alpha_i$ *and, possibly, an arc of* r_0 *that connects* G *to* α_i, *in its interior.*

I6 *For* $0 < i < k$, $f_i \subset f_{i-1}$ *and* $f_{i-1} \backslash f_i$ *is an empty triangular face in* Γ_i *bounded by the following three arcs:*
- *the arc of* q_i *between* x_i *and the special vertex of* q_{i-1},
- *the arc of* q_{i-1} *between* x_i *and* x_{i-1},
- *the arc of* q_{i-2} *between* x_{i-1} *and the special vertex of* q_{i-1}
where $q \in \{r, b\}$.

Remark 1. Note that invariant I5 implies that in Γ_i, g crosses only b and possibly r_0. Moreover, the arcs of q_i and q_{i-1} connecting R and B via x_i are uncrossed in Γ_i.

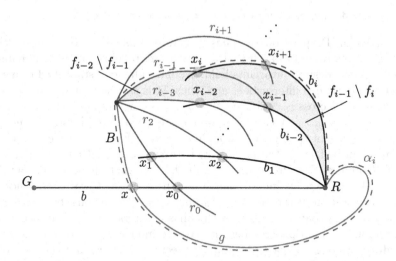

Fig. 6. An example of the sequence of edges described in Lemma 3. The face f_i is the unbounded region delimited by the dashed curve, the face $f_{i-1} \backslash f_i$ is depicted in blue and the face $f_{i-2} \backslash f_{i-1}$ is depicted in green. (Color figure online)

Proof. It follows from the preconditions that B is the special vertex of b and G is the special vertex of g. We will construct the sequence of edges inductively.

Base Case. For the induction base case, we show how to determine r_0 and b_1 such that all invariants are satisfied with respect to r_0. For b_1, we will only establish the invariants I2, I4, I5 and I6.

We traverse from R along b until we encounter an edge r_0 that crosses b and denote its crossing by x_0. If $x_0 = x$ and, hence, $r_0 = g$, we can redraw the part of g that leads from R to x along b such that the crossing at x is removed. Moreover, since the redrawn part is crossing-free, the total number of crossings is decreased and fan-planarity is maintained. Hence, if $x_0 = x$, the statement of the lemma holds.

So assume that $x_0 \neq x$. It follows that, $r_0 \neq g$ since g cannot cross b multiple times. Moreover, $r_0 \neq b$ since edges are realized as simple curves. Since r_0 intersects b, it is incident to B.

Now, we traverse r_0 from B towards x_0 until we encounter a crossing x_1 with an edge b_1. If $x_1 = x_0$ and, hence, $b_1 = b$, we redraw g along the part of b between R and x_0 and the part of r_0 between x_0 and B. The redrawn version of g is crossing-free. Hence, we have eliminated at least one crossing (namely x) while maintaining fan-planarity and, thus, the statement of the lemma holds if $x_1 = x_0$.

So assume that $x_1 \neq x_0$. It follows that b_1 is distinct from b since b has no multiple crossings with r_0. Moreover, $b_1 \neq r_0$ since edges are simple curves. Finally, we show that $b_1 \neq g$. In fact, we actually claim something stronger and prove it in the full version [19].

Proposition 3. *The part of r_0 between B and x_0 cannot cross g.*

In particular, Proposition 3 implies $b_1 \neq g$, as claimed. Thus, we have determined two edges r_0 and b_1 such that b, g, r_0, b_1 are pairwise distinct. It remains to show that the desired invariants hold. We have already established that r_0 is incident to B (since it intersects b) and, thus, I1 is satisfied for r_0.

Since b_1 and b cross r_0, it follows that b_1 shares an endpoint with b, which is the special vertex of r_0. Accordingly, we consider two cases. First, assume that the special vertex of r_0 is G, which is illustrated in Fig. 7. Consider the closed curve α_0 described by g, the part of r_0 between x_0 and B and the part of b between R and x_0. By Proposition 3 and the fact that there are no multiple crossings, the curve α_0 is indeed simple. Orient b and b_1 towards G. Since the resulting orientation of the crossings x_0 and x_1 has to be consistent, it follows that the part of b_1 that connects x_1 with G has to intersect α_0. More specifically, since there are no multiple crossings, it needs to intersect g in some point z. We now redraw g along the part of b that connects R with x_0 and the part of r_0 that connects x_0 with B. The redrawn version of g only has crossings along the part between x_0 and B. In particular, it crosses b_1 at x_1, but the orientation of this crossing is consistent with the orientation of z in the original drawing of g. The same argument applies for all other intersected edges. Consequently, we introduce no additional crossings, eliminate the crossing x, and maintain fan-planarity. Hence, the statement of the lemma holds if the special vertex of r_0 is G. It remains to consider the case where the special vertex of r_0 is R and, hence, b_1 is incident to R. It follows that invariant I2 is satisfied for both r_0 and b_1.

Invariant I3 for r_0 is satisfied by construction (and for b_1 there is nothing to show). Invariant I4 is also satisfied for r_0 and b_1 by construction.

The edge r_0 cannot cross b or b_1 a second time. If it crosses g, then it is incident to G, the special vertex of g. In any case, this implies invariant I5 for Γ_0.

We observe that b_1 cannot cross b or g since this would imply that b_1 is incident to B or G (the special vertex of b and g, respectively) and hence b_1 is parallel to g or b, respectively. Moreover, b_1 cannot cross r_0 a second time. Hence, the part of b_1 that leads from R to x_1 is crossing-free in the drawing Γ_1. Together with invariant I5 for Γ_0, the invariant I5 holds for Γ_1 and invariant I6 holds, which concludes the base case.

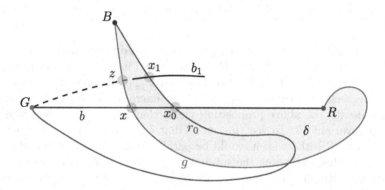

Fig. 7. r_0 can be incident to G. b_1 is drawn as if G is the special vertex of r_0.

Inductive Step: Now, assume the first $j+1$ edges, $r_0, b_1, \ldots, q_j$, have been determined and $j < k$. We assume all invariants hold for $r_0, \ldots q_{j-1}$. Additionally, we assume that I1, I4, I5 and I6 hold for q_j if j is even (and hence $q_j = r_j$ is red), or I2, I4, I5 and I6 hold for q_j if j is odd (and hence $q_j = b_j$ is black).

We will now determine the edge q_{j+1}. If j is even, we need to prove the invariants I2 and I3 for q_j and the invariants I2, I4, I5 and I6 for q_{j+1}. If j is odd, we need to prove the invariants I1 and I3 for q_j and the invariants I1, I4, I5 and I6 (and I2 and I3 if $j+1 = k$) for q_{j+1}.

Case 1: $q_j = r_j$. Note that in this case, we have nothing to prove for invariant I1.

If the edge r_j has no crossings between B and x_j, then we could redraw g along this part of r_j and the arc of b_{j-1} from x_j to R. The redrawn version of g would then be uncrossed by invariant I3 for b_{j-1} and the lemma is proved.

Otherwise r_j has at least one crossing between B and x_j. We determine edge b_{j+1} as follows: traverse along r_j from B towards x_j until we encounter the first edge that crosses r_j, let this edge be b_{j+1}.

Invariant I3 for r_j is satisfied by construction. To prove the remaining invariants, we establish several propositions, the proofs of which can be found in the full version [19]. First we prove invariant I2 for r_j and b_{j+1}.

Proposition 4. *Edge b_{j+1} is incident to R.*

Since r_j crosses both edges b_{j+1} and b_{j-1}, which are both incident to R, the special vertex of r_j is R, which proves invariant I2.

Next we prove invariant I4. We first need to prove b_{j+1} is distinct from all the previous edges.

Proposition 5. *b_{j+1} is distinct from all edges of Γ_j and does not cross edge r_0.*

Lastly, to prove invariants I5 and I6, we have to prove the following proposition.

Proposition 6. *The arc of b_{j+1} between R and x_{j+1} is uncrossed in the drawing Γ_{j+1}.*

So the arc of b_{j+1} between x_{j+1} and R is uncrossed in the drawing Γ_{j+1}. Further, the arc of r_j between x_j and B was uncrossed in Γ_j as noted in Remark 1. Since b_{j+1} is the only new edge introduced for Γ_{j+1}, the arcs of r_j between x_j and x_{j+1} as well as between x_{j+1} and B are uncrossed in Γ_{j+1}. The latter in conjunction with the above proposition yields that indeed, there is a face f_{j+1} admitting invariant I5. To see this, note that b_{j+1} cannot cross g, since it is not incident to G (otherwise it would be parallel to b). Therefore, no additional edges cross g while extending the subdrawing from Γ_j to Γ_{j+1}. Invariant I5 can be combined with the fact that the arc of b_{j-1} from R to x_j is uncrossed by invariant I3 to conclude that the triangular region $f_j \setminus f_{j+1}$ is indeed empty and invariant I6 is established.

This concludes the proof of the lemma in the case when $q_j = r_j$. The second case, $q_j = b_j$, is similar to the first one and can be found in the full version of this paper [19]. $\square$

Now that we concluded the proof of Lemma 3, we have all the tools to prove Lemma 4.

Lemma 4. *Let Γ be a non-simple fan-planar drawing with the properties established by Corollary 4. If there is an edge $b = (G, R)$ in Γ that crosses an edge g at x and g is incident to R, then we can redraw an edge such that the total number of crossings in Γ decreases, and the drawing remains fan-planar.*

Proof. Let $b = (G, R)$ and $g = (B, R)$ be two adjacent edges which cross at x. Their common endpoint is not the special vertex of either of the edges by Corollary 4. Thus, the special vertices of b and g must be B and G, respectively. We apply Lemma 3 on b and g. If g can be redrawn using Lemma 3, this concludes the proof of Lemma 4. Assume that g cannot be redrawn. Then we can determine a sequence of edges $r_0, b_1, r_2, \ldots, r_k$ with the properties described in Lemma 3. We now describe how the edge b can be redrawn to eliminate the crossing x while maintaining fan-planarity and decreasing the overall number of crossings.

Let the other endpoint of edge r_k be W. By invariant I4, r_k has a crossing with edge b. First assume this crossing occurs between x_k and W, i.e., after r_k enters the triangular region $f_{k-2} \backslash f_{k-1}$ at x_k. Since b does not enter this region, r_k has to leave it. It cannot cross b_{k-1} again, nor can it cross r_{k-2}, because it is not incident to its special vertex R (note that $W \neq R$ since otherwise r_k would be parallel to g). Finally, it cannot cross b_{k-3}, because this is the part of b_{k-3} that is uncrossed by invariant I3. Hence, the crossing of r_k and b cannot lie between x_k and W and must instead lie between B and x_k along r_k. In this case, we claim that edge b can be redrawn. Redraw edge b to follow g from R until x, and then follow its previous drawing from x until G while avoiding crossing g at x, as illustrated in Fig. 8. We now prove that this redrawing does not introduce any new crossings on b.

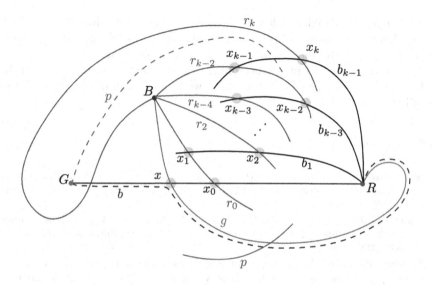

Fig. 8. Redrawing of edge b.

Proposition 7. *Redrawing b does not introduce any new crossings on b.*

Proof. Assume a new crossing with an edge p is introduced on b by the redrawing operation. Since the redrawn part of b is parallel to a part of g, the edge p crosses g as well. Consequently, edge p is incident to G, the special vertex of g.

Consider the closed curve δ formed by the arc of r_k between B and x_k, the arc of r_{k-2} between B and x_{k-1}, and the arc of b_{k-1} between x_{k-1} and x_k. The edge b crosses r_k exactly once, does not cross r_{k-2} due to invariant I4, and also does not cross b_{k-1} since the special vertex of b_{k-1} is B due to invariant I1 and b is not incident to B. This implies that b crosses δ exactly once and thus R and G have to lie on distinct sides of δ. This is illustrated in Fig. 8. Edge g does not cross any of the edges on the boundary of δ since b is the only edge crossed by

g by Remark 1 except possibly for r_0, and even if $r_{k-2} = r_0$ the part of r_{k-2} on δ is still uncrossed by invariant I3, and therefore g is contained in the same side of δ as its endpoint R.

The edge p crosses g and is incident to G, and thus must cross the curve δ since g and G lie on distinct sides of δ. Edge p cannot be incident to R since then p would be parallel to b. Since R is the special vertex of r_k and r_{k-2} and p is not incident to R, p cannot cross the edges r_k and r_{k-2}. Hence, p must cross edge b_{k-1} to cross the curve δ. Then the other endpoint of p must be B, the special vertex of b_{k-1}. However, the part of p connecting G with b_{k-1} is on the same side as the part of r_{k-2} between B and b_{k-1}. Consequently, the part of p connecting B to b_{k-1} and the part of r_{k-2} between B and b_{k-1} lie on distinct sides of b_{k-1}, which contradicts the fan-planarity. Overall, we have shown that p cannot cross δ and, by extension, it cannot cross g; a contradiction.   □

The only redrawn edge is b and no new crossing is introduced on b, which ensures that fan-planarity is maintained. Additionally, we eliminate the crossing x, which decreases the total number of crossings in the drawing.   □

We already described in the beginning of Sect. 3 how Lemmata 1–4 can be combined to obtain a proof of Theorem 1; we formally summarize the proof in the full version [19].

References

1. Ackerman, E., Tardos, G.: On the maximum number of edges in quasi-planar graphs. J. Comb. Theory, Ser. A **114**(3), 563–571 (2007). https://doi.org/10.1016/j.jcta.2006.08.002
2. Agarwal, P.K., Aronov, B., Pach, J., Pollack, R., Sharir, M.: Quasi-planar graphs have a linear number of edges. Combinatorica **17**(1), 1–9 (1997). https://doi.org/10.1007/BF01196127
3. Angelini, P., Bekos, M.A., Kaufmann, M., Kindermann, P., Schneck, T.: 1-fan-bundle-planar drawings of graphs. Theor. Comput. Sci. **723**, 23–50 (2018). https://doi.org/10.1016/j.tcs.2018.03.005
4. Bae, S.W., Baffier, J., Chun, J., Eades, P., Eickmeyer, K., Grilli, L., Hong, S., Korman, M., Montecchiani, F., Rutter, I., Tóth, C.D.: Gap-planar graphs. Theor. Comput. Sci. **745**, 36–52 (2018). https://doi.org/10.1016/j.tcs.2018.05.029
5. Bekos, M.A., Cornelsen, S., Grilli, L., Hong, S.-H., Kaufmann, M.: On the recognition of fan-planar and maximal outer-fan-planar graphs. Algorithmica **79**(2), 401–427 (2016). https://doi.org/10.1007/s00453-016-0200-5
6. Bekos, M.A., Grilli, L.: Fan-planar graphs. In: Hong, S.-H., Tokuyama, T. (eds.) Beyond Planar Graphs, pp. 131–148. Springer, Singapore (2020). https://doi.org/10.1007/978-981-15-6533-5_8
7. Bekos, M.A., Kaufmann, M., Raftopoulou, C.N.: On optimal 2- and 3-planar graphs. In: Aronov, B., Katz, M.J. (eds.) 33rd International Symposium on Computational Geometry, SoCG 2017, July 4–7, 2017, Brisbane, Australia. LIPIcs, vol. 77, pp. 16:1–16:16. Schloss Dagstuhl - Leibniz-Zentrum für Informatik (2017). https://doi.org/10.4230/LIPIcs.SoCG.2017.16

8. Binucci, C., Chimani, M., Didimo, W., Gronemann, M., Klein, K., Kratochvíl, J., Montecchiani, F., Tollis, I.G.: Algorithms and characterizations for 2-layer fan-planarity: from caterpillar to stegosaurus. J. Graph Algorithms Appl. **21**(1), 81–102 (2017). https://doi.org/10.7155/jgaa.00398
9. Binucci, C., Giacomo, E.D., Didimo, W., Montecchiani, F., Patrignani, M., Symvonis, A., Tollis, I.G.: Fan-planarity: properties and complexity. Theor. Comput. Sci. **589**, 76–86 (2015). https://doi.org/10.1016/j.tcs.2015.04.020
10. Brandenburg, F.J.: On fan-crossing graphs. Theor. Comput. Sci. **841**, 39–49 (2020). https://doi.org/10.1016/j.tcs.2020.07.002
11. Cheong, O., Har-Peled, S., Kim, H., Kim, H.-S.: On the number of edges of fan-crossing free graphs. Algorithmica **73**(4), 673–695 (2014). https://doi.org/10.1007/s00453-014-9935-z
12. Didimo, W., Eades, P., Liotta, G.: Drawing graphs with right angle crossings. Theor. Comput. Sci. **412**(39), 5156–5166 (2011). https://doi.org/10.1016/j.tcs.2011.05.025
13. Didimo, W., Liotta, G., Montecchiani, F.: A survey on graph drawing beyond planarity. ACM Comput. Surv. **52**(1), 4:1-4:37 (2019). https://doi.org/10.1145/3301281
14. Holten, D.: Hierarchical edge bundles: visualization of adjacency relations in hierarchical data. IEEE Trans. Vis. Comput. Graph. **12**(5), 741–748 (2006). https://doi.org/10.1109/TVCG.2006.147
15. Holten, D., van Wijk, J.J.: Force-directed edge bundling for graph visualization. Comput. Graph. Forum **28**(3), 983–990 (2009). https://doi.org/10.1111/j.1467-8659.2009.01450.x
16. Hong, S.-H., Tokuyama, T. (eds.): Beyond Planar Graphs. Springer, Singapore (2020). https://doi.org/10.1007/978-981-15-6533-5
17. Kaufmann, M., Ueckerdt, T.: The density of fan-planar graphs. CoRR abs/1403.6184v1 (2014). http://arxiv.org/abs/1403.6184v1
18. Kaufmann, M., Ueckerdt, T.: The density of fan-planar graphs. CoRR abs/1403.6184v2 (2014). http://arxiv.org/abs/1403.6184v2
19. Klemz, B., Knorr, K., Reddy, M.M., Schröder, F.: Simplifying non-simple fan-planar drawings. CoRR abs/2108.13345 (2021). https://arxiv.org/abs/2108.13345
20. Pach, J., Tóth, G.: Graphs drawn with few crossings per edge. Combinatorica **17**(3), 427–439 (1997). https://doi.org/10.1007/BF01215922
21. Telea, A.C., Ersoy, O.: Image-based edge bundles: simplified visualization of large graphs. Comput. Graph. Forum **29**(3), 843–852 (2010). https://doi.org/10.1111/j.1467-8659.2009.01680.x

RAC-Drawability is ∃ℝ-Complete

Marcus Schaefer[✉]

DePaul University, Chicago, IL 60604, USA
mschaefer@cdm.depaul.edu

Abstract. A RAC-*drawing* of a graph is a straight-line drawing in which every crossing occurs at a right-angle. We show that deciding whether a graph has a RAC-drawing is as hard as the existential theory of the reals, even if we know that every edge is involved in at most ten crossings and even if the drawing is specified up to isomorphism.

Keywords: RAC-drawing · Right-angle drawing · Straight-line drawing · Existential theory of the reals · Computational complexity

1 Introduction

It is generally admitted that if we cannot avoid crossings in drawings, then it is advantageous to draw the crossings with large angles. This simplifies reading the drawing as edges become easier to follow individually. In a 2009 paper, Didimo, Eades and Liotta [12] formalized this idea by introducing RAC-drawings of graphs, in which only the largest possible angle, the right-angle, is allowed. In other words, a drawing of a graph is a RAC-*drawing* if all crossings in the drawing occur at right angles.

RAC-drawings have become an increasingly popular subject in the graph drawing literature, see, for example, a recent survey by Didimo [11] summarizing our knowledge. We are specifically interested in the computational complexity of the recognition problem.

Argyriou, Bekos, and Symvonis [3] showed early on that it is **NP**-hard to recognize whether a graph has a RAC-drawing. Why **NP**-hard, and not **NP**-complete? The issue is that a realization of a RAC-drawing may require real coordinates, and, a priori, we do not have any bounds on the precision and we do not even know whether the graph can be realized on a grid. Bieker [6] showed that the problem lies in ∃ℝ, the complexity class associated with deciding the truth of the existential theory of the reals.[1] The exact complexity remained open (as mentioned, for example, in [13, p. 4:11/12]).

Also open, not even known to be **NP**-hard, was the complexity of the *fixed embedding* variant of RAC-drawability, in which we are given a drawing of the graph and have to decide whether the graph has a RAC-drawing isomorphic to the given drawing.

---

[1] For a thorough introduction to the existential theory of the reals, see [17]. For a quick intro, the Wikipedia page [25] will serve.

ⓒ Springer Nature Switzerland AG 2021
H. C. Purchase and I. Rutter (Eds.): GD 2021, LNCS 12868, pp. 72–86, 2021.
https://doi.org/10.1007/978-3-030-92931-2_5

Theorem 1. *Testing whether a graph (with or without fixed embedding) has a RAC-drawing is ∃ℝ-complete, even if each edge has at most ten crossings.*

∃ℝ-hardness implies **NP**-hardness [24], so the fixed embedding variant is **NP**-hard. What does ∃ℝ-hardness add that **NP**-hardness does not give us? Perhaps nothing, since it is possible (if considered unlikely) that **NP** = ∃ℝ. Nevertheless, our ∃ℝ-hardness reduction shows that RAC-drawings can require arbitrarily complex algebraic integers in any realization. And even RAC-drawings that can be realized on a grid, may require double-exponential precision. We discuss these results on area and precision in Sect. 5.

1.1   More on RAC-Drawings

As far as we know the double-exponential lower bound on the area of a RAC-drawing is new, but there have been (single) exponential lower bounds in constrained settings, e.g. for upward RAC-drawings [2], RAC-drawings in which a given horizontal order of the vertices must be realized, and for 1-plane RAC-drawings [8], drawings in the plane with at most one crossing per edge.

If we allow bends along edges, the situation changes dramatically. A RAC_k drawing of a graph is a RAC-drawing in which every edge has at most k bends. RAC-drawings are just the RAC_0-drawings.

Every graph has a RAC_3-drawing [12], but not necessarily a RAC_2-drawing: any graph with a RAC_2-drawing has at most linearly many edges [4]. The complexity of recognizing graphs with RAC_1- and RAC_2-drawings remains tantalizingly open [13, Problem 6].

There are polynomial upper bounds on the area of RAC_k drawings for $k \geq 3$ [15], but no bounds seem to be known for $k = 1, 2$. ∃ℝ-hardness of these cases would likely imply double-exponential lower bounds.

1.2   Overview

Our proof that the RAC-drawability problem is ∃ℝ-complete consists of a sequence of reductions. There is no convenient graph drawing problem to start with, so in Sect. 2 we present an ∃ℝ-complete algebraic problem.

The main idea then is to enrich the RAC-drawing model with a special feature: vertices with a specific type of angle constraints. The proof of Theorem 1 then breaks into two major parts. The first part shows that this additional feature leads to an ∃ℝ-complete problem (even for crossing-free drawings), since it is powerful enough to encode the algebraic problem, see Sect. 3. The second part shows how to simulate the special feature within the RAC-model. This second part can be found in the arXiv version of this paper [21, Appendix B]. In Sect. 5 we discuss issues of precision, area, and universality, before closing with a short section of open questions.

2   The Existential Theory of the Reals

The *existential theory of the reals* is the set of existentially quantified, true statements over the real numbers. The corresponding complexity class is $\exists\mathbb{R}$. (Defined just like **NP** is defined from Boolean formula satisfiability, though there also is a machine model [14].) A problem is $\exists\mathbb{R}$-*hard* if every problem in $\exists\mathbb{R}$ reduces to it; it is $\exists\mathbb{R}$-*complete*, if it is $\exists\mathbb{R}$-hard and lies in $\exists\mathbb{R}$.

$\exists\mathbb{R}$ captures the complexity of many natural problems in graph drawing, particularly if real coordinates are involved. Bieker's thesis [6] surveys many of the relevant graph drawing results. Some recent problems shown $\exists\mathbb{R}$-complete relevant to graph drawing include: visibility graphs of triangulated irregular networks [7], the local rectilinear crossing number [22], simultaneous geometric embeddings of paths [23], and covering polygons by triangles [1].[2]

For our reduction we will be working with an $\exists\mathbb{R}$-complete problem, which, as is often the case, is tailor-made for the situation we find ourselves in. The proof of the theorem can be found in [21, Appendix A]; it combines ideas from Mněv [18], Shor [24], and Richter-Gebert [19].

Theorem 2. *The following problem is* $\exists\mathbb{R}$-*complete: Given equations of the form* $x_i = 2$, $x_i = x_j$, $x_i = x_j + x_k$, *and* $x_i = x_j \cdot x_k$ *for variables* $x_1, \ldots, x_n$, *decide whether the equations have a solution with* $x_i > 1$ *for all* $i \in [n]$.

As we mentioned, **NP** $\subseteq \exists\mathbb{R}$ [24], in particular, $\exists\mathbb{R}$-hard problems are also **NP**-hard. On the other hand, $\exists\mathbb{R} \subseteq$ **PSPACE** [9], so $\exists\mathbb{R}$-complete problems are solvable in polynomial space (and, therefore, exponential time).

3   RAC-Drawings with Angle Constraints

We introduce two types of special vertices that come with angle and rotation constraints. Recall that the rotation at a vertex in a drawing is the clockwise permutation of edges incident to the vertex.

- a $\top$-*junction* is a vertex v which is incident to three special edges e_1, e_2, e_3 (v may be incident to additional edges). In a straight-line drawing we require that the rotation of the special edges at v is $e_1 e_2 e_3$, *or the reverse*, and there are right angles between e_1 and e_2 and e_2 and e_3 at v; additional edges at v can occur, at any angle, between e_1 and e_3 (opposite of e_2),
- a $\times$-*junction* is a vertex v which is incident to four special edges e_1, e_2, e_3, e_4 (v may be incident to additional edges). In a straight-line drawing we require that the rotation of the special edges at v is $e_1 e_2 e_3 e_4$, *or the reverse*, and that there are right angles between e_i and e_{i+1}, for $1 \le i \le 3$; additional edges can occur, at any angle, inside one of the quadrants, e.g. between e_3 and e_4.

Figure 1 shows these junctions, and how we symbolize them in drawings.

---

[2] These results are from 2021. The Wikipedia page mentioned earlier [25] is host to a growing list of complete problems, many from the areas of graph drawing and computational geometry.

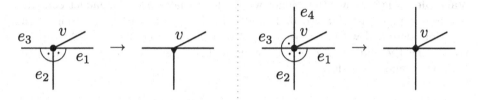

Fig. 1. T- and ×-junctions, and how we draw them in graphs using a ▼ and a ♦. Each of the junctions is shown with one additional edge.

If special edge e_2 in a T-junction ends in a leaf, we can think of the junction as a straight-line, $e_3 e_1$, with a vertex on it.

Two drawings of a graph (with or without junctions) are *isomorphic* if there is a homeomorphism of the plane (which may be orientation-reversing) that maps the graphs to each other.

Theorem 3. *Testing whether a graph G with T- and ×-junctions has a RAC-drawing, is ∃ℝ-complete. This remains true even if we are given a crossing-free drawing of G and G either has no RAC-drawing, or it has a crossing-free RAC-drawing which is isomorphic to the given drawing.*

We will later see that ×-junctions can be simulated by T-junctions, so they are not, strictly speaking, necessary, but they do simplify the constructions.

For the promise version note that T- and ×-junctions are vertices (and not crossings), and that a RAC-drawing does not have to contain any crossings. The theorem implies that testing whether a graph with T- and ×-junctions has a crossing-free RAC-drawing is also ∃ℝ-complete, but we need the stronger promise version for the main proof.

We prepare the proof of Theorem 3 by constructing gadgets to simulate arithmetic.

3.1 Gadgets

We describe the gadgets we use to enforce equations $x_i = 2$, $x_i = x_j + x_k$, $x_i = x_j \cdot x_k$ and $x_i = x_j$. Some additional gadgets will be useful.

We start by defining how a drawing of a graph encodes a real number. Given three collinear points (these will be vertices of the graph) labeled 0, 1 and x, we say x *represents* $\pm d(0, x)/d(0, 1)$, where d is the Euclidean distance of two points. If x lies on the same side of 0 as does 1 (on the common line), we choose the positive value, otherwise we choose the negative value. With this definition we can build our first gadget.

Variables. Fig. 2 shows the gadget we use for variable x_i. The gadget consists of three ×-junctions (labeled 0, 1, and x in this order) with the e_3-edge of a junction identified with the e_1-edge of the next junction. We also add a path connecting the ends a, b, c of the e_2-edges. The e_4-edges can then be used to connect to another gadget.

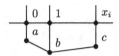

Fig. 2. A gadget for variable x_i.

In a RAC-drawing of the x_i-gadget, all e_4-edges must lie on the same side of the line through 0, 1, x_i. This is because ab and bc cannot cross the line, since otherwise they would overlap with the e_4-edges they are incident to. Thus, the x_i-gadget can be used to represent any number $x_i > 1$. By relabeling the junctions, we can also obtain gadgets for variables between 0 and 1, and variables less than 0.

Copying and Moving Information. Our next gadget will allow us to duplicate information. More precisely, we have points $p_1, \ldots, p_\ell$ along a line (in this order). If two of those points are labeled 0 and 1, then all these points represent numbers. Our gadget allows us to make two copies of these points (along a new common line), that each, by itself, represents the same numbers as the original. See Fig. 3.

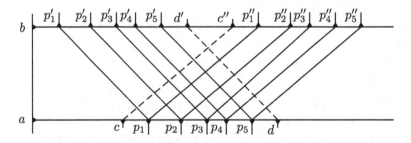

Fig. 3. The copy gadget. The dashed edges have to cross orthogonally, forcing the $p_i p_i'$ as well as the $p_i p_i''$ edges to be parallel as shown.

The two e_2-edges incident to the ⊤-junctions a and b are parallel, since they are both orthogonal to ab, and leave ab in the same direction, since otherwise edge cc'', for example, would have to cross ab at a right angle, and overlap ca. Because of the relative order of the points on the top and bottom line, edge cc'' crosses dd' as well as all edges $p_i p_i'$ at right-angles, so all these edges are parallel.

And since every edge $p_j p_j''$ is crossed by dd', these edges are also all parallel. It follows that $p_1', \ldots, p_\ell'$ and $p_1'', \ldots, p_\ell''$ represent the same numbers as $p_1, \ldots, p_\ell$.

By repeating the copy gadget, we can make any number of copies of a set of points on a line. There are other applications of the copying gadget, some of which we will see below, but here we can show how to use it to test whether $x_i = x_j$ for two variables x_i and x_j which we know to be larger than 1. Figure 4 shows the set-up.

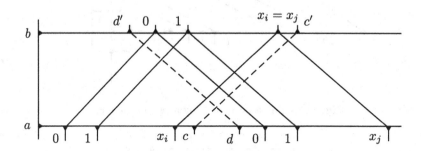

Fig. 4. Testing equality $x_i = x_j$ using the copy gadget.

We move information around the drawing as parallel lines, with the relative distances of the lines encoding the numbers. However, we may have to change direction, and the gadget shown in Fig. 5 allows us to create a copy $p_1', \ldots, p_\ell'$ of a set of points $p_1, \ldots, p_\ell$ at an angle.

The p_i and p_i' are ⊤-junctions, and v is a ×-junction. Edges $p_i p_i'$ must lie between the two lines on which the points lie (and on opposite sides of the e_2-edges of the ⊤-gadgets for p_i and p_i', since otherwise edges would overlap). The ×-junctions surrounding v force vc to also lie between those two lines. Then cv must cross each of the edges $p_i p_i'$ which implies that they are parallel, so the p_i' have the same relative distances from each other as do the p_i, and they represent the same numbers.

By chaining several angling gadget, we can achieve angles of $\pi/2$, π, $3\pi/2$, and 2π. In particular chaining two gadgets, its main axes combined like ⌐Γ, allows us to continue sending the information in the same direction, but at a different (arbitrarily chosen) scale. Hence, we can also use the angling gadget to rescale information. (The resulting horizontal offset can be compensated for by adding two more angling gadgets on top, like ⨽.)

Doing Arithmetic I. Let us start with the number 2. The gadget shown in Fig. 6 consists of ×-junctions combining four squares with diagonals.

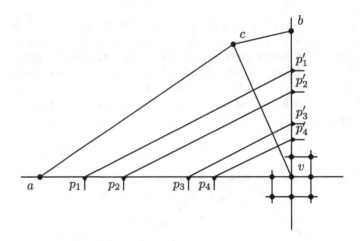

Fig. 5. Making a (scaled) copy at a right angle.

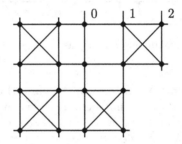

Fig. 6. The number 2 gadget.

In a RAC-drawing, the diagonal edges force the sides of each square to have equal length. This forces the length of 01 and 12 to be the same. (Why do we not place two squares with diagonals right next to each other? The reason is that our ×-junctions only allow additional edges in one quadrant.)

Instead of 0, 1, 2 we can also label the points of the gadget as -1, 0, 1. This gives us a -1 which we can use to build a negation gadget. (While the x_i are all bigger than 1, we will make use of some additional, temporary, variables that are negative.)

We next build a special negation gadget that, for given points 0, 1, and x creates four points, $-x - 1$, $-x$, 0, and 1 that correctly represent their labels. See Fig. 7. On the lower left is a 2-gadget with points relabeled -1, 0, and 1. We added two points to this gadget, labeled x' and $x' + 1$. Using the diagonals, the distance between x' and $x' + 1$ is the same as the distance between 0 and 1 in this gadget. We then use a copy gadget in reverse to combine this information with the x-gadget. The resulting five points then correspond to -1, 0, 1, x, and $x + 1$, or, more useful for us $-x - 1$, $-x$, $-x + 1$, 0 and 1. Of these we only need four as output, namely $-x - 1$, $-x$, 0 and 1.

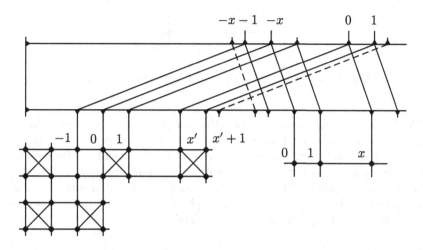

Fig. 7. Negating x gadget.

Doing Arithmetic II. To compute $x_i = x_j + x_k$, we use the negation gadget to create $-x_j - 1$ and $-x_j$, and then perform the addition geometrically, as shown in Fig. 8. The reason for using negation is that this way we do not have to know whether x_j or x_k is the larger one, something that's unavoidable if we place two variables greater than 1 on the same line. (This idea is due to Richter-Gebert [19], also see Matoušek [17].)

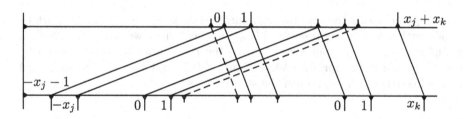

Fig. 8. Addition gadget computing $x_i = x_j + x_k$.

As in the other gadgets, the two dashed edges are orthogonal, and force the other lines to be parallel. If we ignore those two edges and their endpoints, the remaining points on top represent $-x_j - 1$, $-x_j$, 0, 1, x_k. If we relabel the first and second points as 0 and 1, then the fifth point represents $x_j + x_k$, and we have the addition gadget we needed.

To compute $x_i = x_j \cdot x_k$ we use the same trick mentioned above to resolve the issue that we do not know which of x_j and x_k is bigger, so we cannot place them along the same line. Instead of working with negation, here we work with the reciprocal of x_j, which lies (strictly) between 0 and 1. The reciprocal gadget is not much of a gadget, just a relabeling, see Fig. 9.

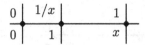

Fig. 9. Computing the reciprocal $1/x$ of a variable $x > 1$.

This is cheating a bit, since for most other gadgets (except the angling gadget) the distance between 0 and 1 did not change, and most gadgets assume they are the same when processing inputs. So before using the result of the reciprocal gadget, we rescale it, so that 0 and 1 have the standard distance.

With that, the product $x_i = x_j \cdot x_k$ can then be calculated using the multiplication gadget shown in Fig. 10. It is simply a copy gadget upside down that allows us to merge $1/x_j$ and x_k into a common scale.

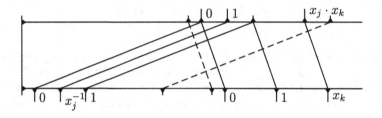

Fig. 10. Multiplication gadget computing the product $x_i = x_j \cdot x_k$.

The points along the top line (ignoring the endpoints of dashed lines which enforce orthogonality) represent $0, 1/x_j, 1$ and x_k, which we can relabel as $0, 1$, x_j and $x_j \cdot x_k$, at which point we can drop the x_j to obtain a gadget computing $x_j \cdot x_k$. As did the reciprocal gadget above, this gadget changes the scale, and we need to rescale to reestablish the standard distance between 0 and 1.

3.2 Proof of Theorem 3

We can assume that we are given a system of equations over variables x_i, $i \in [n]$ as described in Theorem 2. Our goal is to (efficiently) construct a graph G with junctions, so that the system of equations is solvable, if and only if the graph G has a RAC-drawing. Moreover, if the system is solvable, then G has a crossing-free RAC-drawing (remember that junctions do not count as crossings).

We build G in several stages. We first create gadgets for the main operations.

(i) For every variable x_i, $i \in [n]$, we create a variable gadget.
(ii) We have equations of four types: $x_i = 2$, $x_i = x_j$, $x_i = x_j + x_k$, and $x_i = x_j \cdot x_k$, and we create a corresponding gadget for each ($x_i = 2$ can be built by combining an equality gadget with a 2-gadget; the addition and multiplication gadgets include the gadgets for negation and reciprocals).

(*iii*) If variable x_i occurs in ℓ equations, we create ℓ copy-gadgets.

Let use say this gives us m gadgets. Place all m gadgets along a vertical line spaced far apart, and so that incoming and outgoing (information carrying) edges are vertical.

We now need to connect the gadgets using angling gadgets. We need the following connections: the variable x_i gadget needs to be connected as an input to one of the ℓ copy gadgets representing it, and we need to connect these copy gadgets so that we have ℓ outputs corresponding to x_i; finally, if x_i occurs in an equation, we need to connect one of the unused outputs from the bank of copy gadgets representing x_i to the equation.

The connections between the m original gadgets can be built from a chained sequence of at most eight angling gadgets. This is already sufficient to get a G that has a RAC-drawing, but we need one more step to remove all crossings, so we describe how to place the angling gadgets more carefully; the placement only depends on the original equations, not on a specific solution to the equations.

Suppose we are connecting gadget α to gadget β, with $\alpha, \beta < m$. Start at gadget α. At most two angling gadgets let us leave the gadget α horizontally to the right up to a distance of $\alpha + \beta \cdot m$. Another two angling gadgets allow us to angle so we can move vertically downwards or upwards until we have reached the height of gadget β. Two more angling gadgets take us back horizontally to β, and we can connect to β with another two angling gadgets.

This process introduces (orthogonal) crossings between edges of angling gadgets, but we can determine exactly what those crossings are (whether the gadgets are realizable, or not). We replace all crossings within gadgets and the newly introduced crossings between angling gadgets with ×-junctions. The resulting graph is graph G and we have also described a plane embedding of G (which need not satisfy the junction-constraints, of course).

If the system of equations is solvable, we can use a solution to create a RAC-drawing of each gadget, and, following the description above, create an isomorphic RAC-drawing of G. In particular, there are no crossings.

On the other hand, if there is a RAC-drawing of G, all the gadgets work as described, and each occurrence of a variable represents the same value, so there is a solution to the system of equations.

3.3   Forcing Empty Faces

To model ⊤- and ×-junctions as normal vertices we make use of another restricted drawing mode that simplifies the construction. Given a graph G with k pairwise disjoint sets of vertices $V_i \subseteq V(G)$, $i \in [k]$, we are interested in drawings (or RAC-drawings) of G in which the vertices of each V_i lie on the boundary of an empty face, for $i \in [k]$.

The following theorem shows that this type of drawing constraint can be removed, even in RAC-drawings.

Theorem 4. *Let G be a graph, and let $(V_i)_{i \in [k]}$ be k pairwise disjoint sets of vertices of G. We can then construct, in polynomial time, a graph G' so that G*

has a RAC-*drawing in which the vertices of each* V_i *lie on the boundary of an empty face, if and only if* G' *has a* RAC-*drawing.*

For the proof, remember that if a graph has a RAC-drawing, it can have at most $4n - 10$ edges [12].

Proof. Let $v_1, \ldots, v_k$ be k vertices not in G. Connect v_i to each vertex in V_i by a path of length $36n$, where $n = |V(G)|$, for $1 \le i \le k$. Finally, replace each of the newly added edges by $\ell = 146n^2$ paths of length 2 (that is, a $K_{2,\ell}$). See Fig. 11 for an illustration. Call the resulting graph G'. Suppose G' has a RAC-drawing.

Fig. 11. Forcing an empty face; in this example $v = v_i$ and $V_i = \{b_1, b_2, b_3\}$.

Since we added at most n paths of length $36n$ to G, we have at most $36n^2$ of the $K_{2,\ell}$-graphs in G'. We consider them one at a time. Let $E_0 = E(G)$. An edge can cross at most one edge of a $K_{1,\ell}$ at right angles (otherwise, the edges of $K_{1,\ell}$ would overlap). Hence, an edge can cross at most two edges of a $K_{2,\ell}$. Since $\ell > 2n = 2|E_0|$, the first $K_{2,\ell}$ contains a path P_1 of length 2 that crosses none of the edges in E_0. Let $E_1 = E_0 \cup E(P_1)$. If we keep repeating this argument, we obtain $E_i = E_{i-1} \cup E(P_i)$, and $|E_i| = n + 2i$. Then the $(i+1)$-st $K_{2,\ell}$ must contain a path P_{i+1} of length 2 which crosses none of the edges in E_i, since $2|E_i| = 2(n + 2i) < \ell$, for all $1 \le i \le 36n^2$. We conclude that the RAC-drawing of G' contains a RAC-drawing of G together with the v_i and paths from v_i to each vertex in V_i so that none of the paths are involved in any crossings. In other words, for each i there is a crossing-free (subdivided) wheel with center v_i and a perimeter containing V_i. Removing all edges not belonging to G gives us a RAC-drawing of G in which all vertices of V_i lie on the boundary of the same face (the one that contained v_i).

For the other direction, suppose G has a RAC-drawing in which all vertices of each V_i lie on the boundary of the same face. For each i, we create a new vertex v_i and a paths of length $72n$ connecting v_i to each vertex in V_i. By Theorem 1 in [10], the additional edges and vertices can be added to the already existing drawing of G without creating any new crossings. We can then duplicate appropriate sub-paths of length 2 to obtain a RAC-drawing of G'.

4 Proof of Theorem 1

Bieker [6, Section 6.2] shows that the problem lies in ∃ℝ. To prove ∃ℝ-hardness, we are missing one more ingredient, a way to simulate junctions in RAC-drawings.

Theorem 5. *Let D be a planar drawing of a graph G with some vertices identified as $\top$- and $\times$-junctions. We can efficiently construct a graph G' without junctions, with vertex sets $(V_i)_{i\in[k]}$, and a drawing D' of G' so that:*

(i) *if D is isomorphic to a RAC-drawing of G, then D' is isomorphic to a RAC-drawing of G', and all edges are involved in at most ten crossings.*

(ii) *if G does not have a RAC-drawing, then G' does not have a RAC-drawing in which all the vertices of each V_i lie on a common boundary.*

The proof of Theorem 5 can be found in [21, Appendix B]. Let us see how this theorem completes the proof of Theorem 1; we first consider the fixed embedding case. By Theorem 3 it is ∃ℝ-hard to test whether a graph G with junctions has a RAC-drawing, even if we know that the graph either has no RAC-drawing, or that it has a RAC-drawing isomorphic to a given planar drawing D. Using Theorem 5, we construct a graph G' and a drawing D' so that D' is isomorphic to a RAC-drawing if and only if G has a RAC-drawing. This implies that the fixed embedding version of RAC-drawability is ∃ℝ-complete.

To show that the problem is ∃ℝ-complete without fixing the embedding, we need to take one more step: By (ii) we can reduce to G' having a RAC-drawing in which all the vertices of each V_i lie on a common boundary. Using Theorem 4, this then reduces to RAC-drawability (without a fixed embedding).

5 Precision and Area

Given an arbitrary algebraic number α, we can write it as the unique solution of an integer polynomial equation (adding constraints to achieve uniqueness). E.g. $x = \sqrt{2}$ would be the unique x-value for which there is a solution of $(x^2 - 2)^2 + (x - y^2)^2 = 0$. Using the reduction of polynomial equations to RAC-drawability we have seen, we can build a graph G so that in any realization of G there are three collinear points that represent α. We are not limited to just one number. For any semi-algebraic set S in free variables $x_i, i \in [n]$ we can build a graph G so that the triples $(0_i', 1_i', x_i')$ and $(0_i'', 1_i'', x_i'')$ representing $x_i = x_i' - x_i''$ represent the semi-algebraic set in the sense that S consists of the points $(d(0_i', x_i')/d(0_i', 1_i') - d(0_i'', x_i'')/d(0_i'', 1_i''))_{i\in[n]}$.

Similarly, we can build a graph G representing equations $x_1 = 2$, $x_2 = x_1 \cdot x_1$, $x_3 = x_2 \cdot x^2$, ..., $x_n = x_{n-1} \cdot x_{n-1}$. Then any realization of G contains three points representing 0, 1, and 2^{2^n}. So G is a graph of polynomial size which requires double-exponential area. In this example, the points in the gadgets can even be placed so as to lie on a grid.

6   Open Questions

Can we relax the right-angle restriction? Huang, Eades, and Hong [16] studied the impact of large angles (vs right angles) on the readability of drawings, and concluded (among other things) that large angles improve the readability of drawings, but the angles do not have to be right angles. They express the hope that the computational problem becomes easier if the right-angle restriction is relaxed. Using common tricks for the existential theory of the reals, it is possible to show that the RAC-drawability problem remains $\exists\mathbb{R}$-complete even if we relax the angle constraint, and require the angle to lie in the interval $(\pi/2 - \varepsilon, \pi/2 + \varepsilon)$, where ε depends on $n = |V(G)|$ (doubly exponentially so). Does the problem remain $\exists\mathbb{R}$-hard for a fixed value $\varepsilon > 0$? The current construction very much relies on precision to simulate the existential theory of the reals. It is not clear whether gadgets can be braced to still work if angles are only approximate.

We saw that testing RAC-drawability remains $\exists\mathbb{R}$-hard, even if there is a RAC-drawing with at most 10 crossings per edge. Can that number be lowered? We note that the **NP**-hardness result remains true even for 1-planar drawings (at most one crossing per edge) [5]. Is the problem in **NP**? A somewhat similar situation occurs for the geometric local crossing number, $\overline{\mathrm{lcr}}(G)$, that is the smallest number of crossings along each edge in a straight-line drawing of G. Testing whether $\overline{\mathrm{lcr}}(G) \leq 1$ is **NP**-complete [20], but there is a fixed k so that testing $\overline{\mathrm{lcr}}(G) \leq k$ is $\exists\mathbb{R}$-complete [22].

Does RAC-drawability remain $\exists\mathbb{R}$-hard for bounded-degree graphs? Nearly all of our gadgets have bounded degree, the only exception are the empty-face gadgets, which are based on $K_{2,n}$'s, and require unbounded degree. Can these be replaced with bounded degree gadgets?

The *right-angle crossing number* of a graph is the smallest k so that G has a RAC-drawing with at most k crossings. Our result implies that testing whether the right-angle crossing number is finite is $\exists\mathbb{R}$-complete. What about small fixed values? Can we test whether a graph has a RAC-drawing with one, two, three crossings in polynomial time? What about fixed k?

References

1. Abrahamsen, M.: Covering polygons is even harder. ArXiv e-prints (2021). arXiv:2106.02335. Accessed 22 July 2021
2. Angelini, P., et al.: On the perspectives opened by right angle crossing drawings. J. Graph Algorithms Appl. **15**(1), 53–78 (2011). https://doi.org/10.7155/jgaa.00217
3. Argyriou, E.N., Bekos, M.A., Symvonis, A.: The straight-line RAC drawing problem is NP-hard. J. Graph Algorithms Appl. **16**(2), 569–597 (2012). https://doi.org/10.7155/jgaa.00274
4. Arikushi, K., Fulek, R., Keszegh, B., Morić, F., Tóth, C.D.: Graphs that admit right angle crossing drawings. Comput. Geom. **45**(4), 169–177 (2012). https://doi.org/10.1016/j.comgeo.2011.11.008
5. Bekos, M.A., Didimo, W., Liotta, G., Mehrabi, S., Montecchiani, F.: On RAC drawings of 1-planar graphs. Theoret. Comput. Sci. **689**, 48–57 (2017). https://doi.org/10.1016/j.tcs.2017.05.039

6. Bieker, N.: Complexity of graph drawing problems in relation to the existential theory of the reals. Bachelor's thesis, Karlsruhe Institute of Technology (August 2020)
7. Boomari, H., Ostovari, M., Zarei, A.: Recognizing visibility graphs of triangulated irregular networks. Fundam. Informaticae **179**(4), 345–360 (2021). https://doi.org/10.3233/FI-2021-2027
8. Brandenburg, F.J., Didimo, W., Evans, W.S., Kindermann, P., Liotta, G., Montecchiani, F.: Recognizing and drawing IC-planar graphs. Theoret. Comput. Sci. **636**, 1–16 (2016). https://doi.org/10.1016/j.tcs.2016.04.026
9. Canny, J.: Some algebraic and geometric computations in PSPACE. In: Proceedings of the Twentieth Annual ACM Symposium on Theory of Computing, STOC 1988, pp. 460–469. ACM, New York (1988). https://doi.org/10.1145/62212.62257
10. Chan, T.M., Frati, F., Gutwenger, C., Lubiw, A., Mutzel, P., Schaefer, M.: Drawing partially embedded and simultaneously planar graphs. J. Graph Algorithms Appl. **19**(2), 681–706 (2015). https://doi.org/10.7155/jgaa.00375
11. Didimo, W.: Right angle crossing drawings of graphs. In: Hong, S.-H., Tokuyama, T. (eds.) Beyond Planar Graphs, pp. 149–169. Springer, Singapore (2020). https://doi.org/10.1007/978-981-15-6533-5_9
12. Didimo, W., Eades, P., Liotta, G.: Drawing graphs with right angle crossings. Theoret. Comput. Sci. **412**(39), 5156–5166 (2011). https://doi.org/10.1016/j.tcs.2011.05.025
13. Didimo, W., Liotta, G., Montecchiani, F.: A survey on graph drawing beyond planarity. ACM Comput. Surv. **52**(1), 1–37 (2019). https://doi.org/10.1145/3301281
14. Erickson, J., van der Hoog, I., Miltzow, T.: Smoothing the gap between NP and ER. In: 61st Annual Symposium on Foundations of Computer Science–FOCS 2020, pp. 1022–1033. IEEE Computer Society, Los Alamitos (2020). https://doi.org/10.1109/FOCS46700.2020.00099
15. Förster, H., Kaufmann, M.: On compact RAC drawings. In: Grandoni, F., Herman, G., Sanders, P. (eds.) 28th Annual European Symposium on Algorithms, ESA 2020, September 7–9, 2020, Pisa, Italy (Virtual Conference). LIPIcs, vol. 173, pp. 53:1–53:21. Schloss Dagstuhl - Leibniz-Zentrum für Informatik (2020). https://doi.org/10.4230/LIPIcs.ESA.2020.53
16. Huang, W., Eades, P., Hong, S.: Larger crossing angles make graphs easier to read. J. Vis. Lang. Comput. **25**(4), 452–465 (2014). https://doi.org/10.1016/j.jvlc.2014.03.001
17. Matoušek, J.: Intersection graphs of segments and ∃ℝ. ArXiv e-prints (2014). arXiv:1406.2636. Accessed 10 June 2020
18. Mnëv, N.E.: The universality theorems on the classification problem of configuration varieties and convex polytopes varieties. In: Viro, O.Y., Vershik, A.M. (eds.) Topology and Geometry—Rohlin Seminar. LNM, vol. 1346, pp. 527–543. Springer, Heidelberg (1988). https://doi.org/10.1007/BFb0082792
19. Richter-Gebert, J.: Mnëv's universality theorem revisited. Sém. Lothar. Combin., vol. 34 (1995)
20. Schaefer, M.: Picking planar edges; or, drawing a graph with a planar subgraph. In: Duncan, C., Symvonis, A. (eds.) GD 2014. LNCS, vol. 8871, pp. 13–24. Springer, Heidelberg (2014). https://doi.org/10.1007/978-3-662-45803-7_2
21. Schaefer, M.: RAC-drawability is ∃ℝ-complete. ArXiv e-prints (2021). arXiv:2107.11663v2. Accessed 28 July 2021
22. Schaefer, M.: Complexity of geometric k-planarity for fixed k. J. Graph Algorithms Appl. **25**(1), 29–41 (2021). https://doi.org/10.7155/jgaa.00548

23. Schaefer, M.: On the complexity of some geometric problems with fixed parameters. J. Graph Algorithms Appl. **25**(1), 195–218 (2021). https://doi.org/10.7155/jgaa.00557
24. Shor, P.W.: Stretchability of pseudolines is NP-hard. In: Applied Geometry and Discrete Mathematics, DIMACS Ser. Discrete Math. Theoret. Comput. Sci., vol. 4, pp. 531–554. Amer. Math. Soc., Providence, RI (1991)
25. Wikipedia Contributors: Existential theory of the reals – Wikipedia, the free encyclopedia (2021). https://en.wikipedia.org/w/index.php?title=Existential_theory_of_the_reals. Accessed 28 May 2021

Recognizing and Embedding Simple Optimal 2-Planar Graphs

Henry Förster[1](✉) , Michael Kaufmann[1] ,
and Chrysanthi N. Raftopoulou[2]

[1] WSI, Tübingen, Germany
{foersth,mk}@informatik.uni-tuebingen.de
[2] NTUA, Athens, Greece
crisraft@mail.ntua.gr

Abstract. In the area of beyond-planar graphs, i.e. graphs that can be drawn with some local restrictions on the edge crossings, the recognition problem is prominent next to the density question for the different graph classes. For 1-planar graphs, the recognition problem has been settled, namely it is NP-complete for the general case, while optimal 1-planar graphs, i.e. those with maximum density, can be recognized in linear time. For 2-planar graphs, the picture is less complete. As expected, the recognition problem has been found to be NP-complete in general. In this paper, we consider the recognition of simple optimal 2-planar graphs. We exploit a combinatorial characterization of such graphs and present a linear time algorithm for recognition and embedding.

Keywords: 2-planar graphs · Recognition algorithms · Beyond-planarity

1 Introduction

In the field of graph drawing beyond planarity, researchers study graphs that admit drawings with crossings only in restrictive local configurations [6,14,15,19,23]. Mostly studied are *1-planar graphs*, which admit drawings where each edge is crossed at most once. Classic results on these graphs concern mostly the density [7,26], recently, other problems such as generation [27], characterization [21], recognition [2,9–11,13], coloring [8] and page number [3,16] have been studied. The most natural extension of 1-planar graphs, is the family of *k-planar graphs*, that is graphs that admit drawings where each edge is crossed at most k times. While there are plenty of results on 1-planarity, k-planarity turns out to be more challenging. The first outstanding results on the density of simple k-planar graphs came with the work of Pach and Tóth [25] for $k \leq 4$. In particular, they proved that simple 2-planar graphs have at most $5n - 10$ edges. Later on, improved upper

The work of C. Raftopoulou is funded by the National Technical University of Athens research program ΠΕΒΕ 2020. H. Förster and M. Kaufmann are supported by DFG grant KA812-18/2.

© Springer Nature Switzerland AG 2021
H. C. Purchase and I. Rutter (Eds.): GD 2021, LNCS 12868, pp. 87–100, 2021.
https://doi.org/10.1007/978-3-030-92931-2_6

bounds of $5.5n - 11$ [24] and $6n - 12$ [1] have been shown for simple 3- and 4-planar graphs, respectively. The bounds for 2- and 3-planar graphs are tight even for non-simple graph without homotopic parallel edges and self-loops [5].

While testing for planarity can be done in linear time, and several classes of beyond-planar graphs have a linear number of edges, the corresponding recognition problems are NP-hard, which impedes the practical application of related results. In particular, it was recently proven that the recognition of k-planar graphs is NP-hard for every value of $k \geq 1$ [28]. Still, optimal 1-planar graphs, which form a prominent subclass of 1-planar graphs, can be recognized in linear time [11], while recognition of maximal 1-planar graphs and 4-maps takes cubic time [12,17]. Previously, efficient recognition algorithms have been developed for more restricted subclasses of 1-planar graphs, see e.g. [2,10,13,20].

For $k \geq 2$, there are significantly fewer results. A notable result is the linear time recognition algorithm for full outer-2-planar graphs [22], while for $k \geq 3$ the literature still lacks similar results.

Our Contribution. In this paper, we extend the study on the recognition of subclasses of k-planar graphs. Namely, we prove that simple optimal 2-planar graphs, that is simple 2-planar graphs with $5n - 10$ edges, can be efficiently recognized in $\mathcal{O}(n)$ time. In our strategy we first remove a set of edges that must be involved in crossings in any optimal 2-planar drawing of the given graph. We unwrap specific properties of the remaining graph that allow us to find a unique drawing and insert the deleted crossing edges.

Paper Organization. In Sect. 2 we give preliminary notations and properties of optimal 2-planar graphs. Section 3 presents CROSS-BAT, an important 2-planar substructure, and examines its structural properties. Section 4 focuses on the insertion of the deleted crossing edges. Our recognition algorithm in Sect. 5 combines all previous results. We conclude in Sect. 6 with further research directions. Due to space constraints, several proofs and details, indicated by (∗), are omitted and can be found in the full version [18].

2   Preliminaries

The degree of a vertex $v \in V$ is denoted by $d(v)$ and its set of neighbors by $N(v)$. For $u, v \in V$ we write $N(u, v)$ to denote the set of common neighbors of u and v, that is $N(u, v) = N(u) \cap N(v)$. A *drawing* of a graph maps the vertices to points in the plane and the edges to simple open Jordan curves connecting the points corresponding to their endpoints. A drawing is k-*planar* if each curve representing an edge is crossed at most k times by the other edges together. A graph G is k-planar if it admits a k-planar drawing.

The topology of a planar drawing is described by a *rotation scheme*. For k-planarity, a k-*planar rotation scheme* specifies the counter-clockwise circular order of edges around each vertex and, additionally, a sequence of at most k crossings along each edge. Hence, a k-planar rotation scheme defines an equivalence class of k-planar drawings.

For a graph G, let H a subgraph of G and let $\mathcal{R}_H$ be a (k-planar) rotation scheme of H. We say that $\mathcal{R}_H$ is a *partial (k-planar) rotation scheme* of G. We further say that $\mathcal{R}_H$ *can be extended* to a (k-planar) rotation scheme $\mathcal{R}$ of G if the restriction of $\mathcal{R}$ on H is $\mathcal{R}_H$. Conversely, we say that $\mathcal{R}$ *extends* $\mathcal{R}_H$. Let C be a cycle of graph G, whose edges define a simple closed curve in a drawing Γ of G. We refer to the region to the left of this curve in a counter-clockwise walk along C in Γ as the *interior of the region* bounded by cycle C. Note that given drawing Γ, the interior of a region may be unbounded.

A 2-planar graph that reaches the upper bound of $5n - 10$ edges is called *optimal 2-planar*. Note that the value of $5n - 10$ edges can be reached only when $n \equiv 2 \mod 3$. Each induced subgraph H of a 2-planar graph G is also 2-planar and thus has at most $5|V(H)| - 10$ edges. This implies that if G is 2-planar then it is also *9-degenerate*, i.e. there is a vertex ordering $v_1, \ldots, v_n$ such that, for each $i = 1, \ldots, n$, vertex v_i of the induced subgraph $G_i = G[v_1, \ldots, v_i]$ has degree at most 9 in G_i. We call such a vertex ordering a *9-degenerate* sequence of G.

In this work, we consider only *simple* graphs which we assume from now on without explicitly stating it. The structure of simple optimal 2-planar graphs is well established. Namely, an optimal 2-planar rotation scheme $\mathcal{R}$ of a simple optimal 2-planar graph specifies a crossing-free 3-connected[1] spanning pentangulation $\mathcal{P}$ and a set of crossing edges [4,5]. The edges around each vertex are cyclically ordered so that every crossing-free edge (belonging to $\mathcal{P}$) is followed by two crossing edges, which are succeeded by one crossing-free edge, and so on. Hence the degree of every vertex is a multiple of three. Moreover, the minimum degree of a simple optimal 2-planar graph is 9 since $\mathcal{P}$ is 3-connected, i.e., every vertex is incident to at least 3 crossing-free edges. Now, each pentangular face f of $\mathcal{P}$ contains exactly five crossing edges of $\mathcal{R}$ which form a 5-clique in G with the boundary edges of f; we call such a 5-clique a *facial 5-clique*. Note that not all 5-cliques of G are facial in $\mathcal{R}$. This is a major difficulty for recognizing optimal 2-planar graphs. We denote a facial 5-clique of $\mathcal{R}$ with outer cycle $(c_0, \ldots, c_4)$ as $\langle c_0, \ldots, c_4 \rangle$ and say that the crossing edges (c_i, c_j) with $(i - j) \mod 5 \in \{2, 3\}$ *belong to* $\langle c_0, \ldots, c_4 \rangle$. As $\mathcal{P}$ is 3-connected, it suffices to detect all facial 5-cliques to find the optimal 2-planar rotation scheme of a given input graph, if one exists.

In the following, we state some preliminary observations and properties that we will use in the remainder of the paper. In particular, Properties 2.1–2.4 immediately arise from the combinatorial characterization of simple optimal 2-planar graphs in [5] also summarized above. So, let G be an optimal 2-planar graph with an optimal 2-planar rotation scheme $\mathcal{R}$.

Property 2.1. A pair of crossing edges in $\mathcal{R}$ belongs to the same facial 5-clique.

Property 2.2. If in $\mathcal{R}$, edge e is crossed by two edges e_1 and e_2, then e_1 and e_2 share a common endpoint. Also, e, e_1 and e_2 belong to the same facial 5-clique.

Property 2.3. Any two facial 5-cliques in $\mathcal{R}$ share at most two vertices.

---

[1] The 3-connectivity derives from the fact that the graph is simple.

Fig. 1. Forbidden configuration excluded by Corollary 2.2.

Property 2.4. A crossing edge of $\mathcal{R}$ belongs to exactly one facial 5-clique and a crossing-free edge of $\mathcal{R}$ belongs to exactly two facial 5-cliques.

Combining the above properties we can further prove the following:

Corollary 2.1 (∗). *For two edges (u, v) and (u, w) of an optimal 2-planar graph G let $S = N(u, v) \cup N(u, w) \cup \{u, v, w\}$. If $|S| < 8$, then (u, v) and (u, w) belong to the same facial 5-clique in any optimal 2-planar rotation scheme of G.*

Corollary 2.2. *Let G be an optimal 2-planar graph with an optimal 2-planar rotation scheme $\mathcal{R}$. Let (u, v) be a crossing edge inside a facial 5-clique C in $\mathcal{R}$. Further, let $u' \notin C$ be a neighbor of u and $v' \notin C$ be a neighbor of v. Then, edges (u, u') and (v, v') do not cross in $\mathcal{R}$; see Fig. 1.*

Consider an optimal 2-planar rotation scheme $\mathcal{R}$ of G. If edge (u, v) is crossing-free in $\mathcal{R}$, then by Property 2.4 u and v belong to two facial 5-cliques. Hence, $|N(u, v)| \geq 6$ holds. Although this condition is not sufficient to conclude that an edge is crossing-free, we can use it to identify some crossing edges:

Corollary 2.3. *Let G be an optimal 2-planar graph and $(u, v) \in E(G)$. Edge (u, v) is crossing in any optimal 2-planar rotation scheme of G if $|N(u, v)| < 6$.*

Our recognition algorithm relies on the identification of crossing edges based on Corollary 2.3. We say that an edge (u, v) is *clearly crossing* if and only if $|N(u, v)| < 6$. Otherwise (u, v) is *potentially planar*. Note that in an optimal 2-planar rotation scheme, a potentially planar edge is not necessarily crossing-free. However, every crossing-free edge of any optimal 2-planar rotation scheme is potentially planar. We define as G_p the subgraph of G formed by all potentially planar edges. Graph G_p is 3-connected and spans G as the corresponding crossing-free pentangulation of each optimal 2-planar rotation scheme is a subgraph of G_p.

3   The CROSS-BAT Configuration

In this section, we study the 3-connected spanning subgraph G_p formed by the potentially planar edges of G. We want to compute a rotation scheme of G_p which is extendable to an optimal 2-planar rotation scheme of G, if it exists. If in each optimal 2-planar rotation scheme of G subgraph G_p is plane, then the rotation scheme of G_p is unique and easy to compute. Though we cannot assure this property, we prove that in any optimal 2-planar rotation scheme of G, crossings

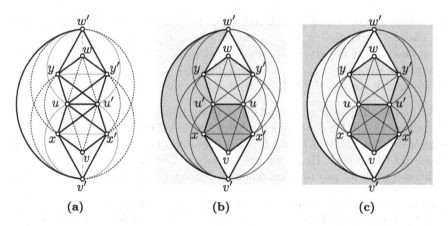

Fig. 2. (a) Graph G_{CB}: black dotted and black bold edges are clearly crossing and potentially planar, resp., gray solid edges might be both, gray dashed edges may be absent. (b)–(c) the possible rotation schemes: in (b) (u, x'), (u, y'), (u', x) and (u', y) are crossing-free; in (c) (u, x), (u, y), (u', x') and (u', y') are crossing-free.

between edges of G_p occur in restricted configurations: A CROSS-BAT instance is an induced subgraph H of G isomorphic to the graph G_{CB} shown in Fig. 2a, so that (i) for edge (u_H, u'_H) of H isomorphic to (u, u') of G_{CB}, it holds that $V(H) = N(u_H) \cup N(u'_H)$, and (ii) the isomorphism between H and G_{CB} preserves the classification of edges to clearly crossing or potentially planar.

In particular, CROSS-BAT has ten vertices, named as in Fig. 2a. Vertices u and u' have degree 9 in G and they have 8 common neighbors (the remaining vertices of CROSS-BAT). The edges of CROSS-BAT form four 5-cliques that pairwise share two vertices, and are shown as facial 5-cliques in Fig. 2a. We call edge (u, u') the *base edge* of CROSS-BAT. No other edges with both endpoints in CROSS-BAT exist, except possibly for edges (v, v') and (w, w'). Figure 2a shows the classification of edges of CROSS-BAT as potentially planar and clearly crossing.

Let $\mathcal{R}$ be an optimal 2-planar rotation scheme of G, such that two potentially planar edges cross each other. These two edges must belong to the same facial 5-clique and, in particular, also to an instance of CROSS-BAT.

Lemma 3.1 (∗). *Let $\mathcal{R}$ be an optimal 2-planar rotation scheme of G and let $C = \langle c_0, \ldots, c_4 \rangle$ be a facial 5-clique in $\mathcal{R}$ such that (c_1, c_3) and (c_2, c_4) are potentially planar edges. Then, vertices c_2 and c_3 have degree 9 in G, and the induced subgraph H of G with vertex set $V(H) = N(c_2) \cup N(c_3)$ is an instance of CROSS-BAT where C is the 5-clique $\langle v, x', u', u, x \rangle$ of Fig. 2a.*

Next, we show that CROSS-BAT has only two rotation schemes:

Lemma 3.2 (∗). *Any instance of CROSS-BAT in an optimal 2-planar rotation scheme $\mathcal{R}$ has one of the two rotation schemes shown in Figs. 2b and 2c.*

Sketch. Vertices are named as in Fig. 2a. First, we prove that there are four facial 5-cliques, namely (i) C_v that contains v, (ii) C_w that contains w, (iii) $C_{x,y}$

that contains both x and y, and, (iv) $C_{x',y'}$ that contains both x' and y'. It is then easy to argue that $x, x' \in C_v$, while $y, y' \in C_w$. Finally, we show that w is contained in the crossing-free cycle (u_1, u_2, y', w', y) and v is contained in the crossing-free cycle (u_2, u_1, x, v', x') where $\{u_1, u_2\} = \{u, u'\}$. The two choices for u_1 and u_2 give the two different rotation schemes of Figs. 2b and 2c. $\qquad\square$

As it is evident in Figs. 2b and 2c, if an optimal 2-planar graph G contains an instance of CROSS-BAT as subgraph, then it admits two different optimal 2-planar rotation schemes $\mathcal{R}$ and $\mathcal{R}'$, that only differ in the choice of the rotation scheme of CROSS-BAT. Hence for any instance of CROSS-BAT, we may arbitrarily choose one of its two rotation schemes. Next, we formalize this observation:

Lemma 3.3. *Let $\mathcal{R}$ be an optimal 2-planar rotation scheme of G, and let H be an instance of CROSS-BAT in G. If H has the rotation scheme of Fig. 2b in $\mathcal{R}$, then there exists another optimal 2-planar rotation scheme $\mathcal{R}'$ of G, in which H has the rotation scheme of Fig. 2c, and vice-versa.*

Both rotation schemes of a CROSS-BAT instance contain two crossings between potentially planar edges:

Lemma 3.4 ($*$). *Let $\mathcal{R}$ be an optimal 2-planar rotation scheme of G. If G contains an instance H of CROSS-BAT, then there exist exactly two pairs of potentially planar edges that belong to H and cross in $\mathcal{R}$.*

Using Lemma 3.2 we can fix a rotation scheme for each instance H of CROSS-BAT identified in G by reclassifying two crossing potentially planar edges of H to clearly crossing edges. After performing this reclassification for all instances, if G is optimal 2-planar, then the subgraph G_p induced by the potentially planar edges is planar. Furthermore, in this case, Lemma 3.3 guarantees that the unique planar embedding of G_p is part of an optimal 2-planar rotation scheme of G.

4   Identifying Facial 5-Cliques

Assume that we have fixed the rotation scheme for every identified instance of CROSS-BAT. Let G_p be the spanning subgraph of G formed by all potentially planar edges after the reclassification process. As discussed in Sect. 3, G_p is planar and 3-connected, i.e. it has a unique planar rotation scheme $\mathcal{R}_p$. Furthermore, by Lemma 3.3, if G is optimal 2-planar, it admits an optimal 2-planar rotation scheme that extends $\mathcal{R}_p$ which we call $\mathcal{R}_p$-*compliant*. For any $\mathcal{R}_p$-compliant optimal 2-planar rotation scheme $\mathcal{R}$, G_p contains the corresponding spanning pentangulation $\mathcal{P}$ as a subgraph. Hence each face of $\mathcal{R}_p$ has length 3, 4 or 5 and is part of a facial 5-clique. Hence, we can arbitrarily triangulate them and assume from now on that G_p is triangulated. Triangulating a face of $\mathcal{R}_p$ corresponds to reclassifying some chords from clearly crossing to potentially planar.

Let (f_1, f, f_2) be a path in the dual G_p^* of G_p so that $f_1 = (u, w_1, v_1)$, $f = (u, w_2, w_1)$ and $f_2 = (u, v_2, w_2)$, as shown in Fig. 3. If the subgraph induced by the vertices $\{u, v_2, w_2, w_1, v_1\}$ is a 5-clique in G, we call $T = \langle f_1, f, f_2 \rangle$ a *triplet*.

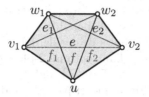

Fig. 3. Illustration of a triplet $\langle f_1, f, f_2 \rangle$, along with its clearly crossing edges e_1, e_2 and e. Edges of G_p are solid black, clearly crossing edges of G are dotted black.

T contains vertices u, v_2, w_2, w_1, v_1, faces f_1, f, f_2 and the three clearly crossing edges $e_1 = (v_1, w_2)$, $e_2 = (v_2, w_1)$ and $e = (v_1, v_2)$, as shown in Fig. 3. We say that e_1, e_2 and e *belong* to triplet T.

In any $\mathcal{R}_p$-compliant optimal 2-planar rotation scheme $\mathcal{R}$ (if it exists), faces and clearly crossing edges of $\mathcal{R}_p$ are partitioned into triplets, such that every face and clearly crossing edge belongs to exactly one of these triplets. Furthermore, the triplets are in 1-1 correspondence with the facial 5-cliques of $\mathcal{R}$. We say that e, e_1 and e_2 are *assigned* to the triplet $T = \langle f_1, f, f_2 \rangle$ w.r.t. $\mathcal{R}$ if T induces a facial 5-clique in $\mathcal{R}$. Similarly, we say that faces f_1, f and f_2 of T are *assigned* to T. If such an assignment is not possible, then G is not optimal 2-planar.

Let $\mathcal{T}$ be a set of triplets such that any two triplets in $\mathcal{T}$ are face-disjoint and contain different clearly crossing edges. Consider the partial 2-planar rotation scheme $\mathcal{R}_\mathcal{T}$ of G that (i) extends $\mathcal{R}_p$, (ii) the clearly crossing edges of each $T \in \mathcal{T}$ are assigned to T, and (iii) there is no other assignment of clearly crossing edges. We say that $\mathcal{T}$ is *bad* if and only if $\mathcal{R}_\mathcal{T}$ cannot be extended to a $\mathcal{R}_p$-compliant optimal 2-planar rotation scheme $\mathcal{R}$ of G. Our goal is to find an assignment of all clearly crossing edges of G to a set of triplets $\mathcal{T}$ such that $\mathcal{T}$ is not bad if G is optimal 2-planar. We actually prove a stronger result, namely that the set $\mathcal{T}$ is unique. We will use two observations that follow from the simplicity of G:

Observation 4.1. *Let $f = (u, v, w)$ be a triangular face of $\mathcal{R}_p$ and let T be a triplet that contains vertices u, v and w. Then T contains face f.*

Observation 4.2. *Let $f = (u, v, w)$ and $f' = (u', w, v)$ be two adjacent faces of $\mathcal{R}_p$ and (u, u') a clearly crossing edge. Let $\mathcal{R}$ be a $\mathcal{R}_p$-compliant optimal 2-planar rotation scheme and T be the triplet that (u, u') is assigned to w.r.t. $\mathcal{R}$. Then, either (i) (u, u') is drawn inside f and f' in $\mathcal{R}$ and T contains both f and f', or (ii) (u, u') is drawn outside f and f' in $\mathcal{R}$ and T contains neither f nor f'.*

If any of e_1, e_2 or e belongs only to one triplet T, then T is a facial 5-clique in any $\mathcal{R}_p$-compliant optimal 2-planar rotation scheme $\mathcal{R}$ and e_1, e_2 and e can be assigned to T. So, we assume that each of e_1, e_2 and e belong to at least one triplet different from T. Let T_1 and T_2 be two triplets that contain edges e_1 and e_2 respectively and are different from T. Note that $T_1 = T_2$ might hold. Let $\mathcal{T} = \{T\}$. If for each such pair of triplets T_1 and T_2 we can conclude that the set $\mathcal{T}' = \{T_1, T_2\}$ is bad, then in any $\mathcal{R}_p$-compliant optimal 2-planar rotation scheme $\mathcal{R}$ of G (if it exists) edges e_1, e_2 and e must be assigned to T. In the following, we compare T against all possible sets T' and prove that at least one of T and T' is bad. This allows to decide, for each triplet T, if T forms a facial 5-clique in every $\mathcal{R}_p$-compliant optimal 2-planar rotation scheme of G or in none. Note that when we write $T = \{T\}$ and $T' = \{T_1, T_2\}$, we assume that $T = \langle f_1, f, f_2 \rangle$ as shown in Fig. 3 and triplets T_1 and T_2 contain edges e_1 and e_2, respectively.

We first restrict how triplets T_1 and T_2 relate to triplet T. Observation 4.2 applied to T_1 implies that if T_1 shares either f or f_1 with T, then, in the presence of edge e_1, T_1 shares both f and f_1 with T. A symmetric argument applies for T_2. It follows that one of the following cases holds for $\{i, j\} = \{1, 2\}$:

C.1 T_i is face-disjoint with T, or
C.2 T_i shares only face f_j with T, or,
C.3 T_i shares only faces f and f_i with T.

In the next two lemmas, we show that every $T' = \{T_1, T_2\}$ is bad if Case C.1 applies either for none or for both of T_1 and T_2.

Lemma 4.1 (*). *Let $T' = \{T_1, T_2\}$. If none of T_1 and T_2 is face-disjoint from T, then T' is bad.*

Lemma 4.2 (*). *Let $T' = \{T_1, T_2\}$. If both T_1 and T_2 are face-disjoint from T, then T' is bad.*

Sketch. Assuming that T' is not bad, we arrive at the configuration shown in Fig. 4a. Since (w_1, w_2) is crossing, Corollary 2.2 is violated. □

By Lemmas 4.1 and 4.2, Case C.1 applies for exactly one of the two triplets T_1 and T_2 and $T_1 \neq T_2$. Then, the other triplet complies with Case C.2 or with Case C.3. For each possible combination we first prove structural properties arising from the assumption that T' is not bad, and then show that these new restrictions make T bad. In the next two lemmas, we consider the setting, where one of T_1 and T_2 complies with Case C.1 while the other one complies with C.2.

Lemma 4.3 (*). *Let $T' = \{T_1, T_2\}$, such that T_1 is face-disjoint from T and T_2 shares only face f_1 with T. For $u \in V(T)$, if (i) $d(u) > 9$, or, (ii) for every vertex $x \in S = N(u) \setminus V(T)$ it holds $|N(x) \cap S| \geq 2$, then T' is bad.*

Sketch. Assuming that T' is not bad and $d(u) = 9$, we arrive at the configuration shown in Fig. 4a. We can conclude that not all conditions of the lemma hold. □

Lemma 4.4 (*). *Let $T = \{T\}$ and let $v \in V(T)$ such that $d(v) = 9$. If there exists a vertex $x \in S = N(v) \setminus V(T)$ such that $|N(x) \cap S| \leq 1$, then T is bad.*

By Lemmas 4.3 and 4.4 it remains to consider the case where one of T_1 and T_2 complies with Case C.1 while the other complies with Case C.3. So, we assume w.l.o.g. that T_1 is face-disjoint with T, while T_2 shares faces f and f_2 with T. Recall that clearly crossing edge e belongs to T. If $T' = \{T_1, T_2\}$ is not bad e

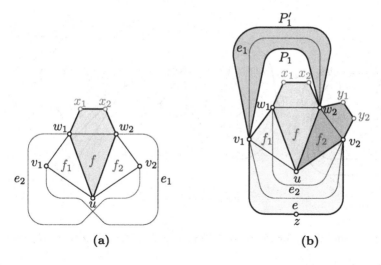

Fig. 4. Illustrations for the proofs of (a) Lemma 4.2, and, (b) Lemma 4.3.

belongs to a triplet $T_e \neq T$ such that $T' \cup \{T_e\}$ is not bad. Note that $T_e \neq T_2$ as otherwise vertex v_1 belongs to T_2 and $T_2 = T$. Hence, either $T_e \neq T_1$ or $T_e = T_1$.

The following four lemmas investigate the subcase $T_e \neq T_1$. Note that T_1 and T_e are face-disjoint as otherwise $T' = \{T_1, T_2, T_e\}$ is bad. The first two lemmas examine the scenario where T_1 contains vertex w_1 or T_e contains vertex u of T.

Lemma 4.5 (*). *Let* $T' = \{T_1, T_2, T_e\}$, *such that* T_1 *is face-disjoint from* T, T_2 *shares faces* f *and* f_2 *with* T, *and* T_e *is face-disjoint from* T_1. *Then:*

- *If* T_1 *contains vertex* w_1 *of* T *and additionally (i)* $d(w_1) > 9$, *or, (ii) there is a vertex* $x \in S = N(w_1) \setminus V(T)$ *such that* $|N(x) \cap S| \geq 4$, *then* T' *is bad.*
- *If* T_e *contains vertex* u *of* T *and additionally (i)* $d(u) > 9$, *or, (ii) there is a vertex* $x \in S = N(u) \setminus V(T)$ *such that* $|N(x) \cap S| \geq 4$, *then* T' *is bad.*

Sketch. Assuming that T' is not bad and $d(w_1) = 9$, in the first case, we arrive at the configuration shown in Fig. 5a. We conclude that not all conditions hold. □

Lemma 4.6 (*). *Let* $\mathcal{T} = \{T\}$ *and let* $v \in V(T)$ *with* $d(v) = 9$. *If for every vertex* $x \in S = N(v) \setminus V(T)$ *it holds* $|N(x) \cap S| \leq 3$, *then* $\mathcal{T}$ *is bad.*

The next two lemmas consider the scenario, where T_1 does not contain vertex w_1 and T_e does not contain vertex u of triplet T.

Lemma 4.7 (*). *Let* $T' = \{T_1, T_2, T_e\}$, *such that* T_1 *is face-disjoint from* T, T_2 *shares faces* f *and* f_2 *with* T, *and* T_e *is face-disjoint from* T_1. *Assume that* T_1 *does not contain vertex* w_1 *and* T_e *does not contain vertex* u *of* T. *If there is no triplet* T_{f_1} *that contains* f_1 *such that (i)* T_{f_1} *shares only vertices of* f_1 *with* T, *and, (ii)* T_{f_1} *is face-disjoint from all of* T_1, T_2 *and* T_e, *then* T' *is bad.*

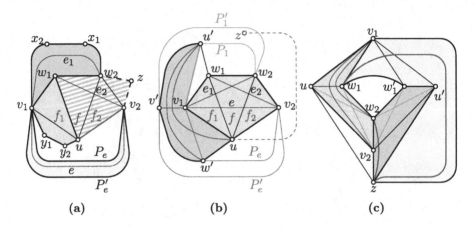

Fig. 5. Illustration for the proofs of Lemmas (a) 4.5, (b) 4.8, and, (c) 4.9.

Lemma 4.8 (∗). *Let triplets T_1, T_2, T_e be pairwise disjoint such that T_1 is face-disjoint from T, T_2 shares faces f and f_2 with T, T_e is face-disjoint from T_1, T_1 does not contain vertex w_1 and T_e does not contain vertex u of T. If there is a triplet T_{f_1} that contains f_1 such that (i) T_{f_1} shares only vertices of f_1 with T, and, (ii) T_{f_1} is face-disjoint from all of T_1, T_2 and T_e, then $T = \{T\}$ is bad.*

Sketch. Assuming that $\mathcal{T}$ is not bad, we arrive at the configuration shown in Fig. 5b where the position of z is not fixed. Edges (u, z) and (w_1, z) belong to T_{f_1} but at least one has three crossings, e.g. edge (u, z) in Fig. 5b.   □

For the second subcase, where $T_1 = T_e$, Property 2.3 assures that T_1 and T_2 share only vertices v_2 and w_2, as otherwise $\mathcal{T}' = \{T_1 = T_e, T_2\}$ would be bad. As indicated by the next lemma, no further restrictions (imposed by the assumption that $\mathcal{T}'$ is not bad) are needed to prove that $\mathcal{T} = \{T\}$ is bad.

Lemma 4.9 (∗). *Let $\mathcal{T} = \{T\}$. Assume that T_1 is face-disjoint from T, that T_2 shares faces f and f_2 with T, and $T_e = T_1$. If T_2 has exactly two common vertices with T_1, then $\mathcal{T}$ is bad.*

Sketch. Assuming that $\mathcal{T}$ is not bad, we obtain the configuration in Fig. 5c. Here, two facial 5-cliques have three common vertices contradicting Property 2.3.   □

In the following two lemmas, we summarize our findings from this section.

Lemma 4.10 (∗). *Let G be optimal 2-planar and let $\mathcal{T} = \{T\}$. $\mathcal{T}$ is not bad if and only if every set $\mathcal{T}' = \{T_1, T_2\}$ is bad.*

Sketch. As G is optimal 2-planar, at least one of $\mathcal{T}$ or $\mathcal{T}' = \{T_1, T_2\}$ for some triplets T_1 and T_2 is not bad. If $\mathcal{T}$ is bad, the lemma holds. For the reverse direction, assume there is a set $\mathcal{T}' = \{T_1, T_2\}$ that is not bad, i.e. none of Lemmas 4.3, 4.5 and 4.7 applies for $\mathcal{T}' \cup \{T_e\}$. If $\mathcal{T}$ is not bad from the corresponding Lemmas 4.4, 4.6 and 4.8, the conditions of Lemma 4.9 are satisfied and $\mathcal{T}$ is bad.   □

Lemma 4.11 (*). *Let G be optimal 2-planar and let $\mathcal{T} = \{T\}$. $\mathcal{T}$ is bad if and only if the conditions of at least one of Lemmas 4.4, 4.6, 4.8 and 4.9 are met.*

Sketch. For a contradiction we assume that $\mathcal{T}$ is bad. Since G is optimal 2-planar, by Lemma 4.10, there exists a set $\mathcal{T}' = \{T_1, T_2\}$ that is not bad. We go through all lemmas and check their conditions. For all the cases, we argue that if the specific lemma can not apply on $\mathcal{T}$, then $\mathcal{T}'$ is bad. □

5 The Recognition and Embedding Algorithm

Now that we have all required ingredients, we are ready to state our main theorem for the recognition of optimal 2-planar graphs.

Theorem 1. *Let G be a simple graph on n vertices. It can be decided in $\mathcal{O}(n)$ time whether G is optimal 2-planar. If the instance is positive, an optimal 2-planar rotation scheme of G is reported.*

The remainder of this section contains the proof of Theorem 1 which is split into two parts. In Sect. 5.1 we describe our algorithm and prove its correctness, while an efficient implementation is discussed in Sect. 5.2.

5.1 Recognition Algorithm

Our recognition algorithm is formalized in Algorithm 1. There are four main steps in the process. The first step is the classification of all edges of G as potentially planar and clearly crossing (line 1). Second, is the identification of the CROSS-BAT instances and the creation of G_p and its planar rotation scheme $\mathcal{R}_p$ (lines 2–10). Third, we identify all triplets of $\mathcal{R}_p$ (line 11). Finally, we decide which of the triplets are not bad (lines 12–13) determining an optimal 2-planar rotation scheme of G if one exists. Having the set of triplets that are facial 5-cliques in any $\mathcal{R}_p$-compliant optimal 2-planar rotation scheme of G, we decide if G is optimal 2-planar (lines 14–15) and compute an optimal 2-planar rotation scheme of G if it exists (line 16). The details of the steps are given in [18].

5.2 Implementation

First, we check that the input graph has at most $5n - 10$ edges and that it is 9-degenerate. This process takes $\mathcal{O}(n)$ time and is described in [18], together with standard techniques and data structures that we use. In the following, we discuss in more detail the time complexity of the involved steps of Algorithm 1. The first step of the process, described in line 1 of Algorithm 1, can be performed in $\mathcal{O}(n)$ time as stated in the following lemma:

Lemma 5.1 (*). *All edges of G can be classified as potentially planar or clearly crossing in $\mathcal{O}(n)$ time.*

Algorithm 1: Recognition of Simple Optimal 2-planar Graphs

Input: A 9-degenerate graph $G = (V, E)$ with $|E| = 5|V| - 10$
Output: An optimal 2-planar rotation scheme of G if it exists, otherwise, `false`

1   Classify each edge $e \in E$ as potentially planar or clearly crossing
2   Identify all instances of CROSS-BAT in G and fix their partial rotation schemes
3   Create subgraph G_p
4   **if** G_p is not 3-connected or is not planar **then**
5 | **return** `false`
6   Compute the unique planar rotation scheme $\mathcal{R}_p$ of G_p
7   **if** $\mathcal{R}_p$ contains a face of length greater than 5
8   **or if** a face does not induce a complete subgraph in G **then**
9 | **return** `false`
10   Augment G_p to maximal planar by triangulating the faces of $\mathcal{R}_p$
11   Compute the set of triplets of $\mathcal{R}_p$
12   **for** every triplet T of $\mathcal{R}_p$ **do**
13 | Label T as facial 5-clique or as non-facial 5-clique
14   Let $\mathcal{T}^*$ be the set of triplets labelled as facial 5-cliques
15   **if** $\mathcal{T}^*$ covers all faces of $\mathcal{R}_p$ and all clearly crossing edges of G exactly once **then**
16 | **return** rotation scheme obtained by making each $T \in \mathcal{T}^*$ a facial 5-clique
17   **return** `false`

We proceed with the second step given in lines 2–10 of Algorithm 1. First, we compute the CROSS-BAT instances of G (line 2).

Lemma 5.2 (∗). *All CROSS-BAT instances of G can be identified in $\mathcal{O}(n)$ time if all edges are already classified as clearly crossing or potentially planar.*

For each CROSS-BAT instance, we choose the rotation scheme of Fig. 2b where (u, y) crosses (u', y') and (u, x) crosses (u', x') by reclassifying (u, y) and (u, x) as clearly crossing. We proceed with lines 3–10 of Algorithm 1. We check the necessary conditions for G_p and compute the planar rotation scheme $\mathcal{R}_p$ and the dual G_p^* after augmenting G_p to maximal planar; see [18].

Next, we compute the set of triplets of $\mathcal{R}_p$ (line 11) in $\mathcal{O}(n)$ time; see [18]. During this process, for each face and each clearly crossing edge of G, we store the set of triplets that contain it. The next lemma describes how to implement the labeling of each triplet as facial 5-clique or not; see line 13.

Lemma 5.3 (∗). *Let $\mathcal{T} = \{T\}$ and $\mathcal{S}_1$, $\mathcal{S}_2$ and $\mathcal{S}_e$ be the set of triplets containing e_1, e_2 and e, respectively. It can be decided in $\mathcal{O}(1)$ time if $\mathcal{T}$ is bad or not.*

After labeling all triplets as facial 5-cliques or non-facial 5-cliques, we consider the set $\mathcal{T}^*$ of triplets labelled as facial 5-cliques. Checking whether $\mathcal{T}^*$ covers all faces and clearly crossing edges of G exactly once, in line 15 of Algorithm 1, can be done in $\mathcal{O}(|\mathcal{T}^*|) = \mathcal{O}(n)$ time. Finally, we augment the planar rotation scheme $\mathcal{R}_p$ of G_p to an optimal 2-planar rotation scheme of G in $\mathcal{O}(n)$ time by inserting the clearly crossing edges of G and identifying the crossings; refer to [18]. The overall time required for the last step is in $\mathcal{O}(n)$.

Every step described above takes time $\mathcal{O}(n)$, hence Algorithm 1 can be implemented to run in linear time. We point out that after fixing $\mathcal{R}_p$, the optimal 2-planar rotation scheme of G is uniquely defined if it exists. In particular, if there are c instances of CROSS-BAT, there exist at most 2^c optimal 2-planar rotation schemes which can be easily enumerated.

6 Conclusions

We showed that simple optimal 2-planar graphs can be recognized and embedded in linear time. It remains open to extend our result also for simple 2-planar graphs with the maximum number of edges when $n \not\equiv 2 \mod 3$, as well as to non-simple optimal 2-planar graphs. Another reasonable attempt would be recognizing 5-maps, similarly to [12]. Finally, a natural question would be if a recognition algorithm for non-simple optimal 2-planar graphs could be adopted for optimal 3-planar graphs that are non-simple, or if our approach could work for simple optimal 3-planar graphs.

Acknowledgement. We thank Michael Bekos for his valuable suggestions and endless discussions on this topic.

References

1. Ackerman, E.: On topological graphs with at most four crossings per edge. CoRR 1509.01932 (2015). http://arxiv.org/abs/1509.01932
2. Auer, C., et al.: Outer 1-planar graphs. Algorithmica **74**(4), 1293–1320 (2016). https://doi.org/10.1007/s00453-015-0002-1
3. Bekos, M.A., Bruckdorfer, T., Kaufmann, M., Raftopoulou, C.: 1-planar graphs have constant book thickness. In: Bansal, N., Finocchi, I. (eds.) ESA 2015. LNCS, vol. 9294, pp. 130–141. Springer, Heidelberg (2015). https://doi.org/10.1007/978-3-662-48350-3_12
4. Bekos, M.A., Giacomo, E.D., Didimo, W., Liotta, G., Montecchiani, F., Raftopoulou, C.N.: Edge partitions of optimal 2-plane and 3-plane graphs. Discret. Math. **342**(4), 1038–1047 (2019). https://doi.org/10.1016/j.disc.2018.12.002
5. Bekos, M.A., Kaufmann, M., Raftopoulou, C.N.: On optimal 2- and 3-planar graphs. In: Aronov, B., Katz, M.J. (eds.) 33rd International Symposium on Computational Geometry, SoCG 2017, 4–7 July 2017, Brisbane, Australia. LIPIcs, vol. 77, pp. 16:1–16:16. Schloss Dagstuhl - Leibniz-Zentrum für Informatik (2017). https://doi.org/10.4230/LIPIcs.SoCG.2017.16
6. Binucci, C., et al.: Fan-planarity: properties and complexity. Theor. Comput. Sci. **589**, 76–86 (2015). https://doi.org/10.1016/j.tcs.2015.04.020
7. Bodendiek, R., Schumacher, H., Wagner, K.: Über 1-optimale Graphen. Math. Nachr. **117**(1), 323–339 (1984)
8. Borodin, O.V.: A new proof of the 6 color theorem. J. Graph Theory **19**(4), 507–521 (1995). https://doi.org/10.1002/jgt.3190190406

9. Brandenburg, F.J.: Recognizing optimal 1-planar graphs in linear time. CoRR 1602.08022 (2016). http://arxiv.org/abs/1602.08022
10. Brandenburg, F.J.: Recognizing IC-planar and NIC-planar graphs. J. Graph Algorithms Appl. **22**(2), 239–271 (2018). https://doi.org/10.7155/jgaa.00466
11. Brandenburg, F.J.: Recognizing optimal 1-planar graphs in linear time. Algorithmica **80**(1), 1–28 (2018). https://doi.org/10.1007/s00453-016-0226-8
12. Brandenburg, F.J.: Characterizing and recognizing 4-map graphs. Algorithmica **81**(5), 1818–1843 (2019). https://doi.org/10.1007/s00453-018-0510-x
13. Brandenburg, F.J., Didimo, W., Evans, W.S., Kindermann, P., Liotta, G., Montecchiani, F.: Recognizing and drawing IC-planar graphs. Theor. Comput. Sci. **636**, 1–16 (2016). https://doi.org/10.1016/j.tcs.2016.04.026
14. Cheong, O., Har-Peled, S., Kim, H., Kim, H.: On the number of edges of fan-crossing free graphs. Algorithmica **73**(4), 673–695 (2015). https://doi.org/10.1007/s00453-014-9935-z
15. Didimo, W., Eades, P., Liotta, G.: Drawing graphs with right angle crossings. Theor. Comput. Sci. **412**(39), 5156–5166 (2011)
16. Dujmovic, V., Joret, G., Micek, P., Morin, P., Ueckerdt, T., Wood, D.R.: Planar graphs have bounded queue-number. J. ACM **67**(4), 22:1-22:38 (2020). https://doi.org/10.1145/3385731
17. Eades, P., Hong, S., Katoh, N., Liotta, G., Schweitzer, P., Suzuki, Y.: A linear time algorithm for testing maximal 1-planarity of graphs with a rotation system. Theor. Comput. Sci. **513**, 65–76 (2013). https://doi.org/10.1016/j.tcs.2013.09.029
18. Förster, H., Kaufmann, M., Raftopoulou, C.N.: Recognizing and embedding simple optimal 2-planar graphs (2021). https://arxiv.org/abs/2108.00665v2
19. Giacomo, E.D., Didimo, W., Liotta, G., Meijer, H., Wismath, S.K.: Planar and quasi-planar simultaneous geometric embedding. Comput. J. **58**(11), 3126–3140 (2015). https://doi.org/10.1093/comjnl/bxv048
20. Hong, S., Eades, P., Katoh, N., Liotta, G., Schweitzer, P., Suzuki, Y.: A linear-time algorithm for testing outer-1-planarity. Algorithmica **72**(4), 1033–1054 (2015). https://doi.org/10.1007/s00453-014-9890-8
21. Hong, S.-H., Eades, P., Liotta, G., Poon, S.-H.: Fáry's theorem for 1-planar graphs. In: Gudmundsson, J., Mestre, J., Viglas, T. (eds.) COCOON 2012. LNCS, vol. 7434, pp. 335–346. Springer, Heidelberg (2012). https://doi.org/10.1007/978-3-642-32241-9_29
22. Hong, S., Nagamochi, H.: A linear-time algorithm for testing full outer-2-planarity. Discret. Appl. Math. **255**, 234–257 (2019). https://doi.org/10.1016/j.dam.2018.08.018
23. Kaufmann, M., Ueckerdt, T.: The density of fan-planar graphs. CoRR 1403.6184 (2014). http://arxiv.org/abs/1403.6184
24. Pach, J., Radoicic, R., Tardos, G., Tóth, G.: Improving the crossing lemma by finding more crossings in sparse graphs. Discrete Comput. Geom. **36**(4), 527–552 (2006). https://doi.org/10.1007/s00454-006-1264-9
25. Pach, J., Tóth, G.: Graphs drawn with few crossings per edge. Combinatorica **17**(3), 427–439 (1997). https://doi.org/10.1007/BF01215922
26. Ringel, G.: Ein Sechsfarbenproblem auf der Kugel. Abhandlungen aus dem Mathematischen Seminar der Universität Hamburg (in German) **29**, 107–117 (1965)
27. Suzuki, Y.: Re-embeddings of maximum 1-planar graphs. SIAM J. Discrete Math. **24**(4), 1527–1540 (2010). https://doi.org/10.1137/090746835
28. Urschel, J.C., Wellens, J.: Testing gap k-planarity is NP-complete. Inf. Process. Lett. **169**, 106083 (2021). https://doi.org/10.1016/j.ipl.2020.106083

A Short Proof of the Non-biplanarity of K_9

Ahmad Biniaz[⊠]

School of Computer Science, University of Windsor, Windsor, Canada
ahmad.biniaz@uwindsor.ca

Abstract. Battle, Harary, and Kodama (1962) and independently Tutte (1963) proved that the complete graph with nine vertices is not biplanar. Aiming towards simplicity and brevity, in this note we provide a short proof of this claim.

Keywords: Biplanar graph · Biplanar drawing · Edge crossing

1 Introduction

An embedding (or drawing) of a graph in the Euclidean plane is a mapping of its vertices to distinct points in the plane and its edges to smooth curves between their corresponding vertices. A planar embedding of a graph is a drawing of the graph such that no two edges cross. A graph that admits such a drawing is called planar. A *biplanar embedding* of a graph $H = (V, E)$ is a decomposition of H into two planar graphs $H_1 = (V, E_1)$ and $H_2 = (V, E_2)$ such that $E_1 \cup E_2 = E$ and $E_1 \cap E_2 = \emptyset$, together with planar embeddings of H_1 and H_2. In this case, H is called *biplanar*. In other words, a graph is called biplanar if it is the union of two planar graphs; that is, if its thickness[1] is 1 or 2. The *complete graph* with n vertices, denoted by K_n, is a graph that has an edge between every pair of its vertices. Let G be a subgraph of K_n that has n vertices. The *complement* of G, denoted by $\overline{G}$, is the graph obtained by removing all edges of G from K_n.

As early as 1960 it was known that K_8 is biplanar and K_{11} is not biplanar. There exist several biplanar embeddings of K_8; see e.g. [2] for a self-complementary drawing. The non-biplanarity of K_{11} is easily seen, since it has 55 edges while a planar graph with eleven vertices cannot have more than 27 edges, by Euler's formula. Finding the smallest integer n, for which K_n is non-biplanar, was a challenging question for some time [7]. The following fundamental theorem due to Battle, Harary, and Kodama ([1], 1962) and independently proved by Tutte ([15], 1963) answers this question and implies that K_9 is non-biplanar.

---

[1] The thickness of a graph G is the minimum number of planar subgraphs whose union equals to G.

Supported by NSERC.

H. C. Purchase and I. Rutter (Eds.): GD 2021, LNCS 12868, pp. 101–106, 2021.
https://doi.org/10.1007/978-3-030-92931-2_7

Theorem 1. *Every planar graph with at least nine vertices has a nonplanar complement.*

Both proofs of Theorem 1 involve a thorough case analysis. Battle, Harary, and Kodama gave an outline of a proof through six propositions. Some of these propositions require detailed case analysis, which is not given in the original paper. For example, the authors write: "There are several cases to discuss in order to establish Propositions 4 and 5. In each case, we can prove that $\overline{G}$ contains a subgraph homeomorphic to $K_{3,3}$ or K_5." A detailed proof of these propositions is appeared in the master's thesis of Hearon [9]. Tutte's proof is a 13-page paper, and enumerates all simple triangulations (with no separating triangles) with up to 9 vertices and verifies that the complement of each triangulation is nonplanar (the connection to triangulations will become clear shortly). It seems that Harary was not quite satisfied with any of these proofs as he noted in his Graph Theory book [8] that "this result was proved by exhaustion; no elegant or even reasonable proof is known." We are still unaware of any short proof of this result. (See [10] for a recent attempt towards a new proof.)

The non-biplanarity of K_9 has the same flavor as the well-known theorem of Kuratowski on non-planar graphs (stated in Theorem 3). The *biplanar crossing number* of a graph is the minimum number of crossings over all drawings of the graph in two planes [3]. It is known that K_9 can be drawn in two planes with one crossing (see e.g. [6]). This and Theorem 1 imply that the biplanar crossing number of K_9 is 1. Determining biplanar crossing numbers of K_n for small values of n is important as they lead to better bounds for biplanar crossing numbers of K_n for large values of n; see e.g. [3,4,13], and [6,14] for more recent progress.

2 Our Proof

In this section we present a short proof of Theorem 1. Our proof is complete, self-contained, and only uses Kuratowski's theorem for non-planar graphs. Towards our proof we show (in Theorem 2) that a particularly restricted drawing of K_8 cannot be biplanar (see Fig. 2(a) for an illustration).

Theorem 2. *Let H be an embedded planar graph with eight vertices such that the boundary of its outer face is a 5-cycle and there are no edges between the three vertices that are not on the outer face. Then the complement of H is nonplanar.*

Proof of Theorem 1. Consider a planar graph G with nine vertices. For the sake of contradiction assume that its complement $\overline{G}$ is also planar. Fix a planar embedding of G and a planar embedding of $\overline{G}$. For convenience we use G and $\overline{G}$ for referring to planar graphs and to their planar embeddings. If there are two vertices in G that lie on the same face and are not connected by an edge, then we transfer the corresponding edge from $\overline{G}$ to G and connect the two vertices by a curve in that face. After this operation both G and $\overline{G}$ remain planar. Repeating this process converts G to an edge-maximal planar graph. In particular G becomes a triangulation in which the boundary of every face (including the outer face) is a triangle (i.e. a 3-cycle).

Claim 1. At least one vertex on the outer face of G has degree larger than four. To prove this claim we use contradiction. Assume that all three vertices on the outer face of G are of degree at most 4. The removal of these three vertices from G results in a 6-vertex graph G'. The region, that is between the boundaries of the outer face of G and the outer face of G' is a polygon with a hole, that is triangulated by at most six edges of G (because every vertex on the outer face of G has at most two edges in the interior of this polygon). The boundary of the outer face of G',

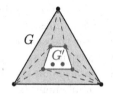

Fig. 1. Seven edges needed to triangulate the shaded polygon.

i.e. the hole, has three vertices because otherwise (if it has at least four vertices) the polygon would require at least seven edges to be triangulated, as in Fig. 1; this can be verified by a simple counting argument using Euler's formula for planar graphs, see also [12, Proof of Lemma 5.2]. Thus the outer face of G' is a 3-cycle. In this case the other three vertices of G' which are in the interior of this 3-cycle together with the three removed vertices from G form a $K_{3,3}$ in $\overline{G}$, which contradicts its planarity. This proves Claim 1.

In view of Claim 1 we assume that at least one vertex, say r, on the outer face of G has degree $k \geq 5$. Remove r from G and $\overline{G}$ and denote the resulting graphs by H and $\overline{H}$, respectively. Notice that $(H, \overline{H})$ is a biplanar embedding of K_8. Let f and $\overline{f}$ be the faces of H and $\overline{H}$, respectively, that contain the removed vertex r, as in Fig. 2(b). Notice that f is the outer face of H. Since $(G, \overline{G})$ was a biplanar embedding of K_9, in which r was connected to all other 8 vertices, we have the following observation.

Observation 1. Every vertex of the resulting graph K_8 lies on f or on $\overline{f}$.

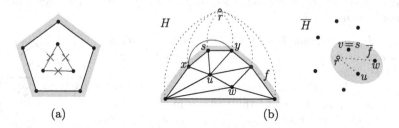

(a)   (b)

Fig. 2. Illustration of (a) the statement of Theorem 2 (b) the proof of Theorem 1.

Since $\overline{G}$ was a simple graph (no multiedges and no loops), the face $\overline{f}$ has at least three vertices; these vertices are not necessarily connected in $\overline{H}$. Since G was a triangulation, the boundary of the outer face f of H is a k-cycle. If $k > 5$ then let s be a vertex of f that also lies on $\overline{f}$; such a vertex exists because $\overline{f}$ has at least three vertices and we have eight vertices in total. Let x and y be the neighbors of s on f. If xy is an edge of H then draw it as a curve in f. If xy is not an edge of H then transfer it from $\overline{H}$ to H and draw it in f, as in Fig. 2(b). Now, the new outer face f of H has $k - 1$ vertices. Repeat the above process until the outer face of H has exactly five vertices.

At this point f has five vertices. Let u, v, w be the vertices of K_8 that are not on f. These three vertices lie on $\overline{f}$, because of Observation 1 and our choices of s (for the case $k > 5$). If any of the edges uv, uw, and vw are not in $\overline{H}$ then transfer them from H to $\overline{H}$ and draw in $\overline{f}$ without crossing other edges. We obtain a planar graph H that satisfies the constraints of Theorem 2 and so that its complement $\overline{H}$ is planar. This contradicts Theorem 2. $\quad\square$

To prove Theorem 2 we use the theorem of Kuratowski [5,11] that "a finite graph is non-planar if and only if it contains a subgraph that is homeomorphic to K_5 or $K_{3,3}$." The following is an alternative statement for Kuratowski's theorem, which is given in [15].

Theorem 3. *A graph G is nonplanar if one of the following conditions hold: (i) G has six disjoint connected subgraphs $A_1, A_2, A_3, B_1, B_2, B_3$ such that for each A_i and B_j there is an edge with one end in A_i and the other in B_j. (ii) G has five disjoint connected subgraphs A_1, A_2, A_3, A_4, A_5 such that for each A_i and A_j, with $i \neq j$, there is an edge with one end in A_i and the other in A_j.*

Proof of Theorem 2. Let the 5-cycle $C = (a_1, a_2, a_3, a_4, a_5)$ be the boundary of the outer face of H, and let u, v, and w be the three vertices that are not on the outer face, i.e., lie on internal faces of H. By the statement of the theorem uv, uw, and vw are edges of the complement graph $\overline{H}$. Except for the three pairs (u, v), (u, w), (v, w), if a pair of vertices lie on the same internal face of H and are not connected by an edge, then we transfer the corresponding edge from $\overline{H}$ to H and connect the two vertices by a curve in the face. After this operation H remains planar. Repeating this process makes H edge-maximal (in the above sense).

Let H' be the embedded planar subgraph of H that is induced by the five vertices of C. The graph H' consists of the cycle C together with zero, one, or two chords as in Fig. 4.

Claim 2. If an internal face f of H' contains u, v, or w then one of them is connected to all boundary vertices of f in H. The shaded region in the figure to the right represents f. To verify the claim, first observe that (by edge-maximality of H) one of the vertices in f, say v, is connected to at least three boundary vertices of f, i.e., v's degree in H is at least three. We argue that v should be connected to all boundary vertices of f. For a contradiction assume that v is not connected to some vertex a_i on f. Let a_j and a_k be the neighbors of v on f that are visited first while walking on boundary of f in clockwise and counterclockwise directions starting from a_i. Since v is not connected to other

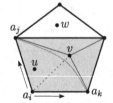

Fig. 3. Moving $a_j a_k$ from $\overline{H}$ to H.

vertices in the interior of f, we could have moved the edge $a_j a_k$ from $\overline{H}$ to H and draw it in f, as in Fig. 3. This means that H is not edge-maximal, which is a contradiction.

Now we consider three cases depending on the number of chords of H'. In each case we show that $\overline{H}$ is nonplanar.

- H' has no chords. Let v be the vertex of H that (by Claim 2) is connected to each a_i; see Fig. 4(a). By planarity of H, each of u and w can only be adjacent to two consecutive vertices of C. Hence there exists a vertex of C (say a_1) that is adjacent to neither u nor w. In this setting, regardless of the locations of u and w, the five connected subgraphs u, w, a_1, $\{a_2, a_4\}$ and $\{a_3, a_5\}$ from $\overline{H}$ satisfy condition (ii) of Theorem 3. Thus $\overline{H}$ is nonplanar.
- H' has one chord. After a suitable relabeling assume that this chord is (a_2, a_5). Let f denote the face of H' whose boundary is the 4-cycle (a_2, a_3, a_4, a_5); this face is shaded in Fig. 4(b). This face contains some vertices of $\{u, v, w\}$ because otherwise H' should have a chord in f (by maximality of H) which contradicts our assumption that H' has one chord. Let v be the vertex in f that (by Claim 2) is connected to all its boundary vertices. By planarity of H, each of u and w can only be adjacent to two consecutive vertices of f. Therefore, the six connected subgraphs u, w, a_1, v, $\{a_2, a_4\}$, and $\{a_3, a_5\}$ from $\overline{H}$ (partitioned into $\{u, w, a_1\}$ and $\{v, \{a_2, a_4\}, \{a_3, a_5\}\}$) satisfy condition (i) of Theorem 3. Thus $\overline{H}$ is nonplanar.
- H' has two chords. Let a_1 be the vertex that is incident to the two chords as in Fig. 4(c). By planarity of H, each of u, v, and w can only be adjacent to one vertex in $\{a_2, a_4\}$ and to one vertex in $\{a_3, a_5\}$. Thus, the five connected subgraphs u, v, w, $\{a_2, a_4\}$, and $\{a_3, a_5\}$ from $\overline{H}$ satisfy condition (ii) of Theorem 3, and hence $\overline{H}$ is nonplanar.   □

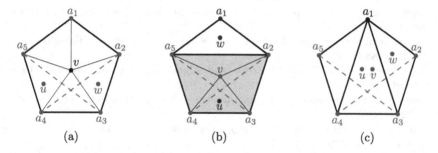

Fig. 4. Solid edges belong to H, bold edges belong to H', dashed edges belong to $\overline{H}$.

3   Conclusions

For any integer $k \geq 1$ let $\nu(k)$ be the smallest integer for which the (edges of the) complete graph with $\nu(k)$ vertices cannot be drawn in k planes without creating a crossing. As the maximum number of (noncrossing) edges that can be drawn in a plane is $3\nu(k) - 6$ and the number of edges of the complete graph is $\binom{\nu(k)}{2}$, a counting argument implies that

$$\nu(k) \leq \left\lfloor \frac{6k + 1 + \sqrt{36k^2 - 36k + 1}}{2} \right\rfloor + 1.$$

This bound implies that $\nu(1) \leq 5$ and $\nu(2) \leq 11$, however for $k \in \{1,2\}$ we already know that $\nu(1) = 5$ and $\nu(2) = 9$. It would be interesting to find exact value of $\nu(k)$ for larger values of k.

References

1. Battle, J., Harary, F., Kodama, Y.: Every planar graph with nine vertices has a nonplanar complement. Bull. Am. Math. Soc. **68**, 569–571 (1962)
2. Beineke, L.: Biplanar graphs: a survey. Comput. Math. Appl. **34**(11), 1–8 (1997)
3. Czabarka, É., Sýkora, O., Székely, L.A., Vrt'o, I.: Biplanar crossing numbers I: a survey of results and problems. In: More Sets, Graphs and Numbers, pp. 57–77 (2006)
4. Czabarka, É., Sýkora, O., Székely, L.A., Vrt'o, I.: Biplanar crossing numbers. II. comparing crossing numbers and biplanar crossing numbers using the probabilistic method. Random Struct. Algorithms **33**(4), 480–496 (2008)
5. Dirac, G.A., Schustér, S.: A theorem of Kuratowski. Nederl. Akad. Wetensch. Proc. Ser. A **57**, 343–348 (1954)
6. Durocher, S., Gethner, E., Mondal, D.: On the biplanar crossing number of K_n. In: Proceedings of the 28th Canadian Conference on Computational Geometry (CCCG), pp. 93–100 (2016)
7. Harary, F.: Problem 28. Bull. Am. Math. Soc. **67**, 542 (1961)
8. Harary, F.: Graph Theory. Addison-Wesley, Boston (1969)
9. Hearon, S.M.: Planar graphs, biplanar graphs and graph thickness. Master's thesis, California State University-San Bernardino (2016)
10. Kuila, S.K.: Algebraic approach to prove non-coplanarity of K_9. Int. J. Eng. Invent. **4**(6), 19–23 (2014)
11. Kuratowski, K.: Sur le problème des courbes gauches en topologie. Fundam. Math. **15**, 271–283 (1930)
12. O'Rourke, J.: Art Gallery Theorems and Algorithms. Oxford University Press, Oxford (1987)
13. Owens, A.: On the biplanar crossing number. IEEE Trans. Circuit Theory **18**(2), 277–280 (1971)
14. Shavali, A., Zarrabi-Zadeh, H.: New bounds on the biplanar and k-planar crossing numbers. arXiv: 1911.06403 (2019)
15. Tutte, W.T.: On non-biplanar character of the complete 9-graph. Can. Math. Bull. **6**, 319–330 (1963)

Morphing and Graph Abstraction

From Tutte to Floater and Gotsman:
On the Resolution of Planar Straight-Line Drawings and Morphs

Giuseppe Di Battista[✉] and Fabrizio Frati

Roma Tre University, Rome, Italy
{giuseppe.dibattista,fabrizio.frati}@uniroma3.it

Abstract. The algorithm of Tutte for constructing convex planar straight-line drawings and the algorithm of Floater and Gotsman for constructing planar straight-line morphs are among the most popular graph drawing algorithms. Surprisingly, little is known about the resolution of the drawings they produce. In this paper, focusing on maximal plane graphs, we prove tight bounds on the resolution of the planar straight-line drawings produced by Floater's algorithm, which is a broad generalization of Tutte's algorithm. Further, we use such a result to prove a lower bound on the resolution of the drawings of maximal plane graphs produced by Floater and Gotsman's morphing algorithm. Finally, we show that such an algorithm might produce drawings with exponentially-small resolution, even when morphing drawings with polynomial resolution.

1 Introduction

In 1963 Tutte [29] presented an algorithm for constructing convex planar straight-line drawings of 3-connected plane graphs. The algorithm is very simple: Given any convex polygon representing the outer cycle of the graph, place each internal vertex at the barycenter of its neighbors. This results in a system of linear equations, whose variables are the coordinates of the internal vertices, which has a unique solution; quite magically, this solution corresponds to a convex planar straight-line drawing of the graph. We call any drawing obtained by an application of Tutte's algorithm a *T-drawing*. Tutte's algorithm is one of the most famous graph drawing algorithms; notably, it has spurred the research on the practical graph drawing algorithms called *force-directed methods* [10,13,23].

A far-reaching generalization of the algorithm of Tutte was presented by Floater [15,16] (and, in a similar form, by Linial, Lovász, and Wigderson [24]). Namely, one can place each internal vertex at *any* convex combination (with positive coefficients) of its neighbors; the resulting system of equations still has a unique solution that corresponds to a convex planar straight-line drawing of the graph. Formally, let G be a 3-connected plane graph and let $\mathcal{P}$ be a convex polygon representing the outer cycle of G. Further, for each internal vertex v of G

This research was supported in part by MIUR Project "AHeAD" under PRIN 20174LF3T8 and by H2020-MSCA-RISE project 734922 – "CONNECT".

© Springer Nature Switzerland AG 2021
H. C. Purchase and I. Rutter (Eds.): GD 2021, LNCS 12868, pp. 109–122, 2021.
https://doi.org/10.1007/978-3-030-92931-2_8

and for each neighbor u of v, let $\lambda_{vu} > 0$ be a real value such that $\sum_{u \in \mathcal{N}(v)} \lambda_{vu} = 1$, where $\mathcal{N}(v)$ (sometimes written as $\mathcal{N}_G(v)$) denotes the set of neighbors of v. For each internal vertex v of G, consider the two equations:

$$x(v) = \sum_{u \in \mathcal{N}(v)} (\lambda_{vu} \cdot x(u)) \quad (1) \qquad y(v) = \sum_{u \in \mathcal{N}(v)} (\lambda_{vu} \cdot y(u)) \quad (2)$$

where $x(v)$ and $y(v)$ denote the x and y-coordinates of a vertex v, respectively. This results in a system of $2N$ equations in $2N$ variables, where N is the number of internal vertices of G, which has a unique solution. This corresponds to a convex planar straight-line drawing, which can hence be represented by a pair $(\Lambda, \mathcal{P})$, where $\mathcal{P}$ is a convex polygon and Λ is a *coefficient matrix*. This matrix has a row for each internal vertex of G and a column for each (internal or external) vertex of G; an element of Λ whose row corresponds to a vertex v and whose column corresponds to a vertex u is the coefficient λ_{vu} if (v, u) is an edge of G and 0 otherwise. We call any drawing resulting from an application of Floater's algorithm an *F-drawing*. F-drawings are extensively used in computer graphics, for surface parameterization and reconstruction, and for texture mapping; see, e.g., [20,21,30,31]. Similar types of drawings have been studied for constructing three-dimensional representations of polytopes [9,25]. Further, *every* convex planar straight-line drawing of a 3-connected plane graph (and thus every planar straight-line drawing of a maximal plane graph) is an F-drawing $(\Lambda, \mathcal{P})$, for suitable Λ and $\mathcal{P}$ [15–18].

Floater and Gotsman [18] devised a powerful application of F-drawings to the construction of planar straight-line morphs. Given a graph G and two convex planar straight-line drawings Γ_0 and Γ_1 of G with the same polygon $\mathcal{P}$ representing the outer cycle, construct two coefficient matrices Λ_0 and Λ_1 such that $\Gamma_0 = (\Lambda_0, \mathcal{P})$ and $\Gamma_1 = (\Lambda_1, \mathcal{P})$. Now, a morph $\mathcal{M}$ between Γ_0 and Γ_1, that is, a continuous transformation of Γ_0 into Γ_1, can be obtained as follows. For each $t \in [0, 1]$, construct a coefficient matrix Λ_t as $(1-t) \cdot \Lambda_0 + t \cdot \Lambda_1$; thus, for each internal vertex v of G and for each neighbor u of v, the element λ_{vu}^t of Λ_t whose row and column correspond to v and u, respectively, is $\lambda_{vu}^t = (1-t) \cdot \lambda_{vu}^0 + t \cdot \lambda_{vu}^1$. Then the morph $\mathcal{M}$ is simply defined as $\{\Gamma_t = (\Lambda_t, \mathcal{P}) : t \in [0, 1]\}$; since Γ_t is an F-drawing, for any $t \in [0, 1]$, this algorithm guarantees that every drawing in $\mathcal{M}$ is a convex planar straight-line drawing. We call any morph constructed by an application of Floater and Gotsman's algorithm an *FG-morph*. The algorithm of Floater and Gotsman is perhaps the most popular graph morphing algorithm; its extensions, refinements, and limits have been discussed, e.g., in [1,22,27,28].

Despite the fame of the above algorithms, little is known about the resolution of T-drawings, F-drawings, and FG-morphs. The *resolution* is perhaps the most studied aesthetic criterion for the readability of a graph drawing. Measuring the resolution of a drawing can be done in many ways. Here, we adopt a natural definition of resolution, namely the ratio between the smallest and the largest distance between two (distinct, non-incident, and non-adjacent) geometric objects representing vertices or edges. The goal of this paper is to study the resolution of T-drawings, F-drawings, and FG-morphs of maximal plane graphs. The only related result we are aware of is the one by Eades and Garvan [14] (independently observed by Chambers et al. [5]), who proved that T-drawings of n-vertex maximal plane graphs might have $1/2^{\Omega(n)}$ resolution. It is even unclear, a priori, whether the worst-case resolution of T-drawings, F-drawings, and FG-morphs can be expressed as *any* function of natural parameters representing the input size and resolution.

We prove the following results[1]. First, we show a lower bound on the resolution of F-drawings (and thus on the resolution of T-drawings).

Theorem 1. *Let $\Gamma = (\Lambda, \Delta)$ be an F-drawing of an n-vertex maximal plane graph G, where $n \geq 4$. The resolution of Γ is larger than or equal to $\frac{r}{2} \cdot \left(\frac{\lambda}{3}\right)^n \in r \cdot \lambda^{O(n)}$, where λ is the smallest positive coefficient in the coefficient matrix Λ and r is the resolution of the prescribed triangle Δ.*

Second, we prove that the bound in Theorem 1 is asymptotically tight.

Theorem 2. *There is a class of maximal plane graphs $\{G_n : n = 5, 6, \dots\}$, where G_n has n vertices, with the following property. For any $0 < \lambda \leq \frac{1}{4}$ and $0 < r \leq \frac{\sqrt{3}}{2}$, there exist a triangle Δ with resolution r and a coefficient matrix Λ for G_n whose smallest positive coefficient is λ such that the F-drawing (Λ, Δ) of G_n has resolution in $r \cdot \lambda^{\Omega(n)}$.*

We remark that algorithms are known for constructing planar straight-line drawings of maximal plane graphs [8,26] and even convex planar straight-line drawings of 3-connected plane graphs [2,6] with polynomial resolution.

Third, we use Theorem 1 to prove a lower bound on the resolution of FG-morphs.

Theorem 3. *Let Γ_0 and Γ_1 be any two planar straight-line drawings of an n-vertex maximal plane graph G such that the outer faces of Γ_0 and Γ_1 are delimited by the same triangle Δ. There is an FG-morph $\mathcal{M} = \{\Gamma_t : t \in [0,1]\}$ between Γ_0 and Γ_1 such that, for each $t \in [0,1]$, the resolution of Γ_t is larger than or equal to $(r/n)^{O(n)}$, where r is the minimum between the resolution of Γ_0 and Γ_1.*

Finally, we prove that FG-morphs might have exponentially small resolution, even if they transform drawings with polynomial resolution.

Theorem 4. *For every $n \geq 6$ multiple of 3, there is an n-vertex maximal plane graph G and two planar straight-line drawings Γ_0 and Γ_1 of G such that: (R1) the outer faces of Γ_0 and Γ_1 are delimited by the same triangle Δ; (R2) the resolution of both Γ_0 and Γ_1 is larger than c/n^2, for a constant c; and (R3) any FG-morph between Γ_0 and Γ_1 contains a drawing whose resolution is in $1/2^{\Omega(n)}$.*

---

[1] At a first glance, our use of the $O(\cdot)$ and $\Omega(\cdot)$ notation seems to be inverted. For example, Theorem 1 shows a bound of $r \cdot \lambda^{O(n)}$ on the resolution of F-drawings. This is a lower bound, and not an upper bound, given that $\lambda < 1$. Indeed, the $O(\cdot)$ notation indicates that the exponent has a value which is *at most something*, hence the entire power has a value which is *at least something* (smaller than one).

The construction of planar straight-line morphs with high resolution has been attracting increasing attention [3,4,7]. A main open question in the area is whether polynomial resolution can be guaranteed in a planar straight-line morph between any two given drawings of a plane graph. Theorem 4 shows that Floater and Gotsman's algorithm cannot be used to settle this question in the positive.

A full version of the paper can be found in [11].

2 Preliminaries

We introduce some definitions, properties and lemmata. Throughout the paper, we assume that every considered graph has $n \geq 4$ vertices.

A biconnected *plane graph* G is a planar graph with a prescribed order of the edges incident to each vertex and a prescribed outer cycle; G is *maximal* if no edge can be added to it without losing planarity or simplicity. We often talk about "faces of G", meaning faces of any planar drawing that respects the prescribed order of the edges incident to each vertex and the prescribed outer cycle. We say that G is *internally-triangulated* if every internal face of G is delimited by a 3-cycle. The sets of internal and external vertices of G are denoted by $\mathcal{I}_G$ and $\mathcal{O}_G$, respectively. Let $\mathcal{C}$ be a cycle of G. An *external chord* of $\mathcal{C}$ is an edge of G that connects two vertices of $\mathcal{C}$, that does not belong to $\mathcal{C}$, and that lies outside $\mathcal{C}$ in G. The *subgraph of G inside $\mathcal{C}$* is composed of the vertices and edges that lie inside or on the boundary of $\mathcal{C}$. The following is easy to observe.

Property 1. Let G be a maximal plane graph and let $\mathcal{C}$ be a cycle of G. The subgraph of G inside $\mathcal{C}$ is biconnected and internally-triangulated.

In a planar straight-line drawing Γ of a graph G, by *geometric object* we mean a point representing a vertex or a straight-line segment representing an edge. We often call "vertex" or "edge" both the combinatorial and the geometric object. Two geometric objects in Γ are *separated* if they share no point. By the planarity of Γ, two geometric objects are hence separated if and only if they are distinct vertices, or non-adjacent edges, or a vertex and a non-incident edge. The *distance* $d_\Gamma(o_1, o_2)$ between two separated geometric objects o_1 and o_2 in Γ is the minimum Euclidean distance between any point of o_1 and any point of o_2. We denote by $d_\Gamma^\updownarrow(u, v)$ the vertical distance between two vertices u and v, i.e., the absolute value of the difference between their y-coordinates. The *resolution* of Γ is the ratio between the distance of the closest separated geometric objects and the distance of the farthest separated geometric objects in Γ. Let $\mathcal{R}$ be a finite connected subset of $\mathbb{R}^2$. The *x-extent* (the *y-extent*) of $\mathcal{R}$ is given by the maximum x-coordinate (resp. y-coordinate) of any point of $\mathcal{R}$ minus the minimum x-coordinate (resp. y-coordinate) of any point of $\mathcal{R}$.

We present a tool that we often use to translate and rotate F-drawings.

Lemma 1. Let (Λ, Δ) be an F-drawing of a maximal plane graph G and let Δ' be the triangle obtained by a proper rigid transformation σ of Δ. The F-drawing (Λ, Δ') coincides with the drawing obtained by applying σ to the drawing (Λ, Δ).

We sometimes use the following elementary property.

Property 2. Let (Λ, Δ) be an F-drawing of a maximal plane graph with at least four (five) vertices. The smallest positive coefficient in Λ is at most $1/3$ $(1/4)$.

We now state some properties on the resolution of triangles.

Property 3. The resolution of a triangle is smaller than or equal to $\frac{\sqrt{3}}{2}$ and there exist triangles with this resolution.

Lemma 2. *Let Δ be a triangle with resolution equal to r and y-extent equal to Y. Then the x-extent X of Δ is at most Y/r.*

3   Lower Bound on the Resolution of F-Drawings

In this section, we prove Theorem 1. Let $\Gamma = (\Lambda, \Delta)$ be an F-drawing of an n-vertex maximal plane graph G with $n \geq 4$; let λ be the smallest positive coefficient in Λ, r be the resolution of Δ, and δ be the smallest distance between any two separated geometric objects in Γ. We have that the smallest distance δ in Γ is achieved "inside" an internal face of G, as in the following.

Lemma 3. *There exist an internal vertex v and an edge $e = (u_e, v_e)$ of G such that: (i) $d_\Gamma(v, e) = \delta$; (ii) v, u_e, and v_e are the vertices of a triangle T delimiting an internal face of G in Γ; and (iii) the altitude of T through v lies inside T.*

By Lemma 1, we can assume that $y(v) = 0$, that e is horizontal, and that v is above e. We show that v's neighbors are not "too high" or "too low" in Γ.

Lemma 4. *For every neighbor u of v, we have that $d_\Gamma^\updownarrow(u, v) \leq \frac{\delta}{\lambda}$.*

Proof Sketch: By Lemma 3, the altitude through v of the triangle with vertices v, u_e, and v_e intersects (u_e, v_e). Hence, $y(u_e) = y(v_e) = -\delta$, which implies the statement for $u \in \{u_e, v_e\}$, as $\lambda < 1$. For every $u \in \mathcal{N}(v) \setminus \{u_e, v_e\}$, we have $y(u) \geq 0$, as otherwise the distance between u and one of (v, u_e) and (v, v_e), or the distance between (v, u) and one of u_e and v_e would be smaller than δ. By Eq. 2, we have $\sum_{u \in \mathcal{N}(v)}(\lambda_{vu} \cdot y(u)) = y(v) = 0$, hence $\sum_{u \in \mathcal{N}(v) \setminus \{u_e, v_e\}}(\lambda_{vu} \cdot y(u)) = (\lambda_{vu_e} + \lambda_{vv_e}) \cdot \delta$. Since $y(u) \geq 0$, for every vertex $u \in \mathcal{N}(v) \setminus \{u_e, v_e\}$, we have $\lambda_{vu} \cdot y(u) \leq (\lambda_{vu_e} + \lambda_{vv_e}) \cdot \delta$, and hence $y(u) < \frac{\delta}{\lambda_{vu}} \leq \frac{\delta}{\lambda}$.   □

We outline the proof of Theorem 1. By Lemma 4, the vertex v and its neighbors lie in Γ in a "narrow" horizontal strip (see Fig. 1(a)). Using that as a starting point, the strategy is now to define a sequence of subgraphs of G, each one larger than the previous one, so that each subgraph is contained in a narrow horizontal strip. The larger the considered graph, the larger the height of the horizontal strip, however this height only depends on λ, on δ, and on the number of vertices of the considered graph. Eventually, this argument leads to a bound on the y-extent of Γ, and from that bound the resolution r of Δ provides

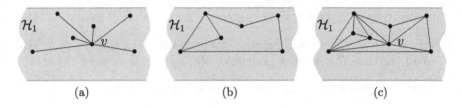

Fig. 1. (a) The edges incident to v. (b) G^v. (c) G_1.

a bound on the maximum distance between two separated geometric objects of Γ. The comparison of such distance with the minimum distance δ between two separated geometric objects of Γ allows us to derive the bound of Theorem 1.

We now formalize the proof. For $i \in \mathbb{N}^+$, denote by $\mathcal{H}_i$ the horizontal strip of height $h(i) := \delta \cdot \left(\frac{3}{\lambda}\right)^i$ bisected by the horizontal line through v.

We prove the existence of a sequence $G_1, \ldots, G_k = G$ of graphs such that, for $i = 1, \ldots, k$, the graph G_i is a biconnected internally-triangulated plane graph that is a subgraph of G satisfying Properties (P1)–(P4) below. Let Γ_i be the restriction of Γ to G_i and let $\mathcal{C}_i$ be the outer cycle of G_i. (P1) G_i has at least $i+3$ vertices; (P2) $\mathcal{C}_i$ does not have any external chord; (P3) G_i is the subgraph of G inside $\mathcal{C}_i$; and (P4) Γ_i is contained in the interior of the horizontal strip $\mathcal{H}_i$.

We define G_1 as follows. Let G^v be the subgraph of G induced by the neighbors of v (see Fig. 1(b)). Since G is a maximal plane graph, G^v is biconnected; let $\mathcal{C}_1$ be the outer cycle of G^v. Then G_1 is the subgraph of G inside $\mathcal{C}_1$ (see Fig. 1(c)). Property (P3) is satisfied by construction. Property (P1) is satisfied, since G_1 contains v and its at least three (as $n \geq 4$) neighbors. Property (P2) is satisfied because $\mathcal{C}_1$ is the outer cycle of G^v and G^v is an induced subgraph of G. Property (P4) is satisfied by Lemma 4, as the distance from v to the top or bottom side of $\mathcal{H}_1$ is $\frac{3}{2} \cdot \frac{\delta}{\lambda}$ and all the vertices of G_1 are in the convex hull of v's neighbors. Finally, G_1 is biconnected and internally-triangulated, by Property 1.

Assume that $G_i \neq G$; we will deal with the case $G_i = G$ later. We describe how to construct G_{i+1} from G_i, so to satisfy Properties (P1)–(P4). In order to do that, we introduce $\updownarrow$-connected vertices and prove some lemmata about them. A vertex v of G_i is $\updownarrow$-connected if it satisfies at least one of the following properties: (i) v has a neighbor above or on the top side of $\mathcal{H}_{i+1}$ and has a neighbor below the bottom side of $\mathcal{H}_i$; (ii) v has a neighbor below or on the bottom side of $\mathcal{H}_{i+1}$ and has a neighbor above the top side of $\mathcal{H}_i$. We have the following.

Lemma 5. *Let u be a vertex of G_i in $\mathcal{I}_G$. If u has a neighbor above or on the top side of $\mathcal{H}_{i+1}$, then it is $\updownarrow$-connected. Analogously, if u has a neighbor below or on the bottom side of $\mathcal{H}_{i+1}$, then it is $\updownarrow$-connected.*

Proof. We prove the first part of the statement. The proof of the second part is analogous. Suppose, for a contradiction, that there exists a vertex u of G_i in $\mathcal{I}_G$ that has a neighbor w above or on the top side of $\mathcal{H}_{i+1}$ and that has no neighbor below $\mathcal{H}_i$. By Eq. 2, we have that $\sum_{z \in \mathcal{N}_G(u)} (\lambda_{uz} \cdot y(z)) = y(u)$.

Fig. 2. Proof of Lemma 6. The fat lines are $\ell(u_1)$, $\ell(u_2)$, and $\ell(u_3)$.

Since $\sum_{z \in \mathcal{N}_G(u)} \lambda_{uz} = 1$, it follows that $\sum_{z \in \mathcal{N}_G(u)} (\lambda_{uz} \cdot (y(z) - y(u))) = 0$, hence

$$\lambda_{uw} \cdot (y(w) - y(u)) = \sum_{z \in \mathcal{N}_G(u) \setminus \{w\}} (\lambda_{uz} \cdot (y(u) - y(z))). \tag{3}$$

The distance between the top side of $\mathcal{H}_{i+1}$ and the top side of $\mathcal{H}_i$ is $\frac{h(i+1)-h(i)}{2}$, hence $\lambda_{uw} \cdot (y(w) - y(u)) \geq \lambda \cdot \frac{h(i+1)-h(i)}{2}$. Since every neighbor of u is above or on the bottom side of $\mathcal{H}_i$ and since u is in the interior of $\mathcal{H}_i$, we have that $y(u) - y(z) < h(i)$, for every neighbor z of u. Hence, $\sum_{z \in \mathcal{N}_G(u) \setminus \{w\}} (\lambda_{uz} \cdot (y(u) - y(z))) < h(i) \cdot \sum_{z \in \mathcal{N}_G(u) \setminus \{w\}} \lambda_{uz} < h(i)$. Furthermore, $\lambda \cdot \frac{h(i+1)-h(i)}{2} - h(i) > \lambda \cdot \frac{h(i+1)}{2} - 3 \cdot \frac{h(i)}{2} = \frac{\delta}{2} \left(\lambda \cdot \left(\frac{3}{\lambda}\right)^{i+1} - 3 \cdot \left(\frac{3}{\lambda}\right)^i \right) = 0$. This implies that $\lambda_{uw} \cdot (y(w) - y(u)) > \sum_{z \in \mathcal{N}_G(u) \setminus \{w\}} (\lambda_{uz} \cdot (y(u) - y(z)))$, which contradicts Eq. 3. $\square$

The second lemma states that few vertices of G_i are $\updownarrow$-connected.

Lemma 6. *The following statements hold true: (S1) G_i contains at most two vertices that are in $\mathcal{I}_G$ and that are $\updownarrow$-connected; and (S2) if G_i contains a vertex in $\mathcal{O}_G$, then it contains at most one vertex that is in $\mathcal{I}_G$ and that is $\updownarrow$-connected.*

Proof Sketch: We only prove (S1), as the proof of (S2) is similar. For each $\updownarrow$-connected vertex u of G_i, we define a polygonal line $\ell(u)$ as follows. Let w (let z) be a neighbor of u that lies above (below) the top side of $\mathcal{H}_i$. Let p_w and p_z be the intersection points of (u, w) and (u, z) with the top side of $\mathcal{H}_i$ and the bottom side of $\mathcal{H}_i$, respectively. Then $\ell(u)$ consists of the segments $\overline{up_w}$ and $\overline{up_z}$.

Suppose, for a contradiction, that G_i contains three $\updownarrow$-connected vertices u_1, u_2, and u_3 in $\mathcal{I}_G$; see Fig. 2. By the planarity of Γ, the lines $\ell(u_1)$, $\ell(u_2)$, and $\ell(u_3)$ do not cross each other. Assume that $\ell(u_2)$ is in-between $\ell(u_1)$ and $\ell(u_3)$ in $\mathcal{H}_i$. Consider any path P of G_i connecting u_1 and u_3. By the planarity of Γ and since G_i lies in the interior of $\mathcal{H}_i$, we have that P does not cross $\ell(u_2)$, except at u_2. However, this implies that the removal of u_2 from G_i separates u_1 from u_3, a contradiction to the fact that G_i is biconnected. This proves (S1). $\square$

Finally, we prove the following main lemma.

Lemma 7. *There exists a vertex u in $\mathcal{O}_{G_i} \cap \mathcal{I}_G$ that is not $\updownarrow$-connected.*

Proof. Since G_i contains at least 4 vertices, by Property (P1), and is biconnected, we have $|\mathcal{O}_{G_i}| \geq 3$. Further, $|\mathcal{O}_{G_i} \cap \mathcal{O}_G| \leq 2$, as if $|\mathcal{O}_{G_i} \cap \mathcal{O}_G| = 3$, then

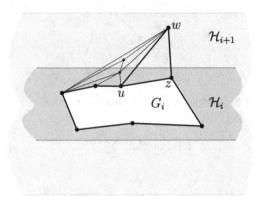

Fig. 3. Construction of G_{i+1} from G_i.

G_i would contain the outer cycle of G, by Property (P2), and we would have $G_i = G$, by Property (P3). We now distinguish three cases. If $|\mathcal{O}_{G_i} \cap \mathcal{O}_G| = 0$, then $|\mathcal{O}_{G_i} \cap \mathcal{I}_G| \geq 3$. By statement (S1) of Lemma 6, a vertex in $\mathcal{O}_{G_i} \cap \mathcal{I}_G$ is not $\updownarrow$-connected. If $|\mathcal{O}_{G_i} \cap \mathcal{O}_G| = 1$, then $|\mathcal{O}_{G_i} \cap \mathcal{I}_G| \geq 2$. By statement (S2) of Lemma 6, a vertex in $\mathcal{O}_{G_i} \cap \mathcal{I}_G$ is not $\updownarrow$-connected. If $|\mathcal{O}_{G_i} \cap \mathcal{O}_G| = 2$, then $|\mathcal{O}_{G_i} \cap \mathcal{I}_G| \geq 1$ and G_i contains two vertices u and w of the outer cycle of G; by Property (P4), these lie in the interior of $\mathcal{H}_i$, hence either every vertex of G_i lies above the bottom side of $\mathcal{H}_i$ or every vertex of G_i lies below the top side of $\mathcal{H}_i$. In both cases, every vertex in $\mathcal{O}_{G_i} \cap \mathcal{I}_G$ is not $\updownarrow$-connected.   □

We describe how to construct G_{i+1} from G_i; see Fig. 3. By Lemma 7, there is a vertex u in $\mathcal{O}_{G_i} \cap \mathcal{I}_G$ that is not $\updownarrow$-connected. Let (u, z) be any of the two edges incident to u in $\mathcal{C}_i$ and let (u, z, w) be the cycle delimiting the face of G incident to (u, z) outside $\mathcal{C}_i$; note that w does not belong to G_i. Let G^w be the biconnected subgraph of G induced by $\{w\} \cup V(\mathcal{C}_i)$. Let $\mathcal{C}_{i+1}$ be the outer cycle of G^w and let G_{i+1} be the subgraph of G inside $\mathcal{C}_{i+1}$. We prove that G_{i+1} satisfies the required properties. Property (P3) is satisfied by construction. Property (P1) is satisfied, since G_i contains at least $i + 3$ vertices (by Property (P1) of G_i) and G_{i+1} also contains w. Property (P2) is satisfied because $\mathcal{C}_{i+1}$ is the outer cycle of G^w and G^w is an induced subgraph of G. Property (P4) is also satisfied by Γ_{i+1}. Namely, the vertices of G_i lie in $\mathcal{H}_i \subset \mathcal{H}_{i+1}$, by Property (P4) of G_i. Further, w lies in the interior of $\mathcal{H}_{i+1}$, since u is not $\updownarrow$-connected and by Lemma 5, and all the vertices of G_{i+1} lie in the convex hull of $\{w\} \cup V(\mathcal{C}_i)$. Finally, G_{i+1} is biconnected and internally-triangulated, by Property 1.

By Property (P1), for some $k \leq n - 3$, we have that G_k contains n vertices, that is, $G_k = G$. By Property (P4), the y-extent Y of Δ is smaller than $\delta \cdot \left(\frac{3}{\lambda}\right)^n$. By Lemma 2, the x-extent X of Δ is smaller than $\frac{\delta}{r} \cdot \left(\frac{3}{\lambda}\right)^n$. The largest distance between two separated geometric objects in Γ is then smaller than $X + Y < \delta \cdot \left(\frac{3}{\lambda}\right)^n \cdot \left(1 + \frac{1}{r}\right) < \frac{2\delta}{r} \cdot \left(\frac{3}{\lambda}\right)^n$. The resolution of Γ is larger than or equal to the ratio between δ and the upper bound on the largest distance between two

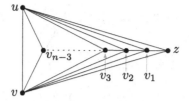

Fig. 4. The graph G_n in the proof of Theorem 2.

separated geometric objects in Γ obtained above. This ratio is $\frac{r}{2} \cdot \left(\frac{\lambda}{3}\right)^n$, which is indeed the bound in Theorem 1. By Property 2, we have that $\frac{r}{2} \cdot \left(\frac{\lambda}{3}\right)^n \in r \cdot \lambda^{O(n)}$.

4   Upper Bound on the Resolution of F-Drawings

In this section, we prove Theorem 2. The theorem is proved by analyzing a class of graphs introduced by Eades and Garvan [14] and depicted in Fig. 4. Consider any values $0 < \lambda \leq \frac{1}{4}$ and $0 < r \leq \frac{\sqrt{3}}{2}$. We remark that these upper bounds on λ and r follow by Properties 2 and 3. Let Δ be the triangle with vertices $u := (0, 0.5)$, $v := (0, -0.5)$, and $z := (r, 0)$; note that the resolution of Δ is r. Further, let $v_0 := z$ and let Λ be the coefficient matrix such that $\lambda_{v_i v_{i+1}} = \lambda$, for $i = 1, \ldots, n - 4$, $\lambda_{v_{i+1} v_i} = \lambda$, for $i = 0, \ldots, n - 4$, $\lambda_{v_i u} = \lambda_{v_i v} = 0.5 - \lambda$, for $i = 1, \ldots, n - 4$, and $\lambda_{v_{n-3} u} = \lambda_{v_{n-3} v} = 0.5 - \lambda/2$.

Easy calculations show that $y(v_i) = 0$, for $i = 1, \ldots, n - 3$. By Eq. 1 and since $x(u) = x(v) = 0$, for $i = 1, \ldots, n - 4$ we have $x(v_i) = \lambda \cdot x(v_{i-1}) + \lambda \cdot x(v_{i+1})$. By the planarity of the drawing (Λ, Δ), we have $x(v_{i+1}) < x(v_i)$, for $i = 0, \ldots, n - 4$. By $x(v_i) = \lambda \cdot x(v_{i-1}) + \lambda \cdot x(v_{i+1})$ and $x(v_{i+1}) < x(v_i)$, we get $x(v_i) \leq \lambda \cdot x(v_{i-1}) + \lambda \cdot x(v_i)$, hence $x(v_i) \leq \frac{\lambda}{1-\lambda} \cdot x(v_{i-1})$. By repeatedly using this, we get $x(v_{n-4}) \leq x(v_0) \cdot \left(\frac{\lambda}{1-\lambda}\right)^{n-4} = r \cdot \left(\frac{\lambda}{1-\lambda}\right)^{n-4}$, which is in $r \cdot \lambda^{\Omega(n)}$. Hence, the distance between v_{n-4} and (u, v) is in $r \cdot \lambda^{\Omega(n)}$. Theorem 2 then follows from the fact that the largest distance between any two separated geometric objects in the drawing is 1.

5   Lower Bound on the Resolution of FG-Morphs

In this section, we prove Theorem 3. Given Γ_0 and Γ_1, we first compute coefficient matrices Λ_0 and Λ_1 such that $\Gamma_0 = (\Lambda_0, \Delta)$, such that $\Gamma_1 = (\Lambda_1, \Delta)$, and such that the smallest positive coefficient in each of Λ_0 and Λ_1 is "not too small".

Lemma 8. *Let Γ be a planar straight-line drawing of an n-vertex maximal plane graph G, let Δ be the triangle delimiting the outer face of Γ, and let r be the resolution of Γ. There exists a coefficient matrix Λ such that $\Gamma = (\Lambda, \Delta)$ and such that the smallest positive coefficient in Λ is larger than r/n.*

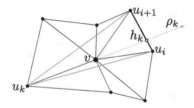

Fig. 5. Illustration for the proof of Lemma 8.

Proof. We employ and analyze a method proposed by Floater and Gotsman [18, Section 5]; refer to Fig. 5. Consider any internal vertex v of G and let $u_0, \ldots, u_{d-1}$ be the clockwise order of the neighbors of v in G. In the following, consider indices modulo d. For $k = 0, \ldots, d-1$, shoot a ray ρ_k starting at u_k and passing through v; this hits either a vertex u_i or the interior of an edge (u_i, u_{i+1}). We have $v = \mu_{k,k} \cdot u_k + \mu_{i,k} \cdot u_i + \mu_{i+1,k} \cdot u_{i+1}$, for some $\mu_{k,k} > 0$, $\mu_{i,k} > 0$, $\mu_{i+1,k} \geq 0$. For every $j \notin \{i, i+1, k\}$, set $\mu_{j,k} = 0$. After all the values $\mu_{j,k}$ have been computed, let $\lambda_{vu_k} = \frac{1}{d} \sum_{j=0,\ldots,d-1} \mu_{k,j}$, for $k = 0, \ldots d-1$. Let δ (let D) be smallest (resp. largest) distance between two separated geometric objects in Γ. In order to prove $\lambda_{vu_k} > r/n$, it suffices to prove $\mu_{k,k} \geq r$; indeed, $\lambda_{vu_k} \geq \mu_{k,k}/d > \mu_{k,k}/n$. Let h_k be the intersection point between ρ_k and (u_i, u_{i+1}). Then $\mu_{k,k} = \frac{|\overline{u_k v}|}{|\overline{u_k h_k}|}$. Further, since (u_k, v) is an edge of G, we have that $|\overline{u_k v}| \geq \delta$. Finally, $|\overline{u_k h_k}| \leq D$, since $|\overline{u_k h_k}| \leq d_\Gamma(u_k, (u_i, u_{i+1}))$. Hence $\mu_{k,k} \geq \delta/D = r$ and $\lambda_{vu_k} > r/n$. $\qquad\square$

The proof of Theorem 3 is as follows. For $i = 0, 1$, let r_i be the resolution of Γ_i. As in Lemma 8, compute a coefficient matrix Λ_i such that $\Gamma_i = (\Lambda_i, \Delta)$ and such that the smallest positive coefficient in Λ_i is larger than $r_i/n \geq r/n$. Let $\mathcal{M} = \{\Gamma_t = (\Lambda_t, \Delta) : t \in [0,1]\}$ be the FG-morph between Γ_0 and Γ_1 such that, for any $t \in [0,1]$, $\Lambda_t = (1-t) \cdot \Lambda_0 + t \cdot \Lambda_1$. For any edge (u, v) of G where $u \in \mathcal{I}_G$, we have $\lambda_{uv}^t = (1-t) \cdot \lambda_{uv}^0 + t \cdot \lambda_{uv}^1 \geq (1-t) \cdot r/n + t \cdot r/n = r/n$. By Theorem 1, the resolution of Γ_t is in $(r/n)^{O(n)}$. This concludes the proof of Theorem 3.

6 Upper Bound on the Resolution of FG-Morphs

In this section, we prove Theorem 4. We employ a triangulated "nested triangles graph" (see, e.g., [12,19]). Let $k = n/3$ and observe that k is an integer. Then G consists of (refer to Fig. 6) the 3-cycles (u_i, v_i, z_i), for $i = 1, \ldots, k$, the paths $(u_1, \ldots, u_k)$, $(v_1, \ldots, v_k)$, and $(z_1, \ldots, z_k)$, and the edges (u_i, z_{i+1}), (z_i, v_{i+1}), and (v_i, u_{i+1}), for $i = 1, \ldots, k-1$. The outer cycle of G is (u_k, v_k, z_k). The planar straight-line drawings Γ_0 and Γ_1 of G for the proof of the theorem are depicted in Fig. 6(a) and Fig. 6(b), respectively.

The construction clearly satisfies Property (R1).

We prove Property (R2). For $i = 0, 1$, the drawing Γ_i lies on an $O(n) \times O(n)$ grid, hence the largest distance between any two separated geometric objects is in $O(n)$. By Lemma 3, the smallest distance between any two separated geometric

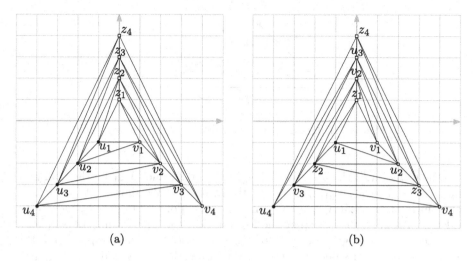

Fig. 6. The graph G in the proof of Theorem 4 (with $n = 12$). (a) shows Γ_0 and (b) shows Γ_1.

objects is the one between a vertex v and an edge e. By Pick's theorem, the area of the triangle T defined by v and e is at least 0.5. Since the length of e is in $O(n)$, the height of T with respect to e, which is with the distance between v and e, is in $\Omega(1/n)$.

We prove Property (R3). To do that, we first show the following.

Claim 1. *For any coefficient matrices Λ_0 and Λ_1 such that $\Gamma_0 = (\Lambda_0, \Delta)$ and $\Gamma_1 = (\Lambda_1, \Delta)$, and for any $i = 2, \ldots, k-1$, we have $\lambda^0_{u_i u_{i+1}} > 0.5$, $\lambda^0_{v_i v_{i+1}} > 0.5$, $\lambda^0_{z_i z_{i+1}} > 0.5$, $\lambda^1_{u_i z_{i+1}} > 0.5$, $\lambda^1_{v_i u_{i+1}} > 0.5$, and $\lambda^1_{z_i v_{i+1}} > 0.5$.*

Proof. We prove that, for any coefficient matrix Λ_0 such that $\Gamma_0 = (\Lambda_0, \Delta)$, we have $\lambda^0_{z_i z_{i+1}} > 0.5$. The other proofs are analogous. By Eq. 2, we have that $y(z_i) = \sum_{w \in \mathcal{N}(z_i)} \lambda^0_{z_i w} \cdot y(w)$, where $\mathcal{N}(z_i) = \{u_i, v_i, z_{i-1}, z_{i+1}, u_{i-1}, v_{i+1}\}$. Since the coefficients $\lambda^0_{z_i w}$ with $w \in \mathcal{N}(z_i)$ are all positive and since the values $y(u_i)$, $y(v_i)$, $y(u_{i-1})$, and $y(v_{i+1})$ are all smaller than $y(z_{i-1}) = i - 1$, we get $y(z_i) = i < (\lambda^0_{z_i u_i} + \lambda^0_{z_i v_i} + \lambda^0_{z_i u_{i-1}} + \lambda^0_{z_i v_{i+1}} + \lambda^0_{z_i z_{i-1}}) \cdot (i - 1) + \lambda^0_{z_i z_{i+1}} \cdot (i + 1)$. From this, it follows that $\lambda^0_{z_i z_{i+1}} > \lambda^0_{z_i u_i} + \lambda^0_{z_i v_i} + \lambda^0_{z_i u_{i-1}} + \lambda^0_{z_i v_{i+1}} + \lambda^0_{z_i z_{i-1}}$, which gives us $\lambda^0_{z_i z_{i+1}} > 0.5$, given that $\sum_{w \in \mathcal{N}(z_i)} \lambda^0_{z_i w} = 1$. $\square$

Consider now any coefficient matrices Λ_0 and Λ_1 such that $\Gamma_0 = (\Lambda_0, \Delta)$ and $\Gamma_1 = (\Lambda_1, \Delta)$, and the FG-morph $\mathcal{M} = \{\Gamma_t = (\Lambda_t, \Delta) : t \in [0, 1]\}$. We will prove that the resolution of $\Gamma_{0.5}$ is exponentially small. By Claim 1, we have $\lambda^0_{u_i u_{i+1}} > 0.5$ and $\lambda^1_{u_i z_{i+1}} > 0.5$. This, together with $\lambda^{0.5}_{u_i u_{i+1}} = 0.5 \cdot \lambda^0_{u_i u_{i+1}} + 0.5 \cdot \lambda^1_{u_i u_{i+1}}$ and $\lambda^{0.5}_{u_i z_{i+1}} = 0.5 \cdot \lambda^0_{u_i z_{i+1}} + 0.5 \cdot \lambda^1_{u_i z_{i+1}}$, implies that $\lambda^{0.5}_{u_i u_{i+1}} > 0.25$ and $\lambda^{0.5}_{u_i z_{i+1}} > 0.25$. Analogously, $\lambda^{0.5}_{v_i v_{i+1}} > 0.25$, $\lambda^{0.5}_{v_i u_{i+1}} > 0.25$, $\lambda^{0.5}_{z_i z_{i+1}} > 0.25$, and $\lambda^{0.5}_{z_i v_{i+1}} > 0.25$.

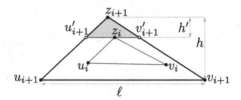

Fig. 7. Triangles Δ_i, Δ_{i+1}, and Δ'_{i+1} (which is gray).

Denote by $\mathcal{A}(T)$ the area of a triangle T. Let Δ_i and Δ_{i+1} be the triangles (u_i, v_i, z_i) and $(u_{i+1}, v_{i+1}, z_{i+1})$, respectively, in $\Gamma_{0.5}$. We show that $\mathcal{A}(\Delta_i)$ is a constant fraction of $\mathcal{A}(\Delta_{i+1})$, which implies the exponential upper bound. Refer to Fig. 7. Assume, w.l.o.g., that the longest side s of Δ_{i+1} connects u_{i+1} and v_{i+1}. Let ℓ be its length and let h be the height of Δ_{i+1} with respect to s. By Lemma 1, we can assume that the x-axis passes through s and that z_{i+1} lies above s. Now consider the vertex t of Δ_i with the highest y-coordinate. We have that t is either z_i or u_i. Assume that $t = z_i$, as the other case is analogous.

We bound $y(z_i)$ in terms of $y(z_{i+1})$. By Eq. 2, in $\Gamma_{0.5}$ we have $y(z_i) = \sum_{w \in \mathcal{N}(z_i)} \lambda_{z_i w}^{0.5} \cdot y(w)$, where $\mathcal{N}(z_i) = \{u_i, v_i, z_{i-1}, z_{i+1}, u_{i-1}, v_{i+1}\}$. Since z_i is the highest vertex of Δ_i, every neighbor of z_i different from z_{i+1} lies below z_i (possibly u_i lies on the same horizontal line as z_i). Hence, $y(z_i) < (\lambda_{z_i u_i}^{0.5} + \lambda_{z_i v_i}^{0.5} + \lambda_{z_i z_{i-1}}^{0.5} + \lambda_{z_i u_{i-1}}^{0.5}) \cdot y(z_i) + \lambda_{z_i z_{i+1}}^{0.5} \cdot y(z_{i+1}) + \lambda_{z_i v_{i+1}}^{0.5} \cdot y(v_{i+1})$. Since $y(v_{i+1}) = 0$, this is $y(z_i) < \frac{\lambda_{z_i z_{i+1}}^{0.5}}{1 - (\lambda_{z_i u_i}^{0.5} + \lambda_{z_i v_i}^{0.5} + \lambda_{z_i z_{i-1}}^{0.5} + \lambda_{z_i u_{i-1}}^{0.5})} \cdot y(z_{i+1})$. Since $\lambda_{z_i v_{i+1}}^{0.5} > 0.25$, we have $\lambda_{z_i z_{i+1}}^{0.5} + \lambda_{z_i u_i}^{0.5} + \lambda_{z_i v_i}^{0.5} + \lambda_{z_i z_{i-1}}^{0.5} + \lambda_{z_i u_{i-1}}^{0.5} < 0.75$. Hence, by setting $\rho := \lambda_{z_i u_i}^{0.5} + \lambda_{z_i v_i}^{0.5} + \lambda_{z_i z_{i-1}}^{0.5} + \lambda_{z_i u_{i-1}}^{0.5}$, we get that $y(z_i) < \frac{0.75 - \rho}{1 - \rho} \cdot y(z_{i+1})$, where $\rho \in (0, 1)$. As the function $f(\rho) := \frac{0.75 - \rho}{1 - \rho}$ decreases as ρ increases over the interval $(0, 1)$, we get that $\frac{0.75 - \rho}{1 - \rho} < 0.75$ and hence $y(z_i) < 0.75 \cdot y(z_{i+1})$ in $\Gamma_{0.5}$.

Consider a horizontal line through z_i and let u'_{i+1} and v'_{i+1} be its intersection points with $\overline{z_{i+1}u_{i+1}}$ and $\overline{z_{i+1}v_{i+1}}$, respectively. Let Δ'_{i+1} be the triangle with vertices z_{i+1}, u'_{i+1}, and v'_{i+1}. Since $y(z_i) < 0.75 \cdot y(z_{i+1})$, the height h' of Δ'_{i+1} with respect to $\overline{u_{i+1}v_{i+1}}$ is larger than $0.25 \cdot h$. By the similarity of Δ_{i+1} and Δ'_{i+1}, the length ℓ' of $u'_{i+1}v'_{i+1}$ is larger than $0.25 \cdot \ell$. Hence, $\mathcal{A}(\Delta'_{i+1}) = h' \cdot \ell'/2 \geq 0.0625 \cdot \mathcal{A}(\Delta_{i+1})$. Since the interiors of Δ_i and Δ'_{i+1} are disjoint, $\mathcal{A}(\Delta_i) \leq \mathcal{A}(\Delta_{i+1}) - \mathcal{A}(\Delta'_{i+1}) \leq 0.9375 \cdot \mathcal{A}(\Delta_{i+1})$. As this holds for $i = 2, \ldots, k-1$, we have that $\mathcal{A}(\Delta_2)/\mathcal{A}(\Delta_k)$ is in $1/2^{\Omega(n)}$. This bound directly transfers to the resolution of $\Gamma_{0.5}$. This concludes the proof of Theorem 4.

7   Conclusions and Open Problems

We studied the resolution of popular algorithms for the construction of planar straight-line graph drawings and morphs. With a focus on maximal plane graphs, we discussed the resolution of the drawing algorithm by Floater [15], which is a broad generalization of Tutte's algorithm [29], and of the morphing algorithm by Floater and Gotsman [18]. Many problems are left open by our research.

1. The lower bounds on the resolution of F-drawings and FG-morphs presented in Theorems 1 and 3 apply to maximal plane graphs. A major objective is to extend such bounds to 3-connected plane graphs.
2. The lower bound on the resolution of F-drawings presented in Theorem 1 is tight, by Theorem 2. However, when applied to T-drawings, the bounds in such theorems do not coincide. Namely, Theorem 1 gives an $r/n^{O(n)} \subseteq r/2^{O(n \log n)}$ lower bound (as the smallest positive term λ of a coefficient matrix is in $1/\Omega(n)$ for graphs with maximum degree in $\Omega(n)$), while Theorem 2 gives an $r/2^{\Omega(n)}$ upper bound (as every internal vertex of the graph in the proof of the theorem has degree in $\Theta(1)$, and hence every element of the coefficient matrix is in $\Theta(1)$). We find it interesting to close this gap.
3. The lower and upper bounds for the resolution of FG-morphs in Theorems 3 and 4 also leave a gap. First, the dependency on n in the lower bound is $1/2^{O(n \log n)}$, while the one in the upper bound is $1/2^{\Omega(n)}$. Second, the resolution r of the input drawings appears in the exponential lower bound of Theorem 3, while it does not appear in the exponential upper bound of Theorem 4; while it is clear that a dependency on r is needed, it is not clear to us whether r should be part of the exponential function.

References

1. Alexa, M.: Recent advances in mesh morphing. Comput. Graph. Forum **21**(2), 173–196 (2002)
2. Bárány, I., Rote, G.: Strictly convex drawings of planar graphs. Doc. Math. **11**, 369–391 (2006)
3. Barrera-Cruz, F., et al.: How to morph a tree on a small grid. In: Friggstad, Z., Sack, J.-R., Salavatipour, M.R. (eds.) WADS 2019. LNCS, vol. 11646, pp. 57–70. Springer, Cham (2019). https://doi.org/10.1007/978-3-030-24766-9_5
4. Barrera-Cruz, F., Haxell, P.E., Lubiw, A.: Morphing Schnyder drawings of planar triangulations. Discrete Comput. Geom. **61**, 161–184 (2019)
5. Chambers, E.W., Eppstein, D., Goodrich, M.T., Löffler, M.: Drawing graphs in the plane with a prescribed outer face and polynomial area. J. Graph Algorithms Appl. **16**(2), 243–259 (2012)
6. Chrobak, M., Goodrich, M.T., Tamassia, R.: Convex drawings of graphs in two and three dimensions (preliminary version). In: Whitesides, S. (ed.) SoCG 1996, pp. 319–328. ACM (1996)
7. Da Lozzo, G., Di Battista, G., Frati, F., Patrignani, M., Roselli, V.: Upward planar morphs. Algorithmica **82**(10), 2985–3017 (2020)
8. de Fraysseix, H., Pach, J., Pollack, R.: How to draw a planar graph on a grid. Combinatorica **10**(1), 41–51 (1990)
9. Demaine, E.D., Schulz, A.: Embedding stacked polytopes on a polynomial-size grid. Discrete Comput. Geom. **57**(4), 782–809 (2017)
10. Di Battista, G., Eades, P., Tamassia, R., Tollis, I.G.: Graph Drawing: Algorithms for the Visualization of Graphs. Prentice-Hall, Hoboken (1999)

11. Di Battista, G., Frati, F.: From Tutte to Floater and Gotsman: on the resolution of planar straight-line drawings and morphs. CoRR abs/2108.09483 (2021)
12. Dolev, D., Leighton, F.T., Trickey, H.: Planar embedding of planar graphs. In: Advances in Computing Research, vol. 2 (1984)
13. Eades, P.: A heuristic for graph drawing. Congr. Numer. **42**(11), 149–160 (1984)
14. Eades, P., Garvan, P.: Drawing stressed planar graphs in three dimensions. In: Brandenburg, F.J. (ed.) GD 1995. LNCS, vol. 1027, pp. 212–223. Springer, Heidelberg (1996). https://doi.org/10.1007/BFb0021805
15. Floater, M.S.: Parametrization and smooth approximation of surface triangulations. Comput. Aided Geom. Des. **14**(3), 231–250 (1997)
16. Floater, M.S.: Parametric tilings and scattered data approximation. Int. J. Shape Model. **4**(3–4), 165–182 (1998)
17. Floater, M.S.: Mean value coordinates. Comput. Aided Geom. Des. **20**(1), 19–27 (2003)
18. Floater, M.S., Gotsman, C.: How to morph tilings injectively. J. Comput. Appl. Math. **101**(1), 117–129 (1999)
19. Frati, F., Patrignani, M.: A note on minimum-area straight-line drawings of planar graphs. In: Hong, S.-H., Nishizeki, T., Quan, W. (eds.) GD 2007. LNCS, vol. 4875, pp. 339–344. Springer, Heidelberg (2008). https://doi.org/10.1007/978-3-540-77537-9_33
20. Gortler, S.J., Gotsman, C., Thurston, D.: Discrete one-forms on meshes and applications to 3D mesh parameterization. Comput. Aided Geom. Des. **23**(2), 83–112 (2006)
21. Groiss, L., Jüttler, B., Mokris, D.: 27 variants of Tutte's theorem for plane near-triangulations and an application to periodic spline surface fitting. Comput. Aided Geom. Des. **85**, 101975 (2021)
22. Ilinkin, I.: Visualization of Floater and Gotsman's morphing algorithm. In: Cheng, S., Devillers, O. (eds.) SoCG 2014, pp. 94–95. ACM (2014)
23. Kobourov, S.G.: Force-directed drawing algorithms. In: Tamassia, R. (ed.) Handbook on Graph Drawing and Visualization, pp. 383–408. Chapman and Hall/CRC (2013)
24. Linial, N., Lovász, L., Wigderson, A.: Rubber bands, convex embeddings and graph connectivity. Combinatorica **8**(1), 91–102 (1988)
25. Mor, A.R., Rote, G., Schulz, A.: Small grid embeddings of 3-polytopes. Discrete Comput. Geom. **45**(1), 65–87 (2011)
26. Schnyder, W.: Embedding planar graphs on the grid. In: Johnson, D.S. (ed.) SODA 1990, pp. 138–148. SIAM (1990)
27. Siu, A.M.K., Wan, A.S.K., Lau, R.W.H., Ngo, C.: Trifocal morphing. In: Banissi, E., et al. (eds.) IV 2003, pp. 24–29. IEEE (2003)
28. Surazhsky, V., Gotsman, C.: Controllable morphing of compatible planar triangulations. ACM Trans. Graph. **20**(4), 203–231 (2001)
29. Tutte, W.T.: How to draw a graph. Proc. Lond. Math. Soc. **3**(13), 743–767 (1963)
30. de Verdière, É.C., Pocchiola, M., Vegter, G.: Tutte's barycenter method applied to isotopies. Comput. Geomet.: Theory Appl. **26**(1), 81–97 (2003)
31. Weiss, V., Andor, L., Renner, G., Várady, T.: Advanced surface fitting techniques. Comput. Aided Geom. Des. **19**(1), 19–42 (2002)

Planar and Toroidal Morphs Made Easier

Jeff Erickson$^{(\boxtimes)}$ and Patrick Lin$^{(\boxtimes)}$

University of Illinois, Urbana-Champaign, Champaign, USA
{jeffe,plin15}@illinois.edu

Abstract. We present simpler algorithms for two closely related morphing problems, both based on the barycentric interpolation paradigm introduced by Floater and Gotsman, which is in turn based on Floater's asymmetric extension of Tutte's classical spring-embedding theorem.

First, we give a very simple algorithm to construct piecewise-linear morphs between planar straight-line graphs. Specifically, given isomorphic straight-line drawings Γ_0 and Γ_1 of the same 3-connected planar graph G, with the same convex outer face, we construct a morph from Γ_0 to Γ_1 that consists of $O(n)$ unidirectional morphing steps, in $O(n^{1+\omega/2})$ time. Our algorithm entirely avoids the classical edge-collapsing strategy dating back to Cairns; instead, in each morphing step, we interpolate the pair of weights associated with a single edge.

Second, we describe a natural extension of barycentric interpolation to geodesic graphs on the flat torus. Barycentric interpolation cannot be applied directly in this setting, because the linear systems defining intermediate vertex positions are not necessarily solvable. We describe a simple scaling strategy that circumvents this issue. Computing the appropriate scaling requires $O(n^{\omega/2})$ time, after which we can compute the drawing at any point in the morph in $O(n^{\omega/2})$ time. Our algorithm is considerably simpler than the recent algorithm of Chambers *et al.* and produces more natural morphs. Our techniques also yield a simple proof of a conjecture of Connelly *et al.* for geodesic torus triangulations.

1 Introduction

Computing morphs between geometric objects is a fundamental problem that has been well studied, with many applications in graphics, animation, modeling, and more. A particularly well-studied setting is that of morphing between planar straight-line graphs. Formally, a morph between two isomorphic planar straight-line graphs Γ_0 and Γ_1 consists of a continuous family of planar straight-line graphs Γ_t starting at Γ_0 and ending at Γ_1.

We describe an extremely simple morphing algorithm for planar graphs, which simultaneously obtains properties of two earlier approaches: Floater and Gotsman's barycentric interpolation method [24,26,43–45] results in morphs that are natural and visually appealing but are represented implicitly; variations on Cairns' edge-collapse method [1,7,8,31,46] result in efficient explicit representations of morphs that are not useful for visualization. Our new algorithm efficiently computes an explicit piecewise-linear representation of a morph

© Springer Nature Switzerland AG 2021
H. C. Purchase and I. Rutter (Eds.): GD 2021, LNCS 12868, pp. 123–137, 2021.
https://doi.org/10.1007/978-3-030-92931-2_9

between drawings of the same 3-connected planar graph, that are potentially more useful for visualization than morphs based on Cairns' method.

We also extend Floater and Gotsman's planar morphing algorithm to geodesic graphs on the flat torus. Recent results of Luo et al. [37] imply that Floater and Gotsman's method directly generalizes to morphs between geodesic triangulations on surfaces of negative curvature, but a direct generalization to the torus generically fails [42]. Our extension is based on simple scaling strategy, and it yields more natural morphs than previous algorithms based on edge collapses [9]. Finally, our arguments yield a straightforward proof of a conjecture of Connelly et al. [15] about the deformation space of geodesic triangulations.

1.1   Related Work

Planar Morphs. Cairns [7,8] was the first to prove the existence of morphs between arbitrary isomorphic planar straight-line triangulations, using an inductive argument based on the idea of collapsing an edge from a low-degree vertex to one of its neighbors. Thomassen [46] extended Cairns' proof to arbitrary planar straight-line graphs. Cairns and Thomassen's proofs are constructive, but yield morphs consisting of an exponential number of steps.

Floater and Gotsman [24] proposed a more direct method to construct morphs between planar graphs, based on an extension by Floater [22] of Tutte's classical spring embedding theorem [49]. Let Γ be a straight-line drawing of a planar graph G, such that the boundary of every face of Γ is a strictly convex polygon. Then every interior vertex in Γ is a strict convex combination of its neighbors; that is, we can associate a positive weight $\lambda_{u \to v}$ with each half-edge or *dart* $u \to v$ in G, such that the vertex positions p_v in Γ satisfy the linear system

$$\sum_{u \to v} \lambda_{u \to v}(p_v - p_u) = (0,0) \qquad \text{for every interior vertex } u \qquad (1)$$

Floater [22] proved that given arbitrary[1] positive weights $\lambda_{u \to v}$ and an arbitrary convex outer face, solving linear system (1) yields a straight-line drawing of G with convex faces. Tutte's original spring-embedding theorem [49] is the special case of this result where every dart has weight 1, but his proof extends verbatim to arbitrary symmetric weights, where $\lambda_{u \to v} = \lambda_{v \to u}$ for every edge uv [29,41,47].

Floater and Gotsman [24] construct a morph between two convex drawings of the same planar graph G, with the same outer face, by linearly interpolating between weights $\lambda_{u \to v}$ consistent with the initial and final drawings. Appropriate initial and final weights can be computed in $O(n)$ time using, for example, Floater's mean-value coordinates [23,30]. The resulting morphs are natural and visually appealing. However, the motions of the vertices are only computed implicitly; vertex positions at any time can be computed in $O(n^{\omega/2})$ time by solving a linear system via nested dissection [4,34], where $\omega < 2.37286$ is the matrix

---

[1] Floater's presentation assumes that $\sum_{u \to v} \lambda_{u \to v} = 1$ for every interior vertex v, but this assumption is clearly unnecessary.

multiplication exponent [3,33]. Gotsman and Surazhsky generalized Floater and Gotsman's technique to arbitrary planar straight-line graphs [26,43–45].

Alamdari *et al.* [1] describe an efficient algorithm to construct planar morphs with explicit piecewise-linear vertex trajectories, based on Cairns' inductive edge-collapsing strategy. Given any two isomorphic straight-line drawings (with the same rotation system and nesting structure) of the same n-vertex planar graph, the algorithm constructs a morph consisting of $O(n)$ *unidirectional* morphing steps, in which all vertices move along parallel lines at fixed speeds. Thus, each vertex moves along a piecewise-linear path of complexity $O(n)$, and the entire morph has complexity $O(n^2)$. Recent results of Klemz [32] imply that this algorithm can be implemented to run in $O(n^2 \log n)$ time. The resulting morph contracts all vertices into an exponentially small neighborhood and then expand them again, so it is not useful for visualization.

Angelini *et al.* [5] consider the setting of *convexity-preserving* morphs between convex drawings; Kleist *et al.* [31] consider morphing to convexify any 3-connected planar drawing. Both describe algorithms that produce piecewise-linear morphs consisting of $O(n)$ steps, and that can be implemented to run in time $O(n^{1+\omega/2})$. (Klemz [32] conjectures that both running times can be improved to $O(n^2 \log n)$.) Combining these algorithms results in an alternative piecewise-linear morph between 3-connected planar drawings.

Toroidal Morphs. Until recently, very little was known about morphing graphs on the torus or other more complex surfaces.

Tutte's spring-embedding theorem was generalized to simple triangulations of surfaces with non-positive curvature by Colin de Verdière [14] and independently by Hass and Scott [27]. Delgado-Friedrichs [19], Lovász [35], and Gortler *et al.* [25] also independently proved an extension of Tutte's theorem to graphs on the flat torus whose universal covers are simple and 3-connected. For any toroidal graph and any assignment of positive *symmetric* weights to the darts, solving a linear system similar to (1) yields vertex positions of a geodesic drawing with strictly convex faces [20,25]; see Sect. 2 for details. Thus, if two isotopic geodesic torus graphs Γ_0 and Γ_1 can both be described by symmetric dart weights, linearly interpolating those weights yields a morph from Γ_0 to Γ_1 [13].

The restriction to symmetric weights is both nontrivial and significant. In a torus graph with convex faces, every vertex can be described as a convex combination of its neighbors, but not necessarily with symmetric weights. Moreover, the linear system expressing vertex positions as convex combinations of its neighbors is rank-deficient, and therefore is not solvable in general; see the full version [21] for an example. Thus, Floater's asymmetric extension of Tutte's theorem does *not* directly generalize to the flat torus.

For similar reasons, Floater and Gotsman's planar morphing algorithm also does not generalize. Suppose we are given two isotopic geodesic torus graphs Γ_0 and Γ_1, each with dart weights that express their vertices as convex combinations of their neighbors. Unfortunately, in general, interpolating those weights yields linear systems that have no solution; we give a simple example in the full version [21].

Steiner and Fischer [42] modify the system by fixing a single vertex, restoring full rank. However, solving the modified system does not necessarily yield a crossing-free drawing, because the fixed vertex may not lie in the convex hull of its neighbors. Moreover, even though the initial and final weights are consistent with crossing-free drawings, averages of those weights may not be. We give an example of this bad behavior in the full version [21].

Chambers et al. [9] described the first algorithm to morph between arbitrary essentially 3-connected geodesic torus graphs. Their algorithm uses a combination of Cairns' edge-collapsing strategy and spring embeddings to construct a morph consisting of $O(n)$ unidirectional morphing steps, in $O(n^{1+\omega/2})$ time. Like planar morphs built from edge collapses, these toroidal morphs contract vertices into small neighborhoods and thus are not suitable for visualization.

Recently, Luo et al. [37] generalized Floater's theorem to geodesic triangulations of arbitrary closed Riemannian 2-manifolds with strictly negative curvature, extending the spring-embedding theorems of Colin de Verdière [14] and Hass and Scott [27] to asymmetric weights. Their result immediately implies that if two geodesic triangulations of such a surface are homotopic, then linearly interpolating the dart weights yields a continuous family of crossing-free geodesic drawings, or in other words, a morph. Their result applies only to surfaces with negative Euler characteristic; the torus has Euler characteristic 0.

1.2   New Results

We describe two applications of Floater and Gotsman's barycentric interpolation strategy, which yield simpler algorithms for morphing planar and toroidal graphs.

First we describe a very simple algorithm to construct piecewise-linear morphs between planar straight-line graphs. Given two isomorphic planar straight-line graphs Γ_0 and Γ_1 with strictly convex faces and the same outer face, we construct a morph from Γ_0 to Γ_1 that consists of $O(n)$ unidirectional morphing steps, in $O(n^{1+\omega/2})$ time. Our morphing algorithm computes barycentric weights for the darts in Γ_0 and Γ_1 in a preprocessing phase, and then for each morphing step, interpolates only the pair of weights associated with a single edge. Our key observation is that changing the weights for a single edge e moves all vertices in the Floater drawing along lines parallel to e. (The same observation was made for *symmetric* edge weights by Chambers et al. [9].) Our algorithm is significantly simpler than that of Angelini et al. [5] for computing convexity-preserving morphs. We then extend our algorithm to drawings with non-convex faces, using a simpler approach than Kleist et al. [31]. Figure 1 shows a morph computed by our algorithm; in each frame, the weights of the bold edge are about to change.

Next, we describe a natural extension of Floater and Gotsman's method to geodesic graphs on the flat torus. Our key observation is that barycentric dart weights can be *scaled* so that barycentric interpolation works. Specifically, we call a weight assignment *morphable* if every *column* of the resulting Laplacian linear system sums to zero; averages of morphable weights are morphable. Given

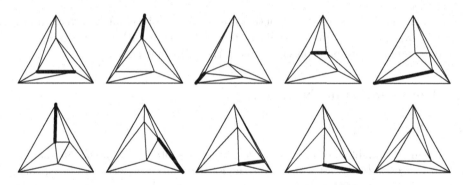

Fig. 1. Incrementally morphing between planar graphs.

any weight assignment consistent with any convex drawing, we can guarantee morphability by scaling the weights of all darts leaving each vertex v—or equivalently, scaling each *row* of the linear system—by a common positive scalar α_v. This scaling obviously has no effect on the solution space of the system. Positivity of the scaling vector α follows from a weighted directed version of the matrix-tree theorem [6,17,48]. We can computing the appropriate scaling in $O(n^{\omega/2})$ time, after which we can compute any intermediate drawing in $O(n^{\omega/2})$ time, matching the performance of Floater and Gotsman exactly. The resulting morphs are natural and visually appealing, and our proofs of correctness are considerably simpler than those of Chambers *et al.* [9]. However, unlike Chambers *et al.*, our new morphing algorithm does not compute explicit vertex trajectories. Figure 2 shows a morph computed by our algorithm between two randomly shifted 6×6 toroidal grids. (The authors' Python implementation is available on request.)

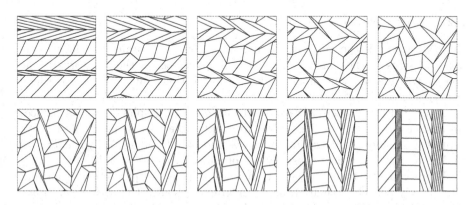

Fig. 2. Morphing between randomly shifted toroidal grids.

It remains an open question whether our results can be combined to compute explicit low-complexity piecewise-linear toroidal morphs without edge collapses. We offer some preliminary observations in the full version [21].

2   Definitions and Notation

2.1   Planar Graphs

Any planar straight-line drawing Γ can be represented by a **position** matrix $P \in \mathbb{R}^{n \times 2}$, each row p_v of which gives the location of some vertex v. Thus, each edge uv is drawn as the straight-line segment $p_u p_v$. We call a planar drawing **convex** if it is crossing-free, every bounded face is a convex polygon, and the outer face is the complement of a convex polygon.

Formally, we regard each edge of any graph as a pair of opposing half-edges or *darts*, each directed from its *tail* to its *head*. We write **rev**(d) to denote the reversal of any dart d. For simple graphs, we write $u{\to}v$ to denote the dart with tail u and head v. A **barycentric weight vector** for Γ assigns a positive real number $\lambda_{u \to v}$ to every dart $u{\to}v$ of a graph, so that the vertex positions p_v satisfy Floater's linear system (1). Conversely, for a fixed graph G with a fixed convex outer face, the **Floater drawing** Γ^λ of G with respect to a positive weight vector λ is the unique drawing whose vertex positions p_v satisfy system (1).

A **morph** between two planar drawings Γ_0 and Γ_1 is a continuous family of crossing-free drawings Γ_t parametrized by time, starting at Γ_0 and ending at Γ_1. A morph is *linear* if each vertex moves along a straight line at uniform speed, and *piecewise-linear* if it is the concatenation of linear morphs. Any piecewise-linear morph can be described by a finite sequence of straight-line drawings. A linear morph is *unidirectional* if vertices move along parallel lines.

2.2   Torus Graphs

The **flat torus** is the quotient space $\mathbb{T} = \mathbb{R}^2 / \mathbb{Z}^2$, also obtained by identifying opposite sides of the unit square $[0, 1]^2$. A **geodesic** on the flat torus is the image of a line segment in $\mathbb{R}^2$ under the projection map $\pi \colon \mathbb{R}^2 \to \mathbb{T}$ where $\pi(x, y) = (x \bmod 1, y \bmod 1)$.

A (crossing-free) **geodesic torus drawing** Γ of a graph G maps its vertices to distinct points in $\mathbb{T}$ and its edges to simple, interior-disjoint geodesics. We explicitly consider graphs containing loops and parallel edges. We write $d : u{\to}v$ to declare that d is a dart (possibly one of many) with tail u and head v.

Every geodesic torus drawing Γ of a graph G is the projection of an infinite, doubly-periodic planar straight-line graph $\widetilde{\Gamma}$, called the **universal cover** of Γ [9]. We call Γ **essentially simple** if its universal cover $\widetilde{\Gamma}$ is simple, and **essentially 3-connected** if $\widetilde{\Gamma}$ is 3-connected [39,40]. Finally, we call Γ a **convex** drawing if every face of $\widetilde{\Gamma}$ is strictly convex. Every convex torus drawing is both essentially simple and essentially 3-connected, since every infinite planar graph with strictly convex faces is 3-connected [18].

Coordinate Representations. Following Chambers *et al.* [9], we use a **coordinate representation** (P, τ) for geodesic torus drawings that records

- a **position** vector $p_v \in \mathbb{R}^2$ for each vertex v, and
- a **translation** vector $\tau_d \in \mathbb{Z}^2$ for each dart d, such that $\tau_{rev(d)} = -\tau_d$.

These vectors indicate that each dart $d : u \rightarrow v$ is drawn as the projection of a line segment from p_u to $p_v + \tau_d$ in the universal cover $\widetilde{\Gamma}$. In particular, if we normalize all vertex positions to the half-open unit square $[0, 1)^2$, then each translation vector τ_d indicates the number of times d crosses the vertical boundary of the unit square to the right, and the number of times d crosses the horizontal boundary of the unit square upward.

Two crossing-free drawings of the same graph on the torus are *isotopic* if one can be deformed into the other through a continuous family of (not necessarily geodesic) crossing-free drawings; such a deformation is called an *isotopy*. Two crossing-free drawings are isotopic if and only if their coordinate representations can be **normalized** so that their translation vectors agree; this condition can be tested in $O(n)$ time [9, Theorem A.1], [12]. A *geodesic isotopy* or *morph* is an isotopy in which all intermediate drawings are geodesic.

Barycentric Weights. In any convex torus drawing Γ, the position p_v of each vertex v can be expressed as a convex combination of its neighbors, as follows. We can assign a weight $\lambda_d > 0$ to each dart d such that any coordinate representation (P, τ) of Γ satisfies the linear system

$$\sum_v \sum_{d:u \rightarrow v} \lambda_d(p_v - p_u + \tau_d) = (0, 0) \qquad \text{for every vertex } u. \qquad (2)$$

We can express this linear system in matrix notation as $L^\lambda P = H^\lambda$, where

$$L_{ij}^\lambda = \begin{cases} \sum_k \sum_{d:i \rightarrow k} \lambda_d & \text{if } i = j \\ \sum_{d:i \rightarrow j} -\lambda_d & \text{otherwise} \end{cases} \qquad \text{and} \quad H_i^\lambda = \sum_j \sum_{d:i \rightarrow j} \lambda_d x_d \qquad (2')$$

The (unnormalized, asymmetric) Laplacian matrix L^λ has rank $n - 1$ [42]. We call any positive weight vector λ satisfying system (2) **barycentric** for Γ. Barycentric weights for any convex torus drawing can be computed in $O(n)$ time using, for example, Floater's mean-value coordinates [23, 30].

On the other hand, suppose we fix the graph G and translation vectors τ_d consistent with an essentially 3-connected (but not necessarily geodesic) drawing of G. Then for any positive weight vector λ, any solution to linear system (2) gives the vertex positions p_v of a convex drawing Γ^λ of G [25]. In this case, we say that the **Floater drawing** Γ^λ **realizes** the weight vector λ, and we call the weight vector λ **realizable** for the graph G. Every realizable weight vector is realized by a two-dimensional family of drawings that differ by translation.

Every *symmetric* positive weight vector (where $\lambda_d = \lambda_{rev(d)}$) is realizable: for any assignment of positive weights to the *edges* of G, there is a corresponding convex torus drawing [14,19,25,27,35]. Realizable weights are not necessarily symmetric: there are convex torus drawings with only asymmetric barycentric weights. Conversely, positive asymmetric weights are not always realizable.

3   Morphing Planar Graphs Edge by Edge

We describe a very simple algorithm to morph planar straight-line graphs that combines the benefits of both the Floater and Gotsman approach [24,26,43–45] and the Cairns approach [1,7,8,31,46]. Our algorithm constructs a morph consisting of $O(n)$ unidirectional morphing steps, in $O(n^{1+\omega/2})$ time. Because our morphs do not use edge collapses, they are also potentially good for visualization.

Fix a planar graph G and a convex outer face. Let p_v^λ denote the position of vertex v in the Floater drawing Γ^λ with respect to weight vector λ. The following lemma is a planar asymmetric version of Lemma 5.1 of Chambers *et al.* [9]. Intuitively, it states that changing the weights of the darts of a single edge e moves each vertex in the Floater drawing along lines parallel to e.

Lemma 1. *Let λ and μ be arbitrary positive weight vectors such that $\lambda_d \neq \mu_d$ or $\lambda_{rev(d)} \neq \mu_{rev(d)}$ for some dart d, but $\lambda_{d'} = \mu_{d'}$ for all darts $d' \notin \{d, rev(d)\}$. For each vertex w, the vector $p_w^\mu - p_w^\lambda$ is parallel to the drawing of d in Γ^λ.*

Proof: Suppose d has tail u and head v, and (by rotating the drawing if necessary) that d is drawn parallel to the x-axis. For each vertex i, let y_i^λ and y_i^μ be the y-coordinates of points p_i^λ and p_i^μ, respectively, so that $y_u^\lambda = y_v^\lambda$. We need to prove that $y_w^\lambda = y_w^\mu$ for every vertex w.

Projecting linear system (1) for λ onto the y-axis gives us

$$\sum_{i \to j} \lambda_{i \to j}(y_j^\lambda - y_i^\lambda) = 0 \qquad \text{for each vertex } i. \tag{3}$$

Swapping entries of λ with corresponding entries of μ in the system (3) changes at most two constraints, corresponding to the two endpoints u and v of d. Moreover, in each changed constraint, the single changed coefficient is multiplied by $y_u^\lambda - y_v^\lambda = y_v^\lambda - y_u^\lambda = 0$, so the y_i^λ's also solve the corresponding system for μ. Since the system (3) and its counterpart for μ each have a unique solution, we conclude that $y_w^\lambda = y_w^\mu$ for every vertex w. $\square$

Under the assumptions of Lemma 1, linearly interpolating the vertex positions from Γ^λ to Γ^μ yields a unidirectional linear morph [1, Corollary 7.2], [9, Lemma 5.2]. It follows that we can morph between isomorphic convex drawings through a sequence of at most $3n-9$ unidirectional linear morphing steps, one for each internal edge, following the algorithm in Fig. 3. Initial and final barycentric weight vectors can be found in $O(n)$ time using, for example, Floater's mean-value method [23,30]. Each intermediate drawing can be computed in $O(n^{\omega/2})$ time using nested dissection [4,34], for a total running time of $O(n^{1+\omega/2})$.

```
MORPHCONVEX(Γstart, Γend):
    λ ← barycentric weights for Γstart
    μ ← barycentric weights for Γend
    k ← 0
    for each internal edge e
        k ← k + 1
        d ← a dart of e
        λd ← μd
        λrev(d) ← μrev(d)
        Γk ← Γλ
    return Γstart, Γ1, Γ2, . . . , Γk (= Γend)
```

Fig. 3. Algorithm for morphing between convex planar drawings.

Because all Floater drawings are convex, Lemma 5.2 of Chambers *et al.* [9] implies that MORPHCONVEX actually produces a *convexity-preserving* piecewise-linear morph; all faces remain convex throughout the morph. Our algorithm is significantly simpler than that of Angelini *et al.* [5].

We can extend the previous algorithm to non-convex drawings by first morphing to convex drawings, as follows. We first add edges to the initial and final drawings to decompose every face into convex polygons, compute barycentric weights for the resulting drawing, and then reduce the weights of each added edge (one-by-one) to zero, effectively deleting that edge. Dropping the added edges yields a piecewise-linear morph from each input drawing to a convex drawing. Again, each intermediate drawing can be computed in $O(n^{\omega/2})$ time. Our complete morphing algorithm is shown in Fig. 4. Our algorithm CONVEXIFY is considerably simpler than that of Kleist *et al.* [31]; however, unlike Kleist *et al.*, our algorithm is *not* necessarily convexity-increasing.

```
CONVEXIFY(Γ):
    Γ' ← convex decomposition of Γ
    λ ← barycentric weights for Γ'
    k ← 0
    for each new edge e in Γ'
        k ← k + 1
        d ← a dart of e
        λd ← 0
        λrev(d) ← 0
        Γk ← Γλ without new edges
    return Γ, Γ1, Γ2, . . . , Γk
```

```
MORPH(Γstart, Γend):
    Sbefore ← CONVEXIFY(Γstart)
    Γbefore ← last drawing in Sbefore
    Safter ← CONVEXIFY(Γend)
    Γafter ← last drawing in Safter
    Sconvex ← MORPHCONVEX(Γbefore, Γafter)
    return Sbefore, Sconvex, reverse(Safter)
```

Fig. 4. Algorithm for morphing between general planar straight-line drawings.

In total, we perform one morphing step for each internal edge of G, plus at most $2(k-3)$ morphing steps for each bounded face with degree k. Euler's formula implies that a 3-connected planar graph has between $1.5n$ and $3n - 6$ edges, and thus at most $3n - 9$ internal edges. Thus, we need to add at most $1.5n - 6$ edges to convexify the initial and final faces, so our morph consists of at most $4.5n - 15$ linear morphing steps. In summary:

Theorem 1. *Given any two isomorphic 3-connected planar straight-line drawings with n vertices and the same convex outer face, we can compute a morph between them consisting of at most $4.5n-15$ unidirectional linear morphing steps, in $O(n^{1+\omega/2})$ time.*

4   Morphable Weight Vectors on the Flat Torus

As observed by Steiner and Fischer [42], Floater and Gotsman's morphing algorithm does not directly generalize to the toroidal setting, since not all positive weight vectors λ are realizable. In particular, given arbitrary barycentric weights $\lambda(0)$ and $\lambda(1)$ of two isotopic convex torus drawings, intermediate weights $\lambda(t) := (1 - t)\lambda(0) + t\lambda(1)$ are not necessarily realizable; see the full version [21].

To bypass this issue, we identify a subspace of *morphable* weight vectors, such that every convex torus drawing has a morphable barycentric weight vector, every morphable weight vector is realizable, and convex combinations of morphable weights are morphable. Specifically, a positive weight vector λ is **morphable** if each *column* of the matrices L^λ and H^λ sums to 0. The following lemma is immediate:

Lemma 2. *Convex combinations of morphable weight vectors are morphable.*

Lemma 3. *Every morphable weight vector is realizable.*

Proof: If λ is a morphable weight vector, then the nth row of the linear system $L^\lambda P = H^\lambda$ is implied by the other $n-1$ rows, so we can remove it. The resulting abbreviated linear system still has rank $n - 1$, so it has a (unique) solution.   □

Lemma 4. *Given a barycentric weight vector λ for a convex torus drawing Γ, a morphable barycentric weight vector for Γ can be computed in $O(n^{\omega/2})$ time.*

Proof: The matrix L^λ has rank $n - 1$, so there is a one-dimensional space of (row) vectors $\alpha = (\alpha_1, \ldots, \alpha_n)$ such that $\alpha L^\lambda = (0, \ldots, 0)$. We can compute a non-zero vector α in $O(n^{\omega/2})$ time using nested dissection [2,4,34].

A directed version of the matrix tree theorem [6,17,48] implies that we can choose all α_i to be positive. Specifically, let $G^\pm$ be the weighted directed graph whose weighted arcs correspond to the weighted darts of G. An *inward directed spanning tree* is an acyclic spanning subgraph of $G^\pm$ where every vertex except one (called the *root*) has out-degree 1. The weight of an inward directed spanning tree is the product of the weights of its arcs. For each i, let α_i be the sum of the weights of all inward directed spanning trees rooted at vertex i; we have $\alpha_i > 0$ because all dart weights are positive. The directed matrix tree theorem implies that $\alpha L = 0$, as required; for an elementary proof, see De Leenheer [17, Theorem 3]. (See also Cohen *et al.* [11, Lemma 1].)

Define a new weight vector μ by setting $\mu_d := \alpha_{tail(d)}\lambda_d$ for each dart d. For each index i, we immediately have $L_i^\mu P = \alpha_i L_i^\lambda P = \alpha_i H_i^\lambda = H_i^\mu$, where P is the position matrix for Γ, so μ is in fact a barycentric weight vector for Γ. Finally, we observe that $(1,\ldots,1)L^\mu = \alpha L^\lambda = (0,\ldots,0)$ and $(1,\ldots,1)H^\mu = \alpha H^\lambda = \alpha L^\lambda P = (0,\ldots,0)P = (0,0)$, which imply that μ is morphable. $\qquad\square$

Theorem 2. *Given coordinate representations of two isotopic essentially 3-connected geodesic torus drawings Γ_0 and Γ_1, we can efficiently compute a morph from Γ_0 to Γ_1. Specifically, after $O(n^{\omega/2})$ preprocessing time, we can compute any intermediate drawing during the morph in $O(n^{\omega/2})$ time.*

Proof: Suppose Γ_0 and Γ_1 are convex drawings. First, if necessary, we normalize the given coordinate representations so that their translation vectors agree, in $O(n)$ time [9, Theorem A.1]. Then we find barycentric weight vectors $\lambda(0)$ and $\lambda(1)$ for Γ_0 and Γ_1, respectively, in $O(n)$ time, for example using Floater's mean-value coordinates [23, 30]. Following Lemma 4, we derive morphable weights $\mu(0)$ and $\mu(1)$ from $\lambda(0)$ and $\lambda(1)$, respectively, in $O(n^{\omega/2})$ time. Finally, given any real number $0 < t < 1$, we set $\mu(t) := (1-t)\mu(0) + t\mu(1)$ and solve the linear system $L^{\mu(t)}P^{(t)} = H^{\mu(t)}$ for the position matrix $P^{(t)}$ of an intermediate drawing $\Gamma^{\mu(t)}$; Lemmas 2 and 3 imply that this system is solvable. The function $t \mapsto \Gamma^{\mu(t)}$ is a convexity-preserving morph between Γ_0 and Γ_1.

If the faces of Γ_0 or Γ_1 are not convex, we morph through an intermediate convex drawing, similarly to Chambers *et al.* [9, Theorem 8.1]. Let Γ_* be the Floater drawing of G obtained by setting every dart weight to 1. Compute any triangulation T_0 of Γ_0, and then triangulate the convex faces Γ_* using the same diagonals, to obtain a triangulation T_* isotopic to T_0. Assign weight 0 to the darts of the diagonals in $T_* \setminus \Gamma_*$ to obtain a barycentric weight vector μ_* for T_*, which is symmetric and therefore morphable. Derive morphable weights μ_0 for T_0 using mean-value coordinates [23, 30] and Lemma 4. Then we can morph from T_0 to T_* by weight interpolation, using the weight vector $\mu(t) := (1-2t)\mu_0 + 2t\mu_*$ for any $0 \le t \le 1/2$. Ignoring the diagonal edges gives us a morph from Γ_0 to Γ_*. A symmetric procedure yields a morph from Γ_* to Γ_1. $\qquad\square$

In the full version [21], we use morphable weights to prove a conjecture of Connelly *et al.* [15] about the deformation space of geodesic torus triangulations.

5   Open Questions

It is natural to ask whether our "best-of-both-worlds" planar morph can be extended to graphs on the flat torus. In the full version [21], we prove a toroidal analog of Lemma 1 for *realizable* weight vectors; unfortunately, the main roadblock is that not all weight vectors are realizable. In particular, given a realizable weight vector (morphable or not), it is not clear when changing the weights for a single edge results in another realizable weight vector.

Several previous planar morphing algorithms [1,5,16,31] rely on a certain convexifying procedure [10,28,31,32], and are (potentially) faster than our algorithm via the implementation recently described by Klemz [32]. It is an open question whether the procedure can be extended to geodesic torus graphs.

One can also ask if the result can be extended to surfaces of higher genus. The recent results of Luo et al. [37] imply that Floater and Gotsman's planar morphing algorithm [24] extends to geodesic triangulations on higher-genus surfaces of negative curvature; however, the existence of (any reasonable analog of) piecewise-linear morphs on such surfaces remains unknown.

Acknowledgments. We thank Anna Lubiw for asking questions about Lemma 5.1 of Chambers et al. [9], whose answers ultimately led to the discovery of Theorem 1, and for other helpful feedback. We also thank Yanwen Luo for making us aware of his recent work [36–38]. Finally, we thank the anonymous reviewers for their comments and helpful suggestions for improvement.

References

1. Alamdari, S., et al.: How to morph planar graph drawings. SIAM J. Comput. **46**(2), 824–852 (2017). https://doi.org/10.1137/16M1069171
2. Aleksandrov, L., Djidjev, H.: Linear algorithms for partitioning embedded graphs of bounded genus. SIAM J. Discrete Math. **9**(1), 129–150 (1996). https://doi.org/10.1137/S0895480194272183
3. Alman, J., Williams, V.V.: A refined laser method and faster matrix multiplication. In: Proceedings of the 32nd Annual ACM-SIAM Symposium Discrete Algorithms, October 2021. https://doi.org/10.1137/1.9781611976465.32
4. Alon, N., Yuster, R.: Matrix sparsification and nested dissection over arbitrary fields. J. ACM **60**(4), 25:1–25:8 (2013). https://doi.org/10.1145/2508028.2505989
5. Angelini, P., Lozzo, G.D., Frati, F., Lubiw, A., Patrignani, M., Roselli, V.: Optimal morphs of convex drawings. In: Proceedings of 31st International Symposium on Computational Geometry, pp. 126–140. No. 34 in Leibniz International Proceedings in Informatics (2015)
6. Borchardt, C.W.: Ueber eine der interpolation entsprechende Darstellung der Eliminations-Resultante. J. Reine Angew. Math. **57**, 111–121 (1860). https://doi.org/10.1515/crll.1860.57.111
7. Cairns, S.S.: Deformations of plane rectilinear complexes. Amer. Math. Monthly **51**(5), 247–252 (1944). https://doi.org/10.2307/2304300
8. Cairns, S.S.: Isotopic deformations of geodesic complexes on the 2-sphere and on the plane. Ann. Math. **45**(2), 207–217 (1944). https://doi.org/10.2307/1969263
9. Chambers, E.W., Erickson, J., Lin, P., Parsa, S.: How to morph graphs on the torus. In: Proceedings of 32nd Annual ACM-SIAM Symposium Discrete Algorithms, pp. 2759–2778 (2021). https://doi.org/10.1137/1.9781611976465.164
10. Chrobak, M., Goodrich, M.T., Tamassia, R.: Convex drawings of graphs in two and three dimensions (preliminary version). In: Proceedings of 12th Annual Symposium on Computational Geometry, pp. 319–328 (1996). https://doi.org/10.1145/237218.237401

11. Cohen, M.B., Kelner, J., Peebles, J., Peng, R., Sidford, A., Vladu, A.: Faster algorithms for computing the stationary distribution, simulating random walks, and more. In: Proceedings of 57th Annual IEEE Symposium on Foundations of Computer Science, pp. 583–592 (2016). https://doi.org/10.1109/FOCS.2016.69
12. Colin de Verdière, É., de Mesmay, A.: Testing graph isotopy on surfaces. Discrete Comput. Geom. **51**(1), 171–206 (2013). https://doi.org/10.1007/s00454-013-9555-4
13. Colin de Verdière, É., Pocchiola, M., Vegter, G.: Tutte's barycenter method applied to isotopies. Comput. Geom. Theory Appl. **26**(1), 81–97 (2003). https://doi.org/10.1016/S0925-7721(02)00174-8
14. Colin de Verdière, Y.: Comment rendre géodésique une triangulation d'une surface? L'Enseignment Mathématique **37**, 201–212 (1991). https://doi.org/10.5169/seals-58738
15. Connelly, R., Henderson, D.W., Ho, C.W., Starbird, M.: On the problems related to linear homeomorphisms, embeddings, and isotopies. In: Continua, Decompositions, Manifolds: Proceedings of Texas Topology Symposium, 1980, pp. 229–239. University of Texas Press (1983)
16. Da Lozzo, G., Di Battista, G., Frati, F., Patrignani, M., Roselli, V.: Upward planar morphs. Algorithmica **82**(10), 2985–3017 (2020). https://doi.org/10.1007/s00453-020-00714-6
17. De Leenheer, P.: An elementary proof of a matrix tree theorem for directed graphs. SIAM Rev. **62**(3), 716–726 (2020). https://doi.org/10.1137/19M1265193
18. Delgado-Friedrichs, O.: Barycentric drawings of periodic graphs. In: Liotta, G. (ed.) GD 2003. LNCS, vol. 2912, pp. 178–189. Springer, Heidelberg (2004). https://doi.org/10.1007/978-3-540-24595-7_17
19. Delgado-Friedrichs, O.: Equilibrium placement of periodic graphs and convexity of plane tilings. Discrete Comput. Geom. **33**(1), 67–81 (2004). https://doi.org/10.1007/s00454-004-1147-x
20. Erickson, J., Lin, P.: A toroidal Maxwell-Cremona-Delaunay correspondence. In: Proceedings of 36th International Symposium on Computational Geometry, pp. 40:1–40:17. No. 164 in Leibniz International Proceedings in Informatics, Schloss Dagstuhl-Leibniz-Zentrum für Informatik (2020). https://doi.org/10.4230/LIPIcs.SoCG.2020.40
21. Erickson, J., Lin, P.: Planar and toroidal morphs made easier. Preprint, August 2021. https://arxiv.org/abs/2106.14086v2
22. Floater, M.S.: Parametric tilings and scattered data approximation. Int. J. Shape Modeling **4**(3–4), 165–182 (1998). https://doi.org/10.1142/S021865439800012X
23. Floater, M.S.: Mean value coordinates. Comput. Aided Geom. Design **20**(1), 19–27 (2003). https://doi.org/10.1016/S0167-8396(03)00002-5
24. Floater, M.S., Gotsman, C.: How to morph tilings injectively. J. Comput. Appl. Math. **101**(1–2), 117–129 (1999). https://doi.org/10.1016/S0377-0427(98)00202-7
25. Gortler, S.J., Gotsman, C., Thurston, D.: Discrete one-forms on meshes and applications to 3D mesh parameterization. Comput. Aided Geom. Des. **23**(2), 83–112 (2006). https://doi.org/10.1016/j.cagd.2005.05.002
26. Gotsman, C., Surazhsky, V.: Guaranteed intersection-free polygon morphing. Comput. Graph. **25**(1), 67–75 (2001). https://doi.org/10.1016/S0097-8493(00)00108-4
27. Hass, J., Scott, P.: Simplicial energy and simplicial harmonic maps. Asian J. Math. **19**(4), 593–636 (2015). https://doi.org/10.4310/AJM.2015.v19.n4.a2
28. Hong, S.H., Nagamochi, H.: Convex drawings of hierarchical planar graphs and clustered planar graphs. J. Discrete Algorithms **8**(3), 282–295 (2010). https://doi.org/10.1016/j.jda.2009.05.003

29. Hopcroft, J.E., Kahn, P.J.: A paradigm for robust geometric algorithms. Algorithmica **7**(1–6), 339–380 (1992). https://doi.org/10.1007/BF01758769
30. Hormann, K., Floater, M.S.: Mean value coordinates for arbitrary planar polygons. ACM Trans. Graph. **25**(4), 1424–1441 (2006). https://doi.org/10.1145/1183287.1183295
31. Kleist, L., Klemz, B., Lubiw, A., Schlipf, L., Staals, F., Strash, D.: Convexity-increasing morphs of planar graphs. Comput. Geom. Theory Appl. **84**, 69–88 (2019). https://doi.org/10.1016/j.comgeo.2019.07.007
32. Klemz, B.: Convex drawings of hierarchical graphs in linear time, with applications to planar graph morphing. In: Proceedings of the 29th Annual European Symposium on Algorithms. Leibniz International Proceedings in Informatics, vol. 204, pp. 57:1–57:15. Schloss Dagstuhl-Leibniz-Zentrum für Informatik (2021). https://doi.org/10.4230/LIPIcs.ESA.2021.57
33. Le Gall, F.: Powers of tensors and fast matrix multiplication. In: Proceedings of 25th International Symposium on Algebraic Computation, pp. 296–303 (2014). https://doi.org/10.1145/2608628.2608664
34. Lipton, R.J., Rose, D.J., Tarjan, R.E.: Generalized nested dissection. SIAM J. Numer. Anal. **16**, 346–358 (1979). https://doi.org/10.1137/0716027
35. Lovász, L.: Discrete analytic functions: an exposition. In: Grigor'yan, A., Yau, S.T. (eds.) Eigenvalues of Laplacians and other geometric operators, pp. 241–273. No. 9 in Surveys in Differential Geometry, International Press (2004). https://doi.org/10.4310/SDG.2004.v9.n1.a7
36. Luo, Y.: Spaces of geodesic triangulations of surfaces. Preprint, August 2020. https://arxiv.org/abs/1910.03070v3
37. Luo, Y., Wu, T., Zhu, X.: The deformation space of geodesic triangulations and generalized Tutte's embedding theorem. Preprint, May 2021. https://arxiv.org/abs/2105.00612
38. Luo, Y., Wu, T., Zhu, X.: The deformation space of geodesic triangulations of flat tori. Preprint, July 2021. https://arxiv.org/abs/2107.05159
39. Mohar, B.: Circle packings of maps–the Euclidean case. Rend. Sem. Mat. Fis. Milano **67**(1), 191–206 (1997). https://doi.org/10.1007/BF02930499
40. Mohar, B.: Circle packings of maps in polynomial time. Europ. J. Combin. **18**(7), 785–805 (1997). https://doi.org/10.1006/eujc.1996.0135
41. Richter-Gebert, J.: Realization Spaces of Polytopes. Lecture Notes in Mathematics, vol. 1643. Springer, Heidelberg (1996). https://doi.org/10.1007/BFb0093761
42. Steiner, D., Fischer, A.: Planar parameterization for closed 2-manifold genus-1 meshes. In: Proceedings of the 9th ACM Symposium Solid Modeling Applications, pp. 83–91 (2004)
43. Surazhsky, V., Gotsman, C.: Controllable morphing of compatible planar triangulations. ACM Trans. Graph. **20**(4), 203–231 (2001). https://doi.org/10.1145/502783.502784
44. Surazhsky, V., Gotsman, C.: Morphing stick figures using optimal compatible triangulations. In: Proceedings of the 9th Pacific Conference on Computer Graphics Applications, pp. 40–49 (2001). https://doi.org/10.1109/PCCGA.2001.962856
45. Surazhsky, V., Gotsman, C.: Intrinsic morphing of compatible triangulations. Int. J. Shape Modeling **9**(2), 191–201 (2003). https://doi.org/10.1142/S0218654303000115
46. Thomassen, C.: Deformations of plane graphs. J. Comb. Theory Ser. B **34**(3), 244–257 (1983). https://doi.org/10.1016/0095-8956(83)90038-2
47. Thomassen, C.: Tutte's spring theorem. J. Graph Theory **45**(4), 275–280 (2004). https://doi.org/10.1002/jgt.10163

48. Tutte, W.T.: The dissection of equilateral triangles into equilateral triangles. Math. Proc. Cambridge Phil. Soc. **44**(4), 463–482 (1948)
49. Tutte, W.T.: How to draw a graph. Proc. London Math. Soc. **13**(3), 743–768 (1963). https://doi.org/10.1112/plms/s3-13.1.743

Visualizing JIT Compiler Graphs

HeuiChan Lim$^{(\boxtimes)}$ ⓘ and Stephen Kobourov$^{(\boxtimes)}$ ⓘ

Department of Computer Science, University of Arizona, Tucson, USA
hlim1@email.arizona.edu, kobourov@cs.arizona.edu

Abstract. Just-in-time (JIT) compilers are used by many modern programming systems in order to improve performance. Bugs in JIT compilers provide exploitable security vulnerabilities and debugging them is difficult as they are large, complex, and dynamic. Current debugging and visualization tools deal with static code and are not suitable in this domain. We describe a new approach for simplifying the large and complex intermediate representation, generated by a JIT compiler and visualize it with a metro map metaphor to aid developers in debugging.

1 Introduction

Many modern programming systems, such as JavaScript engines that are running our web browsers, use just-in-time (JIT) compilers to improve performance. Examples include Google Chrome, Microsoft Edge, Apple Safari, and Mozilla Firefox, which are used by 2.65 billion, 600 million, 446 million, and 220 million, respectively [4]. JIT compiler bugs can lead to exploitable security vulnerabilities [1,6–9]. Such a bug in Google Chrome could be used to hijack passwords and to navigate to other sites and execute malicious programs, as reported by the Microsoft Offensive Security Research team (CVE-2017-5121 [1]). Thus, the ability to quickly analyze, localize and fix JIT compiler problems is important. However, existing work and available tools focus on static code [15,16,23], and so they are not suitable for developers in debugging JIT compilers, which generates code at run-time. Additionally, the size and complexity of JIT-based systems [12] combined with the dynamic nature of JIT compiler optimizations, make it challenging to analyze and locate bugs quickly. For example, Google V8 has more than 2,000 source files and more than 1 million lines of code.

Traditional debuggers rely on text even though the main feature of a JIT compiler is building a graph-like structure to translate bytecode into optimized machine code. With this in mind, we propose a new debugging tool, which visualizes the JIT compiler's intermediate representation (IR). Our approach uses IR identification and generation techniques described by Lim and Debray [26], where the compiler-related half of the visualization tool's pipeline are described in detail. In this paper we focus on the visualization half, which includes: merging multiple IR graphs into a single graph, simplifying the merged graph, converting the simplified graph into a hypergraph, simplifying the hypergraph, and visualizing the hypergraph using a metro map metaphor. Visualizing the JIT compiler's IR allows us to answer questions such as:

H. C. Purchase and I. Rutter (Eds.): GD 2021, LNCS 12868, pp. 138–146, 2021.
https://doi.org/10.1007/978-3-030-92931-2_10

1. What optimizations took place to generate the machine code?
2. What is the relationship among the optimization phases?
3. Which optimization phase was most active?
4. What optimizations affected a specific node?
5. Which optimization phases are likely to be buggy?

Related Work. There are many methods and tools for debugging static code compilers and optimized code, but little on using the intermediate representation and visualizing it to show the explicit information about the compilation and optimization processes. Google V8's Turbolizer [5,13] is one of very few IR visualization tools. It shows the final IR graph after each optimization process and provides interactive features to view the control-flow graphs for each optimization phase. Although Turbolizer provides some information about the IR nodes and their relationships, it does not provide enough information about the optimization process and cannot answer several of our initial set of questions.

Dux *et al.* [21] visualize dynamically modified code at run-time with call graphs and control-flow graphs by showing the graph changes with animation, allowing end-to-end play, pause, and forward/backward step-by-step animation. CFGExplorer [20] visualizes the control-flow graph of a program to represent the program structure for dynamic binary analysis. It provides interactive features allowing developers to find specific memory addresses, loops, and functions to analyze the system. CcNav [19] analyzes and visualizes a C++ compiler's optimization process with a call graph, control-flow graph, and loop hierarchies.

Control-flow graphs and call graphs are popular in program analysis, especially for analyzing static code. However, they are different from dynamically generated IR graphs. Tools for visualizing and interacting with control-flow graphs and call graphs (such as those above) are not sufficient for visualizing the IR graph as, e.g., they cannot capture the optimization phases.

Background. We briefly introduce several concepts relevant to JIT compilers.

Interpreter. A computer program that converts input source code into bytecode and executes it without compiling it into a machine code [22].

Bytecode. Instructions generated from input source code by an interpreter; bytecode is portable, unlike compiled programs, and used in many modern languages and systems, such as JavaScript, Python, and Java [17].

Instruction-level Trace. A file that holds all the instructions that a programming system, such as a JIT compiler, has generated and executed at run-time. The instructions are in a machine-level code with symbol information (e.g., function names) and are used for performance analysis and debugging.

Just-in-Time (JIT) Compiler. A program that turns bytecode into instructions that are sent to a computer's processor, to improve performance [24]; see Fig. 1(a) for an example of JIT compiler in Google's V8 pipeline.

Optimized Code. Machine code generated from bytecode by a JIT compiler that can be directly executed by a processor.

Intermediate Representation (IR). A type of graph also known as sea-of-nodes [11,14,18]. Unlike other graphs used in program analysis, such as control-flow or data-flow graphs which have specific types of nodes, nodes in the

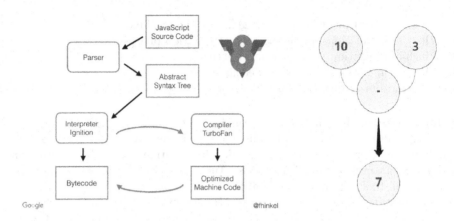

Fig. 1. (a) V8 Pipeline [12] (b) Example of constant folding optimization.

sea-of-nodes graph represent different types: from scalar values and arithmetic operators to variables and control-flow nodes and function entry nodes. Similarly, edges represent different relationships (e.g., semantic and syntax relationships).

Optimization. Adding, removing, and merging nodes and edges in the graph during execution. In a single JIT compilation, the compiler executes several different optimization phases (inlining, loop peeling, constant propagation) to generate efficient machine code, which modify the IR graph and correspond to new hyperedges (the set of all nodes generated or optimized in this phase); see Fig. 1(b) for an example of constant propagation.

Proof-of-Concept Program. An input program that is used to trigger the buggy behavior in the JIT compiler, i.e., a valid program (without any bugs) which when run can reveal bugs in the JIT compiler. In our experiment, we are targeting JavaScript engine V8, so the PoC is a JavaScript program.

2   Visualizing the Intermediate Representation

Our approach for capturing and visualizing the IR of a JIT compiler below uses compiler-related steps 1–4 [26], and steps 5–9 are described in brief below.

1. Modify the input program, P_0, to create similar programs, $\{P_1, ...P_N\}$, by generating the abstract syntax tree for P_0 and then randomly modifying nodes in the tree with allowable edits (passing semantic/syntactic checks). The newly created programs either still contain the code that triggers a bug in the JIT compiler, or the buggy code is replaced and no bug is triggered. In the first case, the execution output of the optimized code is different from the interpreted code (as with P_0).
2. Run each program P_i and collect the instruction-level traces.
3. Analyze traces to check if P_i triggers a bug in the JIT compiler and to identify P_i's IR and the optimization phases executed while optimizing P_i.

4. Select candidate hyperedges, suspected to be buggy, from the information gathered in step 3.
5. Merge all selected candidate hyperedges into the original IR from P_0.
6. Simplify the merged IR by reducing the number of nodes and edges.
7. Convert the simplified graph into a hypergraph by extracting the hyperedges from step 4 and analyzing each node's optimization status.
8. Simplify the hypergraph by reducing the number of hyperedges and nodes.
9. Visualize the simplified hypergraph with MetroSets [25].

2.1   Intermediate Representation

Recall that the intermediate representation (IR) of a JIT compiler is a sea-of-nodes graph that the compiler generates at the beginning of its execution by parsing the bytecode and optimizing it with several optimization phases. Formally, the IR is a simple, undirected graph $G = (V, E)$, where V represents the nodes optimized by the JIT compiler and E contains pairs of nodes connected by different relationships (e.g., semantic and syntax relationships, such as math expressions). By keeping track of the optimization information for each node we construct the hypergraph $H = (V, S)$ from G, where V is a set of nodes optimized by the JIT compiler and each hyperedge in S represents an optimization phase.

Two important node features are phases and opcodes. Phases are the optimization phases where a node was generated and optimized (and which later correspond to hyperedges). Opcodes represent node operations (e.g., add, sub, return). A node also has two different attribute groups: (1) *basic*, such as a node id, address, list of neighbors, opcode, and IR ID; and (2) *optimization*, such as hyperedge (phase) ID, generated hyperedge name, and optimized hyperedge names. Note that a node is generated at one hyperedge, but can be present in multiple different hyperedges, due to different optimization phases.

Recall that given one JavaScript code we generate N similar versions to see if any of them trigger bugs. We generate the IRs for all of these versions (typically about 20). In the real-world examples we work with, each such IR graph has about 300–500 nodes and 30–40 optimization phase executions.

2.2   Merging Intermediate Representation Hyperedges

We now merge the N similar but different intermediate representations into one single graph. There are two main reasons to do this. First, we want to see the differences among the graphs in one single view. Second, by comparing hyperedges from a buggy program IR to hyperedges from a non-buggy program IR, we can find differences in some hyperedges due to different optimizations, and thus find the bug. Consider, for example, a hyperedge α in both buggy and non-buggy program IRs and suppose that an additional node (the result of incorrect optimization) makes a buggy program's α different from the non-buggy program's α. A merged hyperedge will show this additional node, and its attributes will identify the buggy IR. A developer can now see that there was an optimization difference in α and find the bug.

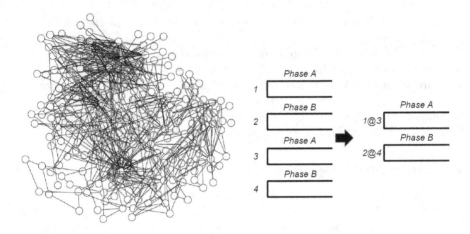

Fig. 2. (a) Example of an IR graph; (b) Example of hypergraph simplification.

Let R_0 be the IR from the original program and $\{R'_1, ..., R'_N\}$ the IRs from the modified programs. Let $\{r'_1, ..., r'_n\}$ be sub-IRs, where r'_i is a subgraph of R'_i when $R'_i \neq R_0$, i.e., $r'_i \subseteq R'_i$, and n is the number of IRs different from R_0 ($n \leq N$). Each r'_i holds buggy candidate hyperedges: R'_i hyperedges are different from R_0's hyperedges. We traverse all sub-IRs, comparing each to R_0, and update the merged IR; see Algorithm 1 in [27] for detail.

2.3 Intermediate Representation Simplification

Although the resulting merged graph may be useful for debugging, its complexity makes it difficult for developers to use; see Fig. 2(a). Therefore, we simplify the graph, convert it into a hypergraph, and simplify the hypergraph (hopefully without losing much information in these simplifications). The main goal is to end up with an interactive visualization that allows developers to debug.

Reducing the IR Graph. We remove dead nodes (nodes with no adjacent edges) as they are not translated into machine code and do not affect other nodes. We then identify nodes that can be merged without losing important information. A pair of nodes is merged if they have the same opcode, the same optimization information, belong to the same IR (which can be identified by the IR id attribute), and share the same neighbors; see Algorithm 2 in [27] for detail.

Reducing the IR Hypergraph. We convert the simplified graph $G = (V, E)$ into a hypergraph $H = (V, S)$, by extracting hyperedges based on the optimization phases; see Algorithm 3 in [27]. Recall that a node v generated in phase/hyperedge α and optimized in phases/hyperedges ϕ and γ now belongs to all three hyperedges. We reduce hypergraph H by merging suitable pairs of hyperedges. Different nodes can have the same hyperedge names as attributes, but different hyperedge IDs, as IDs are assigned based on the execution order.

Therefore, we merge hyperedges with the same name into a single hyperedge while assigning a new unique identifier generated from the original IDs. We use ID concatenation to obtain unique identifiers. Consider two hyperedges A and B executed twice in the order shown in Fig. 2(b). We use the order to create unique IDs by merging the 4 hyperedges into 2 hyperedges and assigning new IDs, generated by concatenating two IDs delimited with a special character '@'; see Algorithm 4 in [27].

This reduces the number of hyperedges but increases the number of nodes in each hyperedge. Next, we traverse each hyperedge $s \in S$, and we use node opcodes to see if they can be merged; see Algorithm 5 and Table 1 in [27] for more details and results.

2.4   Visualizing the Hypergraph with MetroSets

MetroSets [25] uses the metro map metaphor to visualize medium-size hypergraphs. It clearly shows the relationships between hyperedges, which in our case captures the relationships among the optimizations. MetroSets provides simple and intuitive interactions that make it possible to quickly identify hyperedges (metro lines) that contain suspicious nodes (metro stations), or hyperedges that intersect with a particular suspicious hyperedge. Each node in the MetroSet map is labeled with its unique ID (representing the node generation timeline). The attributes shown when hovering over a node are phase, opcode, address, graph ID, and phase ID. A phase attribute tells the user where the node was generated and it is useful when nodes belong to multiple sets. A developer can distinguish the phase that generated a node and phases where it was optimized.

3   Evaluation

We work with Google's JavaScript engine and its JIT compiler, using a dynamic analysis tool built on top of Intel's Pin software [28] to collect instruction-level traces, XED [3] for instruction decoding [3], esprima-python [10] to generate the syntax-tree from JavaScript code, and escodegen [2] to regenerate JavaScript from the syntax-tree. Our data comes from the Chromium bug report site; see [26] for details. We can identify the bugs in all listed bug reports, including Chromium bug report 5129. This version of the compiler has a bug in the *EarlyOptimization* phase. We generate 19 additional modified JavaScript programs from the original and run all 20. The instruction traces are used to generate the IR graph shown in Fig. 2(a) and our visualization is shown in Fig. 3. We can now attempt to answer some of the questions from Sec. 1.

"What optimizations took place to generate the machine code?" The map and the "Key to Lines" legend show all optimization phases.

"What is the relationship among the optimization phases?" We can examine the corresponding lines and use the interactive exploration modes (intersection, union, complement, etc.) to see the relationships among the phases.

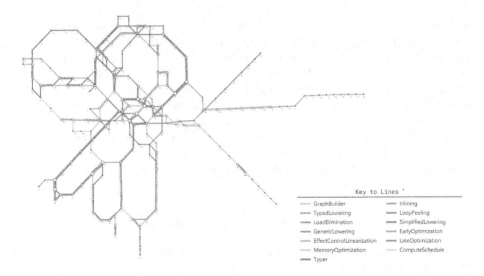

Fig. 3. Metro map of the IR graph from bug report 5129.

"Which optimization phase was most active?" We can visually identify the longest line, or hover over each line and see the number of nodes in it; see Fig. 9 in [27] for an example of the most active optimization phase.

"What optimizations affected a specific node" We can hover over the node of interest, which grays out the lines that don't contain the node. We can then examine each of the corresponding lines and look at the displayed node attributes.

"Which optimization phases are likely to be buggy?" One natural way to do this is to find parts that differ in the IR graphs with the bug and those without. In other words, a program is buggy because either it has additional optimizations or missing optimizations, and this information is captured in the IRs. Any line that has many non-original IRs represents a significant difference between buggy and non-buggy programs. In this case study, the majority of nodes (9 out of 11) in the EarlyOptimization line are from different IRs, indicating a difference in optimization between buggy and non-buggy programs; see the full paper [27] for more detail of the figures and more examples.

Our prototype is available at https://hlim1.github.io/JITCompilerIRViz/.

Acknowledgements. This research was supported in part by the National Science Foundation under grants CNS-1908313 and DMS-1839274.

References

1. Browser security beyond sandboxing. https://www.microsoft.com/security/blog/2017/10/18/browser-security-beyond-sandboxing/ (2017). Accessed 22 Jan 2021
2. Estools/escodegen. https://github.com/estools/escodegen (2012). Accessed 03 Feb 2021

3. Intel xed. https://intelxed.github.io/ (2019). Accessed 03 Feb 2021
4. Internet browser market share 2012–2021. https://www.statista.com/statistics/26 8254/market-share-of-internet-browsers-worldwide-since-2009/ (2021). Accessed 23 May 2021
5. Intro to chrome's v8 from an exploit development angle. https://sensepost. com/blog/2020/intro-to-chromes-v8-from-an-exploit-development-angle/ (2020). Accessed 21 Feb 2021
6. Issue 1072171: security: missing the -0 case when intersecting and computing the type::range in numbermax. https://bugs.chromium.org/p/chromium/issues/ detail?id=1072171 (2020). Accessed 01 Feb 2021
7. Issue 5129: turbofan changes $x - y<0$ to $x<y$ which is not equivalent when (x - y) overflows. https://bugs.chromium.org/p/v8/issues/detail?id=5129 (2016). Accessed 01 Feb 2021
8. Issue 8056: [turbofan] optimized array indexof and array includes ignore a prototype that is not initial. https://bugs.chromium.org/p/v8/issues/detail?id=8056 (2018). Accessed 01 Feb 2021
9. Issue 961237: security: jit difference on comparison in d8. https://bugs.chromium. org/p/chromium/issues/detail?id=961237 (2019). Accessed 01 Feb 2021
10. Kronuz/esprima-python. https://github.com/Kronuz/esprima-python (2017). Accessed 03 Feb 2021
11. Turbofan IR. https://docs.google.com/presentation/d/1Z9iIHojKDrXvZ27gRX51 UxHD-bKf1QcPzSijntpMJBM/edit?usp=embed_facebook (2016). Accessed 21 Jan 2021
12. Understanding v8's bytecode. https://medium.com/dailyjs/understanding-v8s-bytecode-317d46c94775 (2017). Accessed 21 Jan 2021
13. Using turbolizer to inspect the v8 jit compiler. https://lukeolney.me/posts/v8-turbolier/ (2019). Accessed 22 Feb 2021
14. V8: behind the scenes. https://benediktmeurer.de/2016/11/25/v8-behind-the-scenes-november-edition (2016). Accessed 21 Jan 2021
15. Adl-Tabatabai, A., Gross, T.R.: Source-level debugging of scalar optimized code. In: Fischer, C.N. (ed.) Proceedings of the ACM SIGPLAN 1996 Conference on Programming Language Design and Implementation (PLDI), Philadephia, Pennsylvania, USA, 21–24, May 1996, pp. 33–43. ACM (1996). https://doi.org/10.1145/ 231379.231388
16. Brooks, G., Hansen, G.J., Simmons, S.: A new approach to debugging optimized code. In: Feldman, S.I., Wexelblat, R.L. (eds.) Proceedings of the ACM SIGPLAN 1992 Conference on Programming Language Design and Implementation (PLDI), San Francisco, California, USA, 17–19, June 1992, pp. 1–11. ACM (1992). https:// doi.org/10.1145/143095.143108
17. Dahm, M.: Byte code engineering. In: Cap, C.H. (ed.) JIT 1999, Java-Informations-Tage 1999, Düsseldorf 20./21. September 1999, pp. 267–277. Informatik Aktuell, Springer (1999). https://doi.org/10.1007/978-3-642-60247-4_25
18. Demange, D., de Retana, Y.F., Pichardie, D.: Semantic reasoning about the sea of nodes. In: Dubach, C., Xue, J. (eds.) Proceedings of the 27th International Conference on Compiler Construction, CC 2018, Vienna, Austria, 24–25, February 2018, pp. 163–173. ACM (2018). https://doi.org/10.1145/3178372.3179503
19. Devkota, S., Aschwanden, P., Kunen, A., LeGendre, M.P., Isaacs, K.E.: Ccnav: understanding compiler optimizations in binary code. IEEE Trans. Vis. Comput. Graph. **27**(2), 667–677 (2021). https://doi.org/10.1109/TVCG.2020.3030357

20. Devkota, S., Isaacs, K.E.: Cfgexplorer: designing a visual control flow analytics system around basic program analysis operations. Comput. Graph. Forum **37**(3), 453–464 (2018). https://doi.org/10.1111/cgf.13433

21. Dux, B., Iyer, A., Debray, S.K., Forrester, D., Kobourov, S.G.: Visualizing the behavior of dynamically modifiable code. In: 13th International Workshop on Program Comprehension (IWPC 2005), St. Louis, MO, USA, 15–16 May 2005, pp. 337–340. IEEE Computer Society (2005). https://doi.org/10.1109/WPC.2005.45

22. Gregg, D., Ertl, M.A., Krall, A.: Implementing an efficient java interpreter. In: Hertzberger, B., Hoekstra, A., Williams, R. (eds.) HPCN-Europe 2001. LNCS, vol. 2110, pp. 613–620. Springer, Heidelberg (2001). https://doi.org/10.1007/3-540-48228-8_70

23. Hölzle, U., Chambers, C., Ungar, D.M.: Debugging optimized code with dynamic deoptimization. In: Feldman, S.I., Wexelblat, R.L. (eds.) Proceedings of the ACM SIGPLAN 1992 Conference on Programming Language Design and Implementation (PLDI), San Francisco, California, USA, 17–19, June 1992, pp. 32–43. ACM (1992). https://doi.org/10.1145/143095.143114

24. Ishizaki, K., et al.: Design, implementation, and evaluation of optimizations in a javatm just-in-time compiler. Concurr. Pract. Exp. **12**(6), 457–475 (2000). https://doi.org/10.1002/1096-9128(200005)12:6⟨457::AID-CPE485⟩3.0.CO;2-0

25. Jacobsen, B., Wallinger, M., Kobourov, S.G., Nöllenburg, M.: Metrosets: Visualizing sets as metro maps. IEEE Trans. Vis. Comput. Graph. **27**(2), 1257–1267 (2021). https://doi.org/10.1109/TVCG.2020.3030475

26. Lim, H., Debray, S.: Automated bug localization in JIT compilers. In: Titzer, B.L., Xu, H., Zhang, I. (eds.) VEE 2021: 17th ACM SIGPLAN/SIGOPS International Conference on Virtual Execution Environments, Virtual USA, 16, April 2021, pp. 153–164. ACM (2021). https://doi.org/10.1145/3453933.3454021

27. Lim, H., Kobourov, S.: Visualizing the intermediate representation of just-in-time compilers (2021). https://arxiv.org/abs/2107.00063

28. Luk, C., et al.: Pin: building customized program analysis tools with dynamic instrumentation. In: Sarkar, V., Hall, M.W. (eds.) Proceedings of the ACM SIGPLAN 2005 Conference on Programming Language Design and Implementation, Chicago, IL, USA, 12–15, June 2005, pp. 190–200. ACM (2005). https://doi.org/10.1145/1065010.1065034

Geometric Constraints

Upward Planar Drawings with Three and More Slopes

Jonathan Klawitter[ID] and Johannes Zink[(✉)][ID]

Universität Würzburg, Würzburg, Germany
{klawitter,zink}@informatik.uni-wuerzburg.de

Abstract. We study upward planar straight-line drawings that use only a constant number of slopes. In particular, we are interested in whether a given directed graph with maximum in- and outdegree at most k admits such a drawing with k slopes. We show that this is in general NP-hard to decide for outerplanar graphs ($k = 3$) and planar graphs ($k \geq 3$). On the positive side, for cactus graphs deciding and constructing a drawing can be done in polynomial time. Furthermore, we can determine the minimum number of slopes required for a given tree in linear time and compute the corresponding drawing efficiently.

Keywords: Upward planar · Slope number · NP-hardness

1 Introduction

One of the main goals in graph drawing is to generate clear drawings. For visualizations of directed graphs (or digraphs for short) that model hierarchical relations, this could mean that we explicitly represent edge directions by letting each edge point upward. We may also require a planar drawing and, if possible, we would thus get an upward planar drawing. For schematic drawings, we try to keep the visual complexity low, for example by using only few different geometric primitives [28] – in our case few slopes for edges. If we allow two different slopes we get orthogonal drawings [14], with three or four slopes we get hexalinear and octilinear drawings [37], respectively. Here, we combine these requirements and study upward planar straight-line drawings that use only few slopes.

Upward Planarity. An *upward planar drawing* of a digraph G is a planar drawing of G where every edge is drawn as a monotonic upward curve. We call G *upward planar* if it admits an upward planar drawing and *upward plane* if it is equipped with an upward planar embedding. Note that an upward planar embedding, given by the edge order around each vertex, is necessarily *bimodal*, that is, each cyclic sequence can be split into two contiguous subsequences of incoming edges and outgoing edges [14]. Di Battista and Tamassia [15] have shown that if a digraph is upward planar, then it also admits an upward planar straight-line drawing.

While upward planarity testing is an NP-complete problem for general digraphs [23], there exist several FPT algorithms [9,19,25] and polynomial-time

© Springer Nature Switzerland AG 2021
H. C. Purchase and I. Rutter (Eds.): GD 2021, LNCS 12868, pp. 149–165, 2021.
https://doi.org/10.1007/978-3-030-92931-2_11

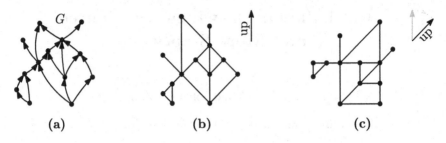

Fig. 1. (a) A digraph G with (b) upward planar 3-slope drawing; (c) drawing rotated by 45°. For readability, edge directions are now given implicitly.

algorithms for special classes, e.g., for single source digraphs [6], outerplanar digraphs [39], series-parallel digraphs [19], and triconnected digraphs [5]. If the embedding is given, upward planarity can be tested in polynomial time [5].

k-slope Drawings. A *k-slope drawing* of a (not necessarily directed) graph G is a straight-line drawing of G where every edge is drawn with one of at most k different slopes; see Fig. 1a and b. The *slope number* of G is the smallest k such that G admits a k-slope drawing. If only (upward) planar drawings are allowed, the number is called the *(upward) planar slope number* of G. The general and planar slope number have been studied extensively in the past for a variety of classes [8, 16, 17, 20, 21, 26, 27, 31, 33–35, 38, 42]. Recently, also the interest in upward planar drawings on few slopes has grown. For example, allowing one bend per edge, Bekos et al. [3] studied so-called bitonic *st*-graphs and complementarily Di Giacomo et al. [18] considered series-parallel digraphs. Brückner et al. [8] studied level-planar drawings with a fixed slope set. Older works include results by Czyzowicz et al. [12, 13] on lattices and several results for trees [1, 2, 7, 10].

In a companion paper to this one, Klawitter and Mchedlidze [29] show that it can be decided in linear time whether a given upward plane digraph admits an upward planar 2-slope drawing. For the variable embedding scenario and two slopes, they give a linear-time algorithm for single-source digraphs, a quartic-time algorithm for series-parallel digraphs, and an FPT algorithm for general digraphs.

Here, we study the problem of whether a digraph admits an upward planar k-slope drawing for any k – with a special focus on the next natural case $k = 3$. Clearly, we can presume that G has maximum in- and outdegree at most k. Note that a 2-slope drawing can be sheared in the direction of one slope without affecting the length of edges drawn with the other slope. The fact that this does not hold for three or more slopes introduces interesting new geometric aspects.

For the choice of k specific slopes, we propose three settings. In the *general* setting, any set of k distinct slopes can be chosen. In the *uniform (angles)* setting, we use slopes with angles in $\{i \cdot \pi/k \mid i \in \{0, \ldots, k-1\}\}$ clockwise (cw) from the x-axis. In the *regular grid* setting, we define a set of slopes as follows. Let c be the middle grid point of a $W \times W$ square grid, where $W = 2\lceil \log_2 k \rceil - 1$. Pick any k distinct slopes that you get from connecting c to any of the other grid points.

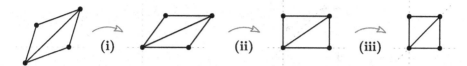

Fig. 2. Given a drawing using any set of three slopes, we can **(i)** Rotate, **(ii)** Shear, and **(iii)** Scale it to only use the slope set $\{\uparrow, \nearrow, \rightarrow\}$.

We remark that these slopes are contained in (an extended version of) the W-th Farey sequence[1]. Uniform angles naturally lead to more balanced drawings with more rotational symmetry, which we find more visually appealing. The downside of this setting is that we cannot always use grid points of the regular 2D grid. E.g., for $k = 6$, the third slope is $\tan(\frac{2}{6}\pi) = \sqrt{3}$, which is an irrational number. Therefore, we assume henceforth for uniform angles a computation and representation model that can handle implicit coordinates or alternatively real numbers. On the other hand in the regular grid setting, we get unbalanced edge angles and irrational edge lengths. Since all of these settings have their natural justification, we consider all of them. However, note that $k = 3$ is a special case because no matter which three slopes we pick, they can be affinely transformed to the slopes of the angles $\{45°, 90°, 135°\}$ as illustrated in Fig. 2. Hence, we restrict considerations to this slope set. For illustrative purposes however, we often rotate drawings by 45° cw and thus use the slope set $\{\uparrow, \nearrow, \rightarrow\}$; see Fig. 1c.

A *k-slope assignment* of a digraph G assigns each edge of G one of k slopes. If G is upward plane, we call a k-slope assignment of G *consistent* if the assignment complies with the cyclic edge order around each vertex; e.g., for $k = 3$, if a vertex has three incoming edges, they need to be assigned the slopes $\rightarrow$, $\nearrow$, and $\uparrow$ in counterclockwise (ccw) order. Clearly, if an upward plane embedding does not admit a consistent k-slope assignment, it also does not admit an upward planar k-slope drawing.

Contribution. We mainly contribute three results to the study of upward planar k-slope drawings. Firstly, we classify the upward planar slope numbers of directed ordered and unordered trees and show how to construct a drawing. Secondly, we show that for cactus graphs we can construct an upward planar k-slope drawing in polynomial time. Thirdly, we show that it is NP-hard to decide whether a given upward outerplanar digraph admits an upward planar 3-slope drawing. We extend the NP-hardness to $k > 3$ but restrict the graph class to upward planar (except for $k = 4$ if no embedding is given).

For statements marked with "$\star$", a proof is available in the full version [30].

---

[1] The W-th Farey sequence is the sequence of all completely reduced fractions where the nominator and denominator is at most W in order of increasing size.

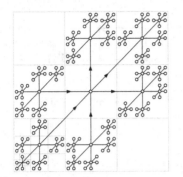 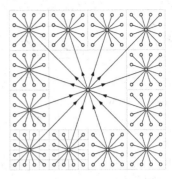

Fig. 3. Upward k-slope drawings of unordered tree $T_{3,3}$ with $k = 3$ and $T_{6,2}$ with $k = 6$, respectively, on the grid.

2   Trees

In this section, we consider upward planar k-slope drawings of directed trees. While our trees are in general not rooted, results for rooted trees can be derived or are partially already known [2,7]. For drawings of trees in the fixed and variable embedding scenario, the terms *ordered* tree, where a planar embedding is specified, and *unordered* tree, where it is not, are used. Note that naturally every unordered tree is upward planar, while an ordered tree is upward planar if and only if its embedding is bimodal. Not surprisingly, the upward planar slope number of an unordered directed tree T equals the maximum in- and outdegree of T; compare this to the planar slope number of $\lceil \Delta/2 \rceil$ of an unordered undirected tree with max. Degree Δ [20]. To show this, we draw T as subgraph of a larger, regular tree $T_{k,h}$ for $h \geq 1$ where every non-leaf vertex has in- and outdegree k and each leaf has distance h to a central vertex. To draw $T_{k,h}$ on a grid with k slopes, we adopt the strategy of Bachmaier et al. [1] for complete rooted trees; see Fig. 3. Alternatively, $T_{k,h}$ can be drawn with k uniform angles; see Fig. 4.

Theorem 1 ($\star$). *Let T be an unordered directed tree with maximum indegree and outdegree k. Then T admits an upward planar k-slope drawing on the regular grid and another upward planar k-slope drawing with uniform angles.*

Note that this recursive drawing procedure requires an exponential-size drawing area (or an exponential edge-length ratio). Different from ordered directed trees (see below), it is not clear whether exponential area is necessary for some unordered directed trees when restricting to k slopes. A $\Theta(n \log n)$ area suffices for an arbitrary number of slopes [22].

Next, let T be a bimodally ordered directed tree. With the drawing approach from unordered trees, it is clear that to determine the upward planar slope number of T it suffices to find a consistent k-slope assignment for T with minimal k. In this regard, note that the maximum in- and outdegree are natural lower bounds but that the choice of the (minimal) slope for an edge uv cannot be determined locally at u and v. For example, the edge vw in Fig. 5a is the third incoming edge at w

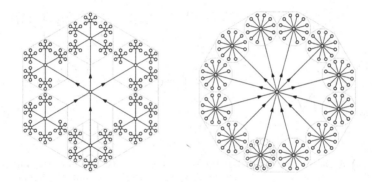

Fig. 4. Upward k-slope drawings of unordered tree $T_{3,3}$ with $k = 3$ and $T_{6,2}$ with $k = 6$, respectively, with uniform angles.

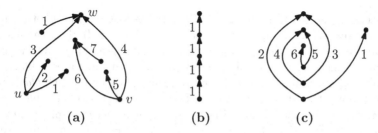

Fig. 5. (a) The minimal slope of an edge is not determined locally; (b) A directed path requires only one slope; (c) An alternating path requires $n - 1$ slopes.

but requires at least slope 4, since its preceding edge uw already requires slope 3 at u. This effect only appears along alternating intervals of incoming and outgoing edges. Hence we have the following observation – see also Fig. 5b and c.

Observation 2. *The upward planar slope number of ordered directed trees with n vertices, $n \geq 2$, is bounded within 1 and $n - 1$ and these bounds are tight.*

However, a simple greedy algorithm finds a consistent k-slope assignment for T, where k is minimal. The algorithm first identifies all edges that can have slope 1, e.g., if an edge uv is the sole incoming edge at v and the ccw first outgoing edge at u. Then, it gives any subsequent edge xy the maximum of the slope of its ccw preceding outgoing edge at x and its ccw preceding incoming edge at y plus one. Since linear time suffices for this and any additional bookkeeping, we get:

Theorem 3. *The upward planar slope number of an ordered directed tree can be determined in linear time.*

Corollary 1. *Let T be an ordered directed tree with maximum in-/outdegree k. We can decide in linear time whether T admits an upward planar k-slope drawing.*

Regarding the area requirement for k-slope drawings of ordered directed trees, we want to remark that Quapil and Jungeblut [40] have shown that there are ordered trees with a spiral structure that require exponential area for 3 slopes.

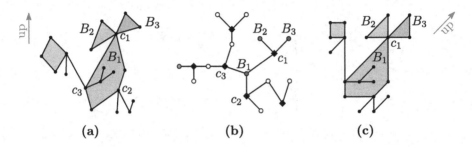

Fig. 6. (a) A cactus graph G; (b) Block-cut tree of G; (c) 3-slope drawing of G.

3   Cactus Graphs

In this section, we show that it can be decided in polynomial time whether a given cactus digraph admits an upward planar k-slope drawing, both in the fixed and the variable embedding scenario. We use a dynamic program on the block-cut tree of the cactus that computes combinable k-slope assignments for each block.

Recall that a block-cut tree T of a graph G has a vertex for each *block* (biconnected component) and each cut vertex of G and an edge between a block B and a cut vertex c if c is part of B; see Fig. 6b. Let G be a cactus graph. Note that in a block-cut tree T of G each block vertex is either a cycle or an edge – we thus distinguish between *cycle blocks* and *edge blocks*. The block-cut tree of G can be computed in linear time [41]. For G to admit an upward planar k-slope drawing, each block of T must be drawable under constraints imposed by other blocks. For example in Fig. 6a–c and $k = 3$, under a fixed embedding the two edges of the block B_1 incident to the cut vertex c_1 need the slopes $\nearrow$ and $\uparrow$ because of the blocks B_2 and B_3. Our strategy is thus as follows.

Algorithm. In the first phase we run a dynamic program (described below) on the blocks of T to find a consistent slope assignment for each block such that the blocks are combinatorially combinable. If successful, we enter the second phase, where we compute drawings of the blocks that are geometrically combinable. In the last phase we put all block drawings together.

Let G be a cactus and let T be the block-cut tree of G. We pick an arbitrary block vertex, say B', of T as root and direct all edges towards B_ℓ. As a result, each block vertex B (except B') has one outgoing edge towards a cut vertex c. We then say c is the *anchor* of B. Let B be a cycle block with anchor c and let e and e' be the edges of B incident to c. Suppose we have a slope assignment for B. Then the *anchor type* $t_c(B)$ of c for B is defined as the slopes of e and e' and if they are incoming or outgoing edges at c; see Fig. 7. For an edge block B with edge e, the *anchor type* $t_c(B)$ describes the slope of e and if e is incoming or outgoing at c. For cycle blocks and edge blocks, there are $2k \cdot (2k - 1)$ and $2k$ different anchor types, respectively.

For a block vertex B with anchor c, a *feasible tuple* $\tau_B = \langle \phi_B, t_c(B) \rangle$ consists of a consistent k-slope assignment ϕ_B of B and an anchor type $t_c(B)$ of c. A

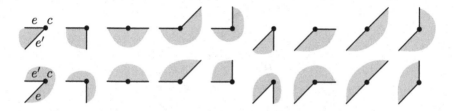

Fig. 7. A subset of the anchor types of a cycle block for $k = 3$.

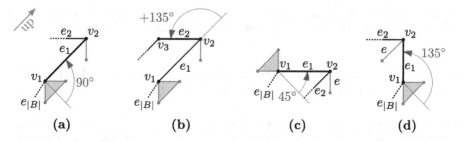

Fig. 8. Computing the possible slopes and rotations for each edge of a cycle.

feasible set for B is a maximal set of feasible tuples for B that have pairwise different anchor types. We process T in a post-order traversal. For each block we compute the feasible set based on the feasible tuples of its descendant blocks.

Combinatorial Realization. Computing the feasible set of a cycle block B with anchor c works as follows. Let B be the cycle $(c = v_1, e_1, v_2, e_2, \ldots, v_{|B|}, e_{|B|}, v_1)$ – if an embedding is given, let the order be cw around the inner face. For every possible slope of e_1, we walk around B once and store all tuples of possible slopes and how far we rotated from the start. We start with e_1 and consider the $\mathcal{O}(1)$ feasible tuples of descendant blocks of B anchored at v_1 and v_2. In the example of Fig. 8a for $k = 3$, assuming a fixed embedding, the edge e_1 can only have slope ↗ and we have thus rotated 90° (starting from the original x-axis). For this tuple (↗, 90°), the edge e_2 in Fig. 8b has also only one possible slope, namely →, and the rotation increases by 135°. However, in the variable embedding scenario, e_1 can also have slopes → and ↑, see Fig. 8c and d. In general, for an edge e_i, $i \in \{2, \ldots, |B|\}$, we consider for all tuples of e_{i-1} how e_i can proceed; again we consider the feasible tuples of descendant blocks of B at v_i and v_{i+1}. For each found tuple of e_i we store a pointer to the tuple(s) of e_{i-1} it is based on.

When we handle $e_{|B|}$, we reject all tuples that do not result in a 2π rotation if the embedding is given or with $\pm 2\pi$ if no embedding is given. This ensures that the cycle has a geometric realization [11]. Combining the slope of e_1 and $e_{|B|}$ as well as whether the rotation is $+2\pi$ or -2π yields an anchor type of B at c. We backtrack from the tuple of $e_{|B|}$ to find a consistent slope assignment of B.

Since the edge e_{i-1} can have at most $\mathcal{O}(k|B|)$ possible rotation values, which imply a slope each, we can compute all possible tuples of e_i in $\mathcal{O}(k|B|)$ time.

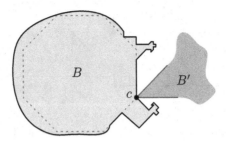

Fig. 9. When drawing a single block, we make sure that the anchor point c lies at a $2k$-gon edge within the algorithm by Culberson and Rawlins [11]; here $k = 4$.

Thus, a single feasible tuple of the whole block B can be computed in $\mathcal{O}(k|B|^2)$ time and all $\mathcal{O}(k)$ feasible tuples of B in $\mathcal{O}(k^2|B|^2)$ time.

Geometric Realization. Suppose we have found a consistent k-slope assignment for every cycle. In the variable embedding scenario, we now know whether and how cycles nest. We thus re-root T such that the root block lies on the outer face. Next, we describe how to obtain a drawing of a cycle block B as a polygon that does not intersect the edges of its parent block B' at its anchor point c.

We describe this only for the uniform angles setting and leave it as an open question for the regular grid setting. Given any sequence σ of rational angles (i.e., a rational number times π) that sum up to $\pm 2\pi$, Culberson and Rawlins [11] describe an algorithm that outputs a polygon with σ as turning angles. Internally, their so-called *Turtlegon* algorithm works as follows. It defines a base angle α as the greatest common divisor of π and all angles in σ; in our case this is π/k. Larger angles are split into sequences of $\pm\alpha$ resulting in a new angle sequence σ'. W.l.o.g. let σ' contain more angles $+\alpha$ than $-\alpha$. Using some of the αs, their algorithm draws a regular $(2\pi/\alpha)$-gon (in our case $2k$-gon). To accommodate additional angles in between, it inserts exponentially shrinking detours at the corners of the $(2\pi/\alpha)$-gon. In the end, we get the original larger angles from merging the smaller angles [11].

The difficulty for us when employing this $\mathcal{O}(k|B|)$ time algorithm, is to ensure that the edges of the parent block B' can reach the anchor point c without intersecting the polygon of B. This might be impossible if c lies within a spiral inside a detour. However, we can avoid this if we let an incident edge of c be a side of the $2k$-gon (this is always possible because we can pick an appropriate set of α angles of σ' for the $2k$-gon) and if we let each detour edge shrink by a sufficiently large factor (e.g., $k|B|$); see Fig. 9.

The running time of this step is in $\mathcal{O}(k|B|)$. Since each vertex is in at most k blocks, we have that $\sum_{i=1}^{\ell} |B_i| \leq kn$. Hence, the total running time is in $\mathcal{O}(k^2 n)$.

Putting Blocks Together. We start with a drawing of the root block. We then recursively draw each child (in a BFS-like order) such that its anchor point coincides with the corresponding vertex of the parent polygon and scale down the drawing of the child block such that the appended polygon does not intersect the existing drawing. Note that it always suffices to scale down each child to the size of the minimum distance of any two vertices within in the parent polygon. We can determine vertex pairs of minimum and maximum distance for a block B in $\mathcal{O}(|B|\log|B|)$ time and then place and scale each polygon in linear time.

The total running time is dominated by the dynamic program, which runs in $\mathcal{O}(k^2|B|^2)$ time for one block B and, hence, in $\mathcal{O}(k^4 n^2)$ time for all blocks.

Theorem 4. *Let G be an upward planar (or plane) cactus graph with maximum in- and outdegree k. It can be constructively tested in $\mathcal{O}(k^4 n^2)$ time whether G admits an upward planar k-slope drawing in the uniform angles setting.*

For the regular grid setting, we cannot use the algorithm by Culberson and Rawlins [11] because we have irrational multiples of π as turning angles. For a sequence of general turning angles, the algorithm by Hartley [24] computes a polygon realizing that sequence. However, it is not immediately clear how to guarantee that the edges of the parent polygon at the anchor point are not intersected. For general polygons, we believe that we can iteratively shrink the spikes to resolve potential intersections. Since such a procedure involves some more technicalities, we leave it as an open question for now.

4   Outerplanar and Planar Graphs

In this section, we show that for any constant $k \geq 3$ deciding whether an upward planar (for $k = 3$, outerplanar) digraph admits an upward planar k-slope drawing is NP-hard. Except for $k = 4$, this hardness holds true regardless of whether we prescribe an embedding or not. However, it remains open if the problem is also NP-complete. Containment in NP is not immediately clear, since it is open whether some graphs require irrational (or super-polynomial precise) coordinates for any k-slope drawing. We first describe our NP-hardness reduction for embedded outerplanar graphs for 3 slopes. Afterwards, we show how this extends to the variable embeddings and to larger k.

We reduce from PLANAR MONOTONE 3-SAT [4], an NP-complete version of 3-SAT, where the three literals of each clause are all either negated or unnegated – from now on called *negative* and *positive* clauses, respectively. Moreover, the incidence graph[2] has a planar drawing where the vertices are rectangles, the edges are vertical straight-line segments, the variables are arranged on a horizontal line, the positive clauses are above, and the negative clauses are below this line; see Fig. 11a. For a given formula F and a rectangular drawing of its incidence graph, we construct a corresponding upward outerplanar digraph G_F, which can only be drawn upward planar with 3 slopes if F is satisfiable. Our construction follows ideas of Nöllenburg [36] and Kraus [32] and utilizes the following observations.

---

[2] The incidence graph of a SAT formula has a vertex for each variable and clause and an edge for each occurrence of a variable in a clause between the corresponding vertices.

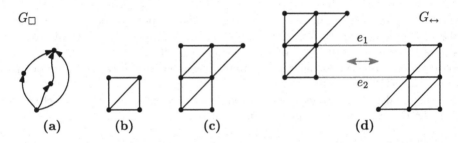

Fig. 10. (a) The digraph $G_\square$ admits only an upward 3-slope drawing as square (b); (c) By combining copies of $G_\square$ and triangles we can build larger rigid structures; (d) Upward planar 3-slope drawing of the digraph $G_{\leftrightarrow}$.

Up to scaling and mirroring diagonally, $G_\square$ in Fig. 10a admits an upward planar 3-slope drawing only as an outerplanar square as in Fig. 10b. We can attach multiple squares (and triangles) to each other as in Fig. 10c. The drawing of such a bigger digraph is unique up to scaling and mirroring diagonally. If the squares form a tree, the drawing is outerplanar. We refer to these squares as *unit squares*, since, once set, the side lengths for all attached squares are the same. To allow a certain small degree of freedom, we exploit the following.

Lemma 1 (⋆)**.** *In any upward planar 3-slope drawing of $G_{\leftrightarrow}$ (see Fig. 10d)*

- *the edges e_1 and e_2 are parallel and have the same arbitrary length $\ell > 0$,*
- *all edges are oriented as in Fig. 10d up to mirroring along a diagonal axis,*
- *and all vertical and horizontal edges (besides e_1 and e_2) have the same lengths, as well as all diagonal edges.*

With this construction kit of useful (sub)graphs in hand, we build a graph whose upward planar drawings represent the satisfying truth assignments for F. The **high-level construction** is depicted in Fig. 11b. We construct, for each variable x_i, a specific digraph – the *variable gadget for x_i*. Similarly, for each clause c_j, there is a specific digraph – the *clause gadget for c_j*. All gadgets mainly consist of chains of $G_\square$s. For a drawing, this enforces a rigid frame structure built from unit squares. We glue all variable gadgets together in a row and connect variable and clause gadgets by *edge gadgets* such that the composite graph remains upward outerplanar (see Fig. 11b) and $G_\square$s are drawn as unit squares.

A **variable gadget** is depicted in Fig. 12. Its base structure is the (violet) frame composed of chains of unit squares. The core element is the (red) central chain of unit squares (with a few side-arms), which has one degree of flexibility, namely, moving as whole to the left or to the right without leaving the frame structure of the gadget. It looks and behaves a bit like a pipe cleaning brush that is stuck inside the frame but can be moved a bit back and forth. Hence, we call it a *brush*. It is connected via a $G_{\leftrightarrow}$ to the brush of the previous variable gadget (see Fig. 12a/d) and the first brush is connected to the frame via a $G_{\leftrightarrow}$. This allows only a horizontal shift of the brushes, but no vertical movement relative

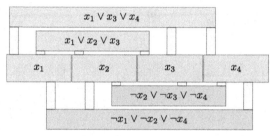

(a) Rectangular incidence graph drawing of a PLANAR MONOTONE 3-SAT formula F.

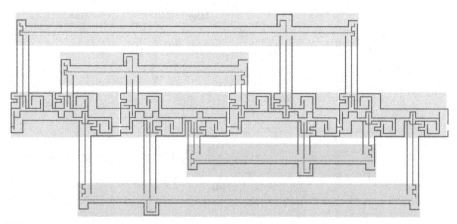

(b) Outerplanar drawing of the digraph G_F obtained from (a). Chains of unit squares are drawn as straight line segments. The variable/clause/edge gadgets occupy the areas of their corresponding rectangles. Here, x_1 and x_4 are set to false (brush on the left), while x_2 and x_3 are set to true (brush on the right)

Fig. 11. Schematic example for our NP-hardness reduction.

to its anchor point at the frame structure. Note that the horizontal position in any variable gadget is independent of those in all other gadgets. If the brush is positioned to the very left (right), the corresponding variable is set to false (true).

For each occurrence of a variable in a positive clause, we have a construction as depicted in Fig. 12b. There, a long chain of (green) $G_\square$s – from now on called *bolt* – is attached to the frame structure via two $G_\leftrightarrow$s, which allow only a vertical, but no horizontal shift. The bolt has on its left side an arm, which can only be placed in one of two pockets of the frame. It can always be placed in the upper pocket, which pushes the bolt outwards with respect to the variable gadget (into an edge and then a clause gadget). It can only be placed in the lower pocket if the brush is shifted to the very right (i.e. set to true) – then the bolt can "fall" into a cove of the brush. For each occurrence of a variable in a negative clause, we have this construction upside-down, such that the bolt can be pulled into the variable gadget only if the brush is shifted to the very left (i.e. set to false).

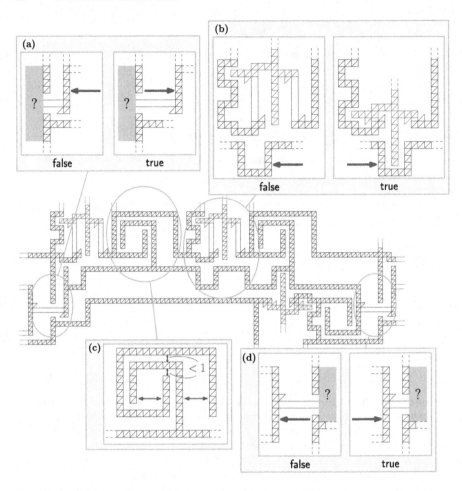

Fig. 12. A variable gadget, which is contained in two positive and one negative clauses. The brush is positioned to the left and, thus, the variable is set to false. (Color figure online)

Note that, to maintain outerplanarity of the whole construction, the frame structure is not contiguous, but connected by $G_{\leftrightarrow}$s and the arms of the bolts. Hence, the frame structure decomposes into many components that have fixed relative horizontal positions and their unit squares have the same side lengths. However, the components can shift up and down relative to each other. To keep this vertical shift small enough not to affect the correct functioning of our reduction, we use, for each such component, the construction depicted in Fig. 12c. The chain of brushes has no vertical flexibility and serves as a base ground for an "anchor" of the frame. The frame can move less than one unit up or down unless it violates planarity. If the frame would be shifted up enough to be completely above the brush, it would get in conflict with the adjacent bolt.

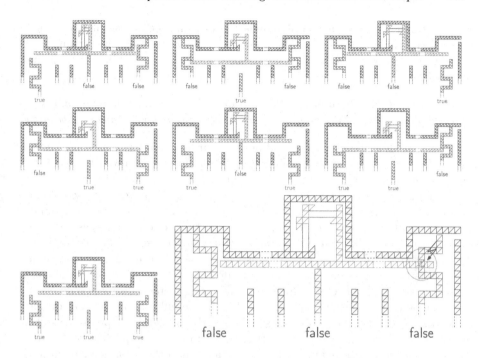

Fig. 13. Positive clause gadget in 8 configurations. (Color figure online)

An **edge gadget** consists of only three straight chains – two frame segments and a bolt in the middle. Their purpose is to synchronize the distance of the clause gadgets to the variable gadgets and to preserve the size of the unit squares. Several edge gadgets are depicted on yellow background color in Fig. 11b.

A **clause gadget** for a positive clause is depicted in Fig. 13. Within a frame, which is connected at six points to the frames of three edge gadgets, there is a horizontal (orange) bar, which is attached via two $G_{\hookleftarrow}$s to the frame – one $G_{\hookleftarrow}$ allows a horizontal, the other allows a vertical shift. It resembles a crane that can move up and extend its arm, while it holds the horizontal bar on a vertical (orange) rope. The three bolts from the corresponding variable gadgets reach into the clause gadget. The lengths of these bolts is chosen such that, if they are pushed out of their variable gadget and into the clause gadget, they only slightly fit inside the gadget. Depending on whether each of the bolts is pushed into the clause gadget or pulled out of it, we have eight possible configurations (with sufficiently small vertical slack). They represent the eight possible truth assignments to a clause. In Fig. 13, we illustrate that in each configuration, we can accommodate the horizontal bar in an upward planar 3-slope drawing of the clause gadget – except for the case when all three bolts push into the clause gadget, which represents the truth assignment false to all contained variables. A negative clause gadget uses the same construction, but mirrored vertically.

Note that, since we have only connected $G_{\square}$s and $G_{\leftrightarrow}$s, the planar embedding of the constructed graph is unique up to mirroring along a diagonal axis. Therefore, our reduction holds true also for the variable embedding scenario and we conclude:

Theorem 5 ($\star$). *Deciding whether an upward outerplanar digraph admits an upward planar 3-slope drawing is NP-hard with and without a given embedding.*

Next, we describe how to extend our NP-hardness reduction to more than 3 slopes. There, however, we use only upward planar instead of upward outerplanar graphs. Observe that, if we fix the embedding and give up outerplanarity, we can add dummy leaves to each vertex to occupy all but the originally used slopes. Since any 3 slopes can be projected to $\{\uparrow, \nearrow, \rightarrow\}$ and we block all other slopes, our arguments work for all sets of k slopes and the reduction remains correct.

Last, we show that our NP-hardness reduction remains applicable for $k > 4$ in the variable embedding setting. This leaves $k = 4$ in the variable embedding setting as the only open case. Again, we do this be extending the graph such that it has only planar embedding up to mirroring along a diagonal axis.

Assume for now that k is an odd number; in the full version [30], we consider otherwise. From the given k slopes, we pick the 3 middle slopes to host the graph of the hardness construction described before. For simplicity, we visualize these 3 middle slopes again as $\{\uparrow, \nearrow, \rightarrow\}$ and the other slopes in quadrants II and IV around a vertex. The key idea is to occupy the unused slopes at each vertex by *fans* and *beaters* as depicted in Fig. 14 instead of simple leaves. Fans are appended to the outside of each vertex if the angle that has been formed in the old construction is $\geq 180°$. For each other remaining slope at each vertex, we add a beater. This is a graph obtained from the wheel graph W_{2k+1} of which one spoke e^* is broken free. This enforces an order on the spokes and, hence, we prescribe the slope of e^*. Note that the whole beater could be mirrored leaving two possible slopes for e^*. However, this is unproblematic since in our construction the "mirrored" slope is also occupied by a beater or a fan. In the full version [30], we prove that this suffices to enforce a desired embedding. Though we lost upward outerplanarity, note that the underlying undirected graph remains outerplanar.

Theorem 6 ($\star$). *Deciding whether an upward planar digraph with maximum in- and outdegree k admits an upward planar drawing with k slopes is NP-hard for $k \geq 3$ if a bimodal embedding is given and is NP-hard for $k \in \mathbb{N}^+ \setminus \{1, 2, 4\}$ if no embedding is given. This holds true for all choices of k slopes.*

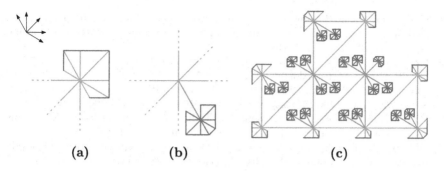

Fig. 14. Example for $k = 5$ slopes: **(a)** A fan and **(b)** a beater. **(c)** We add fans and beaters to each vertex of the graph such that all unused slopes are occupied.

References

1. Bachmaier, C., Brandenburg, F.J., Brunner, W., Hofmeier, A., Matzeder, M., Unfried, Thomas: Tree drawings on the hexagonal grid. In: Tollis, I.G., Patrignani, M. (eds.) GD 2008. LNCS, vol. 5417, pp. 372–383. Springer, Heidelberg (2009). https://doi.org/10.1007/978-3-642-00219-9_36
2. Bachmaier, C., Matzeder, M.: Drawing unordered trees on k-grids. J. Graph Algorithms Appl. **17**(2), 103–128 (2013). https://doi.org/10.7155/jgaa.00287
3. Bekos, M.A., Di Giacomo, E., Didimo, W., Liotta, G., Montecchiani, F.: Universal slope sets for upward planar drawings. In: Biedl, T., Kerren, A. (eds.) GD 2018. LNCS, vol. 11282, pp. 77–91. Springer, Cham (2018). https://doi.org/10.1007/978-3-030-04414-5_6
4. de Berg, M., Khosravi, A.: Optimal binary space partitions for segments in the plane. Int. J. Comput. Geom. Appl. **22**(3), 187–206 (2012). https://doi.org/10.1142/s0218195912500045
5. Bertolazzi, P., Di Battista, G., Liotta, G., Mannino, C.: Upward drawings of triconnected digraphs. Algorithmica **12**(6), 476–497 (1994). https://doi.org/10.1007/BF01188716
6. Bertolazzi, P., Di Battista, G., Mannino, C., Tamassia, R.: Optimal upward planarity testing of single-source digraphs. SIAM J. Comput. **27**(1), 132–169 (1998). https://doi.org/10.1137/S0097539794279626
7. Brunner, W., Matzeder, M.: Drawing ordered $(k - 1)$–any treees $(k$–grids. In: Brandes, U., Cornelsen, S. (eds.) GD 2010. LNCS, vol. 6502, pp. 105–116. Springer, Heidelberg (2011). https://doi.org/10.1007/978-3-642-18469-7_10
8. Brückner, G., Krisam, N.D., Mchedlidze, T.: Level-planar drawings with few slopes. In: Archambault, D., Tóth, C.D. (eds.) GD 2019. LNCS, vol. 11904, pp. 559–572. Springer, Cham (2019). https://doi.org/10.1007/978-3-030-35802-0_42
9. Chan, H.: A parameterized algorithm for upward planarity testing. In: Albers, S., Radzik, T. (eds.) ESA 2004. LNCS, vol. 3221, pp. 157–168. Springer, Heidelberg (2004). https://doi.org/10.1007/978-3-540-30140-0_16
10. Crescenzi, P., Di Battista, G., Piperno, A.: A note on optimal area algorithms for upward drawings of binary trees. Comput. Geom. **2**(4), 187–200 (1992). https://doi.org/10.1016/0925-7721(92)90021-J

11. Culberson, J.C., Rawlins, G.J.E.: Turtlegons: generating simple polygons for sequences of angles. In: Proceedings of 1st Symposium on Computational Geometry (SoCG 1985), pp. 305–310. ACM (1985)
12. Czyzowicz, J.: Lattice diagrams with few slopes. J. Comb. Theory Ser. A **56**(1), 96–108 (1991). https://doi.org/10.1016/0097-3165(91)90025-C
13. Czyzowicz, J., Pelc, A., Rival, I.: Drawing orders with few slopes. Discrete Math. **82**(3), 233–250 (1990). https://doi.org/10.1016/0012-365X(90)90201-R
14. Di Battista, G., Eades, P., Tamassia, R., Tollis, I.G.: Graph Drawing: Algorithms for the Visualization of Graphs. Prentice-Hall, Hoboken (1999)
15. Di Battista, G., Tamassia, R.: Algorithms for plane representations of acyclic digraphs. Theor. Comput. Sci. **61**(2), 175–198 (1988). https://doi.org/10.1016/0304-3975(88)90123-5
16. Di Giacomo, E., Liotta, G., Montecchiani, F.: Drawing outer 1-planar graphs with few slopes. J. Graph Algorithms Appl. **19**(2), 707–741 (2015). https://doi.org/10.7155/jgaa.00376
17. Di Giacomo, E., Liotta, G., Montecchiani, F.: Drawing subcubic planar graphs with four slopes and optimal angular resolution. Theor. Comput. Sci. **714**, 51–73 (2018). https://doi.org/10.1016/j.tcs.2017.12.004
18. Di Giacomo, E., Liotta, G., Montecchiani, F.: 1-bend upward planar slope number of SP-digraphs. Comput. Geom. **90**, 101628 (2020). https://doi.org/10.1016/j.comgeo.2020.101628
19. Didimo, W., Giordano, F., Liotta, G.: Upward spirality and upward planarity testing. SIAM J. Discrete Math. **23**(4), 1842–1899 (2010). https://doi.org/10.1137/070696854
20. Dujmović, V., Eppstein, D., Suderman, M., Wood, D.R.: Drawings of planar graphs with few slopes and segments. Comput. Geom. **38**(3), 194–212 (2007). https://doi.org/10.1016/j.comgeo.2006.09.002
21. Dujmović, V., Suderman, M., Wood, D.R.: Graph drawings with few slopes. Comput. Geom. **38**(3), 181–193 (2007). https://doi.org/10.1016/j.comgeo.2006.08.002
22. Frati, F.: On minimum area planar upward drawings of directed trees and other families of directed acyclic graphs. Int. J. Comput. Geom. Appl. **18**(03), 251–271 (2008). https://doi.org/10.1142/S021819590800260X
23. Garg, A., Tamassia, R.: On the computational complexity of upward and rectilinear planarity testing. SIAM J. Comput. **31**(2), 601–625 (2001). https://doi.org/10.1137/S0097539794277123
24. Hartley, R.I.: Drawing polygons given angle sequences. Inf. Process. Lett. **31**(1), 31–33 (1989)
25. Healy, P., Lynch, K.: Two fixed-parameter tractable algorithms for testing upward planarity. Int. J. Found. Comput. Sci. **17**(05), 1095–1114 (2006). https://doi.org/10.1142/S0129054106004285
26. Jelínek, V., Jelínková, E., Kratochvíl, J., Lidický, B., Tesař, M., Vyskočil, T.: The planar slope number of planar partial 3-trees of bounded degree. Graphs Comb. **29**(4), 981–1005 (2013). https://doi.org/10.1007/s00373-012-1157-z
27. Keszegh, B., Pach, J., Pálvölgyi, D.: Drawing planar graphs of bounded degree with few slopes. SIAM J. Discrete Math. **27**(2), 1171–1183 (2013). https://doi.org/10.1137/100815001
28. Kindermann, P., Meulemans, W., Schulz, A.: Experimental analysis of the accessibility of drawings with few segments. J. Graph Algorithms Appl. **22**(3), 501–518 (2018). https://doi.org/10.7155/jgaa.00474
29. Klawitter, J., Mchedlidze, T.: Upward planar drawings with two slopes. Arxiv report (2021), http://arxiv.org/abs/2106.02839

30. Klawitter, J., Zink, J.: Upward planar drawings with three slopes. Arxiv report (2021). http://arxiv.org/abs/2103.06801
31. Knauer, K., Micek, P., Walczak, B.: Outerplanar graph drawings with few slopes. Comput. Geom. **47**(5), 614–624 (2014). https://doi.org/10.1016/j.comgeo.2014.01.003
32. Kraus, R.: Level-außenplanare zeichnungen mit wenigen steigungen. Bachelor Thesis, University of Würzburg (2020)
33. Lenhart, W., Liotta, G., Mondal, D., Nishat, R.I.: Planar and plane slope number of partial 2-trees. In: Wismath, S., Wolff, A. (eds.) GD 2013. LNCS, vol. 8242, pp. 412–423. Springer, Cham (2013). https://doi.org/10.1007/978-3-319-03841-4_36
34. Mukkamala, P., Pálvölgyi, D.: Drawing cubic graphs with the four basic slopes. In: van Kreveld, M., Speckmann, B. (eds.) GD 2011. LNCS, vol. 7034, pp. 254–265. Springer, Heidelberg (2012). https://doi.org/10.1007/978-3-642-25878-7_25
35. Mukkamala, P., Szegedy, M.: Geometric representation of cubic graphs with four directions. Comput. Geom. **42**(9), 842–851 (2009). https://doi.org/10.1016/j.comgeo.2009.01.005
36. Nöllenburg, M.: Automated drawing of metro maps. Diploma Thesis, University of Karlsruhe (TH) (2010)
37. Nöllenburg, M., Wolff, A.: Drawing and labeling high-quality metro maps by mixed-integer programming. IEEE Trans. Visual. Comput. Graph. **17**(5), 626–641 (2011). https://doi.org/10.1109/TVCG.2010.81
38. Pach, J., Pálvölgyi, D.: Bounded-degree graphs can have arbitrarily large slope numbers. Electron. J. Comb. **13**(1), N1 (2006)
39. Papakostas, A.: Upward planarity testing of outerplanar dags (extended abstract). In: Tamassia, R., Tollis, I.G. (eds.) GD 1994. LNCS, vol. 894, pp. 298–306. Springer, Heidelberg (1995). https://doi.org/10.1007/3-540-58950-3_385
40. Quapil, V.: Bachelor thesis on slope drawings. In: Jungeblut, P (ed.) Karlsruhe Institute of Technology (2021)
41. Tarjan, R.: Depth-first search and linear graph algorithms. SIAM J. Comput. **1**(2), 146–160 (1972). https://doi.org/10.1137/0201010
42. Wade, G.A., Chu, J.H.: Drawability of complete graphs using a minimal slope set. Comput. J. **37**(2), 139–142 (1994). https://doi.org/10.1093/comjnl/37.2.139

Planar Straight-Line Realizations of 2-Trees with Prescribed Edge Lengths

Carlos Alegría, Manuel Borrazzo, Giordano Da Lozzo$^{(\boxtimes)}$,
Giuseppe Di Battista, Fabrizio Frati, and Maurizio Patrignani

Roma Tre University, Rome, Italy
{carlos.alegria,manuel.borrazzo,giordano.dalozzo,giuseppe.dibattista,
fabrizio.frati,maurizio.patrignani}@uniroma3.it

Abstract. We study a classic problem introduced thirty years ago by
Eades and Wormald. Let $G = (V, E, \lambda)$ be a weighted planar graph,
where $\lambda : E \to \mathbb{R}^+$ is a *length function*. The FIXED EDGE-LENGTH PLA-
NAR REALIZATION problem (FEPR for short) asks whether there exists
a *planar straight-line realization* of G, i.e., a planar straight-line drawing
of G where the Euclidean length of each edge $e \in E$ is $\lambda(e)$. Cabello,
Demaine, and Rote showed that the FEPR problem is NP-hard, even
when λ assigns the same value to all the edges and the graph is triconnected. Since the existence of large triconnected minors is crucial to
the known NP-hardness proofs, in this paper we investigate the computational complexity of the FEPR problem for weighted 2-trees, which
are K_4-minor free. We show its NP-hardness, even when λ assigns to
the edges only up to four distinct lengths. Conversely, we show that the
FEPR problem is linear-time solvable when λ assigns to the edges up to
two distinct lengths, or when the input has a prescribed embedding.
Furthermore, we consider the FEPR problem for weighted maximal outerplanar graphs and prove it to be linear-time solvable if their dual tree is
a path, and cubic-time solvable if their dual tree is a caterpillar. Finally,
we prove that the FEPR problem for weighted 2-trees is slice-wise polynomial in the length of the longest path.

1 Introduction and Preliminary Results

The problem of producing drawings of graphs with geometric constraints is a
core topic for Graph Drawing [3–5,11,14,23,24,27,34,43,45]. In this context, a
classic question is the one of testing if a planar graph can be drawn planarly and
straight-line with prescribed edge lengths. The study of such a question is related
to several topics in computational geometry [17,41,47], rigidity theory [16,30,32],
structural analysis of molecules [8,31], and sensor networks [13,38,40]. Formally,
given a weighted planar graph $G = (V, E, \lambda)$, i.e., a planar graph equipped with

This research was partially supported by MIUR Project "AHeAD" under PRIN
20174LF3T8, by H2020-MSCA-RISE project 734922 – "CONNECT", and by Roma
Tre University Azione 4 Project "GeoView".

© Springer Nature Switzerland AG 2021
H. C. Purchase and I. Rutter (Eds.): GD 2021, LNCS 12868, pp. 166–183, 2021.
https://doi.org/10.1007/978-3-030-92931-2_12

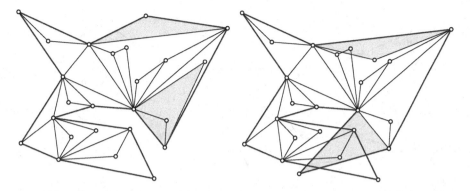

Fig. 1. A planar and a non-planar straight-line realization of the same 2-tree.

a *length function* $\lambda : E \to \mathbb{R}^+$, the Fixed Edge-Length Planar Realization problem (FEPR for short) asks whether there exists a *planar straight-line
realization* of G (PR for short), i.e., a planar straight-line drawing of G where the
Euclidean length of each edge $e \in E$ is $\lambda(e)$. The FEPR problem was first studied
by Eades and Wormald [26], who showed its NP-hardness for triconnected planar
graphs and for biconnected planar graphs with unit lengths. Cabello, Demaine,
and Rote strengthened this result by proving NP-hardness for triconnected planar graphs with unit lengths [12]. Abel et al. [1] proved the $\exists\mathbb{R}$-completeness of
the FEPR problem with unit lengths, solving a problem posed by Schaefer [42].

Since large triconnected minors are essential in the known NP-hardness proofs
of the FEPR problem, we study its complexity for 2-trees, which are the maximal
graphs with no K_4-minor. A *2-tree* is a graph composed of 3-cycles glued together
along edges in a tree-like fashion; see Fig. 1, where we show a planar and a non-
planar realization of a weighted 2-tree. Every 2-tree is planar and biconnected,
and the class of 2-trees is the class of maximal series-parallel graphs. There is a
vast amount of research on 2-trees in Graph Drawing (e.g., in [18,25,28,33,39]).
The edge lengths of 2-trees have been studied in [9,10].

In this paper, we first show that the FEPR problem can be solved in linear
time for 2-trees with prescribed embedding[1]. We note the FEPR problem is
NP-hard for general planar graphs with a prescribed embedding [12]. Second, we
show that, in the variable embedding setting, the FEPR problem is NP-hard
when the number of distinct lengths is at least four, whereas it is linear-time
solvable when the number of distinct lengths is one or two. Note that, for general
planar graphs, the problem is NP-hard even when all the edges are required to
have the same length [26]. Third, we deal with maximal outerplanar graphs.
We show that the FEPR problem can be solved in linear time for maximal
outerpaths, i.e., the maximal outerplanar graphs whose dual tree is a path, and in
cubic time for maximal outerpillars, i.e., the maximal outerplanar graphs whose

---

[1] As in [12], our algorithms adopt the real RAM model, which is customary in computational geometry and supports standard arithmetic operations in constant time.

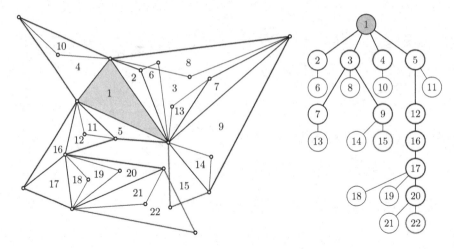

Fig. 2. A 2-tree and its decomposition tree rooted at the 3-cycle with label 1.

dual tree is a caterpillar. Finally, we present a slice-wise polynomial algorithm for 2-trees, parameterized by the length of the longest path.

Because of space limitations, several proofs are omitted. They can be found in the full version of the paper [2].

Preliminaries. We assume familiarity with Graph Drawing (see, e.g., [21]). A planar drawing of a graph G defines a clockwise order of the edges incident to each vertex of G; the set of such orders for all the vertices is a *rotation system* for G. Two planar drawings of G are *equivalent* if (i) they define the same rotation system for G and (ii) their outer faces have the same boundaries. An equivalence class of planar drawings is a *plane embedding* (or simply an *embedding*). When referring to a planar drawing Γ of a graph that has a prescribed embedding $\mathcal{E}$, we always imply that Γ respects $\mathcal{E}$; sometimes, we explicitly stress this.

An *outerplanar drawing* is a planar drawing in which all the vertices are incident to the outer face. An *outerplane embedding* is an equivalence class of outerplanar drawings. An *outerplanar graph* is a graph that admits an outerplanar drawing. The *dual tree* T of a biconnected outerplanar graph G is defined as follows. Consider the (unique) outerplane embedding $\mathcal{O}$ of G. Then T has a node for each internal face of $\mathcal{O}$ and has an edge between two nodes if the corresponding faces of $\mathcal{O}$ are incident to the same edge of G. An *outerpath* is a biconnected outerplanar graph whose dual tree is a path. A *caterpillar* is a tree that becomes a path if its leaves are removed. An *outerpillar* is a biconnected outerplanar graph whose dual tree is a caterpillar.

A *2-tree* is recursively defined as follows. A 3-cycle is a 2-tree. Given a 2-tree G containing an edge (u, w), the graph obtained by adding to G a vertex v and two edges (v, u) and (v, w) is a 2-tree. We observe that the neighbors of any degree-2 vertex are adjacent. The tree-like structure of a 2-tree G is encoded by means of the *decomposition tree* T *rooted* at a 3-cycle of G. Each node in

T represents a 3-cycle of G, and two nodes are adjacent if their corresponding 3-cycles share an edge; see Fig. 2. The decomposition tree of a 2-tree is easily computed in linear time. We adopt the Euclidean metric and assume that the length function of G is such that every 3-cycle satisfies the triangle inequality. This is a necessary condition for the existence of a *straight-line realization* of G, i.e., a (not necessarily planar) drawing of G in which each edge is represented by a line segment with the prescribed length. We often refer to 3-cycles of G, nodes of T, and triangles in a straight-line realization of G interchangeably.

Prescribed Embedding. First, we deal with 2-trees with a prescribed rotation system or embedding. We start by presenting a geometric tool.

Theorem 1. *Let G be an n-vertex weighted 2-tree, $\mathcal{E}$ be a plane embedding (resp. $\mathcal{R}$ be a rotation system) for G, and Γ be a straight-line realization of G. There is an $O(n)$-time algorithm to test whether Γ is a PR respecting $\mathcal{E}$ (resp. $\mathcal{R}$).*

The proof of Theorem 1 is based on an algorithm that: **i.** tests if Γ respects $\mathcal{E}$ (resp. $\mathcal{R}$), **ii.** triangulates the faces of Γ and checks if they are simple polygons [15], and **iii.** tests if the obtained drawing is a convex subdivision [20].

We now present our prescribed embedding result.

Theorem 2. *Let G be an n-vertex weighted 2-tree and $\mathcal{E}$ be a plane embedding (resp. $\mathcal{R}$ be a rotation system) for G. There is an $O(n)$-time algorithm to test whether G admits a PR that respects $\mathcal{E}$ (resp. $\mathcal{R}$) and to construct one, if any.*

The proof of Theorem 2 is based on: **i.** computing a decomposition tree T rooted at the 3-cycle c of G with the largest sum of the edge lengths; **ii.** computing a candidate PR Γ by visiting T in pre-order while greedily adding to Γ the drawing of each 3-cycle t, by exploiting $\mathcal{E}$ (resp. $\mathcal{R}$) and the containment relationship between c and t; and **iii.** testing whether Γ is a PR of G whose plane embedding (resp. rotation system) is $\mathcal{E}$ (resp. $\mathcal{R}$) by means of Theorem 1.

2 NP-Hardness for 2-Trees with 4 Edge Lengths

We sketch a reduction from the NP-complete PLANAR MONOTONE 3-SAT problem [7] (PMS for short) to the FEPR problem with *four* edge lengths.

Theorem 3. *The FEPR problem is NP-hard for weighted 2-trees, even for instances whose number of distinct edge lengths is 4.*

A Boolean CNF formula ϕ is an instance of PMS if the variable-clause incidence graph G_ϕ of ϕ is planar, and each clause of ϕ is either *positive* (it consists of positive literals) or *negative* (it consists of negated literals). The PMS problem is NP-complete even when G_ϕ comes with a *monotone rectilinear representation* [7], i.e., a crossing-free drawing Γ_ϕ of G_ϕ in which **i.** variables and clauses are boxes, **ii.** edges are vertical segments, and **iii.** positive (resp. negative) clauses lie above (resp. below) the horizontal strip containing the variable boxes; see Fig. 3(a).

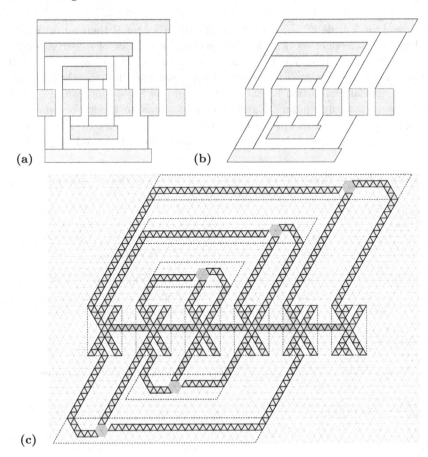

(a)   (b)

(c)

Fig. 3. (a) The monotone rectilinear representation Γ_ϕ of G_ϕ, and (b) Its modified version Γ_ϕ^*. (c) Overview of the reduction showing only the frame triangles (Color figure online).

First, we transform Γ_ϕ into a representation Γ_ϕ^* of G_ϕ that uses segments with slope 0°, 60°, or 90°; see Fig. 3(b). Then we obtain from Γ_ϕ^* a weighted 2-tree H_ϕ that admits a PR if and only if ϕ is satisfiable; see Fig. 3(c). The edges of H_ϕ are assigned the lengths $w_1 = 1$, $w_2 = 0.9$, $w_3 = 0.2$, and $w_4 = 1.61$. To obtain H_ϕ we construct gadgets for the variables, the clauses, and the edges of G_ϕ. Our gadgets exploit two main types of triangles: equilateral triangles with sides of length w_1 (*frame triangles*), and isosceles triangles with base of length w_1 and two sides of length w_2 (*transmission triangles*). The union of the frame triangles of the gadgets representing variables (gray), edges (yellow), and clauses (green), together with a set of frame triangles connecting the variable gadgets (blue), forms a maximal outerplanar graph. Since this graph is formed by frame triangles, it has a unique PR up to rigid transformations.

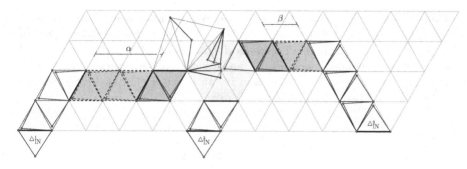

Fig. 4. The (α, β)-clause gadget with $\alpha = 4$ and $\beta = 2$. The values of α and β depend on the relative positions in Γ_ϕ^* of the involved variable gadgets (Color figure online).

Our strategy is to construct H_ϕ from a "rigid" part (mainly formed by the union of the frame triangles of the gadgets), and a part that instead allows for different embedding choices (mainly encoded by the flips of transmission triangles). Consider for example Fig. 4, where we illustrate the PR of a clause gadget, which is the most critical gadget in the construction. Each transmission triangle $\triangle$ (in red) has two possible different embeddings. The choice of this embedding influences the choice of the embeddings of the transmission triangles that "conflict" with $\triangle$, that is, that overlaps with $\triangle$ in one of their embeddings. These chains of conflict relationships allow for "truth values" that come from the variable gadgets to "move" along the gadgets representing edges. The relevant triangles $\triangle_{\mathrm{IN}}^1$, $\triangle_{\mathrm{IN}}^2$, and $\triangle_{\mathrm{IN}}^3$ encode such values: In Fig. 4, $\triangle_{\mathrm{IN}}^1$ and $\triangle_{\mathrm{IN}}^2$ point downward since the corresponding variable is True, and $\triangle_{\mathrm{IN}}^3$ points upward since the corresponding variable is False. A special set of transmission triangles, whose flip depends on the orientation of the relevant triangles, overlap in the pink hexagonal region if all the relevant triangles point upward. Conversely, if at least one relevant triangle points downward, the clause gadget admits a PR.

3   Linear-Time Algorithm for 2-Trees with 2 Edge Lengths

This section is devoted to sketch the proof of the following theorem.

Theorem 4. Let $G = (V, E, \lambda)$ be an n-vertex weighted 2-tree, where $\lambda : E \to \{w_1, w_2\}$ with $w_1, w_2 \in \mathbb{R}^+$. There is an $O(n)$-time algorithm to test whether G admits a PR and to construct one, if any.

Let Γ be a PR of G. If $w_1 = w_2$, then the existence of Γ implies that G is an outerplanar graph, which is a linear-time testable property [19,36,46], and that Γ is outerplanar. Since G is 2-connected, it has a unique outerplane embedding $\mathcal{E}$ which can be constructed in linear time [19,36,37,44,46]. Hence, the problem reduces to the problem of testing whether G has a PR that respects $\mathcal{E}$, which can be solved in linear time by Theorem 2.

Consider now the case in which $w_1 \neq w_2$. W.l.o.g. assume $w_1 < w_2$. Also, let $r = \frac{w_2}{w_1} > 1$. The realization of any 3-cycle of G is one of the following types of triangles: **i.** an equilateral (*small equilateral*) triangle of side w_1, **ii.** an equilateral (*big equilateral*) triangle of side w_2, **iii.** an isosceles (*tall isosceles*) triangle with base w_1 and two sides of length w_2, and **iv.** an isosceles (*flat isosceles*) triangle with base w_2 and two sides of length w_1; refer to Fig. 5.

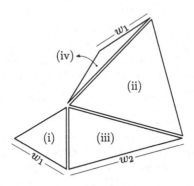

Fig. 5. The four possible types of triangles that represent the 3-cycles of G.

Let $\triangle_1$ and $\triangle_2$ be two triangles realizing two different 3-cycles in a PR of G. We say that $\triangle_1$ is *drawn inside* $\triangle_2$ if all the points of $\triangle_1$ are points of $\triangle_2$ and at least one vertex of $\triangle_1$ is an interior point of $\triangle_2$. Let T be the decomposition tree of G. We have the following.

Lemma 1. *If T is rooted at the 3-cycle with the largest sum of edge lengths and $\triangle_1$ is drawn inside $\triangle_2$, then $\triangle_1$ is a leaf triangle that shares a side with $\triangle_2$.*

By Lemma 1, we assume that T is rooted at a 3-cycle with the largest sum of edge lengths. The *framework* of G is the subgraph $G_F \subseteq G$ obtained, in linear time, as follows: For each leaf triangle $\triangle_i$ that can be drawn inside its parent or a sibling triangle $\triangle_j$, we remove from G the vertex v that $\triangle_i$ does not share with $\triangle_j$, along with the two edges incident to v. Note that **i.** G_F is a 2-tree which may contain any type of triangle, **ii.** by Lemma 1, no triangle of G_F is drawn inside any other triangle in any PR of G, and **iii.** T is rooted at a triangle of the framework. We test in linear time if G_F is outerplanar and, in such case, we compute in linear time its unique outerplanar embedding $\mathcal{E}$. Exploiting Theorem 2, we test if G_F admits a PR respecting $\mathcal{E}$. In the negative case G admits no PR, otherwise we denote by Γ_F the obtained PR of G_F. Hereafter, we assume that Γ_F exists.

Refer to Fig. 6, where we show an example of a PR of a weighted 2-tree. Let $L_\triangle$ be the set of leaf triangles that were removed from G to obtain G_F. Observe that $L_\triangle$ is formed by small equilateral and flat isosceles triangles. We show how to extend Γ_F to a PR of G by embedding the triangles of $L_\triangle$, if possible. Let

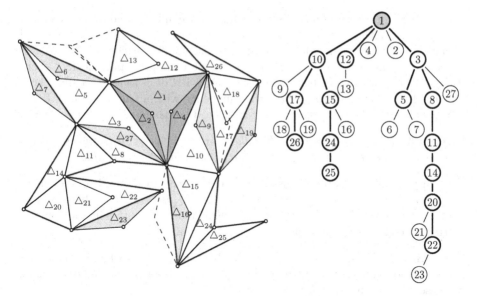

Fig. 6. (Left) A PR of a weighted 2-tree G. Dashed triangles represent alternative embeddings for some 3-cycles, which would generate conflicts. (Right) The decomposition tree of G rooted at the framework triangle $\triangle_1$. Triangles of Γ_F are thick, and edges of the leaf triangles of $L_\triangle$ that are not in Γ_F are thin.

$\triangle$ denote a triangle in $L_\triangle$. The triangle $\triangle$ has a side in G_F that is incident to either two internal faces, or to an internal face and the outer face of Γ_F; hence $\triangle$ has exactly two embedding choices. We say that $\triangle$ has an *internal embedding* if it is embedded inside an internal face of Γ_F, and an *outer embedding* otherwise. We say that $\triangle$ induces a *framework conflict* if it has an outer embedding and $\Gamma_F \cup \triangle$ is not planar; e.g., see the (dashed) outer embedding of $\triangle_{16}$ in Fig. 6. Let $\triangle_i$ and $\triangle_j$ be two triangles of $L_\triangle$. We say that $\triangle_i$ and $\triangle_j$ induce an *internal conflict* if both have an internal embedding and $\Gamma_F \cup \triangle_i \cup \triangle_j$ is not planar; e.g., see the (dashed) internal embedding of $\triangle_9$ and the (solid) internal embedding of $\triangle_{18}$. On the other hand, we say that $\triangle_i$ and $\triangle_j$ induce an *external conflict* if both have an outer embedding and $\Gamma_F \cup \triangle_i \cup \triangle_j$ is not planar; e.g., see the (dashed) outer embeddings of $\triangle_6$ and $\triangle_{13}$.

Lemma 2. *Let $\triangle_i$ and $\triangle_j$ be two leaf triangles of $L_\triangle$ that can both be drawn inside some triangle $\triangle \in \Gamma_F$. The triangles $\triangle_i$ and $\triangle_j$ induce an internal conflict if at least one of the following properties holds true:*

(a) $\triangle_i$ and $\triangle_j$ share an edge;
(b) $\sqrt{3} < r \leq 2\cos(\pi/12)$ and $\triangle$ is a tall isosceles;
(c) $1 < r \leq \sqrt{3}$.

The weighted 2-tree shown in Fig. 6 is such that $\sqrt{3} < r \leq 2\cos(\pi/12)$, that is $r < 2$. By Property b of Lemma 2, two flat isosceles triangles can be drawn inside a big equilateral triangle without inducing conflicts, and any pair of leaf triangles (i.e., either two flat isosceles triangles or a flat isosceles triangle and a small equilateral triangle) induce an internal conflict inside a tall isosceles triangle. On the other hand, by Property a of Lemma 2 and the fact every triangle in $L_\triangle$ has two embedding choices, for a PR of G to exist, it must hold that no three triangles in $L_\triangle$ share the same edge with a triangle of G_F. If $L_\triangle$ satisfies this requirement, then we say that $L_\triangle$ is *consistent*.

Lemma 3. *If $L_\triangle$ is consistent, then there are $O(1)$ pairs of triangles sharing an edge with the same triangle of G_F that induce an internal conflict.*

Proof. By Lemma 1, two triangles of $L_\triangle$ that induce an internal conflict share an edge with a common triangle $\triangle$ of G_F. Since $L_\triangle$ is consistent, there exist at most 6 triangles in $L_\triangle$ incident to $\triangle$. Hence, at most $\binom{6}{2} = 15$ pairs of triangles sharing an edge with $\triangle$ can induce an internal conflict.   □

Extending Γ_F to a PR of G. Next, we show how to test whether there is a choice of embeddings for the triangles in $L_\triangle$ that yields a PR of G. We distinguish two cases, based on whether $r \geq 2$ or $r < 2$.

$\boxed{\textbf{Case } \mathbf{r \geq 2.}}$ In this case there are no flat isosceles triangles, and hence the setting is much simpler than the one depicted in Fig. 6. Every leaf triangle in $L_\triangle$ is a small equilateral triangle whose parent is a tall isosceles triangle. We act as follows. For any two triangles $\triangle_1, \triangle_2 \in L_\triangle$ that share an edge e, we embed $\triangle_1$ and $\triangle_2$ in Γ_F on opposite sides of e. We embed every other leaf triangle in $L_\triangle$ inside its parent triangle. At the end of this process we obtain, in linear time, a straight-line realization Γ of G, and a plane embedding $\mathcal{E}$ of G.

Lemma 4. *G has a PR if and only if Γ is planar.*

Proof Sketch. First, if two leaf triangles share an edge e, then they lie on opposite sides of e in any PR, since they would overlap otherwise. Second, each isosceles triangle may contain only one leaf triangle, since it has only one side of length w_1. Hence, embedding a leaf triangle $\triangle$ inside an isosceles triangle that shares an edge with $\triangle$ cannot cause crossings, since $\triangle$ does not induce internal conflicts. Therefore, a crossing in Γ can only be caused by framework and external conflicts, which are, however, unavoidable.   □

By Lemma 4, in order to test whether G admits a PR, we can apply Theorem 1 to test in $O(n)$ time whether Γ is a PR of G with embedding $\mathcal{E}$.

$\boxed{\textbf{Case } \mathbf{r < 2.}}$ In this case there might be flat isosceles triangles in G which might or might not need to be embedded inside a framework triangle; in Fig. 6 such triangles are shaded light gray. Also, more than one leaf triangle can be drawn inside the same framework triangle which might or might not induce internal conflicts. Recall that the triangles in $L_\triangle$ are small equilateral triangles and/or flat isosceles triangles.

We construct a 2SAT formula ϕ with a Boolean variable for each triangle $\triangle \in L_\triangle$. The values of ϕ are associated with the two possible embeddings of $\triangle$.

If two triangles in $L_\triangle$ induce a conflict for certain embeddings, then ϕ contains a clause that is True if and only if, at least one of the variables representing the triangles does not have the value corresponding to the embedding generating the conflict. Further, for each triangle that induces a framework conflict, ϕ contains a clause that is True if and only if, the variable representing the triangle does not have the value corresponding to an outer embedding. We test in $O(|\phi|)$ time if ϕ is satisfiable [6]. In the positive case, we obtain a PR of G from Γ_F by embedding each triangle in $L_\triangle$ according to the value of the corresponding variable. We reject the instance in the negative case.

It only remains to prove that the number of conflicts (and hence the size of ϕ) is in $O(n)$ and that such conflicts can be found in $O(n)$ time. Detecting internal conflicts is fairly easy: **i.** by Lemma 1, triangles inducing an internal conflict are "close" in T (they share an edge with a common framework triangle); **ii.** by Lemma 3, there exist $O(1)$ leaf triangles sharing an edge with the same framework triangle; and **iii.** since $L_\triangle$ is consistent, the maximum degree of T is bounded by a constant.

Hence, by traversing T we compute in $O(n)$ time the set of $O(n)$ pairs of leaf triangles that induce an internal conflict.

Efficiently detecting external and framework conflicts is more challenging. Let $L'_\triangle$ be the subset of $L_\triangle$ composed of those triangles that are incident to external edges of Γ_F. We give an outer embedding to every triangle in $L'_\triangle$. This results in a (possibly non-planar) straight-line realization Γ'_F of the graph $G'_F := G_F \cup L'_\triangle$. We now construct a bounded-degree graph H whose nodes are associated with sets of vertices of G'_F so that the following properties hold: **(a)** Each node is associated with $O(1)$ degree-2 vertices of G that belong to triangles in $L'_\triangle$, **(b)** if two triangles in $L'_\triangle$ induce an external conflict, then their degree-2 vertices are associated either with the same node or with adjacent nodes, and **(c)** if a triangle $\triangle \in L'_\triangle$ intersects an edge e of G_F (inducing a framework conflict), then the degree-2 vertex of $\triangle$ and the end-vertices of e are associated either with the same node or with adjacent nodes.

After constructing H, the external and framework conflicts can be detected with a linear-time traversal of H.

The graph H is defined as follows. Assume that the bottom-left corner of the bounding box of Γ'_F lies on the origin of the Cartesian axes. Consider a square grid covering the plane whose grid cells have side length $3w_2$; see Fig. 7. Assign a label $l(v) = (\lfloor \frac{x(v)}{3w_2} \rfloor, \lfloor \frac{y(v)}{3w_2} \rfloor)$ to each vertex v of G'_F. Then H has a node for each label assigned to at least one vertex of G'_F, and two distinct nodes (i,j) and (i',j') are connected if and only if $|i - i'| \leq 1$ and $|j - j'| \leq 1$. Note that H has $O(n)$ edges since it has at most n nodes and maximum degree 8.

We now prove that H satisfies Property **(a)**. The number of degree-2 vertices of G that belong to triangles in $L'_\triangle$ and are associated with a node (i,j) of H, is upper bounded by the number k of framework triangles that **i.** are contained in the union of the grid cell (i,j) and its 8 surrounding grid cells, and **ii.** share an edge with a triangle in $L'_\triangle$. Note that k is actually the number of big equilateral and tall isosceles triangles in such nine cells. Since the area of a big equilateral or tall isosceles triangle is at least the area of a small equilateral triangle, then k is upper bounded by the ratio between the area of 9 cells, which is $81w_2^2$, and

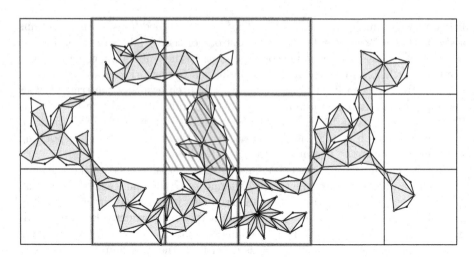

Fig. 7. The straight-line realization Γ'_F of the graph $G'_F := G_F \cup L'_\triangle$, where each triangle in $L'_\triangle$ has an outer embedding. The triangles of G_F are gray and those in $L'_\triangle$ are white. The leaf triangles in the dashed cell can induce conflicts only with the triangles in the eight highlighted cells surrounding such a cell.

the area of a small equilateral triangle, which is $w_1^2\sqrt{3}/4$. Therefore, since $r < 2$, we have that $k \in O(r^2) \subseteq O(1)$.

We next sketch an algorithm to construct H in $O(n)$ time. The vertex set of H is constructed by removing repetitions from the set of labels $l(v)$ computed for the vertices v in G'_F. To this aim, we compute a total order π of the vertices of G'_F such that, for any two vertices u and v with $l(u) = (i_u, j_u)$ and $l(v) = (i_v, j_v)$, we have $u \prec_\pi v$ if and only if **i.** $i_u < i_v$ or **ii.** $i_u = i_v$ and $j_u < j_v$.

Since G'_F is connected and any edge of G'_F has length at most w_2, then $0 \le i, j \le \frac{w_2 n}{3w_2} = \frac{1}{3}n$ for any label (i, j). Hence, we compute π in $O(n)$ time with counting sort. Since vertices with the same label are consecutive in π, repetitions can be removed with a linear scan of π.

The edge set of H consists of four disjoint subsets E_-, $E_|$, $E_/$, and $E_\backslash$. These sets contain the edges that connect nodes of H corresponding to grid cells that are adjacent horizontally, vertically, along the main, and the minor diagonal, respectively; see Fig. 8. We appropriately define four orders π_-, $\pi_|$, $\pi_/$, and $\pi_\backslash$ of the nodes of H such that nodes that are connected by an edge in E_-, $E_|$, $E_/$, and $E_\backslash$ are consecutive in the corresponding order. We compute the four sets of edges with a linear scan of the orders π_-, $\pi_|$, $\pi_/$, and $\pi_\backslash$.

4   Maximal Outerplanar Graphs

In this section we study the FEPR problem for weighted outerplanar 2-trees, i.e., for weighted maximal outerplanar graphs. We prove the following theorems.

Theorem 5. *Let G be an n-vertex weighted maximal outerpath. There is an $O(n)$-time algorithm to test whether G admits a PR and to construct one, if any.*

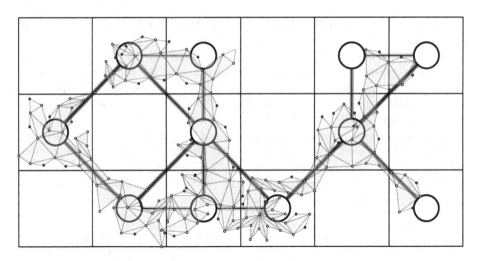

Fig. 8. The edge set of the graph H computed from the drawing Γ'_F in Fig. 7. The sets $E_-, E_|, E_/$, and $E_\backslash$ are shown in yellow, blue, green, and red, respectively (Color figure online).

Theorem 6. *Let G be an n-vertex weighted maximal outerpillar. There is an $O(n^3)$-time algorithm to test whether G admits a PR and to construct one, if any.*

Let G be a weighted 2-tree and e be an edge of G. An *e-outer realization* of G is a PR of G such that e is incident to the outer face. An e-outer realization Γ of G is *e-optimal* if, for every e-outer realization Γ' of G, there is a rigid transformation of Γ such that the segment representing e coincides with the one in Γ' and such that the interior of Γ is a subset of the interior of Γ'.

We sketch the proof of Theorem 5; the proof of Theorem 6 uses similar ideas. Let G be an n-vertex weighted maximal outerpath; see Fig. 9. Let T be the dual tree of the outerplane embedding $\mathcal{O}$ of G; since G is an outerpath, T is a path $(p_1, \ldots, p_k)$. For $i = 1, \ldots, k$, let c_i be the 3-cycle of G bounding the internal face of $\mathcal{O}$ dual to p_i and let $\mathcal{C}_i$ be the unique, up to rigid transformation, PR of c_i. For $i = 1, \ldots, k-1$, let e_i be the edge of G dual to (p_i, p_{i+1}). Let $x \in \{1, \ldots, k\}$

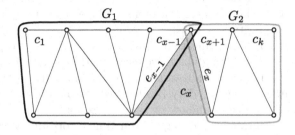

Fig. 9. A maximal outerpath G.

be such that c_x has maximum edge length sum. Let G_1 and G_2 be the subgraphs of G composed of the cycles $c_1, c_2, \ldots, c_{x-1}$ and $c_{x+1}, c_{x+2}, \ldots, c_k$, respectively. Since the length of c_x is maximum, the restrictions of any PR of G to G_1 and G_2 are e_{x-1}-outer and e_x-outer realizations, respectively. We prove that G_1 (resp. G_2) admits an e_{x-1}-outer (resp. e_x-outer) realization if and only if it admits an e_{x-1}-optimal (resp. e_x-optimal) realization. The core of the proof of Theorem 5 is an $O(n)$-time algorithm, called OUTER-CHECKER, that constructs an e_{x-1}-optimal (resp. an e_x-optimal) realization Γ_1 of G_1 (resp. Γ_2 of G_2) and its plane embedding, if any such a realization exists. If OUTER-CHECKER concludes that both G_1 and G_2 admit a PR, then Γ_1 and Γ_2 (as well as their embeddings) can be combined in four ways with $\mathcal{C}_x$ (see Fig. 10) and each resulting straight-line realization can be tested for planarity in $O(n)$ time, by Theorem 1.

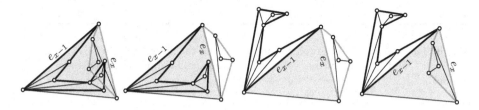

Fig. 10. The four different ways to combine Γ_1 and Γ_2 with $\mathcal{C}_x$.

We describe how OUTER-CHECKER works on G_1. A key observation is that the restriction of any e_{x-1}-optimal realization of G_1 to the graph G_1^i composed of the cycles $c_1, c_2, \ldots, c_i$ is an e_i-optimal realization of G_1^i. This allows OUTER-CHECKER to work by induction on i to decide whether G_1^i has an e_i-optimal realization Γ_1^i. If $i = 1$, the graph G_1^1 is the cycle c_1 whose unique PR $\mathcal{C}_1$ is e_1-optimal. If $i > 1$, an e_i-optimal realization Γ_1^i of G_1^i is constructed, if it exists, by combining Γ_1^{i-1} and $\mathcal{C}_i$ so that e_{i-1} coincides in the two realizations. Three things might happen. First, if Γ_1^{i-1} "fits" inside $\mathcal{C}_i$, as in Fig. 11(left), then the resulting PR Γ_1^i is e_i-optimal. Else, if Γ_1^{i-1} "fits" outside $\mathcal{C}_i$, as in Fig. 11(middle), once cycles $\mathcal{C}_i$ and $\mathcal{C}_{i-1}$ lie on different sides of e_{i-1}, then the resulting PR Γ_1^i is e_i-optimal. Otherwise, G_1^i admits no e_i-optimal realization, as in Fig. 11(right).

A naive implementation of OUTER-CHECKER takes $O(n^2)$ time. Indeed, for each of the $O(n)$ inductive steps, one can check in $O(n)$ time whether Γ_1^{i-1} fits inside and/or outside $\mathcal{C}_i$ using Theorem 1. We achieve $O(n)$ total running time avoiding a planarity test at each step. For $i = 1, \ldots, x - 1$, we compute a "candidate" straight-line realization Γ_1^i of G_1^i, and only test for planarity the final realization Γ_1^{x-1}. By "candidate" we mean that, if G_1^i admits an e_i-optimal realization, then Γ_1^i is such a realization. In order to do that, OUTER-CHECKER dynamically maintains the boundary $\mathcal{B}_1^i$ of the convex hull of Γ_1^i, which is guaranteed to actually be the boundary of the convex hull of Γ_1^i if Γ_1^i is planar. We compute $\mathcal{B}_1^i$ by suitably exploiting a linear-time algorithm by Melkman [35], which incrementally computes the convex hull of a point set spanned by a planar path, provided that the points are given in the order of the path.

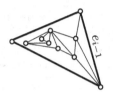 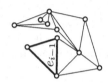 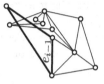

Fig. 11. The three cases in the construction of Γ_1^i. The triangle $\mathcal{C}_i$ is bold. (Left) Γ_1^{i-1} fits inside $\mathcal{C}_i$. (Middle) Γ_1^{i-1} does not fit inside $\mathcal{C}_i$, but it fits outside $\mathcal{C}_i$. (Right) Γ_1^{i-1} fits neither inside nor outside $\mathcal{C}_i$.

After constructing Γ_1^{x-1} (which comes with a plane embedding), we test its planarity in $O(n)$ time using Theorem 1. If the test is successful, Γ_1^{x-1} is an e_{x-1}-optimal PR of G_1^{x-1}, otherwise no e_{x-1}-optimal PR of G_1^{x-1} exists. Each step of OUTER-CHECKER takes $O(1)$ time, except for the computation of the boundary $\mathcal{B}_1^i$. However, the computation of the boundaries $\mathcal{B}_1^1, \mathcal{B}_1^2, \ldots, \mathcal{B}_1^{x-1}$ takes $O(n)$ time in total [35]. Hence, the overall running time of OUTER-CHECKER is in $O(n)$.

5 2-Trees with Short Longest Path

In this section, we sketch a proof of the following theorem.

Theorem 7. *Let G be an n-vertex weighted 2-tree and let ℓ be the length of a longest path of G. There is an $n^{O(4^\ell)}$-time algorithm to test whether G admits a PR and to construct one, if any.*

Theorem 7 is actually a corollary of a stronger theorem, which relates to SPQ-trees; these are a specialization for 2-trees of the well-known *SPQR-trees* [22,29]. The *SPQ-tree* $\mathcal{T}$ of G is a tree that represents a recursive decomposition of G into subgraphs along separation pairs. Each node μ of $\mathcal{T}$ corresponds to a subgraph G_μ of G, which is joined to the rest of the graph via two vertices u_μ and v_μ. Assume that $\mathcal{T}$ is rooted at the neighbor of an edge of G with maximum length and let h be the height of $\mathcal{T}$. We design an $n^{O(2^h)}$-time algorithm that tests whether G admits a PR and, in the positive case, constructs such a realization. Then Theorem 7 follows, as we can prove that $h \le 2\ell - 2$.

The $n^{O(2^h)}$-time algorithm performs a visit of $\mathcal{T}$. When visiting a node μ, the algorithm either concludes that G admits no PR, or constructs a set $\mathcal{R}_\mu$ of "optimal" PRs of G_μ. Here, "optimal" means that, for every PR Γ_μ of G_μ, there is a PR $\Gamma'_\mu \in \mathcal{R}_\mu$ whose interior is a subset of the interior of Γ_μ, after a suitable rigid transformation. The main ingredient needed for bounding the running time of the algorithm is the following. Suppose that G_μ consists of a "parallel" composition of graphs $G_{\nu_1}, \ldots, G_{\nu_k}$. Then "few" of the permutations of $G_{\nu_1}, \ldots, G_{\nu_k}$ need to be considered when constructing $\mathcal{R}_\mu$. Namely, we can sort $G_{\nu_1}, \ldots, G_{\nu_k}$ by increasing length of the 2-edge paths between u_μ and v_μ they contain. Then, in any PR of G, the graph G_{ν_i} is either "to the left" or "to the right" of all the graphs $G_{\nu_1}, \ldots, G_{\nu_{i-1}}$; further, whether a PR of G is optimal only depends on the choice of the "leftmost" and "rightmost" graphs among $G_{\nu_1}, \ldots, G_{\nu_k}$ (and on their drawings, which are taken from $\mathcal{R}_{\nu_1}, \ldots, \mathcal{R}_{\nu_k}$), and not on the permutation of the remaining graphs, as long as planarity is ensured.

6   Open Problems

Our results on the FEPR problem when G is a 2-tree motivate the study of several open questions:

- Determine the computational complexity of the FEPR problem for weighted 2-trees with 3 prescribed edge lengths (we proved it is linear-time solvable for 2 and NP-hard for 4).
- Determine if it is possible to improve our XP algorithm for general 2-trees to an FPT algorithm.
- Study the computational complexity of the FEPR problem for general maximal outerplanar graphs.
- Study the computational complexity of the FEPR problem for graphs with treewidth 2 and for 2-degenerate planar graphs; both these classes generalize the one of 2-trees.

References

1. Abel, Z., Demaine, E.D., Demaine, M.L., Eisenstat, S., Lynch, J., Schardl, T.B.: Who needs crossings? Hardness of plane graph rigidity. In: Fekete, S.P., Lubiw, A. (eds.) 32nd International Symposium on Computational Geometry (SoCG 2016). LIPIcs, vol. 51, pp. 3:1–3:15. Schloss Dagstuhl - Leibniz-Zentrum für Informatik (2016). https://doi.org/10.4230/LIPIcs.SoCG.2016.3
2. Alegría, C., Borrazzo, M., Da Lozzo, G., Di Battista, G., Frati, F., Patrignani, M.: Planar straight-line realizations of 2-trees with prescribed edge lengths. CoRR abs/2108.09483 (2021). https://arxiv.org/abs/2108.09483
3. Alon, N., Feldheim, O.N.: Drawing outerplanar graphs using three edge lengths. Comput. Geom. **48**(3), 260–267 (2015). https://doi.org/10.1016/j.comgeo.2014.10.006
4. Angelini, P., et al.: Anchored drawings of planar graphs. In: Duncan, C., Symvonis, A. (eds.) GD 2014. LNCS, vol. 8871, pp. 404–415. Springer, Heidelberg (2014). https://doi.org/10.1007/978-3-662-45803-7_34
5. Angelini, P., et al.: Windrose planarity: embedding graphs with direction-constrained edges. ACM Trans. Algorithms **14**(4), 54:1–54:24 (2018). https://doi.org/10.1145/3239561
6. Aspvall, B., Plass, M.F., Tarjan, R.E.: A linear-time algorithm for testing the truth of certain quantified boolean formulas. Inf. Process. Lett. **8**(3), 121–123 (1979). https://doi.org/10.1016/0020-0190(79)90002-4
7. de Berg, M., Khosravi, A.: Optimal binary space partitions for segments in the plane. Int. J. Comput. Geom. Appl. **22**(3), 187–206 (2012). https://doi.org/10.1142/S0218195912500045

8. Berger, B., Kleinberg, J.M., Leighton, F.T.: Reconstructing a three-dimensional model with arbitrary errors. J. ACM **46**(2), 212–235 (1999). https://doi.org/10.1145/301970.301972

9. Blažej, V., Fiala, J., Liotta, G.: On the edge-length ratio of 2-trees. In: GD 2020. LNCS, vol. 12590, pp. 85–98. Springer, Cham (2020). https://doi.org/10.1007/978-3-030-68766-3_7

10. Borrazzo, M., Frati, F.: On the planar edge-length ratio of planar graphs. J. Comput. Geom. **11**(1), 137–155 (2020). https://doi.org/10.20382/jocg.v11i1a6

11. Brandes, U., Schlieper, B.: Angle and distance constraints on tree drawings. In: Kaufmann, M., Wagner, D. (eds.) GD 2006. LNCS, vol. 4372, pp. 54–65. Springer, Heidelberg (2007). https://doi.org/10.1007/978-3-540-70904-6_7

12. Cabello, S., Demaine, E.D., Rote, G.: Planar embeddings of graphs with specified edge lengths. J. Graph Algorithms Appl. **11**(1), 259–276 (2007). https://doi.org/10.7155/jgaa.00145

13. Capkun, S., Hamdi, M., Hubaux, J.: GPS-free positioning in mobile ad hoc networks. Clust. Comput. **5**(2), 157–167 (2002). https://doi.org/10.1023/A:1013933626682

14. Chaplick, S.: Recognizing stick graphs with and without length constraints. J. Graph Algorithms Appl. **24**(4), 657–681 (2020). https://doi.org/10.7155/jgaa.00524

15. Chazelle, B.: Triangulating a simple polygon in linear time. Discrete Comput. Geom. **6**(3), 485–524 (1991). https://doi.org/10.1007/BF02574703

16. Connelly, R.: On generic global rigidity. In: Gritzmann, P., Sturmfels, B. (eds.) Proceedings of a DIMACS Workshop on Applied Geometry And Discrete Mathematics. DIMACS Series in Discrete Mathematics and Theoretical Computer Science, vol. 4, pp. 147–156. DIMACS/AMS (1990). https://doi.org/10.1090/dimacs/004/11

17. Coullard, C.R., Lubiw, A.: Distance visibility graphs. Int. J. Comput. Geom. Appl. **2**(4), 349–362 (1992). https://doi.org/10.1142/S0218195992000202

18. Da Lozzo, G., Devanny, W.E., Eppstein, D., Johnson, T.: Square-contact representations of partial 2-trees and triconnected simply-nested graphs. In: Okamoto, Y., Tokuyama, T. (eds.) 28th International Symposium on Algorithms and Computation (ISAAC 2017). LIPIcs, vol. 92, pp. 24:1–24:14. Schloss Dagstuhl - Leibniz-Zentrum für Informatik (2017). https://doi.org/10.4230/LIPIcs.ISAAC.2017.24

19. Deng, T.: On the implementation and refinement of outerplanar graph algorithms. Master's thesis, University of Windsor, Ontario, Canada (2007)

20. Devillers, O., Liotta, G., Preparata, F.P., Tamassia, R.: Checking the convexity of polytopes and the planarity of subdivisions. Comput. Geom. **11**(3–4), 187–208 (1998). https://doi.org/10.1016/S0925-7721(98)00039-X

21. Di Battista, G., Eades, P., Tamassia, R., Tollis, I.G.: Graph Drawing: Algorithms for the Visualization of Graphs. Prentice-Hall, Hoboken (1999)

22. Di Battista, G., Tamassia, R.: On-line planarity testing. SIAM J. Comput. **25**(5), 956–997 (1996). https://doi.org/10.1137/S0097539794280736

23. Di Battista, G., Tamassia, R., Tollis, I.G.: Constrained visibility representations of graphs. Inf. Process. Lett. **41**(1), 1–7 (1992). https://doi.org/10.1016/0020-0190(92)90072-4

24. Di Giacomo, E., Didimo, W., Liotta, G.: Radial drawings of graphs: Geometric constraints and trade-offs. J. Discrete Algorithms **6**(1), 109–124 (2008). https://doi.org/10.1016/j.jda.2006.12.007

25. Di Giacomo, E., Hančl, J., Liotta, Gi.: 2-Colored point-set embeddings of partial 2-trees. In: Uehara, R., Hong, S., Nandy, S.C. (eds.) WALCOM 2021. LNCS, vol. 12635, pp. 247–259. Springer, Cham (2021). https://doi.org/10.1007/978-3-030-68211-8_20

26. Eades, P., Wormald, N.C.: Fixed edge-length graph drawing is NP-hard. Disc. Appl. Math. **28**(2), 111–134 (1990). https://doi.org/10.1016/0166-218X(90)90110-X

27. Geelen, J., Guo, A., McKinnon, D.: Straight line embeddings of cubic planar graphs with integer edge lengths. J. Graph Theory **58**(3), 270–274 (2008). https://doi.org/10.1002/jgt.20304

28. Goodrich, M.T., Johnson, T.: Low ply drawings of trees and 2-trees. In: Durocher, S., Kamali, S. (eds.) Proceedings of the 30th Canadian Conference on Computational Geometry (CCCG 2018), pp. 2–10 (2018), http://www.cs.umanitoba.ca/%7Ecccg2018/papers/session1A-p1.pdf

29. Gutwenger, C., Mutzel, P.: A linear time implementation of SPQR-trees. In: Marks, J. (ed.) GD 2000. LNCS, vol. 1984, pp. 77–90. Springer, Heidelberg (2001). https://doi.org/10.1007/3-540-44541-2_8

30. Hendrickson, B.: Conditions for unique graph realizations. SIAM J. Comput. **21**(1), 65–84 (1992). https://doi.org/10.1137/0221008

31. Hendrickson, B.: The molecule problem: exploiting structure in global optimization. SIAM J. Optim. **5**(4), 835–857 (1995). https://doi.org/10.1137/0805040

32. Jackson, B., Jordán, T.: Connected rigidity matroids and unique realizations of graphs. J. Comb. Theory, Ser. B **94**(1), 1–29 (2005). https://doi.org/10.1016/j.jctb.2004.11.002

33. Lenhart, W., Liotta, G., Mondal, D., Nishat, R.I.: Planar and plane slope number of partial 2-trees. In: Wismath, S., Wolff, A. (eds.) GD 2013. LNCS, vol. 8242, pp. 412–423. Springer, Cham (2013). https://doi.org/10.1007/978-3-319-03841-4_36

34. Lubiw, A., Miltzow, T., Mondal, D.: The complexity of drawing a graph in a polygonal region. In: Biedl, T., Kerren, A. (eds.) GD 2018. LNCS, vol. 11282, pp. 387–401. Springer, Cham (2018). https://doi.org/10.1007/978-3-030-04414-5_28

35. Melkman, A.A.: On-line construction of the convex hull of a simple polyline. Inf. Process. Lett. **25**(1), 11–12 (1987). https://doi.org/10.1016/0020-0190(87)90086-X

36. Mitchell, S.L.: Linear algorithms to recognize outerplanar and maximal outerplanar graphs. Inf. Process. Lett. **9**(5), 229–232 (1979). https://doi.org/10.1016/0020-0190(79)90075-9

37. Moran, S., Wolfstahl, Y.: One-page book embedding under vertex-neighborhood constraints. SIAM J. Discrete Math. **3**(3), 376–390 (1990). https://doi.org/10.1137/0403034

38. Priyantha, N.B., Chakraborty, A., Balakrishnan, H.: The cricket location-support system. In: Pickholtz, R.L., Das, S.K., Cáceres, R., Garcia-Luna-Aceves, J.J. (eds.) 6th Annual International Conference on Mobile Computing and Networking (MOBICOM 2000), pp. 32–43. ACM (2000). https://doi.org/10.1145/345910.345917

39. Rengarajan, S., Veni Madhavan, C.E.: Stack and queue number of 2-trees. In: Du, D.-Z., Li, M. (eds.) COCOON 1995. LNCS, vol. 959, pp. 203–212. Springer, Heidelberg (1995). https://doi.org/10.1007/BFb0030834

40. Savarese, C., Rabaey, J., Beutel, J.: Location in distributed ad-hoc wireless sensor networks. In: 2001 IEEE International Conference on Acoustics, Speech, and Signal Processing. Proceedings, vol. 4, pp. 2037–2040 (2001). https://doi.org/10.1109/ICASSP.2001.940391

41. Saxe, J.: Embeddability of Weighted Graphs in K-space is Strongly NP-hard. CMU-CS-80-102, Carnegie-Mellon University, Department of Computer Science, Pittsburgh (1980). https://books.google.it/books?id=vClAGwAACAAJ
42. Schaefer, M.: Realizability of graphs and linkages. In: Pach, J. (ed.) Thirty Essays on Geometric Graph Theory, pp. 461–482. Springer, New York (2013). https://doi.org/10.1007/978-1-4614-0110-0_24
43. Silveira, R.I., Speckmann, B., Verbeek, K.: Non-crossing paths with geographic constraints. Discret. Math. Theor. Comput. Sci. 21(3), 1–13 (2019). https://doi.org/10.23638/DMTCS-21-3-15
44. Syslo, M.M.: Characterizations of outerplanar graphs. Discret. Math. **26**(1), 47–53 (1979). https://doi.org/10.1016/0012-365X(79)90060-8
45. Tamassia, R.: Constraints in graph drawing algorithms. Constraints Int. J. **3**(1), 87–120 (1998). https://doi.org/10.1023/A:1009760732249
46. Wiegers, M.: Recognizing outerplanar graphs in linear time. In: Tinhofer, G., Schmidt, G. (eds.) WG 1986. LNCS, vol. 246, pp. 165–176. Springer, Heidelberg (1987). https://doi.org/10.1007/3-540-17218-1_57
47. Yemini, Y.: Some theoretical aspects of position-location problems. In: 20th Annual Symposium on Foundations of Computer Science (FOCS 1979), pp. 1–8. IEEE Computer Society (1979). https://doi.org/10.1109/SFCS.1979.39

One-Bend Drawings of Outerplanar Graphs Inside Simple Polygons

Patrizio Angelini[1] , Philipp Kindermann[2(✉)] , Andre Löffler[3],
Lena Schlipf[4] , and Antonios Symvonis[5]

[1] John Cabot University, Rome, Italy
pangelini@johncabot.edu
[2] Universität Trier, Trier, Germany
kindermann@uni-trier.de
[3] Universität Würzburg, Würzburg, Germany
andre.loeffler@uni-wuerzburg.de
[4] Universität Tübingen, Tübingen, Germany
schlipf@informatik.uni-tuebingen.de
[5] National Technical University of Athens, Athens, Greece
symvonis@math.ntua.gr

Abstract. We consider the problem of drawing an outerplanar graph with n vertices with at most one bend per edge if the outer face is already drawn as a simple polygon. We prove that it can be decided in $O(nm)$ time if such a drawing exists, where $m \leq n - 3$ is the number of interior edges. In the positive case, we can also compute such a drawing.

Keywords: Partial embedding · Outerplanar graphs · Visibility graph · Simple polygon

1 Introduction

One of the fundamental problems in graph drawing is to draw a planar graph crossing-free under certain geometric or topological constraints. Many classical algorithms draw planar graphs under the constraint that all edges have to be straight-line segments [6,17,18]. In practical applications, however, we do not always have the freedom of drawing the whole graph from scratch, as some important parts of the graph may already be drawn.

This problem is known as the PARTIAL DRAWING EXTENSIBILITY problem. Formally, given a planar graph $G = (V, E)$, a subgraph $H = (V', E')$ with $V' \subseteq V$ and $E' \subsetneq E$ called the *host* graph, and a planar drawing Γ_H of host H, the problem asks for a planar drawing Γ_G of G such that the drawing of H in Γ_G coincides with Γ_H. This problem was first proposed by Brandenburg et al. [3] in 2003 and has received a lot of attention in the subsequent years.

This work was initiated at the Workshop on Graph and Network Visualization 2019. We thank all the participants for helpful discussions and Anna Lubiw for bringing the problem to our attention.

Lena Schlipf—This research is supported by the Ministry of Science, Research and the Arts Baden-Württemberg (Germany).

H. C. Purchase and I. Rutter (Eds.): GD 2021, LNCS 12868, pp. 184–192, 2021.
https://doi.org/10.1007/978-3-030-92931-2_13

Problem Statement. In this paper, we consider a special drawing extension setting, in which G is a biconnected outerplanar graph with n vertices and m interior edges, the host graph H is the simple cycle bounding the outer face of G, and Γ_H is a 1-bend drawing of H. In other words, we have a simple polygon $P = \Gamma_H$ as input and we want to draw the interior edges of G inside P without crossings. Testing for a straight-line extension is trivial. Moreover, for any integer k, there exists some instances that have a k-bend extension but no $(k-1)$-bend extension; see, e.g., Fig. 1a for $k = 2$. Hence, it is of interest to test for a given k whether a k-bend extension exists. In this paper, we present an algorithm that solves this problem for $k = 1$ in $O(nm)$ time. More formally, we prove the following theorem.

Theorem 1. *Given a biconnected outerplanar graph $G = (V, E)$ with n vertices and m interior edges, and a 1-bend drawing of the outer face of G, we can decide in $O(nm)$ time whether G admits an outerplanar drawing with at most one bend per edge that contains the given drawing of the outer face. In the positive case, we can also compute such a drawing.*

Related Work. For the case of extending a given straight-line drawing of a planar graph using straight-line segments as edges, Patrignani [15] showed the problem to be NP-hard, but he could not prove membership in NP, as a solution may require coordinates not representable with a polynomial number of bits. Recently, Lubiw, Miltzow, and Mondal [10] proved that a generalization of the problem, where overlaps (but not proper crossings) between edges of $E \setminus E'$ and E' are allowed, is hard for the existential theory of the reals ($\exists \mathbb{R}$-complete).

These results motivate allowing bends in the drawing. Angelini et al. [1] presented an algorithm to test in linear time whether there exists any topological planar drawing of G with pre-drawn subgraph, and Jelínek, Kratochvíl, and Rutter [9] gave a characterization via forbidden substructures. Chan et al. [4] showed that a linear number of bends ($72|V'|$) per edge suffices. This number is also asymptotically worst-case optimal as shown by Pach and Wenger [14] for the special case of the host graph not containing edges ($E' = \emptyset$).

Special attention has been given to the case that the host graph H is exactly the outer face of G. Already Tutte's seminal paper [18] showed how to obtain a straight-line convex drawing of a triconnected planar graph with its outer face drawn as a prescribed convex polygon. This result has been extended by Hong and Nagamochi [8] to the case that the outer face is drawn as a star-shaped polygon without chords. Mchedlidze and Urhausen [12] study the number of bends required based on the shape of the drawing of the outer face and show that one bend per edge suffices if the outer face is drawn as a star-shaped polygon. Mchedlidze, Nöllenburg, and Rutter [11] give a linear-time algorithm to test for the existence of a straight-line drawing of G in the case that H is biconnected and Γ_H is a convex drawing. Ophelders et al. [13] characterize the instances in which G is a (planar) graph and H is a cycle that admit a positive solution for *any* straight-line drawing Γ_H as the outer face.

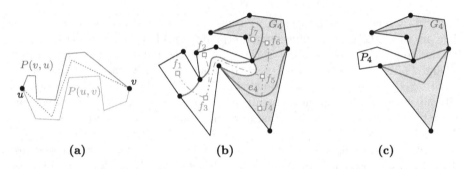

Fig. 1. (a) The edge $e = (u, v)$ requires two bends. (b) A biconnected outerplanar graph and its dual tree (dash-dotted) rooted in f_7. (c) A drawing of G_4 in P_4.

Notation and Preliminaries. For a pair of vertices u and v, we denote by $\overline{uv}$ the straight-line segment between them. Starting at u and following the boundary ∂P of P in counterclockwise order until reaching v, we obtain the interval $P(u, v)$.

The faces of G induce a unique dual tree T [16] where each edge of T corresponds to an *interior* edge of G; see Fig. 1b. In the following, we consider T to be rooted at some degree-1 node f_{m+1}.

We will traverse the dual tree twice – first bottom-up and then top-down. In the bottom-up traversal, we incrementally process the interior edges of G and refine P by cutting off parts that cannot be used anymore. In the top-down traversal, we compute positions for the bend points inside the refined polygons.

We label the faces of G as $f_1, \ldots, f_{m+1}$ according to the bottom-up traversal. This sequence also implies a sequence of subtrees of T. For step $i = 1, \ldots, m$, let T_i be the subtree of T induced by the nodes $f_i, \ldots, f_{m+1}$. We denote by $p(f_i)$ the parent of f_i in T, and by e_i the edge of G corresponding to the edge of T between f_i and its parent. We say that f_i and f_j are *siblings* if $p(f_i) = p(f_j)$. We denote by G_i the subgraph of G induced by the vertices incident to the faces $f_i, \ldots f_{m+1}$, hence $G = G_1$. Similar to the sequence of subtrees, we also have a corresponding sequence of refined polygons $P = P_1 \subseteq \ldots \subseteq P_{m+1}$, where P_i contains (at least) the vertices of G_i.

We draw the edge e_i inside P_i with one bend point b. This drawing of e_i cuts P_i into two parts; the part that contains all the vertices of G_{i+1} is the *unobstructed region* $U_i(b)$ of b, the other part is the *obstructed region* $O_i(b)$ of b; see Fig. 2a. We classify edge e_i based on the type of corner that b will induce in $U_i(b)$. Edge e_i is either (i) a *convex* edge if it is possible to draw e_i such that b is a convex corner in $U_i(b)$, or (ii) a *reflex* edge otherwise, i.e., any drawing of e_i results in a strictly reflex corner at b in $U_i(b)$; see Fig. 2b.

2   Refining the Polygon

During the bottom-up traversal of T, we maintain the invariant that G_{i+1} can be drawn in P_{i+1} if and only if G_i can be drawn in P_i. Observe that, while $P_1 = P$

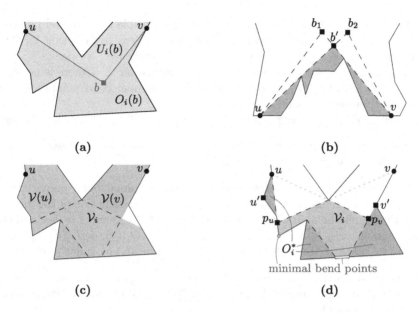

Fig. 2. (a) The region $O_i(b)$ for some $b \in \mathcal{V}_i$; (b) e_i is reflex and b' is the minimal bend point. (c–d) $e_i = (u, v)$ is convex: (c) Construction of $\mathcal{V}_i$; (d) Construction of p_u, p_v, the obstructed region O_i^*, and the set of minimal bend points.

represents a 1-bend drawing of the outer face of $G_1 = G$, in general some edges on the outer face of G_i might be drawn in the interior of P_i; see Fig. 1c. To this end, to obtain P_{i+1}, we want to refine P_i in the least restrictive way – cutting away as little of P_i as possible. In particular, we will choose P_{i+1} as the union of the unobstructed regions for all possible bend points of e_i.

Among all leaves of T_i, we choose the next node f_i to process as follows: If T_i has a leaf corresponding to a reflex edge, then we process the corresponding interior edge next. Otherwise, all leaves in T_i correspond to convex edges, and we choose one of the nodes of the dual tree among them that has the largest distance to the root f_{m+1} in T. We do this to make sure that a convex edge is only chosen if all siblings corresponding to reflex edges have already been processed. The idea is that for a reflex edge we can determine its "best" drawing, namely one that cuts off only a part of the polygon that would be cut off by any valid drawing of this edge (see Lemma 1), while for convex edges this is generally not possible. Furthermore, drawing a reflex edge may restrict the possible drawings for siblings corresponding to convex edges, while drawing convex edges (with a convex bend point) does not (see Lemma 2), so after processing all siblings corresponding to reflex edges, we can compute all possible bend points for a convex edge.

Consider a leaf f_i and the corresponding interior edge $e_i = (u, v)$. W.l.o.g. assume that $P_i(u, v)$ does not contain an edge of the root f_{m+1}. Let $\mathcal{V}(u)$ and $\mathcal{V}(v)$ be the (closed) regions inside P_i visible from u and v, respectively, and let

$\mathcal{V}_i = \mathcal{V}(u) \cap \mathcal{V}(v)$ be their intersection – the region visible by both end points of e_i; see Fig. 2c. Clearly, any valid bend point for e_i needs to be inside $\mathcal{V}_i$; thus, if the interior of $\mathcal{V}_i$ is empty, then we reject the instance. We define a set of *minimal bend points* for e_i: a point $b \in \mathcal{V}_i$ is minimal for e_i if there is no other point $b' \in \mathcal{V}_i$ with $O_i(b') \subsetneq O_i(b)$. Note that all minimal bend points for e_i lie on $\partial \mathcal{V}_i$, so they are not valid bend points.

For reflex edges, we have the following lemma regarding minimal bend points.

Lemma 1. *Let $e_i = (u, v)$ be a reflex edge. If there is a valid drawing of e_i in P_i, then there is a unique minimal bend point b for e_i.*

Proof. For a contradiction, let b_1 and b_2 be two minimal bend points for e_i. Since e_i is a reflex edge, both b_1 and b_2 must lie on the same side of the line uv. Hence, there must be a crossing b' between one of the segments $\overline{ub_1}, \overline{b_1v}$ and one of the segments $\overline{ub_2}, \overline{b_2v}$. Further, $O_i(b') = O_i(b_1) \cap O_i(b_2)$, so $O_i(b') \subseteq O_i(b_1)$ and $O_i(b') \subseteq O_i(b_2)$. Since $O_i(b_1) \neq O_i(b_2)$, one of b_1 and b_2 is not minimal. $\square$

Based on Lemma 1, we construct P_{i+1} as the polygon $P_i(v, u) \circ \overline{ub} \circ \overline{bv}$ for a reflex edge e_i having minimal bend point b. Note that we can efficiently find b since it is a corner of $\mathcal{V}_i$.

For a convex edge $e_i = (u, v)$, we can no longer rely on having a single minimal bend point. Hence, we need to refine our notation.

We define the *obstructed region* $O_i^* = \bigcap_{b \in \mathcal{V}_i} O_i(b)$ of e_i to be the region of P_i obstructed by all valid drawings of e_i – wherever we place the bend point of e_i, all the points of O_i^* will be cut off. Conversely, for every point $p \in P_i \setminus O_i^*$, there is a placement of the bend point of e_i such that p is not cut off; see Fig. 2d. In view of this, we are going to set $P_{i+1} = P_i \setminus O_i^*$. However, we cannot directly compute O_i^* according to its definition, as this would require considering all possible bend points. Thus, in the following we describe an efficient way to compute P_{i+1}.

If u and v lie in $\mathcal{V}_i$, then let $p_u = u$ and $p_v = v$. Otherwise, let (u, u') and (v', v) be the segments of $P_i(u, v)$ incident to u and v, respectively. Shoot a ray from u through u'. Rotate this ray in counterclockwise direction until it hits $\mathcal{V}_i$; let this point be p_u. Do the same with the ray from v through v', rotating it clockwise; let the point where it hits $\mathcal{V}_i$ be p_v. Let $\mathcal{V}_i(p_u, p_v)$ be the path from p_u to p_v along $\partial \mathcal{V}_i$ in counterclockwise order. Then, P_{i+1} is the polygon $P_i(v, u) \circ \overline{up_u} \circ \mathcal{V}_i(p_u, p_v) \circ \overline{p_v v}$.

Note that all vertices of G_{i+1} lie on $P_i(v, u)$, so they are contained in P_{i+1}. The bend point of e_i has to lie in $\mathcal{V}_i$, which is completely contained in P_{i+1}. Hence, no bend point of another edge $e_j, j > i$ can lie in the region $O_i^* = P_i \setminus P_{i+1}$, because then e_i and e_j would cross.

If we manage to construct a non-degenerate polygon P_{m+1}, i.e., its interior is non-empty, then we declare the instance as positive and we compute a drawing of G in P as follows. We first draw G_{m+1} in P_{m+1}, by picking an arbitrary bend point b from the interior of $\mathcal{V}_m$ in P_{m+1} for e_m. Suppose that we have a drawing Γ_{i+1} for G_{i+1} in P_{i+1} where edge $e_{i+1} = (u, v)$ is drawn with bend point b. To draw G_i in P_i, we have to find valid bend points for the edges corresponding to the children of f_i inside the polygon $P_i' = P_i(u, v) \circ \overline{vb} \circ \overline{bu}$. Consider such an

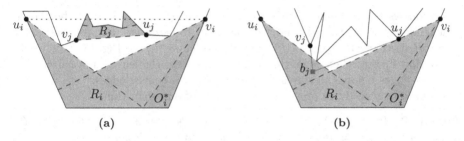

Fig. 3. (a) u_j and v_j lie in $P(u_i, v_i) \circ \overline{v_i u_i}$, but R_i and R_j are interior-disjoint. (b) $P(u_j, v_j)$ forces bend b_j to lie inside R_i. Then, b_j creates a reflex angle.

edge $e_j = (u', v')$. If e_j is reflex, then we place its bend point inside $\mathcal{V}_j$, very close to the minimum bend point. If e_j is convex, then we place its bend point on an arbitrary point in the interior of $\mathcal{V}_j$ that induces a convex corner.

3   Correctness

In this section, we show that G_{i+1} can be drawn in P_{i+1} if and only if G_i can be drawn in P_i. If e_i is a reflex edge, then P_{i+1} contains exactly the points not cut off by placing e_i at its unique minimal bend point b, as Lemma 1 indicates. Hence, G_{i+1} can be drawn in P_{i+1} if and only if G_i can be drawn in P_i with the bend point of e_i arbitrarily close to b inside $\mathcal{V}_i$.

For each convex edge $e_i = (u, v)$, we define the open *restricted region* in P_i as $R_i = \text{interior} \left[\left(\left(\bigcup_{b \in \mathcal{V}_i} O_i(b) \right) \setminus O_i^* \right) \cap \left(P_i(u, v) \circ \overline{vu} \right) \right]$; see Fig. 3a. For each point $r \in R_i$, there are two convex bend points b and b' for e_i such that bending e_i at b cuts off r, whereas bending e_i at b' does not. Note that ∂R_i contains all minimal bend points for e_i. Recall that we only process a convex edge e_i if all siblings of f_i in the dual tree T_i are leaves and correspond to convex edges. We now prove that the restricted region of e_i does not interfere with the restricted regions of these siblings.

Lemma 2. *Let e_i, e_j be convex edges such that f_i and f_j are leaves and siblings in T_i. Then, the restricted regions R_i and R_j are interior-disjoint.*

Proof. Let $e_i = (u_i, v_i)$ and $e_j = (u_j, v_j)$. Since f_i and f_j are siblings, $P(u_i, v_i)$ and $P(u_j, v_j)$ do not share any interior point.

By definition, the restricted regions of e_i and e_j lie in the polygon $P(u_i, v_i) \circ \overline{v_i u_i}$ and the polygon $P(u_j, v_j) \circ \overline{v_j u_j}$, respectively; see Fig. 3a. Hence, if these two polygons are disjoint, the statement follows. W.l.o.g. assume that at least one of u_j, v_j lies in $P(u_i, v_i) \circ \overline{v_i u_i}$. This implies that $R_i(v_i, u_i)$ has exactly one bend that forms a reflex angle in R_i. Further, by construction, there is no vertex in the interior of R_i; in particular, neither u_j nor v_j can lie in the interior of R_i. Therefore, no point of $P(u_j, v_j) \circ \overline{v_j u_j}$ lies in the interior of R_i.

It follows from Lemma 2 that the bend points for e_i and e_j can be chosen independently. Note that the only reason for e_j to be drawn with its bend point inside R_i is that some part of $P(u_j, v_j)$ intersects $\overline{u_j v_j}$; see Fig. 3b. This would imply that e_j is a reflex edge; this is why we process the leaves that correspond to reflex edges before the leaves that correspond to convex edges.

We are now ready to prove the correctness of our algorithm.

Lemma 3. *If* interior$(\mathcal{V}_i) = \emptyset$, *then G_i cannot be drawn in P_i. Otherwise, G_i can be drawn in P_i if and only if G_{i+1} can be drawn in P_{i+1}.*

Proof (Sketch). The first statement of the lemma follows immediately.

For the second statement, first assume that we have a drawing Γ_i of G_i in P_i. Recall that $P_{i+1} = P_i \setminus O_i^*$. The bend point of e_i cannot lie in O_i^* and so e_i is drawn in P_{i+1}. Hence, the edges $e_{i+1}, \ldots, e_m$ also must be drawn in P_{i+1}. Thus, restricting Γ_i to G_{i+1}, we obtain a drawing of G_{i+1} in P_{i+1}.

Now assume that we have a drawing of G_{i+1} in P_{i+1} where e_i is drawn with bend point b. In the top-down traversal of the algorithm, we place the bend points of the edges corresponding to the children of f_i in the interior of the polygon $P_i' = P_i(u, v) \circ \overline{vb} \circ \overline{bu}$. These edges cannot cross any edge $e_i, \ldots, e_m$.

For any child f_h of f_i, we can prove that there is still a valid bend point for e_h inside P_i'. For any pair of children f_h, f_j of f_i, e_h and e_j do not cross. This can be proven using Lemma 1 or 2, depending on whether these edges are convex or reflex. The full proof can be found in the full version of the paper [2]. $\qquad\blacksquare$

We are now ready to prove Theorem 1, which is the main result of this paper. We report the statement for the reader's convenience.

Theorem 1. *Given a biconnected outerplanar graph $G = (V, E)$ with n vertices and m interior edges, and a 1-bend drawing of the outer face of G, we can decide in $O(nm)$ time whether G admits an outerplanar drawing with at most one bend per edge that contains the given drawing of the outer face. In the positive case, we can also compute such a drawing.*

Proof. The correctness follows immediately from Lemma 3.

For the running time, we first argue that, for every $0 \le i \le m$, P_{i+1} has $O(n)$ corners. If e_i is a reflex edge, then obviously P_{i+1} has fewer corners than P_i, so assume that e_i is a convex edge. To create P_{i+1} from P_i, we cut off the obstructed region O_i^* of e_i. Through an easy charging argument, we can charge the new corners on the boundary of P_{i+1} to removed corners of P_i: each of the new introduced edges on the boundary of P_{i+1} has one end point that is a corner of P_i, and it cuts off at least one corner from P_i. Note that two newly introduced edges may form a new vertex, but both of these edges cut off at least one vertex.

In each step of the algorithm, we have to compute the visibility region $\mathcal{V}_i$ for e_i. Since $\mathcal{V}_i$ is a simple polygon with $O(n)$ edges, it can be computed in $O(n)$ time, as demonstrated by Gilbers [7, page 15]. Doing two traversals, the visibility region of each edge needs to be computed at most twice. The remaining parts of the algorithm (computing the dual graph of G, choosing the order of the faces

f_i in which we traverse the graph, computing O_i^*, "cutting off" parts of P_i, and placing the bend points) can clearly be done within this time.

Thus, the total running time is $O(nm)$.

4 Open Problems

There are several interesting open problems related to our work. What if we allow more than one bend per edge? What if we allow some well-behaved crossings, e.g., outer-1-planar drawings [5]? The ∃ℝ-completeness proof by Lubiw, Miltzow, and Mondal [10] for any graph with pre-drawn outer face allows overlaps between interior edges and edges of the outer face. Is the problem still ∃ℝ-complete if overlaps are forbidden?

References

1. Angelini, P., et al.: Testing planarity of partially embedded graphs. ACM Trans. Algorithms **11**(4), 1–42 (2015). https://doi.org/10.1145/2629341
2. Angelini, P., Kindermann, P., Löffler, A., Schlipf, L., Symvonis, A.: One-bend drawings of outerplanar graphs inside simple polygons. Arxiv Report 2108.12321 (2021). https://arxiv.org/abs/2108.12321
3. Brandenburg, F., Eppstein, D., Goodrich, M.T., Kobourov, S., Liotta, G., Mutzel, P.: Selected open problems in graph drawing. In: Liotta, G. (ed.) GD 2003. LNCS, vol. 2912, pp. 515–539. Springer, Heidelberg (2004). https://doi.org/10.1007/978-3-540-24595-7_55
4. Chan, T.M., Frati, F., Gutwenger, C., Lubiw, A., Mutzel, P., Schaefer, M.: Drawing partially embedded and simultaneously planar graphs. J. Graph Algorithms Appl. **19**(2), 681–706 (2015). https://doi.org/10.7155/jgaa.00375
5. Dehkordi, H.R., Eades, P.: Every outer-1-plane graph has a right angle crossing drawing. Int. J. Comput. Geom. Appl. **22**(6), 543–558 (2012). https://doi.org/10.1142/S021819591250015X
6. de Fraysseix, H., Pach, J., Pollack, R.: How to draw a planar graph on a grid. Combinatorica **10**(1), 41–51 (1990). https://doi.org/10.1007/bf02122694
7. Gilbers, A.: Visibility domains and complexity. Ph.D. thesis, Rheinische Friedrich-Wilhelms-Universität Bonn (2014)
8. Hong, S., Nagamochi, H.: Convex drawings of graphs with non-convex boundary constraints. Discrete Appl. Math. **156**(12), 2368–2380 (2008). https://doi.org/10.1016/j.dam.2007.10.012
9. Jelínek, V., Kratochvíl, J., Rutter, I.: A Kuratowski-type theorem for planarity of partially embedded graphs. Comput. Geom. **46**(4), 466–492 (2013). https://doi.org/10.1016/j.comgeo.2012.07.005
10. Lubiw, A., Miltzow, T., Mondal, D.: The complexity of drawing a graph in a polygonal region. In: Biedl, T., Kerren, A. (eds.) GD 2018. LNCS, vol. 11282, pp. 387–401. Springer, Cham (2018). https://doi.org/10.1007/978-3-030-04414-5_28
11. Mchedlidze, T., Nöllenburg, M., Rutter, I.: Extending convex partial drawings of graphs. Algorithmica **76**(1), 47–67 (2016)
12. Mchedlidze, T., Urhausen, J.: β-stars or on extending a drawing of a connected subgraph. In: Biedl, T., Kerren, A. (eds.) GD 2018. LNCS, vol. 11282, pp. 416–429. Springer, Cham (2018). https://doi.org/10.1007/978-3-030-04414-5_30

13. Ophelders, T., Rutter, I., Speckmann, B., Verbeek, K.: Polygon-universal graphs. In: SoCG. LIPIcs, vol. 189, pp. 55:1–55:15. Schloss Dagstuhl - Leibniz-Zentrum für Informatik (2021)
14. Pach, J., Wenger, R.: Embedding planar graphs at fixed vertex locations. Graphs Comb. **17**(4), 717–728 (2001). https://doi.org/10.1007/pl00007258
15. Patrignani, M.: On extending a partial straight-line drawing. Int. J. Found. Comput. Sci. **17**(5), 1061–1069 (2006). https://doi.org/10.1142/s0129054106004261
16. Proskurowski, A., Syslo, M.: Efficient vertex- and edge-coloring of outerplanar graphs. SIAM J. Algebraic Discrete Methods **7**, 131–136 (1986). https://doi.org/10.1137/0607016
17. Schnyder, W.: Embedding planar graphs on the grid. In: Johnson, D.S. (ed.) Proceedings of the First Annual ACM-SIAM Symposium on Discrete Algorithms, pp. 138–148. Society for Industrial and Applied Mathematics (1990). https://doi.org/10.5555/320176.320191, http://dl.acm.org/citation.cfm?id=320176.320191
18. Tutte, W.T.: How to draw a graph. Proc. Lond. Math. Soc. **13**(1), 743–767 (1963). https://doi.org/10.1112/plms/s3-13.1.743

Topological and Upward Drawings

Quasi-upward Planar Drawings
with Minimum Curve Complexity

Carla Binucci$^{(\boxtimes)}$ ⓘ, Emilio Di Giacomo ⓘ, Giuseppe Liotta ⓘ,
and Alessandra Tappini ⓘ

Università degli Studi di Perugia, Perugia, Italy
{carla.binucci,emilio.digiacomo,giuseppe.liotta,
alessandra.tappini}@unipg.it

Abstract. This paper studies the problem of computing quasi-upward planar drawings of bimodal plane digraphs with minimum curve complexity, i.e., drawings such that the maximum number of bends per edge is minimized. We prove that every bimodal plane digraph admits a quasi-upward planar drawing with curve complexity two, which is worst-case optimal. We also show that the problem of minimizing the curve complexity in a quasi-upward planar drawing can be modeled as a min-cost flow problem on a unit-capacity planar flow network. This gives rise to an $\tilde{O}(m^{\frac{4}{3}})$-time algorithm that computes a quasi-upward planar drawing with minimum curve complexity; in addition, the drawing has the minimum number of bends when no edge can be bent more than twice. For a contrast, we show bimodal planar digraphs whose bend-minimum quasi-upward planar drawings require linear curve complexity even in the variable embedding setting.

1 Introduction

Let G be a *plane digraph*, i.e., a directed graph with a given planar embedding. A vertex v of G is *bimodal* if the circular order of the edges around v can be partitioned into two (possibly empty) sets of consecutive edges, one consisting of the incoming edges and the other one consisting of the outgoing edges. If every vertex of G is bimodal, G is a *bimodal plane digraph*. See for example Fig. 1(a).

A planar drawing of a bimodal plane digraph G is *upward planar* if all the edges are represented by curves monotonically increasing in the vertical direction. A digraph that admits an upward planar drawing is *upward planar*. Having a bimodal embedding is a necessary but not sufficient condition for a digraph to be upward planar [12]. For example, the (embedded) digraph in Fig. 1(a) is not upward planar.

This work is partially supported by: (*i*) MIUR, grant 20174LF3T8 "AHeAD: efficient Algorithms for HArnessing networked Data", (*ii*) Dipartimento di Ingegneria - Università degli Studi di Perugia, grants RICBA19FM: "Modelli, algoritmi e sistemi per la visualizzazione di grafi e reti" and RICBA20EDG: "Algoritmi e modelli per la rappresentazione visuale di reti".

© Springer Nature Switzerland AG 2021
H. C. Purchase and I. Rutter (Eds.): GD 2021, LNCS 12868, pp. 195–209, 2021.
https://doi.org/10.1007/978-3-030-92931-2_14

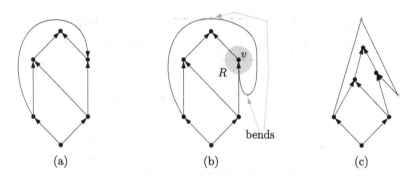

Fig. 1. (a) A bimodal plane digraph G. (b) A quasi-upward planar drawing of G. (c) The same drawing with poly-line edges.

Garg and Tamassia [20] proved that testing a bimodal digraph for upward planarity is NP-hard in the variable embedding setting, i.e., when all possible bimodal planar embeddings must be checked. In this setting, an $O(n^4)$-time algorithm exists for series-parallel digraphs [17], where n is the number of vertices. FPT solutions, SAT formulations, and branch-and-bound approaches have also been proposed for general digraphs (see, e.g., [3,10,17,22]). On the other hand, upward planarity testing can be solved in polynomial time in the fixed embedding setting, i.e., when the input is a bimodal plane digraph G and the algorithm tests whether G admits an upward planar drawing that preserves the given bimodal embedding [4]. See also [16] for a survey on upward planarity.

Motivated by the observation that only restricted families of bimodal digraphs are upward planar, Bertolazzi et al. introduced quasi-upward planar drawings [3]. A drawing Γ of a digraph G is *quasi-upward planar* if it has no edge crossings and for each vertex v there exists a sufficiently small disk R of the plane, properly containing v, such that, in the intersection of R with Γ, the horizontal line through v separates the incoming edges (below the line) from the outgoing edges (above the line); see Fig. 1(b). Intuitively, all the incoming edges enter v from "below" and all the outgoing edges leave v from "above". A digraph that admits a quasi-upward planar drawing is *quasi-upward planar*. An edge of a quasi-upward planar drawing that is not upward has at least one horizontal tangent. Each point of horizontal tangency is called a *bend* (see Fig. 1(b)). This term is justified by the fact that an edge with b points of horizontal tangency can be represented as a poly-line with b bends (substituting each point of tangency with a vertex v and suitable orienting the edges incident to v, we obtain an upward planar digraph which always has a straight-line upward planar drawing [12]); see Fig. 1(c).

Bertolazzi et al. [3] prove that, different from upward planarity, having a planar bimodal embedding is necessary and sufficient for a digraph to be quasi-upward planar. They also study the problem of computing quasi-upward planar drawings with the minimum number of bends and use a suitable flow network to solve it in $\tilde{O}(n^2)$-time in the fixed embedding setting. They also describe a branch-and-bound algorithm in the variable embedding setting. A list of papers about quasi-upward planarity also includes [7–9].

Our Contribution. In this paper we study the problem of computing quasi-upward planar drawings with minimum curve complexity of bimodal plane digraphs possibly having multiple edges. The *curve complexity* is the maximum number of bends along any edge of the drawing. We recall that minimizing the curve complexity is a classical subject of investigation in Graph Drawing (see, e.g., [2,5,7,11,13–15,23,25,26]). Our results can be summarized as follows.

- In Sect. 3 we prove that every bimodal plane digraph admits an embedding-preserving quasi-upward planar drawing with curve complexity two. This is worst-case optimal, since the number of bends per edge in a quasi-upward planar drawing is an even number and not all bimodal plane digraphs are upward planar [12]. This result is the counterpart in the quasi-upward planar setting of a well-known result by Biedl and Kant who prove, in the orthogonal setting, that every plane graph of degree at most four admits an orthogonal drawing with curve complexity two [5].
- In Sect. 4 we study the problem of minimizing the curve complexity in a quasi-upward planar drawing. This problem can be modeled as a min-cost flow problem on a unit-capacity planar flow network. By exploiting a result of Karczmarz and Sankowski [24] we obtain an algorithm to compute embedding-preserving quasi-upward planar drawings that minimize the curve complexity and that have the minimum number of bends when no edge can be bent more than twice that runs in $\tilde{O}(m^{\frac{4}{3}})$ time, where m is the number of edges of the input digraph. We recall that the problem of computing planar drawings that minimize the number of bends while keeping the curve complexity bounded by a constant has already been studied, for example in the context of orthogonal representations (see, e.g., [18,19,27]).
- A quasi-upward planar drawing with minimum curve complexity may have linearly many bends in total. Thus, a natural question to ask is whether these many bends are sometimes necessary if we just minimize the number of bends independent of the curve complexity and, if so, what the curve complexity may be. In Sect. 5 we prove that for every $n \geq 39$ there exists a planar bimodal digraph with n vertices whose bend-minimum quasi-upward planar drawings have at least cn bends on a single edge, for a constant $c > 0$. We show that this bound holds even in the variable embedding setting. This result can be regarded as the counterpart in the quasi-upward planar setting of a result by Tamassia et al. [28] showing a similar lower bound on the curve complexity of bend-minimum planar orthogonal representations.

Preliminaries are in Sect. 2, while Sect. 6 lists some open problems. Proofs marked with (⋆) are omitted or sketched and can be found in [6].

2   Preliminaries

We consider multi-digraphs, that are directed graphs which can have multiple edges. For simplicity we shall call them *digraphs*. We also assume the digraphs to be connected; indeed a digraph G has a quasi-upward drawing if and only if each connected component of G has a quasi-upward drawing. A vertex of a digraph G without incoming (outgoing) edges is a *source* (*sink*) of G. A vertex that is not a source nor a sink is an *internal vertex*.

A *drawing* Γ of a digraph $G = (V, E)$ is a mapping of the vertices of V to points of the plane, and of the edges in E to Jordan arcs connecting their corresponding endpoints but not passing through any other vertex. Drawing Γ is *planar* if any two edges can only meet at common endpoints. A digraph is *planar* if it admits a planar drawing. A planar drawing of a planar digraph G subdivides the plane into topologically connected regions, called *faces*. The infinite region is the *external face*. A *planar embedding* $\mathcal{E}$ of G is an equivalence class of planar drawings that define the same set of faces and have the same external face. A planar embedding of a connected digraph can be uniquely identified by the clockwise circular order of the edges around each vertex and by the external face. A *plane digraph* G is a planar digraph with a given planar embedding. The number of vertices encountered in a closed walk along the boundary of a face f of G is the *degree* of f, denoted as $\delta(f)$. If G is not biconnected, a vertex may be encountered more than once, thus contributing more than once to the degree of the face. The *dual digraph* of G is a plane digraph with a vertex for each face of G and an edge e' between two faces for each edge e of G shared by the two faces. The edge e' is oriented from the face to the left of e to the face to the right of e.

3   Subdivisions of Bimodal Plane Digraphs

In this section we show how to suitably subdivide the edges of a bimodal plane digraph so to obtain an upward plane digraph. We start by recalling the notions of large angles and upward consistent assignments [4].

Bimodality and Upward Consistent Assignments. Let G be a bimodal plane digraph. Let f be a face of G, let e_1 and e_2 be two consecutive edges encountered in this order when walking counterclockwise along the boundary of f, and let v be the vertex shared by e_1 and e_2; the pair (e_1, e_2) is an *angle* of f at vertex v (Fig. 2(a) highlights the angles of a face f). Notice that if v has exactly one incident edge e, then $e_1 = e_2 = e$ and the pair (e, e) is also an angle of f at v. Let f be a face of G, let v be a vertex of f, and let (e_1, e_2) be an angle of f at v. Angle (e_1, e_2) is a *source-switch* of f if e_1 and e_2 are both outgoing edges for v; (e_1, e_2) is a *sink-switch* of f if e_1 and e_2 are both incoming edges for v. An angle of f that is neither a source-switch nor a sink-switch is a *non-switch* of f. It is easy to observe that for any face f the number of source-switches equals the number of sink-switches (in Fig. 2(a) the source- and sink-switches of f are indicated). The number of source-switches in a face f is denoted by $\mathcal{A}(f)$. The *capacity* of f is $\mathcal{A}(f) + 1$ if f is the external face and it is $\mathcal{A}(f) - 1$ otherwise.

Lemma 1 [4]**.** *Let G be a bimodal plane digraph. The number of source and sink vertices of G is equal to the sum of the capacities of the faces of G.*

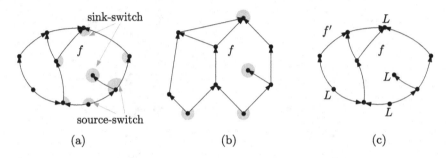

Fig. 2. (a) Angles (shown in gray), source-switches, and sink-switches of a face f of a bimodal plane digraph G; $\mathcal{A}(f) = 2$; (b) An upward planar drawing of G. (c) The assignments of L labels to the angles of G ($L(f)=1$, $L(f')=3$).

Let G be an upward plane digraph and let Γ be an embedding-preserving upward planar drawing of G. The angles of G correspond to geometric angles in Γ. In particular, an angle (e, e) of G corresponds to a 2π angle in Γ. For each face f of G, and for each source- or sink-switch a of f, we assign a label L to a, if a is larger than π in Γ; in this case we say that a is a *large angle*. Figure 2(b) shows an embedding-preserving upward planar drawing of the graph in Fig. 2(a) with the angles larger that π highlighted; the corresponding angles in Fig. 2(c) are labeled with an L. We denote the number of L labels on the angles of f by $L(f)$. Also, if v is a vertex of G, we denote by $L(v)$ the number of L labels on all angles at vertex v. In [4] it is shown that $L(v) = 0$ if v is an internal vertex, and $L(v) = 1$ if v is a source or a sink. Also, $L(f) = \mathcal{A}(f) + 1$ if f is the external face and $L(f) = \mathcal{A}(f) - 1$ otherwise; in other words, the number of large angles inside each face is equal to its capacity (see, e.g., faces f and f' in Fig. 2(c)).

A bimodal plane digraph G is upward planar if and only if it is acyclic and it admits an upward consistent assignment [4]. An *upward consistent assignment* is an assignment of the source and sink vertices of G to its faces such that: (i) Each source or sink v is assigned to exactly one of its incident faces. (ii) For each face f, the number of source and sink vertices assigned to f is equal to the capacity of f. Assigning a source or sink v to a face f corresponds to assigning an L label to an angle that v forms in f. See Fig. 2(c) for an example.

2-Subdivisions of Bimodal Plane Digraphs. Let G be a bimodal plane digraph. A face f of G is *nice* if $\delta(f) = 4$ and each angle of f is either a source-switch or a sink-switch (see, e.g., f_2 in Fig. 3(a)). We augment G by adding edges (possibly creating multiple edges) so that the augmented digraph G' is bimodal and each face of G' either has degree two, or three, or it is nice (note that there can be more than one such augmentations). The resulting digraph is called a *quasi-triangulation* of G and all its faces have degree at most four. See Figs. 3(a) and 3(b). The following lemma, whose proof is reported in [6] for completeness, can also be derived as a special case of [1, Lemma 5].

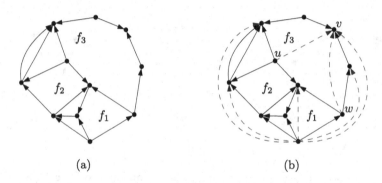

(a)   (b)

Fig. 3. (a) A bimodal plane digraph G; f_2 is nice, while f_1 is not; f_3 is such that $\delta(f_3) \geq 5$. (b) A quasi-triangulation of G: The added edges are dashed.

Lemma 2 ($\star$). *Every bimodal plane digraph admits a quasi-triangulation.*

The *2-subdivision* of a bimodal plane digraph G is the graph $\hat{G}$ obtained from G by replacing each edge $e = (u,v)$ of G with the three edges (u, t_e), (s_e, t_e), (s_e, v), where s_e and t_e are two subdivision vertices. See Fig. 4(a) for an illustration. Notice that s_e is a source and t_e is a sink. Furthermore, the 2-subdivision $\hat{G}$ of G is bimodal, it has no multiple edges, and it is acyclic even if G is not. We prove that $\hat{G}$ is upward planar, which implies that G admits an embedding preserving quasi-upward planar drawing with curve complexity two. Since $\hat{G}$ is bimodal and acyclic, it is sufficient to prove that $\hat{G}$ admits an upward consistent assignment. We model the problem of computing an upward consistent assignment of $\hat{G}$ as a matching problem on a suitably defined bipartite graph.

The *bipartite description* of $\hat{G}$ is the bipartite graph $H_{\hat{G}} = (A, B, E)$ defined as follows. The vertex set A contains for each face $\hat{f}$ of $\hat{G}$ a set of vertices $a_1(\hat{f}), a_2(\hat{f}), \ldots, a_c(\hat{f})$, where c is the capacity of $\hat{f}$. Each vertex $a_i(\hat{f})$ (for $i = 1, \ldots, c$) is called a *representative vertex* of face f. The vertex set B contains the source and sink vertices of $\hat{G}$. There is an edge $(a_i(\hat{f}), v)$ in E if v is a source or sink vertex of face $\hat{f}$ (for $i = 1, \ldots, c$). See Fig. 4(b) for an illustration.

Lemma 3 ($\star$). *The 2-subdivision of a bimodal plane digraph admits an upward consistent assignment if and only if its bipartite description has a perfect matching.*

A bimodal plane digraph is *face-acyclic* if its face boundaries are not cycles. To prove the main result of this section, we first consider face-acyclic bimodal plane digraphs. We then show how to extend the result to the general case.

Lemma 4. *The 2-subdivision of a face-acyclic bimodal plane digraph is upward planar.*

Proof. Let G be a face-acyclic bimodal plane digraph. We assume that G is a quasi-triangulation. If not, by Lemma 2 we can augment G to a quasi-triangulation and the statement follows because the 2-subdivision of G is a

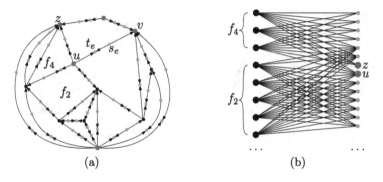

Fig. 4. (a) The 2-subdivision $\hat{G}$ of the graph G in Fig. 3(b). (b) A portion of the bipartite description $H_{\hat{G}}$ of $\hat{G}$.

subgraph of the 2-subdivision of the obtained quasi-triangulation. By Lemma 3, it suffices to prove that $H_{\hat{G}}$ has a perfect matching. According to Hall's theorem $H_{\hat{G}} = (A, B, E)$ has a perfect matching if and only if for each $A' \subseteq A$, we have that $|A'| \leq |N(A')|$, where $N(A') \subseteq B$ is the set of neighbors of the vertices in A' [21]. Let A' be a subset of A and let $\{f_1, \ldots, f_k\}$ be the faces with a representative vertex in A'. A' is *complete* if it contains all representative vertices for each face f_i ($1 \leq i \leq k$).

Claim 1. *Let A' and A'' be two distinct subsets of A that contain the representative vertices of the same set of faces. If A' is complete and $|A'| \leq |N(A')|$, then $|A''| \leq |N(A'')|$.*

Proof. Since $A'' \subseteq A'$ and since all the representative vertices of a face have the same neighbors in B, we have that $|A''| \leq |A'|$ and $N(A') = N(A'')$. ∎

By Claim 1, it is sufficient to prove that Hall's theorem holds for any complete subset A' of A. Let $N_1(A') = \{v \in N(A') \mid v \text{ is a subdivision vertex of } \hat{G}\}$ and let $N_2(A') = N(A') \backslash N_1(A')$. Let F be the set of the faces of G and let $F' \subseteq F$ be the set of faces whose representative vertices are in A'. We denote by $G_{F'}$ the bimodal plane subgraph of G induced by the edges of the boundaries of the faces in F'. Note that $G_{F'}$ can have faces other than those in F' (see Figs. 5(a) and 5(b)). Let F'' be the set of faces of $G_{F'}$ that are not in F'. Each face in F'' is the union of one or more faces of $F \setminus F'$. Further, let $F'_i = \{f \in F' \mid f \text{ has degree } i \text{ in } G\}$, for $i = 2, 3, 4$. Let E_b be the set of edges of G shared by the faces in F' and those in F'' (bold edges in Fig. 5(b)). Finally, we set $\alpha = 1$ if the external face of G belongs to F', and $\alpha = 0$ otherwise.

Claim 2. $|N_1(A')| - |A'| = |E_b| - |F'_4| - 2\alpha$.

Proof. For each edge of $G_{F'}$ there are two vertices in $N_1(A')$ because each edge has two subdivision vertices. Thus $|N_1(A')| = 2|E_b| + 2|E_x|$, where E_x is the set of edges of $G_{F'}$ that are not in E_b.

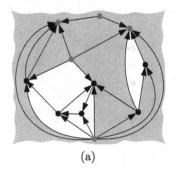
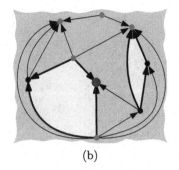

(a)   (b)

Fig. 5. Illustration for Lemma 4. (a) Faces in F' are gray. (b) The graph $G_{F'}$; faces of F'' have a light background. The largest vertex belongs to $S_{F'}$. The edges of E_b are bold.

Since G is face-acyclic, each face of degree 2 (resp. 3 and 4) in G has capacity 2 (resp. 3 and 5) in $\hat{G}$ if it is an internal face, and it has capacity 4 (resp. 5 and 7) if it is the external face. It follows that for each face in F_2' there are two vertices in A', for each face in F_3' there are three vertices in A', and for each face in F_4' there are five vertices in A'. If the external face of G belongs to F', then there are two additional vertices in A'. Thus, $|A'| = 2|F_2'| + 3|F_3'| + 5|F_4'| + 2\alpha$ and $|N_1(A')| - |A'| = 2|E_b| + 2|E_x| - 2|F_2'| - 3|F_3'| - 5|F_4'| - 2\alpha$.

Each face in F_i' has i edges in $G_{F'}$, for $i = 2, 3, 4$. Each edge of E_x belongs to two faces of F', while each edge of E_b belongs to only one face of F'. Thus we have $|E_b| + 2|E_x| = 2|F_2'| + 3|F_3'| + 4|F_4'|$ and therefore $|N_1(A')| - |A'| = |E_b| + (2|F_2'| + 3|F_3'| + 4|F_4'|) - 2|F_2'| - 3|F_3'| - 5|F_4'| - 2\alpha = |E_b| - |F_4'| - 2\alpha$. ∎

Graph $G_{F'}$ can have some source or sink vertices that are not source or sink vertices of G because $G_{F'}$ has a subset of the edges of G (see for example the red vertex in Fig. 5(b)). Let $S_{F'}$ be the set of such vertices. Also, let F_2'' and F_3'' be the set of faces of F'' that have degree 2 and 3, respectively, in $G_{F'}$ and let F_x'' be the set $F'' \setminus (F_3'' \cup F_2'')$.

Claim 3. $|N_2(A')| \geq |F_4'| - \frac{|E_b|}{2} - \frac{|F_3''|}{2} - |F_2''| - |F_x''| + 2.$

Proof. By Lemma 1, the number of source and sink vertices of $G_{F'}$ is equal to the total capacity of the faces of $G_{F'}$. The number of source and sink vertices of $G_{F'}$ is $|N_2(A')| + |S_{F'}|$. The total capacity of the faces of $G_{F'}$ is given by two terms: The total capacity C' of the faces in F' plus the total capacity C'' of the faces in F''. We have $C' = |F_4'| + 2\alpha$. Indeed, let f be a face of F': If f is internal it has capacity 0 if it has degree 2 or 3 and it has capacity 1 if it has degree 4; if f is the external face of $G_{F'}$, it has capacity 2 if it has degree 2 or 3 and it has capacity 3 if it has degree 4. The term C'' is equal to $\sum_{f \in F''}(\mathcal{A}(f) - 1) + 2(1 - \alpha)$, where $2(1 - \alpha)$ takes into account the fact that the external face of $G_{F'}$ may belong to F''. (Recall that the capacity of a face is $\mathcal{A}(f) - 1$ if it is internal and

Quasi-upward Planar Drawings with Minimum Curve Complexity   203

$\mathcal{A}(f) + 1$ if it is external. Also, if $\alpha = 1$ the external face of $G_{F'}$ belongs to F', otherwise it belongs to F''.)

Each vertex in $S_{F'}$ belongs to at least one face of F''. Let v be a vertex of $S_{F'}$ and let f be a face of F'' that contains v. Since v is a source or a sink vertex, f has at least one angle at v that is either a source-switch or a sink-switch. In other words, for each vertex in $S_{F'}$ there is at least one source-switch or a sink-switch in a face of F''. Since at least $\frac{|S_{F'}|}{2}$ of these angles are source-switches or sink-switches, we have that $\sum_{f \in F''} \mathcal{A}(f) \geq \frac{|S_{F'}|}{2}$ (recall that $\mathcal{A}(f)$ is equal to the number of source-switches of f, which is equal to the number of sink-switches of f). It follows that $C'' = \sum_{f \in F''}(\mathcal{A}(f) - 1) + 2(1 - \alpha) \geq \frac{|S_{F'}|}{2} - |F''| + 2(1 - \alpha)$.

Thus, we have $|N_2(A')| + |S_{F'}| = C' + C'' \geq |F_4'| + 2\alpha + \frac{|S_{F'}|}{2} - |F''| + 2(1 - \alpha)$. From $F'' = F_2'' \cup F_3'' \cup F_x''$, we obtain $|N_2(A')| \geq |F_4'| - \frac{|S_{F'}|}{2} - |F_2''| - |F_3''| - |F_x''| + 2$. Since each edge of E_b is shared by a face of F' and a face of F'' and since each face of F'' has only edges of E_b in its boundary, we have that $\sum_{f \in F''} \delta(f) = |E_b|$. Also, each face in F_3'' has at most two switches and therefore at least one of its vertices does not belong to $|S_{F'}|$. It follows that $|S_{F'}| \leq \sum_{f \in F''} \delta(f) - |F_3''| = |E_b| - |F_3''|$, and therefore $|N_2(A')| \geq |F_4'| - \frac{|E_b|}{2} + \frac{|F_3''|}{2} - |F_2''| - |F_3''| - |F_x''| + 2 = |F_4'| - \frac{|E_b|}{2} - \frac{|F_3''|}{2} - |F_2''| - |F_x''| + 2$.   ∎

In order to prove that the condition of Hall's theorem holds for A', we will show that $|N(A')| - |A'| \geq 0$. By Claims 2 and 3, $|N(A')| - |A'| = |N_1(A')| + |N_2(A')| - |A'| \geq |E_b| - |F_4'| - 2\alpha + |F_4'| - \frac{|E_b|}{2} - \frac{|F_3''|}{2} - |F_2''| - |F_x''| + 2 = \frac{|E_b|}{2} - \frac{|F_3''|}{2} - |F_2''| - |F_x''| + 2 - 2\alpha$.

Since the faces in F'' do not share edges, we have that $|E_b| \geq 2|F_2''| + 3|F_3''| + 4|F_x''|$, and therefore $|N(A')| - |A'| \geq \frac{2|F_2''|}{2} + \frac{3|F_3''|}{2} + \frac{4|F_x''|}{2} - \frac{|F_3''|}{2} - |F_2''| - |F_x''| + 2 - 2\alpha = |F_3''| + |F_x''| + 2 - 2\alpha$. Since α is either 0 or 1, we have that $|N(A')| - |A'| \geq |F_3''| + |F_x''| \geq 0$ and the condition of Hall's theorem holds.   ∎

The next theorem extends Lemma 4 to graphs that may not be face-acyclic.

Theorem 1. *The 2-subdivision of a bimodal plane digraph is upward planar.*

Proof. Let G be a bimodal plane digraph. If G is face-acyclic the statement follows from Lemma 4. Otherwise, for every face f of G whose boundary is a cycle, we insert a source vertex v inside f and connect it to every vertex of the boundary of f. Let G' be the resulting digraph. Clearly, G' is a face-acyclic bimodal digraph and by Lemma 4 the 2-subdivision $\hat{G}'$ of G' is upward planar. Since the 2-subdivision of G is a subgraph of $\hat{G}'$, the statement follows.   ∎

4   Computing Minimum Curve Complexity Drawings

To efficiently compute quasi-upward planar drawings with minimum curve complexity, we define a variant of the flow network used by Bertolazzi et al. [3].

204   C. Binucci et al.

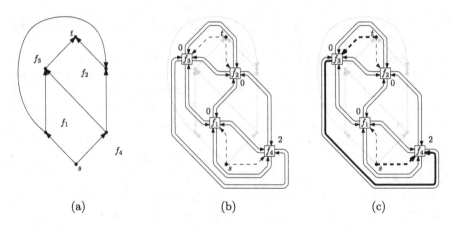

(a)   (b)   (c)

Fig. 6. (a) A bimodal planar embedding of the graph G of Fig. 1(a). (b) The unit-capacity flow network $\mathcal{N}_u(G)$. The dashed arcs (vertex-to-face arcs) have cost 0, while solid arcs (face-to-face arcs) have cost 2. The number close to each face-node indicates the capacity of the face. (c) A feasible flow φ for $\mathcal{N}_u(G)$. A unit of flow traverses the edges highlighted in bold. The cost of the flow is 2. A 2-bend drawing of G corresponding to φ is the one shown in Fig. 1(b).

Feasible flows in this network correspond to quasi-upward planar drawings. Intuitively, each unit of flow represents a large angle; large angles are produced by sources and sinks and are consumed by the faces.

Let G be a bimodal plane digraph. The *unit-capacity flow network of G*, denoted as $\mathcal{N}_u(G)$, is defined as follows (see Fig. 6). For each edge e of $\mathcal{N}_u(G)$, we denote by $\beta(e)$, $\chi(e)$ and $\varphi(e)$ the capacity, the cost, and the flow of e, respectively.

- The nodes of $\mathcal{N}_u(G)$ are all the sources and sinks (*vertex-nodes*), and all the faces (*face-nodes*) of G.
- Each vertex-node v of $\mathcal{N}_u(G)$ supplies a flow equal to 1. This means that exactly one of the angles at v must be large.
- Each face-node of $\mathcal{N}_u(G)$ that corresponds to a face f demands a flow equal to the capacity of f. This means that f must have a number of large angles equal to its capacity. If f is an internal face and it is a directed cycle then $\mathcal{A}(f) = 0$ and the capacity of f is -1, that is, f supplies a flow equal to 1.
- For each source or sink v that belongs to a face f, there is a *vertex-to-face arc* (v, f) in $\mathcal{N}_u(G)$ such that $\beta(v, f) = 1$ and $\chi(v, f) = 0$. Intuitively, a unit of flow on this arc means that f has a large angle at v.
- For each edge e of G shared by two faces f and g, there is a pair of *face-to-face arcs* (f, g) and (g, f) in $\mathcal{N}_u(G)$ such that $\beta(f, g) = \beta(g, f) = 1$ and $\chi(f, g) = \chi(g, f) = 2$. Intuitively, a unit of flow on (f, g) or on (g, f) represents the insertion of two bends along e. This corresponds to two units of cost. The two arcs (f, g) and (g, f) are called the *dual arcs* of edge e.

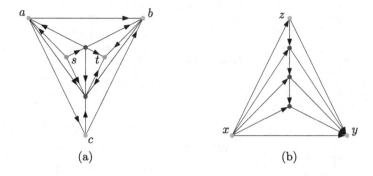

Fig. 7. (a) Supplier gadget. (b) Barrier gadget.

We remark that the main differences between the flow network defined above and the flow network $\mathcal{N}(G)$ defined by Bertolazzi et al. [3] are as follows: *(i)* In $\mathcal{N}_u(G)$ we have two opposite face-to-face arcs for each edge e of G, while in $\mathcal{N}(G)$ there are two opposite face-to-face arcs for each pair of adjacent faces, even when they share more than one edge; *(ii)* the capacity of the face-to-face arcs is one in $\mathcal{N}_u(G)$ and it is unbounded in $\mathcal{N}(G)$. The fact that $\mathcal{N}_u(G)$ is well-defined is a consequence of Lemma 1. The following lemma will be used to prove that a quasi-upward planar drawing with minimum curve complexity can be computed by means of the flow network $\mathcal{N}_u(G)$.

Lemma 5 ($\star$). *Let G be a bimodal plane digraph. For each feasible flow φ_u in $\mathcal{N}_u(G)$ there exists a quasi-upward planar drawing Γ of G such that the number of bends along each edge e of G is equal to the sum of the costs of the flows along the two face-to-face arcs that are the dual arcs of e.*

The next theorem gives the main result of this section and exploits the algorithm by Karczmarz and Sankowski [24] to compute a min-cost flow on a planar unit-capacity flow network.

Theorem 2 ($\star$). *Let G be a bimodal plane digraph with m edges. There exists an $\tilde{O}(m^{\frac{4}{3}})$-time algorithm that computes a quasi-upward planar drawing Γ of G with the following properties: (i) Γ has minimum curve complexity, which is at most two. (ii) Γ has the minimum number of bends among the quasi-upward planar drawings of G with minimum curve complexity.*

If G does not have homotopic multiple edges (i.e., multiple edges that define a face), then $m \in O(n)$ and the time complexity of Theorem 2 is $\tilde{O}(n^{\frac{4}{3}})$.

5   A Lower Bound on the Curve Complexity

By Theorem 2, every bimodal plane digraph admits a quasi-upward planar drawing with curve complexity two, which is worst-case optimal. One may wonder whether curve complexity two and minimum total number of bends can be simultaneously achieved. The next lemma shows that this is not always possible.

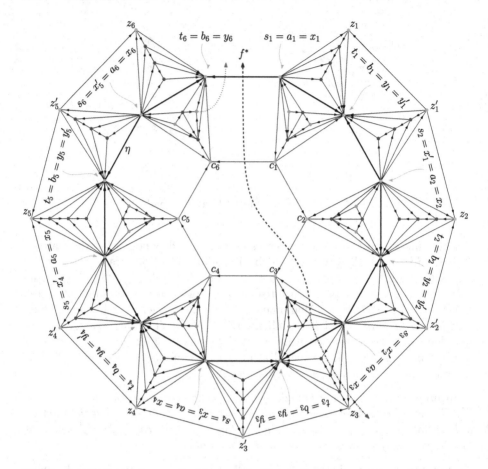

Fig. 8. Graph G_6 in the proof of Lemma 6. The dashed arrow represents a path in the dual that is shorter than the one represented by the dot-dashed arrow. The dotted arrow also represents a shortest path.

Lemma 6 (⋆). *For every integer $k \geq 3$, there exists a bimodal planar acyclic digraph G_k with $14k-3$ vertices and $41k-13$ edges such that every quasi-upward planar drawing of G_k with the minimum number of bends has one edge with at least $2k - 2$ bends.*

Sketch of Proof. For each $k \geq 3$, we construct a graph G_k by suitably combining different copies of two gadgets. The first gadget is shown in Fig. 7(a) and it is called *supplier gadget* because it contains one source and one sink vertex, denoted as s and t in Fig. 7(a), that supply two units of flow. Graph G_k has k copies of the supplier gadget that in total supply $2k$ units of flow; G_k is such that $k - 1$ units of this flow have to reach a specific face. To force these $k - 1$ units of flow to "traverse" the same edge (thus creating $2k - 2$ bends along this edge), we use the second gadget, shown in Fig. 7(b). This gadget is called *barrier gadget* because it is used to prevent the flow to traverse some edges. Graph G_k is shown in Fig. 8 for $k = 6$. See [6] for more details on the construction of G_k.

We now prove that every quasi-upward planar drawing of G_k with the minimum number of bends has at least one edge with at least $2k - 2$ bends. Let Γ be any quasi-upward planar drawing of G_k with the minimum number of bends and let ψ be the planar embedding of Γ. Drawing Γ corresponds to a minimum cost flow on the flow network $\mathcal{N}(G_k)$ defined on the planar embedding ψ [3]. Since G_k is triconnected, all its planar embeddings have the same set of faces and each embedding is defined by the choice of a face as the external one. Let f^* be the face that is external in the embedding of G_k shown in Fig. 8. Face f^* has k source-switches, which implies that its capacity is at least $k - 1$ in $\mathcal{N}(G_k)$: Namely, it is $k + 1$ if f^* is the external face in ψ and $k - 1$ otherwise. Since there are one source and one sink in each of the k supplier gadgets, the total amount of flow consumed by the faces is $2k$. In any planar embedding of G_k the cycle η, highlighted by bold edges in Fig. 8, separates the source and sink vertices from f^*. It follows that at least $k - 1$ units of flow must go through the dual arcs of the edges of η, thus creating at least $2k - 2$ bends along the edges of η. We now prove that all these bends are on the edge (s_1, t_k). Consider a unit flow that goes from a source or a sink node v to f^* following a path π in $\mathcal{N}(G_k)$; the cost of sending this unit of flow along π is equal to the number of face-to-face arcs in π because face-to-face arcs have cost two. Each face-to-face arc of π is the dual arc of an edge in G_k shared by two adjacent faces. Hence, to obtain a minimum cost flow, each unit of flow that goes from a source or a sink node v to f^* in $\mathcal{N}(G_k)$ follows a shortest path π in the dual graph of G_k connecting a face incident to v with f^*. The lemma holds because any shortest path in the dual graph of G_k connecting a face incident to v with f^* includes a dual arc of the edge (s_1, t_k). (See [6] for more details.)   □

The next theorem extends the result of Lemma 6 to the cases when n is not equal to $14k - 3$, for some $k > 0$.

Theorem 3 ($\star$). *For every $n \geq 39$ there exists a bimodal planar digraph with n vertices such that every bend-minimum quasi-upward planar drawing has curve complexity at least $\frac{n-24}{7}$.*

6   Open Problems

We conclude by mentioning some open problems that are naturally suggested by the results in this paper.

- Is it possible to improve the time complexity stated by Theorem 2?
- We showed that every bimodal plane digraph becomes upward planar if every edge is subdivided twice. It would be interesting to minimize the total number of subdivision vertices (with at most two subdivision vertices per edge) such that the resulting graph admits an upward straight-line drawing of polynomial area.

208   C. Binucci et al.

References

1. Angelini, P., Chaplick, S., Cornelsen, S., Da Lozzo, G.: Planar L-drawings of bimodal graphs. CoRR arXiv:2008.07834 (2020)
2. Bekos, M.A., Kaufmann, M., Krug, R.: On the total number of bends for planar octilinear drawings. J. Graph Algorithms Appl. **21**(4), 709–730 (2017)
3. Bertolazzi, P., Di Battista, G., Didimo, W.: Quasi-upward planarity. Algorithmica **32**(3), 474–506 (2002). https://doi.org/10.1007/s00453-001-0083-x
4. Bertolazzi, P., Di Battista, G., Liotta, G., Mannino, C.: Upward drawings of tri-connected digraphs. Algorithmica **12**(6), 476–497 (1994). https://doi.org/10.1007/BF01188716
5. Biedl, T.C., Kant, G.: A better heuristic for orthogonal graph drawings. Comput. Geom. **9**(3), 159–180 (1998). https://doi.org/10.1016/S0925-7721(97)00026-6
6. Binucci, C., Di Giacomo, E., Liotta, G., Tappini, A.: Quasi-upward planar drawings with minimum curve complexity. CoRR arXiv:2108.10784 (2021)
7. Binucci, C., Didimo, W.: Quasi-upward planar drawings of mixed graphs with few bends: heuristics and exact methods. In: Pal, S.P., Sadakane, K. (eds.) WALCOM 2014. LNCS, vol. 8344, pp. 298–309. Springer, Cham (2014). https://doi.org/10.1007/978-3-319-04657-0_28
8. Binucci, C., Didimo, W.: Computing quasi-upward planar drawings of mixed graphs. Comput. J. **59**(1), 133–150 (2016). https://doi.org/10.1093/comjnl/bxv082
9. Binucci, C., Didimo, W., Patrignani, M.: Upward and quasi-upward planarity testing of embedded mixed graphs. Theor. Comput. Sci. **526**, 75–89 (2014). https://doi.org/10.1016/j.tcs.2014.01.015
10. Chan, H.: A parameterized algorithm for upward planarity testing. In: Albers, S., Radzik, T. (eds.) ESA 2004. LNCS, vol. 3221, pp. 157–168. Springer, Heidelberg (2004). https://doi.org/10.1007/978-3-540-30140-0_16
11. Chaplick, S., Lipp, F., Wolff, A., Zink, J.: Compact drawings of 1-planar graphs with right-angle crossings and few bends. Comput. Geom. **84**, 50–68 (2019)
12. Di Battista, G., Eades, P., Tamassia, R., Tollis, I.G.: Graph Drawing. Prentice Hall, Upper Saddle River (1999)
13. Di Giacomo, E., Didimo, W., Liotta, G., Meijer, H.: Area, curve complexity, and crossing resolution of non-planar graph drawings. In: Eppstein, D., Gansner, E.R. (eds.) GD 2009. LNCS, vol. 5849, pp. 15–20. Springer, Heidelberg (2010). https://doi.org/10.1007/978-3-642-11805-0_4
14. Di Giacomo, E., Gasieniec, L., Liotta, G., Navarra, A.: On the curve complexity of 3-colored point-set embeddings. Theor. Comput. Sci. **846**, 114–140 (2020)
15. Di Giacomo, E., Liotta, G., Trotta, F.: Drawing colored graphs with constrained vertex positions and few bends per edge. Algorithmica **57**(4), 796–818 (2010). https://doi.org/10.1007/s00453-008-9255-2
16. Didimo, W.: Upward graph drawing. In: Encyclopedia of Algorithms, pp. 2308–2312 (2016). https://doi.org/10.1007/978-1-4939-2864-4_653
17. Didimo, W., Giordano, F., Liotta, G.: Upward spirality and upward planarity testing. SIAM J. Discrete Math. **23**(4), 1842–1899 (2009). https://doi.org/10.1137/070696854
18. Didimo, W., Liotta, G., Ortali, G., Patrignani, M.: Optimal orthogonal drawings of planar 3-graphs in linear time. In: Chawla, S. (ed.) Proceedings of the 2020 ACM-SIAM Symposium on Discrete Algorithms, SODA 2020, pp. 806–825. SIAM (2020). https://doi.org/10.1137/1.9781611975994.49

19. Didimo, W., Liotta, G., Patrignani, M.: Bend-minimum orthogonal drawings in quadratic time. In: Biedl, T., Kerren, A. (eds.) GD 2018. LNCS, vol. 11282, pp. 481–494. Springer, Cham (2018). https://doi.org/10.1007/978-3-030-04414-5_34

20. Garg, A., Tamassia, R.: On the computational complexity of upward and rectilinear planarity testing. SIAM J. Comput. **31**(2), 601–625 (2001). https://doi.org/10.1137/S0097539794277123

21. Hall, P.: On representatives of subsets. J. London Math. Soc. **s1–10**(1), 26–30 (1935)

22. Healy, P., Lynch, K.: Two fixed-parameter tractable algorithms for testing upward planarity. Int. J. Found. Comput. Sci. **17**(5), 1095–1114 (2006). https://doi.org/10.1142/S0129054106004285

23. Kant, G.: Drawing planar graphs using the canonical ordering. Algorithmica **16**(1), 4–32 (1996). https://doi.org/10.1007/BF02086606

24. Karczmarz, A., Sankowski, P.: Min-cost flow in unit-capacity planar graphs. In: Bender, M.A., Svensson, O., Herman, G. (eds.) 27th Annual European Symposium on Algorithms, ESA 2019, LIPIcs, vol. 144, pp. 66:1–66:17. Schloss Dagstuhl - Leibniz-Zentrum für Informatik (2019). https://doi.org/10.4230/LIPIcs.ESA.2019.66

25. Kaufmann, M., Wiese, R.: Embedding vertices at points: few bends suffice for planar graphs. J. Graph Algorithms Appl. **6**(1), 115–129 (2002). https://doi.org/10.7155/jgaa.00046

26. Kindermann, P., Montecchiani, F., Schlipf, L., Schulz, A.: Drawing subcubic 1-planar graphs with few bends, few slopes, and large angles. J. Graph Algorithms Appl. **25**(1), 1–28 (2021). https://doi.org/10.7155/jgaa.00547

27. Rahman, M.S., Nakano, S., Nishizeki, T.: A linear algorithm for bend-optimal orthogonal drawings of triconnected cubic plane graphs. J. Graph Algorithms Appl. **3**(4), 31–62 (1999). https://doi.org/10.7155/jgaa.00017

28. Tamassia, R., Tollis, I.G., Vitter, J.S.: Lower bounds and parallel algorithms for planar orthogonal grid drawings. In: Proceedings of the Third IEEE Symposium on Parallel and Distributed Processing, SPDP 1991, Dallas, Texas, USA, 2–5 December 1991, pp. 386–393. IEEE Computer Society (1991). https://doi.org/10.1109/SPDP.1991.218215

Non-homotopic Loops with a Bounded Number of Pairwise Intersections

Václav Blažej[1] ⓘ, Michal Opler[2] ⓘ, Matas Šileikis[3](✉) ⓘ, and Pavel Valtr[4] ⓘ

[1] Faculty of Information Technology, Czech Technical University in Prague, Prague, Czech Republic
[2] Computer Science Institute, Charles University, Prague, Czech Republic
[3] Czech Academy of Sciences, Institute of Computer Science, Prague, Czech Republic
[4] Department of Applied Mathematics, Faculty of Mathematics and Physics, Charles University, Prague, Czech Republic

Abstract. Let V_n be a set of n points in the plane and let $x \notin V_n$. An *x-loop* is a continuous closed curve not containing any point of V_n. We say that two x-loops are *non-homotopic* if they cannot be transformed continuously into each other without passing through a point of V_n. For $n = 2$, we give an upper bound $e^{O(\sqrt{k})}$ on the maximum size of a family of pairwise non-homotopic x-loops such that every loop has fewer than k self-intersections and any two loops have fewer than k intersections. The exponent $O(\sqrt{k})$ is asymptotically tight. The previous upper bound $2^{(2k)^4}$ was proved by Pach et al. [6]. We prove the above result by proving the asymptotic upper bound $e^{O(\sqrt{k})}$ for a similar problem when $x \in V_n$, and by proving a close relation between the two problems.

Keywords: Graph drawing · Non-homotopic loops · Curve intersections · Plane

1 Introduction

The *crossing lemma* bounds the number of edge crossings of a graph drawn in the plane where the graph has n vertices and $m \geq 4n$ edges. It was proved independently by Ajtai, Chvátal, Newborn, and Szemerédi [1] and by Leighton [5]. Recently, Pach et al. [6] proved a modification of the crossing lemma for multigraphs with non-homotopic edges. In the proof, they used a bound on the maximum size of collections of so-called non-homotopic loops. We focus on improving the bounds for two settings – one used by the authors of [6] and a slightly altered

MŠ was supported by the Czech Science Foundation, grant number GJ20-27757Y, with institutional support RVO:67985807. VB acknowledges the support of the OP VVV MEYS funded project CZ.02.1.01/0.0/0.0/16_019/0000765 "Research Center for Informatics". This work was supported by the Grant Agency of the Czech Technical University in Prague, grant No. SGS20/208/OHK3/3T/18. PV and MO were supported by project 19-17314J of the Czech Science Foundation.

ⓒ Springer Nature Switzerland AG 2021
H. C. Purchase and I. Rutter (Eds.): GD 2021, LNCS 12868, pp. 210–222, 2021.
https://doi.org/10.1007/978-3-030-92931-2_15

one. We provide an upper bound $e^{O(\sqrt{k})}$ with the asymptotically tight exponent $O(\sqrt{k})$ in both settings (Theorem 1); we also show a new relation between the extremal functions in the two settings (Proposition 1).

For an integer $n \geq 1$, let $V_n = \{v_1, \ldots, v_n\}$ be a set of n distinct points in the plane $\mathbb{R}^2$. Given $x \in \mathbb{R}^2$, an x-*loop* is a continuous function $\ell : [0,1] \to \mathbb{R}^2$ such that $\ell(0) = \ell(1) = x$ and $\ell(t) \notin V_n$ for $t \in (0,1)$. Two x-loops ℓ_0, ℓ_1 are *homotopic*, denoted $\ell_0 \sim \ell_1$, if there is a continuous function $H : [0,1]^2 \to \mathbb{R}^2$ (a *homotopy*) such that

$$H(0,t) = \ell_0(t) \quad \text{and} \quad H(1,t) = \ell_1(t) \quad \text{for all } t \in [0,1],$$
$$H(s,0) = H(s,1) = x \quad \text{for all } s \in [0,1], \text{ and}$$
$$H(s,t) \notin V_n \quad \text{for all } s,t \in (0,1).$$

In the case when $x \in V_n$, we will, without loss of generality, assume $x = v := v_1$, and refer to x-loops as v-loops (dropping the subscript for simplicity). Henceforth, when we use the term x-*loop*, we will tacitly assume that $x \notin V_n$. When x (or v) is clear from the context we will also call an x-loop (v-loop) simply a *loop*.

A *self-intersection* of a loop ℓ is an unordered pair $\{t, u\} \subset (0,1)$ of distinct numbers such that $\ell(t) = \ell(u)$, while an intersection of two loops ℓ_1, ℓ_2 is an *ordered* pair $(t, u) \in (0,1)^2$ such that $\ell_1(t) = \ell_2(u)$.

Given integers $n, k \geq 1$ and $x \notin V_n$ ($v \in V_n$), let $f(n,k)$ (respectively, $g(n,k)$) be the largest number of pairwise non-homotopic x-loops (respectively, v-loops) such that every loop has fewer than k self-intersections and any two loops have fewer than k intersections.

Pach et al. [6] considered x-loops (they also added a convenient restriction that no loop passes through x, which holds trivially in the setting of v-loops). The quantities $f(n,k)$ and $g(n,k)$ are related by the following inequalities. In [3] we proved that for every $n, k \geq 1$ we have

$$g(n,k) \leq f(n,k) \leq g(n+1,k). \tag{1}$$

In the current paper we give the following inequality, which allows us to improve the upper bound from [6] on $f(n,k)$ by proving an upper bound on $g(n,k)$.

Proposition 1. *For every* $n, k \geq 1$ *we have*

$$f(n,k) = O(k^2) \cdot g(n, 5k). \tag{2}$$

Proposition 1 is proved (with a multiplicative constant of 484) in Sect. 6.

Pach et al. [6] showed that for $n \geq 2$

$$f(n,k) \leq 2^{(2k)^{2n}} \tag{3}$$

and

$$f(n,k) \geq \begin{cases} 2^{\sqrt{nk}/3}, & \text{for } n \leq 2k, \\ (n/k)^{k-1}, & \text{for } n \geq 2k. \end{cases} \tag{4}$$

Pach et al. [6] also proved that if $n = 1$, then there are at most $2k + 1$ non-homotopic loops with fewer than k self-intersections (that is, if we do not bound the number of intersections) implying $f(1, k) \leq 2k + 1$.

In our main result we focus on the function g in case $n = 2$.

Inequalities (1) and (3) imply that $g(2, k) \leq 2^{16k^4}$. After submitting this paper to GD2021, the authors became aware that the latter inequality was not the best at that time. From the proofs in the paper of Juvan, Malnič and Mohar [4] it follows that $f(2, k) \leq k^{Ck^2}$ for some absolute constant $C > 0$ (this paper focuses on the generality of spaces in which the loops are drawn rather than on quantitative bounds; it also implies good bounds for $f(n, k)$ for any fixed n).

In [3] the authors proved

$$g(n, k) = 2^{O(k)}.$$

The following theorem (which we prove in Sect. 5) improves the upper bounds on $g(2, k)$ and $f(2, k)$ significantly. Interestingly, the bound on $g(2, k)$ only uses the restriction on self-intersections (this is not enough for $n \geq 3$), while the restriction on intersections is used only in Proposition 1.

Theorem 1. *Let $n = 2$. For any k, the size of any collection of non-homotopic v-loops with fewer than k self-intersections is $e^{O(\sqrt{k})}$. In particular $g(2, k) = e^{O(\sqrt{k})}$, and, in view of (2), we have $f(2, k) = e^{O(\sqrt{k})}$.*

Note that in view of (4) the exponent $O(\sqrt{k})$ is asymptotically tight.

There is still a huge gap between lower and upper bounds on $f(n, k)$ (and $g(n, k)$) for general n; see [6] (also the implicit bounds from [4] are probably better for many pairs (n, k)). In the proof of Theorem 1, we use several lemmas (Lemmas 1–7), which might help to narrow this gap, as they provide useful tools and are usually stated for general n.

2   Setup and Notation

2.1   Obstacles, Equator and Gaps

Depending on the context, we will treat $S := \mathbb{R}^2 \setminus V_n$ either as the plane with n points removed, or as a sphere with $n + 1$ points removed (where n of these points come from the set $V_n = \{v_1, \ldots, v_n\}$ and the last point, denoted by v_0, corresponds to the "point at infinity"). We define $V_n^\infty = \{v_0\} \cup V_n$ and refer to the elements of V_n^∞ as *obstacles*.

Given a finite collection of loops, by infinitesimal perturbations, without creating any new intersections or self-intersections, we can ensure that

1. no two (self-)intersections occur at the same point of S,
2. every (self-)intersection is a *crossing*, that is, one loop "passes to the other side" of the other loop (rather than two loops 'touching').

Given a drawing of the loops satisfying the above conditions, we choose a closed simple curve on the sphere which goes through the obstacles $v_0, \ldots, v_n$ in this order (for x-loops, we choose this curve so that it also avoids x). We call this loop the *equator*. Removing the equator from the sphere, we obtain two connected sets, which we arbitrarily name the *northern hemisphere* and the *southern hemisphere*. We refer to the $n+1$ open curves into which the equator is split by excluding points v_i as *gaps*. We assign label i to the gap between v_i and v_{i+1}, with indices counted modulo $n+1$. Moreover, when talking about v-loops, we treat $v = v_1$ as an additional, special gap, with a label v; see Fig. 1, where the equator is drawn as a circle, for simplicity.

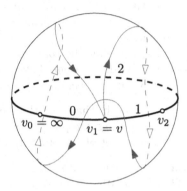

Fig. 1. The equator, the gaps, and a v-loop that induces the word $w = v2102v$.

By a careful choice of the equator, we can assume the following conditions:

3. every loop in the collection intersects the equator a finite number of times,
4. each of these intersections (except for, possibly, the intersection at v) is a crossing, (i.e., no loop touches the equator),
5. no point of self-intersection or intersection lies on the equator.

2.2   Segments and Induced Words

Part of a given loop ℓ between a pair of distinct intersections with the equator (inclusively) is called a *segment*. Treating a loop (respectively, a segment) as a function $\ell : [0,1] \to \infty$ (respectively, the restriction of ℓ to a closed subinterval of $[0,1]$), gives a natural orientation of a loop or a segment. A minimal segment is called an *arc*. If an arc intersects itself, we can remove the part of the arc between these self-intersections without changing the homotopy class of the loop, which allows us to make yet another assumption that

6. there are no self-intersections within any arc.

Consider a segment s that intersects the equator t times (including the beginning and the end). By listing the labels of gaps that the loop crosses as it traverses s, we obtain a word $w = w_1 \ldots w_t$. In this case we say that s is a *w-segment*. If we take the maximal segment of a loop ℓ, that is, from the first to the last crossing of equator (which is the whole loop in the case of v-loops), then we say that word w is the word *induced* by ℓ; see Fig. 1. Given a loop that induces a word w, the segments of the loop correspond to subwords (that is, words consisting of consecutive letters of w) of length at least 2.

Note that the word induced by a v-loop ℓ starts and ends in v. Dropping these vs we obtain a word which we call the *inner word induced by* a v-loop ℓ.

If we reverse the orientation of a segment s, the order of gaps is reversed and hence we obtain the *reverse* of the word w, denoted $\overline{w} := w_t \ldots w_1$. Sometimes we talk about segments as unoriented objects, treating a segment simultaneously as a w-segment and a $\overline{w}$-segment.

Given an oriented segment s, we define the *polarity* of s as the hemisphere to which the first arc of s belongs. We call an oriented w-segment a *w-downsegment* or a *w-upsegment* whenever we want to specify the polarity of the first arc.

Remark 1. For a word w of even length (that is, with an even number of letters) a w-segment is also a $\overline{w}$-segment of the same polarity, while for a word w of odd length a w-segment is also a $\overline{w}$-segment of the opposite polarity.

For example, consider a v-loop with the first arc in the southern hemisphere that induces the word 01201. It has 01-segments of both polarities (and hence 10-segments of both polarities), as well as a 012-downsegment (which is also a 210-upsegment) but, say, there is no 012-upsegment.

2.3 Patterns

To simplify notation we use a concept of *pattern*, that is a finite sequence of symbols, usually, the first letters of the Greek alphabet $\alpha, \beta, \ldots$. Given a pattern T, we say that a word w is a *T-word* if it is obtained by replacing symbols by letters, so that two letters in w are equal if and only if the symbols in T are equal. For example, words $010, 202, 121$ match the pattern $\alpha\beta\alpha$, but 111 does not. Given a pattern T, we say a word is *T-free*, if it contains no subword matching the pattern T. With a slight abuse of notation, by a *T-segment* we also mean a w-segment such that word w matches the pattern T.

3 General n

In this section we state and prove several facts that are valid for general n, including all prerequisites for the proof of Theorem 1.

3.1   Simplifying Words

Part (i) of the following lemma allows to simplify the words induced by a given family of v-loops, in particular simplifying the setting for the proof of Theorem 1. Part (ii) is used in the proof of the inequality (2) of Proposition 1.

Lemma 1. *Assume that $n \geq 1$.*

(i) *Given any family of v-loops, each loop can be replaced by a homotopic v-loop inducing an $\alpha\alpha$-free inner word, with the first and last letters in $\{2, \ldots, n\}$, so that the numbers of pairwise intersections or self-intersections do not increase.*

(ii) *Suppose that a family of x-loops and the equator are such that there is a path connecting x to some $v \in V_n$ which does not intersect any loop or the equator. Then each x-loop can be replaced by a homotopic x-loop inducing an $\alpha\alpha$-free word so that the numbers of pairwise intersections or self-intersections do not increase.*

The part (i) was already proved in [3] and the part (ii) is easily established using the same ideas. The details of the proof can be found in the full version of this paper [2].

3.2   Characterization of Homotopic Loops

Once we can simplify loops using Lemma 1, we can use the following lemma to describe the homotopy classes of loops in terms of induced words.

Recall that v-loops start and end at v_1, which is incident to gaps 0 and 1.

Lemma 2. (i) *Two x-loops inducing $\alpha\alpha$-free words w_1 and w_2 are homotopic if and only if $w_1 = w_2$.*

(ii) *Suppose that two v-loops ℓ_1 and ℓ_2 are such that all four initial/final arcs lie in the same hemisphere. Suppose ℓ_1, ℓ_2 induce $\alpha\alpha$-free inner words*

$$w_1 = x_1 u_1 z_1, \quad \text{and} \quad w_2 = x_2 u_2 z_2, \tag{5}$$

where (possibly zero-length) words x_1, x_2, z_1, z_2 use only letters 0 and 1 and words u_1, u_2 start and end in letters other than 0 or 1 (if u_i is empty, we assume that $x_i = w_i$ and z_i is empty). Then ℓ_1 and ℓ_2 are homotopic if and only if $u_1 = u_2$ and the lengths of x_1 and x_2 have the same parity.

(iii) *If two v-loops ℓ_1 and ℓ_2 induce words w_1 and w_2 that start and end in a letter other than 0 or 1, then ℓ_1 and ℓ_2 are homotopic if and only if ℓ_1 and ℓ_2 start in the same hemisphere, end in the same hemisphere, and $w_1 = w_2$.*

Lemma 2 is proved in the full version of this paper [2]. It is based on the description of the (fundamental) homotopy group of the plane with several points removed and the correspondence between words and the generators of this group.

The idea of the part (ii) is that we may "unwind" the initial and final segments that just "wind around" the obstacle v, without changing the homotopy class. The reason why, say, the prefixes x_1 and x_2 have to have matching parities (and merely $u_1 = u_2$ is not enough), is that otherwise the segments corresponding to the subwords u_1 and u_2 would have opposite polarity. One can see, say, that if two loops start and end in the northern hemisphere and induce words 02 and 20 (so that $u_1 = u_2 = 2$), they are not homotopic (in fact, ℓ_1 is homotopic to ℓ_2 reversed).

3.3 Subwords Forcing Intersections

The equator has two natural orientations. For any ordered triple (a, b, c) of distinct gaps we assign the orientation of the equator such that if we circle the equator starting from the gap a, we encounter the gap b before c. In particular, for any three distinct gaps a, b, c triples $(a, b, c), (b, c, a), (c, a, b)$ have the same orientation while triples (a, b, c) and (c, b, a) have opposite orientations. For example, if $n \geq 2$, recalling our labeling of gaps from Sect. 2 (in particular that the vertex $v = v_1$ is a special gap) we have that $(0, v, 1)$, $(v, 1, 2)$ and $(0, 1, 2)$ have the same orientation.

Lemma 3. *Let $k \geq 0$ be an integer. Consider two segments of the same polarity corresponding to $\alpha\alpha$-free words $a_0 a_1 \ldots a_k a_{k+1}$ and $b_0 b_1 \ldots b_k b_{k+1}$ such that for $i = 1, \ldots, k$ we have $a_i = b_i$, while $a_0 \neq b_0$ and $a_{k+1} \neq b_{k+1}$. Suppose that (a_0, b_0, a_1) and (a_k, b_{k+1}, a_{k+1}) have opposite orientations for even k, and the same orientation for odd k. Then there is $i \in \{0, \ldots, k\}$ such that the $a_i a_{i+1}$-arc of the first word intersects the $b_i b_{i+1}$-arc of the second word.*

Proof Idea. Assume, for a contradiction, that the $a_i a_{i+1}$-arc and the $b_i b_{i+1}$-arc are disjoint for every i. It follows inductively that the orientation of the triple (a_i, b_{i+1}, a_{i+1}) is uniquely determined for every i. In particular, it must be the same as the orientation of (a_0, b_0, a_1) for even i and opposite otherwise. For $i = k$, we arrive at a contradiction. The details of the proof are included in the full version [2].

3.4 Windings in v-Loops

We now focus on v-loops and describe the intersections forced by specific alternating words.

For a word w we write w^k a concatenation of k copies of w, say $(ab)^2 = abab$. Given an obstacle v_i, $i \neq 1$, let a, b be the gaps incident to v_i. For integer $s \geq 1$, an *s-winding around* v_i is a w-segment, where w has a form $tw'u$, where w' is of the form $(ab)^s a$, $(ba)^s b$, $(ab)^{s+1}$ or $(ba)^{s+1}$ and t, u are letters other than $\{a, b\}$. Assuming the gaps incident to v_1 are 0 and 1, an *s-winding around* $v := v_1$ is a w-segment with w of the same form as above (for $\{a, b\} = \{0, 1\}$), but with t, u being letters other than 0, 1, and v (the difference from the first case is that we do not allow $t = v$ or $u = v$), see the left part of Fig. 2.

The proofs of the following lemmas are sketched at the end of this subsection; the detailed proofs are included the full version of this paper [2].

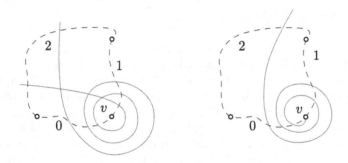

Fig. 2. Left: a 2-winding around v with word $2(01)^3 2$; right: a $(2, 2)$-snail with word $2(01)^2 0v$. The dashed line is the equator.

Lemma 4. *Suppose S and T are an s-winding and a t-winding, respectively, both around the same obstacle $v \in V_n$. Then*

(i) *Segment S has at least s self-intersections.*
(ii) *S and T have at least $2 \cdot \min\{s, t\}$ mutual intersections (provided $S \neq T$).*

We excluded from the definition of windings the case where the obstacle is v_1 and the alternating sequence appears at the beginning/end of the (inner) word. Given a positive integer m, let $(ab)^{-m} := (ba)^m$ and let $(ab)^0$ stand for an empty word. Given an integer s and a letter a other than $0, 1$ (but possibly v), by a (s, a)-*snail* we call a w-segment where w is a $\alpha\alpha$-free word of the form

$$v(01)^s w' a, \quad \text{where } w' \in \{0, 1, 01, 10\}, a \notin \{0, 1\},$$

see the right part of Fig. 2. (Note: since w is $\alpha\alpha$-free, we cannot, say, have $w' \in \{1, 10\}$ if $s > 0$.)

Lemma 5. *Consider a (s, a)-snail and a (t, b)-snail of the same polarity. If $st < 0$, the snails intersect at least $\min\{|s|, |t|\}$ times. If $st > 0$ and $a, b \neq v$, then the snails intersect at least $|s - t| - 1$ times.*

The following lemma is used in the proof of the inequality (2) of Proposition 1.

Lemma 6. *Fix a word u that starts an and ends in a letter other than 0 or 1. If we have a family F with more than $4(2k + 1)^2$ v-loops of the same polarity, each of which induces an even-length $\alpha\alpha$-free word of the form*

$$v(01)^s w' u w'' (10)^t v, \quad t, s \in \mathbb{Z}, \quad w', w'' \in \{0, 1, 01, 10\}, \tag{6}$$

then there are $\ell_1, \ell_2 \in F$ (possibly $\ell_1 = \ell_2$) with at least k (self)-intersections.

We note in passing that Lemmas 4 and 5 are proved by finding sufficiently many segment pairs that satisfy the conditions of Lemma 3 and arguing that each pair implies a distinct intersection. Lemma 6 then follows from Lemma 5 by relatively straightforward pigeonhole-type arguments.

The detailed proofs of Lemmas 4 to 6 are included in the full version of this paper [2].

4   Expansions of Words

Given an $\alpha\alpha$-free word w of length at least two, consider all maximal sub-words that use two letters. For example, if $w = 2010212$, the maximal words are 20,010,02, and 212. Ordering these words by the position of the first letter, it is clear that every two consecutive words overlap in a single letter. We classify these subwords according to the pair of letters they use. For distinct a, b a maximal subword that uses a and b is called an ab-*word* (the ordering of a and b does not matter). For a given pair a, b, the ab-words are disjoint and surrounded by letters other than a and b (we assume that w is extended by adding the letter v at each end).

Each ab-word starts either with ab or ba. If we replace each of these two-letter subwords by its power, say, ab by $(ab)^{s+1}$, $s \geq 0$, we obtain another $\alpha\alpha$-free word, which we call an ab-*expansion* of w. Hence if w has ℓ ab-words, each ab-expansion of w is uniquely described by a vector $s = (s_1, ..., s_\ell)$ of nonnegative integers.

If, in addition, w is $\alpha\beta\alpha\beta$-free, then for each distinct a, b all maximal ab-words have at most three letters. If w is not $\alpha\beta\alpha\beta$-free, it can be obtained by consecutive ab-expansions of a $\alpha\beta\alpha\beta$-free word, one for each pair a, b of distinct letters.

The *self-intersection number* of a loop-word w is defined as the smallest number of self-intersections in a loop that induces w.

Lemma 7. *Let a, b be two gaps adjacent to the same obstacle v. Let $\ell = \ell(k) \geq 2\sqrt{k}$ be a positive integer.*

Let w be an $\alpha\alpha$-free word which contains no subword abab or baba. Suppose that there are at most ℓ maximal ab-words in w. If $v = v_1$, then also assume that none of these words appears at the beginning or the end of w. The number of ab-expansions of w with the self-intersection number smaller than k is

$$\left(\frac{\ell}{\sqrt{k}}\right)^{O(\sqrt{k})} e^{O(\sqrt{k})}.$$

In particular, if also $\ell = O(\sqrt{k})$, then the above estimate becomes $e^{O(\sqrt{k})}$.

4.1   Sketch of the Proof of Lemma 7

Let ℓ' be the number of maximal ab-words in w and consider an ab-expansion of w determined by a vector $s = (s_1, \ldots, s_{\ell'})$. In this expansion the ith maximal ab-word, together with the letters surrounding it, is either an s_i-winding or an $(s_i + 1)$-winding around v. By Lemma 4 such an ab-expansion has at least

$$\sum_i s_i + 2 \sum_{i<i'} \min\{s_i, s_{i'}\} \tag{7}$$

self-intersections.

Our goal is to give an upper bound, in terms of k and ℓ, on the number of vectors s of length ℓ' such that (7) is smaller than k. Since $\ell' \leq \ell$ and the number of such vectors is clearly largest for $\ell' = \ell$, let us further assume that $\ell' = \ell$.

For $i = 0, \ldots, k$, let $m_i = m_i(s)$ denote the multiplicity of i in s, formally,

$$m_i = m_i(s) := |\{j \in [\ell] : s_j = i\}|.$$

Given an integer $\alpha \geq 0$, let us write $m_{\geq \alpha} := \sum_{i \geq \alpha} m_i$ and note that

$$m_{\geq 0} = \ell. \tag{8}$$

Moreover

$$m_{\geq \alpha} \leq \sqrt{k/\alpha}, \qquad \alpha = 1, 2, \ldots, k, \tag{9}$$

since otherwise, noting that $m_{\geq \alpha} = \{j : s_j \geq \alpha\}$, by (7) the self-intersection number is at least

$$m_{\geq \alpha}\alpha + 2\binom{m_{\geq \alpha}}{2}\alpha = m_{\geq \alpha}^2 \alpha > k,$$

giving a contradiction.

An upper bound on the number of vectors s can be obtained by bounding (i) the number of vectors s giving the same vector $m = (m_0, \ldots, m_k)$ and (ii) the number of distinct vectors $m = (m_0, \ldots, m_k)$ with nonnegative integer coordinates satisfying the constraints (8) and (9). The product of the two obtained bounds is then an upper bound on the number of vectors s. The two bounds are obtained by combinatorial methods in the full version of this paper [2], and they are $(\ell/\sqrt{k})^{\sqrt{k}} e^{O(\sqrt{k})}$ for (i) and $e^{O(k^{1/3}(\ln \ell + \ln k))}$ for (ii). Their product is of order $(\ell/\sqrt{k})^{O(\sqrt{k})} e^{O(\sqrt{k})}$, which gives Lemma 7.

5   Proof of Theorem 1

Recall that without loss of generality we assume $v = v_1$ so that the gaps adjacent to v are 0 and 1. By Lemma 1 we can assume that every v-loop in the collection induces an $\alpha\alpha$-free word so that the first and the last letters are 2. By Lemma 2 (iii), the v-loops induce different words, so it is enough to show that the number of words of the such form with self-intersection numbers less than k is $e^{O(\sqrt{k})}$.

Recall that the inner words use letters in $\{0, 1, 2\}$. Whenever we talk about two distinct letters a, b, let c refer to the remaining third letter. The maximal ab-words are disjoint and each of them is surrounded by c or v. We can replace w by an $\alpha\beta\alpha\beta$-free word w' by repeatedly applying an operation which replaces a subword of a form $abab$, for some distinct letters a and b, by a subword ab. Note that this does not change the structure of maximal words and when the procedure terminates, every maximal word is a $\alpha\beta$-word or a $\alpha\beta\alpha$-word. Moreover, w' remains $\alpha\alpha$-free and the first and last letters remain 2 in each intermediate word.

As discussed in Subsect. 4, word w can be reconstructed from an $\alpha\beta\alpha\beta$-free word w' by three consecutive ab-expansions, one for each $ab \in \{01, 02, 12\}$. Note that we can also assume that, as in w, the first and the last letters in both w' and the intermediate expansions are 2.

In view of this claim, it is enough to count $\alpha\beta\alpha\beta$-free words with fewer than k self-intersections, calculate the bound ℓ on the number of maximal ab-words in such a word, and apply Lemma 7 three times.

We claim that an $\alpha\alpha$-free word with fewer than k self-intersections has at most $4\sqrt{k}$ maximal ab-words for each pair ab. Assume the contrary. Let v be the obstacle incident to gaps a and b. If at least $\sqrt{k}$ of the maximal ab-words are s-windings around v with $s \geq 1$, then there are at least k self-intersections by (7). So further we assume there are at least $3\sqrt{k}$ maximal ab-words of the form $cabc$ or $cbac$. Note that each such word corresponds either to a cab-upsegment or a cab-downsegment. If among them there are at least $\sqrt{k}$ of each polarity, then we again have k self-intersections between cab-upsegments and cab-downsegments by Lemma 3. So further assume there are more than $2\sqrt{k}$ words $cabc$ of the same polarity (note that reversing $cabc$ does not change the polarity). In particular there is a pair of such maximal ab-words which has no other maximal ab word between them. Since their polarity is the same, between them there is an even-length word which starts and ends in a letter other than c. We claim that such subword contains ab, giving a contradiction. Say the word has $2m$ letters and argue by induction: if $m = 1$, the first and last letter is not c, so the word is ab or ba. Otherwise either the first two letters are ab or ba (in which case we're done) or one of ac and bc, in which case removing them we are left with a shorter word of even number of letters starting and ending not in c, which by induction hypothesis contains ab.

We have shown that every word w with fewer than k self-intersections is a 'triple' extension of a $\alpha\beta\alpha\beta$-free word w' with at most $\ell := 4\sqrt{k}$ maximal ab-words for each $ab \in \{01, 02, 12\}$. Since the first maximal word in w' has at most three letters, and each subsequent maximal word has at most two additional letters, w' has at most $3 + 2(3\ell - 1) = 6\ell + 1$ letters. Taking into account that w' is $\alpha\alpha$-free and necessarily starts with 2, there are at most

$$\sum_{i=2}^{6\ell+1} 2^{i-1} \leq 2^{6\ell} = e^{O(\sqrt{k})} \tag{10}$$

choices of w'. Now Lemma 7 implies that the number of ab-expansions is $e^{O(\sqrt{k})}$, which, together with (10) implies that the number of different words induced by the v-loops is $e^{O(\sqrt{k})}$. By the remark at the beginning of the proof, this proves the theorem.   □

6   Proof of Proposition 1

For the detailed proof, see the full version of this paper [2].

Proof Sketch. Let $f = f(n, k)$ and choose a family of non-homotopic x-loops $\ell_1, \ldots, \ell_f$ in S of the maximal size. We choose an obstacle v, a point x' on some loop, and a path P from v to x' that does not intersect any loop. We assume without loss of generality that x' lies on the loop ℓ_f.

We first turn the x-loops into non-homotopic x'-loops while keeping the number of pairwise intersections and self-intersections bounded. We choose a path R with no self-intersections that connects x to x' and is contained in the graph of ℓ_f. The x-loop ℓ_f is already an x'-loop and we turn every other x-loop ℓ_i into an x'-loop ℓ_i' by (i) following R from x' to x, (ii) going along ℓ_i, and (iii) returning back to x' via R. Note that to avoid an infinite number of (self-)intersections, the loops cannot follow R precisely. Rather, they use pairwise disjoint paths that run along R in sufficiently small distance so that if any loop intersects them it must also intersect R (and thus ℓ_f). See the left and middle parts of Fig. 3.

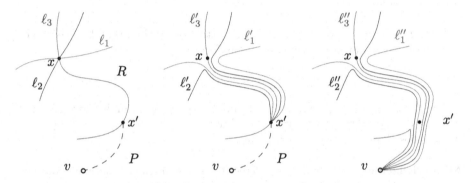

Fig. 3. Transforming x-loops into x'-loops (note that for simplicity of the example we assume $f = 3$, so $\ell_3 = \ell_f$ is unchanged) and then transforming x'-loops into v-loops.

It is easy to see that the obtained x'-loops are pairwise non-homotopic. And since R is a subset of the loop ℓ_f, every newly created crossing between a pair of loops ℓ_i' and ℓ_j' corresponds to some crossing between ℓ_i and ℓ_f, or ℓ_j and ℓ_f. This fact allows us to bound the total number of additional (self-)intersections by $4k$.

Now we choose the equator so that it does not cross the path P (see Fig. 4). Applying Lemma 1.(ii) we modify the x'-loops without increasing the numbers of intersections so that they induce $\alpha\alpha$-free words.

Finally we turn each x'-loop ℓ_i' into a v-loop ℓ_i'' so that no additional intersections are created, and the inner word that ℓ_i'' induces is the same as the word induced by ℓ_i'. This is done similarly as before – the loop ℓ_i'' is obtained by concatenating (i) a path closely following P from v to x', (ii) the x'-loop ℓ_i', and (iii) a path closely following P back to v. See the right part of Fig. 3.

Some resulting v-loops may be homotopic. Partition the v-loops into maximal sets of homotopic v-loops $H_1, \ldots, H_m$ and note that $m \le g(n, 5k)$. Since the first and the last arc of every v-loop lies in the same hemisphere, by Lemma 2 (ii) for

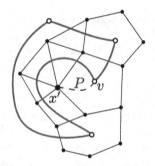

Fig. 4. Drawing the equator (the thicker line) so that it does not separate x' from the obstacle v. Path P is dashed.

each set H_j there is a word u_j starting in letters other than 0 or 1 so that each $\ell \in H_j$ induces an even-length word of the form

$$vw'_\ell u_j w''_\ell v,$$

where words w'_ℓ and w''_ℓ use letters 0 and 1. Applying Lemma 6, we see that $|H_j| \leq 4(2 \cdot 5k + 1)^2 \leq 4(11k)^2 = 484k^2$ for every j. And since $m \leq g(n, 5k)$, this implies that $f(n, k) \leq 484k^2 g(n, 5k)$.

Acknowledgement. We thank the referees for a careful reading and numerous corrections, in particular, one of the referees for pointing out the upper bound by Juvan, Malnič and Mohar.

References

1. Ajtai, M., Chvátal, V., Newborn, M.M., Szemerédi, E.: Crossing-free subgraphs. In: Theory and Practice of Combinatorics, North-Holland Mathematics Studies, vol. 60, pp. 9–12. North-Holland, Amsterdam (1982)
2. Blažej, V., Opler, M., Šileikis, M., Valtr, P.: Non-homotopic loops with a bounded number of pairwise intersections. arXiv:2108.13953v1 (2021)
3. Blažej, V., Opler, M., Šileikis, M., Valtr, P.: On the intersections of non-homotopic loops. In: Mudgal, A., Subramanian, C.R. (eds.) CALDAM 2021. LNCS, vol. 12601, pp. 196–205. Springer, Cham (2021). https://doi.org/10.1007/978-3-030-67899-9_15
4. Juvan, M., Malnič, A., Mohar, B.: Systems of curves on surfaces. J. Combin. Theory Ser. B **68**(1), 7–22 (1996). https://doi.org/10.1006/jctb.1996.0053
5. Leighton, F.T.: Complexity Issues in VLSI. Foundations of Computing. MIT Press, Cambridge (1983)
6. Pach, J., Tardos, G., Tóth, G.: Crossings between non-homotopic edges. In: GD 2020. LNCS, vol. 12590, pp. 359–371. Springer, Cham (2020). https://doi.org/10.1007/978-3-030-68766-3_28

On the Number of Edges of Separated Multigraphs

Jacob Fox[1], János Pach[2], and Andrew Suk[3(✉)]

[1] Stanford University, Stanford, CA, USA
jacobfox@stanford.edu
[2] Rényi Institute, Budapest and MIPT, Moscow, Russia
pach@cims.nyu.edu
[3] University of California at San Diego, La Jolla, CA, USA
asuk@ucsd.edu

Abstract. We prove that the number of edges of a multigraph G with n vertices is at most $O(n^2 \log n)$, provided that any two edges cross at most once, parallel edges are noncrossing, and the lens enclosed by every pair of parallel edges in G contains at least one vertex. As a consequence, we prove the following extension of the Crossing Lemma of Ajtai, Chvátal, Newborn, Szemerédi and Leighton, if G has $e \geq 4n$ edges, in any drawing of G with the above property, the number of crossings is $\Omega\left(\frac{e^3}{n^2 \log(e/n)}\right)$. This answers a question of Kaufmann *et al.* and is tight up to the logarithmic factor.

Keywords: Multigraphs · Lenses · Crossing lemma

1 Introduction

A *topological graph* is a graph drawn in the plane such that its vertices are represented by points, and the edges are represented by simple continuous arcs connecting the corresponding pairs of points. In notation and terminology, we do not distinguish between the vertices and the points representing them and the edges and the arcs representing them. The edges are allowed to intersect, but they cannot pass through any vertex other than their endpoints. If two edges share an interior point, then they must properly *cross* at that point, *i.e.*, one edge passes from one side of the other edge to its other side.

A *multigraph* is a graph in which two vertices can be joined by several edges. Two edges that join the same pair of vertices are called *parallel*.

J. Fox—Supported by a Packard Fellowship and by NSF award DMS-1855635.
J. Pach—Supported by NKFIH grants K-176529, KKP-133864, Austrian Science Fund Z 342-N31, Ministry of Education and Science of the Russian Federation MegaGrant No. 075-15-2019-1926, ERC Advanced Grant "GeoScape."
A. Suk—Supported by an NSF CAREER award, NSF award DMS-1952786, and an Alfred Sloan Fellowship.

© Springer Nature Switzerland AG 2021
H. C. Purchase and I. Rutter (Eds.): GD 2021, LNCS 12868, pp. 223–227, 2021.
https://doi.org/10.1007/978-3-030-92931-2_16

According to the *crossing lemma* of Ajtai, Chvátal, Newborn, Szemerédi [1] and Leighton [4], every topological graph G with n vertices and $e > 4n$ edges has at least $c\frac{e^3}{n^2}$ edge crossings, where $c > 0$ is an absolute constant. In notation, we have

$$\mathrm{cr}(G) \geq c \cdot \frac{e^3}{n^2}. \tag{1}$$

In a seminal paper which was an important step towards the solution of Erdős's famous problem on distinct distances [2], Székely [9] generalized the crossing lemma to multigraphs: for every topological multigraph G with n vertices and $e > 4n$ edges, in which the *multiplicity* of every edge is at most m, we have

$$\mathrm{cr}(G) \geq c\frac{e^3}{mn^2}. \tag{2}$$

As the maximum multiplicity m increases, (2) gets weaker. However, as was shown in [7] and [3], under certain special conditions on the multigraphs, the inequality (1) remains true, independently of m. Some related results were established in [6]. In all of these papers, one of the key elements of the argument was to find an analogue of Euler's theorem for the corresponding classes of "nearly planar" multigraphs.

Throughout this paper, we consider only *single-crossing* topological multigraphs, *i.e.*, we assume that any two edges cross *at most once*. Hence, two edges that share an endpoint may also have a common interior point. Two edges are said to be *independent* if they do not share an endpoint, and they are called *disjoint* if they are independent and do not cross.

Definition 1. *A multigraph G is called* separated *if no two parallel edges of G cross, and the "lens" enclosed by them has at least one vertex in its interior.*

It was conjectured in [3] that any separated single-crossing topological multigraph with n vertices has at most $O(n^2)$ edges. The aim of this note is to verify this conjecture apart from a logarithmic factor.

Theorem 1. *The number of edges of a separated single-crossing topological multigraph G on n vertices satisfies $|E(G)| \leq O(n^2 \log n)$.*

Note that in a separated multigraph, any pair of vertices can be connected by at most $n-1$ edges. This immediately implies the bound $|E(G)| \leq \binom{n}{2}(n-1) = O(n^3)$.

If we plug in Theorem 1 into the machinery of [3] and [7], a routine calculation gives the following.

Corollary 1. *Every separated single-crossing topological multigraph on n vertices and $e \geq 4n$ edges has at least $c\frac{e^3}{n^2 \log(e/n)}$ crossings, where $c > 0$ is a suitable constant.*

For simplicity, we will assume that a multigraph does not have *loops*. It is easy to see that Theorem 1 also holds for topological multigraphs with loops, assuming that each loop contains a vertex.

Theorem 1 does not remain true if we replace the assumption that G is *single-crossing* by the weaker one that any two edges cross at most *twice*. To see this, let the vertices of G lie on the x-axis: set $V(G) = \{1, 2, \ldots, n\}$. Let each edge consist of a semicircle in the upper half-plane and a semicircle below it that meet at a point of the x-axis. More precisely, for any pair of integers $i, j \in V(G)$ with $i < j$, and for any k with $i \leq k < j$, pick a distinct point p_{ikj} in the open interval $(k, k+1)$. Let γ_{ikj} be the union of two semicircles centered at the x-axis: an upper semicircle connecting i to p_{ikj} and a lower one connecting p_{ikj} to j. Let $E(G)$ consist of all arcs γ_{ikj} over all triples $i \leq k < j$. Observe that any two edges of G cross at most twice: once above the x-axis and once below it. No two parallel edges, γ_{ihj} and γ_{ikj} with $h < k$, cross each other, and the region enclosed by them contains the vertex $k \in V(G)$. Therefore, G is a separated topological multigraph with $\sum_{i,j(i<j)} (j - i) = \Omega(n^3)$.

The proof of Theorem 1 is presented in the next section.

All logarithms used in the sequel are of base 2. We omit all floor and ceiling signs wherever they are not crucially important.

2   Proof of Theorem 1

We will need the following simple lemma.

Lemma 1. *Let G be a single-crossing topological graph on n vertices with no parallel edges, in which every pair of independent edges cross. Then we have $|E(G)| \leq 4n$.*

Proof. Let $V(G) = A \cup B$ be a bipartition of the vertex set such that at least half of the edges of G run between A and B. Denote the corresponding bipartite graph by $G(A, B)$. Any pair of independent edges of $G(A, B)$ cross once, that is, an *odd* number of times. Assume without loss of generality that A and B are separated by a horizontal line. By "flipping" one of the half-planes bounded by this line from left to right, we obtain a drawing of $G(A, B)$, in which any pair of independent edges cross an *even* number of times. According to the Hanani-Tutte theorem [8,10], this implies that $G(A, B)$ is a planar graph. Any bipartite planar graph on $n \geq 3$ vertices has at most $2n - 4$ edges. Therefore, $|E(G)| \leq 2|E(G(A, B))| \leq 4n - 8$.

Proof (Proof of Theorem 1). Let $G = (V, E)$ be a separated single-crossing topological multigraph on n vertices. If two vertices, u and v, are joined by $j > 1$ parallel edges, then they cut the plane into j pieces, one of which is unbounded. The bounded pieces are called *lenses*. Each lens is bounded by two adjacent edges joining a pair of vertices. Let L denote the set of lenses determined by G.

If $|L| \leq \frac{|E(G)|}{2}$, then keeping only one edge between every pair of adjacent vertices, we obtain a simple graph G' whose number of edges satisfies

$$\frac{|E(G)|}{2} \leq |E(G')| \leq \binom{n}{2}.$$

This implies that $|E(G)| < n^2$, and we are done.

From now on, we can and will assume that

$$|L| \geq \frac{|E(G)|}{2}. \tag{3}$$

For any lens $\ell \in L$, let $|\ell|$ denote the number of vertices in the interior of ℓ. For $t = \log n$, we partition L into t parts, $L_1 \cup L_2 \cup \cdots \cup L_t$, where $\ell \in L_i$ if and only of $2^{i-1} \leq |\ell| < 2^i$. By the pigeonhole principle, there is an integer $k, 1 \leq k \leq t$, such that

$$|L_k| \geq \frac{|L|}{\log n}. \tag{4}$$

Fix an integer k with the above property, and let $d_k(v)$ denote the number of lenses in L_k that contain vertex v in its interior. Then we have

$$\sum_{v \in V} d_k(v) = \sum_{\ell \in L_k} |\ell| \geq |L_k| 2^{k-1}.$$

Hence, there is a vertex $v \in V$ that lies in the interior of at least $|L_k|\frac{2^{k-1}}{n}$ lenses from L_k. Assume without loss of generality that v is located at the origin o, and let L^o denote the set of lenses in L_k which contain the origin. Hence, we have

$$|L^o| \geq |L_k| \cdot \frac{2^{k-1}}{n}.$$

Combining this with (3) and (4), we obtain

$$|L^o| \geq \frac{|E(G)|}{n \log n} \cdot 2^{k-2}. \tag{5}$$

Let G^o denote the subgraph of G consisting of all vertices and the edges that bound a lens in L^o. Any two vertices of G^o are connected by 0 or 2 edges of G^o.

Now we use the idea of the probabilistic proof of the crossing lemma; see [5]. Let W be a random subset of V in which each vertex is picked independently with probability $p = 2^{-k}$. Let $G^o[W]$ be the subgraph of G^o induced by W. Let $L^o(W)$ denote the set of *empty* lenses in $G^o[W]$ (that is, the set of lenses with empty interiors). For the expected values of $|W|$ and $|L^o(W)|$, we have

$$\mathbb{E}[|W|] = pn$$

and

$$\mathbb{E}[|L^o(W)|] \geq p^2(1-p)^{2^k}|L^o|.$$

By linearity of expectation, there is a subset W' of V such that

$$|L^o(W')| - 4W'| \geq \mathbb{E}[|L^o(W)|] - 4\mathbb{E}[|W|] \geq p^2(1-p)^{2^k}|L^o| - 4pn. \tag{6}$$

For each lens in $\ell \in L^o(W')$, we arbitrarily pick one of the two edges enclosing ℓ, and denote the resulting simple topological graph by G'. We now make the following observation.

Lemma 2. *Any two independent edges of G' cross each other.*

Proof. Suppose, for contradiction, that G' has two independent edges, e and e', which do not cross. Let ℓ and ℓ' be the corresponding empty lenses in $G^o[W']$. Since the interiors of ℓ and ℓ' are empty, neither of them can contain an endpoint of the other. Both of these lenses contain the origin o, which implies that they cannot be disjoint. Therefore, both sides of ℓ must cross both sides of ℓ', contradicting the choice of e and e'. Here, we used the assumption that G and, hence, G' are single-crossing.

In view of Lemma 2, we can apply Lemma 1 to G'. We obtain $|E(G')| = |L^o(W')| \leq 4|W'|$ and hence by (6) we have $p^2(1-p)^{2^k}|L^o| \leq 4pn$. It follows that

$$|L^o| \leq 4p^{-1}(1-p)^{-2^k}n.$$

Substituting $p = 2^{-k}$, we get

$$|L^o| \leq 16 \cdot 2^k n.$$

Comparing this with (5), we conclude that

$$|E(G)| \leq O(n^2 \log n).$$

This completes the proof of Theorem 1.

References

1. Ajtai, M., Chvátal, V., Newborn, M., Szemerédi, E.: Crossing-free subgraphs. In: North-Holland Mathematics Studies, vol. 60, pp. 9–12 (1982)
2. Guth, L., Katz, N.H.: On the erdős distinct distances problem in the plane. Ann. Math. **181**, 155–190 (2015)
3. Kaufmann, M., Pach, J., Tóth, G., Ueckerdt, T.: The number of crossings in multigraphs with no empty lens. In: Biedl, T., Kerren, A. (eds.) GD 2018. LNCS, vol. 11282, pp. 242–254. Springer, Cham (2018). https://doi.org/10.1007/978-3-030-04414-5_17
4. Leighton, F.T.: Complexity Issues in VLSI: Optimal Layouts for the Shuffle-Exchange Graph and Other Networks. MIT Press, Cambridge (1983)
5. Matousek, J.: Lectures on Discrete Geometry. Springer, Heidelberg (2002). https://doi.org/10.1007/978-1-4613-0039-7
6. Pach, J., Tardos, G., Tóth, G.: Crossings between non-homotopic edges. In: GD 2020. LNCS, vol. 12590, pp. 359–371. Springer, Cham (2020). https://doi.org/10.1007/978-3-030-68766-3_28
7. Pach, J., Tóth, G.: A crossing lemma for multigraphs. Discrete Comput. Geom. **63**(4), 918–933 (2020). https://doi.org/10.1007/s00454-018-00052-z
8. Schaefer, M.: Toward a theory of planarity: Hanani-Tutte and planarity variants. J. Graph Algorithms Appl. **17**, 367–440 (2013)
9. Székely, L.A.: Crossing numbers and hard erdős problems in discrete geometry. Comb. Probab. Comput. **6**, 353–358 (1997)
10. Tutte, W.T.: Toward a theory of crossing numbers. J. Comb. Theory Ser. A **8**, 45–53 (1970)

Linear Layouts

On the Queue-Number of Partial Orders

Stefan Felsner[1], Torsten Ueckerdt[2(✉)], and Kaja Wille[1]

[1] Institut für Mathematik, Technische Universität Berlin, Berlin, Germany
felsner@math.tu-berlin.de, wille@campus.tu-berlin.de
[2] Institute of Theoretical Informatics, Karlsruhe Institute of Technology,
Karlsruhe, Germany
torsten.ueckerdt@kit.edu

Abstract. The queue-number of a poset is the queue-number of its cover graph viewed as a directed acyclic graph, i.e., when the vertex order must be a linear extension of the poset. Heath and Pemmaraju conjectured that every poset of width w has queue-number at most w. Recently, Alam et al. constructed posets of width w with queue-number $w + 1$. Our contribution is a construction of posets with width w with queue-number $\Omega(w^2)$. This asymptotically matches the known upper bound.

Keywords: Poset · Queue-number · Width · Lower bounds

1 Introduction

A *queue layout* of a graph consists of a total ordering on its vertices and a partition of its edge set into queues, i.e., no two edges in a single block of the partition are nested. The minimum number of queues needed in a queue layout of a graph G is its queue-number and denoted by $\mathrm{qn}(G)$.

To be more precise, let G be a graph and let L be a linear order of the vertices. A *k-rainbow* is a set of k edges $\{a_i b_i : 1 \leq i \leq k\}$ such that $a_1 < a_2 < \cdots < a_k < b_k < \cdots < b_2 < b_1$ in L. A pair of edges forming a 2-rainbow is said to be *nested*. A *queue* is a set of edges without nesting. Given G and L, the edges of G can be partitioned into k queues if and only if there is no rainbow of size $k+1$ in L. The *queue-number* of G is the minimum number of queues needed to partition the edges of G over all linear orders L.

Queue layouts were introduced by Heath and Rosenberg in 1992 [6] as a counterpart of book embeddings. Queue layouts were implicitly used before and have applications in fault-tolerant processing, sorting with parallel queues, matrix computations, scheduling parallel processes, and in communication management in distributed algorithm (see [4,6,8]). There is a rich literature exploring bounds on the queue-number of different classes of graphs [2,4,6,9].

Here we study the queue-number of posets. This parameter was introduced in 1997 by Heath and Pemmaraju [5], inspired by the older concept of the queue-number of directed acyclic graphs. For a queue layout of a directed acyclic graph,

© Springer Nature Switzerland AG 2021
H. C. Purchase and I. Rutter (Eds.): GD 2021, LNCS 12868, pp. 231–241, 2021.
https://doi.org/10.1007/978-3-030-92931-2_17

it is required that a precedes b in the total vertex ordering whenever there is a directed edge $a \to b$. I.e., it is a topological ordering of the graph.

A *poset* is a pair $P = (X, <)$ of a finite set X of elements, called the ground set, and a transitive (if $a < b$ and $b < c$, then $a < c$) and antisymmetric (if $a < b$, then $b \not< a$) binary relation $<$ on X. Two elements a, b are called comparable if either $a < b$ or $b < a$, and incomparable otherwise. A relation $a < b$ in P is a *cover* if it is not implied by transitivity, i.e., there is no element c such that $a < c < b$. In the context of drawings, embeddings and layouts for posets $P = (X, <)$, it is natural to work with their *directed cover graphs*, having vertex set X and a directed edge $a \to b$ for every cover relation $a < b$ in P. For example a *diagram* of P is an upward drawing of the directed cover graph where the direction on edges is usually omitted as each edge is implicitly directed upwards.

Now, a *linear extension* L of P is simply a topological ordering of its directed cover graph, and we write $a < b$ in L if a precedes b in L (though not necessarily in P). The *queue-number* of P, denoted by $\mathrm{qn}(P)$, is the smallest k such that there is a linear extension L of P for which the resulting linear layout of the directed cover graph contains no $(k+1)$-rainbow. Figure 1 shows an example.

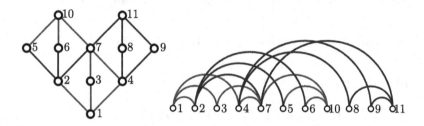

Fig. 1. A poset of width 5 and a queue layout with 2 queues indicated by colors.

Clearly, if G_P denotes the undirected cover graph of P, then $\mathrm{qn}(G_P) \le \mathrm{qn}(P)$, i.e., the queue-number of a poset is at least as large as the queue-number of its (undirected) cover graph. It was shown by Heath and Pemmaraju [5] that even for planar posets P there is no function f such that $\mathrm{qn}(P) \le f(\mathrm{qn}(G_P))$. They also investigated the maximum queue-number of several classes of posets, in particular with respect to bounded width (the maximum number of pairwise incomparable elements) and height (the maximum number of pairwise comparable elements). In particular they gave a nice argument showing that $\mathrm{qn}(P) \le \mathrm{width}(P)^2$ (see Proposition 2 below). The poset P of height 2 and width w whose cover graph is the complete bipartite graph $K_{w,w}$ attains $\mathrm{qn}(P) = \mathrm{width}(P)$. Actually, Heath and Pemmaraju conjectured that $\mathrm{qn}(P) \le \mathrm{width}(P)$ for every poset P.

Knauer, Micek and the second author [7] showed that the inequality $\mathrm{qn}(P) \le \mathrm{width}(P)$ holds for all posets of width 2. Last year, Alam et al. [1] constructed a non-planar poset P_3 of width 3 whose queue-number is 4; thus refuting the conjecture of Heath and Pemmaraju. Using a simple lifting argument from [7],

Alam et al. generalized their example and constructed for every $w > 3$ a poset P_w with $\text{width}(P_w) = w$ and $\text{qn}(P_w) = w + 1$. Figure 2 shows their construction. In fact, consider the lifting construction in the middle of Fig. 2 and a fixed linear extension L. If $a < b$ in L, then the cover edge from the bottommost element to b nests above the lower copy of P_{w-1}. Symmetrically, if $b < a$ in L, the cover edge from b to the topmost element nests above the upper copy of P_{w-1}. In any case, we extend any rainbow in P_{w-1} by one edge. Similarly, in the right of Fig. 2 one of the diagonal cover edges will nest above one of the copies of P_{w-1} in any linear extension.

Let us also mention that a second contribution of Alam et al. consists in a slight improvement of the upper bound: They show $\text{qn}(P) \leq (w - 1)^2 + 1$ for all posets P of width at most w.

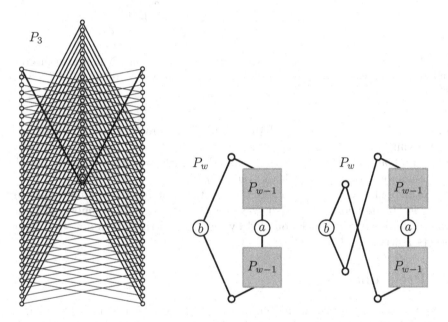

Fig. 2. Left: The construction of Alam et al. of a poset P_3 of width 3 and queue-number 4. Middle and right: Two possibilities of lifting a poset P_{w-1} s.t. $\text{width}(P_w) = \text{width}(P_{w-1}) + 1$ and $\text{qn}(P_w) \geq \text{qn}(P_{w-1}) + 1$.

Our contribution is the following theorem.

Theorem 1. *For every $w > 3$ there is a poset P_w of width w with*

$$\text{qn}(P_w) \geq w^2/8.$$

These examples (asymptotically) match the upper bound. Besides yielding a strong improvement of the lower bound, we also believe that the analysis of our construction is conceptually simpler than the example provided by Alam et

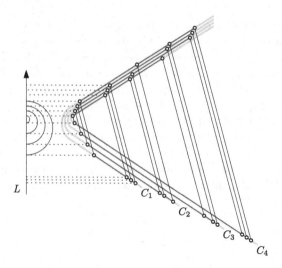

Fig. 3. A poset P of width $w = 4$ (A partition into 4 chains is indicated in grey.) and a linear extension L (ordering the elements by their y-coordinates) of P with a w^2-rainbow.

al. to disprove the conjecture of Heath and Pemmaraju. The key difference is that we improve the lifting step rather than the base case. In particular, we show how to lift any poset of width w so that the width goes up by only 2, but the queue-number goes up by at least $\lceil (w-1)/2 \rceil$.

As an open problem we promote the question whether the original conjecture holds for planar posets. In [7] it was shown that the queue-number of planar posets of width w is upper bounded by $3w - 2$ and that there are such planar posets P with $\mathrm{qn}(P) = \mathsf{width}(P) = w$.

2 Preliminaries

Before presenting our construction, we like to revisit the nice upper bound argument of Heath and Pemmaraju. Let $P = (X, <)$ be a poset of width w. Dilworth's Theorem asserts that X can be decomposed into w chains of P.

Proposition 2 (Heath and Pemmaraju). *For every poset P we have* $\mathrm{qn}(P) \le \mathsf{width}(P)^2$.

Proof. Let $w = \mathsf{width}(P)$, let $C_1, \ldots, C_w$ be a chain partition, and let L be any linear extension of P. Partition the cover edges into w^2 sets $Q_{i,j}$ with $i, j \in [w]$ such that $(u, v) \in Q_{i,j}$ if $u \in C_i$ and $v \in C_j$. We claim that each $Q_{i,j}$ is a queue.

Let $a < b < c < d$ in L support a pair of nesting cover edges and suppose that both edges (a, d) and (b, c) belong to $Q_{i,j}$. By definition $a, b \in C_i$ and $c, d \in C_j$ and from the ordering in L we get $a < b$ and $c < d$ in P. Now we have $a < b$ and $b < c$ and $c < d$ in P and thus the relation $a < d$ is implied by transitivity. This contradicts that (a, d) is a cover edge. □

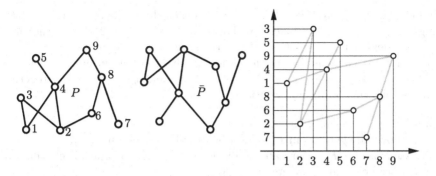

Fig. 4. A poset P, its dual $\bar{P}$, and a 2-dimensional drawing of P.

In fact we have shown a much stronger statement: If P and a chain partition $C_1, \ldots, C_w$ are given, then there is a partition of the edges of the cover graph of P into parts $Q_{i,j}$ with $i, j \in [w]$ such that each $Q_{i,j}$ is a queue for every(!) linear extension L of P. Let us remark that for some posets P and some linear extensions L of P, the resulting queue layout indeed has a $\mathsf{width}(P)^2$-rainbow. An example is indicated in Fig. 3.

2.1 Concepts Needed for the Construction

Let P be a poset. The *dual* of P, denoted $\bar{P}$, is the poset on the same ground set such that: $x < y$ in $P \iff y < x$ in $\bar{P}$. In terms of its diagram, the dual of P is obtained by flipping along a horizontal line.

A poset P is *2-dimensional* if and only if there are two linear extensions L_1 and L_2 such that: $x < y$ in $P \iff x < y$ in L_1 and L_2. Such a pair L_1, L_2 is called a *realizer* of P.

When drawing 2-dimensional posets, it is common to represent each element x by a point with coordinates (x_1, x_2) where x_1 is the position of x in L_1 and x_2 is the position of x in L_2, see Fig. 4. This is also called a *dominance drawing*.

3 Proof of Theorem 1

We define P_w recursively, focusing on the recursive step. As mentioned in the introduction, the recursive step involves lifting a given poset P_{w-2} of width $w-2$ to the desired poset P_w of width w such that $\mathsf{qn}(P_w) \geq \mathsf{qn}(P_{w-2}) + \lceil (w-1)/2 \rceil$. Our lifting can be seen as an extension of the situation on the very right of Fig. 2. Specifically, for $w \geq 3$, the construction of P_w is based on

- a copy of P_{w-2},
- a reinforcement poset R_{w-2} of width $w-2$,
- two linear extensions L_x and L_y of R_{w-2}, and
- the duals $\overline{P_{w-2}}, \overline{R_{w-2}}, \overline{L_x}, \overline{L_y}$ of the above.

We invite the reader to take a look at Fig. 5, which shows the construction of P_w using P_{w-2} and R_{w-2} as a black box. Formally, let $r = r(w-2)$ denote the number of elements in R_{w-2}. Then, P_w contains besides $P_{w-2}, R_{w-2}, \overline{P_{w-2}}, \overline{R_{w-2}}$, two additional elements a and b, and four chains of additional r elements $x_1 < \cdots < x_r, y_1 < \cdots < y_r, \overline{x_r} < \cdots < \overline{x_1}$, and $\overline{y_r} < \cdots < \overline{y_1}$, together with the following additional relations:

- b is below x_1, y_1 and above $\overline{x_1}, \overline{y_1}$.
- a is above all elements in P_{w-2} and below all elements in $\overline{P_{w-2}}$.
- All elements of P_{w-2} are above all elements of R_{w-2}, and all elements of $\overline{P_{w-2}}$ are below all elements of $\overline{R_{w-2}}$.
- x_i is above the i-th element in the linear extension L_x of R_{w-2}, $i = 1, \ldots, r$.
- $\overline{x_i}$ is below the i-th element in the dual $\overline{L_x}$ of $\overline{R_{w-2}}$.
- y_i is above the i-th element in the linear extension L_y of R_{w-2}, $i = 1, \ldots, r$.
- $\overline{y_i}$ is below the i-th element in the dual $\overline{L_y}$ of $\overline{R_{w-2}}$.
- All relations that are transitively implied by the above.

First we observe that $\mathsf{width}(P_w) = \mathsf{width}(P_{w-2}) + 2 = w$, as $\mathsf{width}(P_{w-2}) = \mathsf{width}(R_{w-2}) = w - 2$ and the additional elements (except a, which can be incorporated into an existing chain) can be covered by two chains. Also note that the number $p(w)$ of elements of the poset P_w is given by the recursion $p(w) = 2p(w-2) + 6r(w-2) + 2$. (Recall that $r(w-2)$ is the number of elements of R_{w-2}.) Further note that x_i and the i-th element of L_x in R_{w-2} indeed form a cover edge, as L_x is a linear extension of R_{w-2}, $i = 1, \ldots, r$. Similarly for the edges between R_{w-2} and y_i, as well as between $\overline{R_{w-2}}$ and $\overline{x_i}, \overline{y_i}$, $i = 1, \ldots, r$.

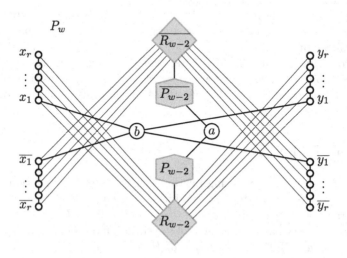

Fig. 5. Recursive construction of P_w.

Furthermore, it can be seen that P_w is self-dual; the reflection $P_w \leftrightarrow \overline{P_w}$ having two fixed points a and b. This shows that when analyzing $\mathrm{qn}(P_w)$, we can restrict the attention to linear extensions L of P_w which have a before b. With this assumption, a rainbow between R_{w-2} and either $X = \{x_1, \ldots, x_r\}$ or $Y = \{y_1, \ldots, y_r\}$ nests above every rainbow of P_{w-2}. See Fig. 6 for an illustration. If we let q_{w-2} be the size of a rainbow between R_{w-2} and either X or Y, then we have the recursion:

$$\mathrm{qn}(P_w) \geq \mathrm{qn}(P_{w-2}) + q_{w-2} \qquad (1)$$

We think of this use of a self-dual construction as the *symmetry trick*. Again, let us mention that constructions given in [7] (proof of Prop. 2) and [1] (proof of Thm. 4) also use a recursion based on two copies of the poset from the previous level of the recursion, as illustrated in the middle of Fig. 2. However, this only forces one edge to nest over the rainbow from the previous level of the recursion. Our lifting forces a rainbow of edges whose size is linear in the width to nest over the previous level construction and its rainbow; thus giving overall a quadratic lower bound.

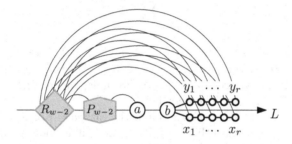

Fig. 6. The general structure of a linear extension L of P_w with a before b.

It remains to construct the poset R_{w-2} together with two linear extensions L_x and L_y such that in any linear extension L having R_{w-2} entirely before $X \cup Y$, a large rainbow between R_{w-2} and either X or Y appears. Recall that q_{w-2} denotes the largest such rainbow and we seek to construct R_{w-2} such that q_{w-2} is at least linear in w.

As the elements in X form a chain $x_1 < \cdots < x_r$ and thus are ordered in this way in L, rainbows between R_{w-2} and X are in bijection with subsets of elements in R_{w-2} that are *oppositely ordered* in L and L_x. Similarly, rainbows between R_{w-2} and Y appear when elements in R_{w-2} are oppositely ordered in L and L_y. Thus our goal is to construct R_{w-2}, L_x and L_y such that for every linear extension L of R_{w-2} there is a long increasing sequence in L which is decreasing in L_x or L_y.

To illustrate this idea, suppose that for each width $u < w$, we choose the poset R_u to be an antichain of size u and the linear extensions L_x and L_y to be a realizer (think of L_x as the identity permutation and of L_y as its reverse).

The Lemma of Erdős-Szekeres asserts that in every linear extension of R_u there is an increasing or a decreasing sequence of size at least $\lceil \sqrt{u} \rceil$, i.e., $q_u = \lceil \sqrt{u} \rceil$. This value of q_u together with Inequality (1) yields

$$\text{qn}(P_w) \geq \sum_{u<w;\ u\equiv w(2)} \lceil \sqrt{u} \rceil \quad \in \Theta(w^{3/2}).$$

For the proof of the theorem we need a better construction for the reinforcement posets R_u. In particular, we seek to have $q_u \geq \lceil \frac{u+1}{2} \rceil$ instead of just $q_u \geq \lceil \sqrt{u} \rceil$. A construction of such a R_u is given in Subsect. 3.1 and based on the following lemma.[1]

Lemma 3. *For each $u \geq 1$, there is a 2-dimensional poset R_u of width u with a realizer L_x, L_y, such that if L is a linear extension of R_u and d_x and d_y denote the maximum lengths of an increasing sequence in L which is decreasing in L_x and L_y respectively, then $d_x + d_y \geq u + 1$.*

The lemma says that we can assume the value $q_u = \lceil \frac{u+1}{2} \rceil$. With Inequality (1) we get:

$$\text{qn}(P_w) \geq \sum_{u<w;\ u\equiv w(2)} \left\lceil \frac{u+1}{2} \right\rceil$$

In the case w odd, $w = 2s+1$, we get $\text{qn}(P_w) \geq \sum_{k=1}^{s} k = \binom{s+1}{2}$. In the case w even, $w = 2s$, we get $\text{qn}(P_w) \geq \sum_{k=2}^{s} k = \binom{s+1}{2} - 1$. A simple computation shows that for $w \geq 4$ we get $\text{qn}(P_w) \geq w^2/8$, independent of the parity of w. This completes the proof of Theorem 1.

The base of our recursive construction is the case $w = 1$ or $w = 2$, depending on the parity of w. For the validity of Theorem 1, it is enough to let P_w with $w \in \{1,2\}$ be any poset of width w. Of course, it is beneficial to start with a higher queue-number, also given that our bound of $w^2/8$ is less than the $w+1$ of Alam et al. [1] for small w. The best results are achieved by starting at width 3 or 4 (depending on the parity of the target width w) with the poset of Alam et al. [1] with queue-number 4, respectively 5.

3.1 The Construction of R_u for Lemma 3

The construction of R_u is again recursive. Let R_1 be a single element. Then clearly $d_x + d_y = 2$. For the construction of R_u for $u \geq 2$ we again use the symmetry trick. We take two copies Q_1, Q_2 of R_u and two additional elements a and b. Then R_u is obtained by a series composition of $Q_1 + a + Q_2$, and a parallel composition of the result with element b. Formally,

– a is above every element of Q_1 and below every element of Q_2, while

---

[1] The lemma with a different proof was discovered (but not yet published) in October 2020 by the first and the second author together with Francois Dross, Piotr Micek, and Michał Pilipczuk.

– b is incomparable to all other elements.

The two linear extensions of the realizer L_x, L_y of R_u are obtained as follows.

– $L_x = b, L_x(Q_1), a, L_x(Q_2)$
– $L_y = L_y(Q_1), a, L_y(Q_2), b,$

where $L_x(Q_i), L_y(Q_i)$ is the realizer of the copy Q_i of R_{u-1}, $i = 1, 2$. We invite the reader to look at Fig. 7 for two illustrations of this recursive construction step for R_u and its realizer L_x, L_y.

First, we observe that $\mathsf{width}(R_u) = \mathsf{width}(R_{u-1}) + 1 = u$, as element b can be covered by a new chain and element a can be incorporated into an existing chain. Also note that the number $r(u)$ of elements in R_u is given by the recursion $r(u) = 2r(u-1) + 2$, which with $r(1) = 1$ solves for $r(u) = \frac{3}{2} \cdot 2^u - 2$. Further observe that R_u is again self-dual. In particular the two copies Q_1 and Q_2 of R_{u-1} are isomorphic. The reflection $R_u \leftrightarrow \overline{R_u}$ has two fixed points a and b.

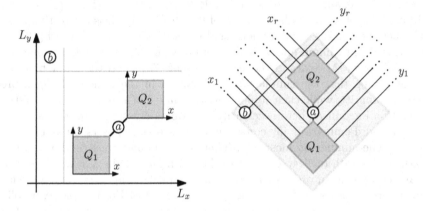

Fig. 7. The recursive construction of R_u with its realizer L_x, L_y.

Now let L be any linear extension of R_u. First suppose that $a < b$ in L. Let L' be the restriction of L to Q_1. By induction the lengths d'_x and d'_y of increasing sequences of L' which are decreasing in the two linear extensions of the realizer $L_x(Q_1), L_y(Q_1)$ of Q_1 satisfy $d'_x + d'_y \geq u$. Since b precedes Q_1 in L_x and comes after Q_1 in L, we have $d_x \geq d'_x + 1$. Together with the trivial $d_y \geq d'_y$, we get $d_x + d_y \geq u + 1$.

If we have $b < a$ in L, then we consider Q_2. As before we get the two values d'_x and d'_y for the restriction L' of L to Q_2 and know by induction that $d'_x + d'_y \geq u$. This time b precedes Q_2 in L but comes after Q_2 in L_y, which gives $d_y \geq d'_y + 1$. Together with the trivial $d_x \geq d'_x$ we again see that $d_x + d_y \geq u + 1$. This completes the proof of Lemma 3.

We remark that in both the construction of P_w and R_u, the element a is used only for the sake of the exposition. It would suffice to add all relations between P_{w-2} and $\overline{P_{w-2}}$, respectively Q_1 and Q_2. While this gives slightly smaller constructions for P_w and R_u, they would still be exponential in their width. (Recall that $r(u) = \frac{3}{2} \cdot 2^u - 2$ and hence $p(w) = 2p(w-2) + 6r(w-2) + 2 = \Theta(2^w)$.)

4   Conclusions

We have made substantial progress in the understanding of queue-numbers of partially ordered sets. We take the opportunity to list and comment on open questions in the field.

- An obvious question is to ask for improved upper and lower bounds. More precisely, we now know that the growth rate of the maximum queue-number of posets of width w is $(C + o(1))w^2$ for some constant C between $1/8$ and 1. What is the precise value of constant C?
- Our reinforcement poset R_u is 2-dimensional for every u. However our entire lower bound example P_w is not (already for $w = 3$), and the same holds for the example of Alam et al. in the left of Fig. 2. We think it is interesting to see whether there exists any 2-dimensional poset P with $\text{qn}(P) \geq \text{width}(P) + 1$.
- What is the maximum queue-number of posets of width w with a planar diagram? Knauer, Micek, and the second author [7] proved the lower bound w by observing that the simple lifting operation in the middle of Fig. 2 preserves planarity, while their upper bound is $3w - 2$. Clearly, the better lifting operation introduced here necessarily introduces crossing cover edges.
- Heath and Pemmaraju [5] conjectured that planar posets on n elements have queue-number at most $\sqrt{n}$. Their lower bound construction is an r-antichain R with realizer L_x, L_y together with an r-chain $X = x_1 < \cdots < x_r$ matched upward in order of L_x and an r-chain $Y = y_1 < \cdots < y_r$ matched downward in order of L_y; see Fig. 8. The Lemma of Erdős-Szekeres implies for this planar poset P with $n = 3r$ elements that $\text{qn}(P) \geq \lceil \sqrt{n/3} \rceil$. It is open whether there is an asymptotically matching upper bound.

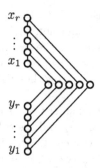

Fig. 8. Heath and Pemmaraju's construction [5] of a planar poset P on $n = 3r$ elements with $\text{qn}(P) \geq \lceil \sqrt{n/3} \rceil$.

- Dujmović and Wood [3] show that a random vertex ordering for an undirected graph G has with positive probability no rainbow of size $\lceil e\sqrt{m} \rceil$, where e is the base of the natural logarithm and m is the number of edges in G. Can a similar result be obtained by considering a random linear extension of a poset P? Note that a positive answer would resolve (up to a constant factor) the previous question of Heath and Pemmaraju about planar posets.
- In [7] is was shown that posets P of width 2 have $\text{qn}(P) \leq 2$. In [1] it was shown that posets P of width 3 may have $\text{qn}(P) \geq 4$ and satisfy $\text{qn}(P) \leq 5$. Is 4 or 5 the best upper bound in this case?

References

1. Alam, J.M., Bekos, M.A., Gronemann, M., Kaufmann, M., Pupyrev, S.: Lazy queue layouts of posets. In: GD 2020. LNCS, vol. 12590, pp. 55–68. Springer, Cham (2020). https://doi.org/10.1007/978-3-030-68766-3_5
2. Dujmović, V., Joret, G., Micek, P., Morin, P., Ueckerdt, T., Wood, D.R.: Planar graphs have bounded queue-number. J. ACM **67**, 22:1–22:38 (2020)
3. Dujmović, V., Wood, D.R.: On linear layouts of graphs. Discrete Math. Theor. Comput. Sci. **6**(2), 339–358 (2004)
4. Heath, L.S., Leighton, F.T., Rosenberg, A.L.: Comparing queues and stacks as machines for laying out graphs. SIAM J. Discrete Math. **5**(3), 398–412 (1992)
5. Heath, L.S., Pemmaraju, S.V.: Stack and queue layouts of posets. SIAM J. Discrete Math. **10**(4), 599–625 (1997)
6. Heath, L.S., Rosenberg, A.L.: Laying out graphs using queues. SIAM J. Comput. **21**(5), 927–958 (1992)
7. Knauer, K., Micek, P., Ueckerdt, T.: The queue-number of posets of bounded width or height. In: Biedl, T., Kerren, A. (eds.) GD 2018. LNCS, vol. 11282, pp. 200–212. Springer, Cham (2018). https://doi.org/10.1007/978-3-030-04414-5_14
8. Nešetřil, J., Ossona de Mendez, P., Wood, D.R.: Characterisations and examples of graph classes with bounded expansion. Eur. J. Combin. **33**(3), 350–373 (2012). https://doi.org/10.1016/j.ejc.2011.09.008
9. Wiechert, V.: On the queue-number of graphs with bounded tree-width. Electron. J. Comb. **24**(1), 1–65 (2017)

On the Upward Book Thickness Problem: Combinatorial and Complexity Results

Sujoy Bhore[1] , Giordano Da Lozzo[2(✉)] , Fabrizio Montecchiani[3] ,
and Martin Nöllenburg[4]

[1] Indian Institute of Science Education and Research, Bhopal, India
`sujoy@iiserb.ac.in`
[2] Roma Tre University, Rome, Italy
`giordano.dalozzo@uniroma3.it`
[3] Department of Engineering, University of Perugia, Perugia, Italy
`fabrizio.montecchiani@unipg.it`
[4] Algorithms and Complexity Group, TU Wien, Vienna, Austria
`noellenburg@ac.tuwien.ac.at`

Abstract. A long-standing conjecture by Heath, Pemmaraju, and Trenk states that the upward book thickness of outerplanar DAGs is bounded above by a constant. In this paper, we show that the conjecture holds for subfamilies of upward outerplanar graphs, namely those whose underlying graph is an internally-triangulated outerpath or a cactus, and those whose biconnected components are st-outerplanar graphs. On the complexity side, it is known that deciding whether a graph has upward book thickness k is NP-hard for any fixed $k \geq 3$. We show that the problem, for any $k \geq 5$, remains NP-hard for graphs whose domination number is $O(k)$, but it is FPT in the vertex cover number.

1 Introduction

A *k-page book embedding* (or *k-stack layout*) of an n-vertex graph $G = (V, E)$ is a pair $\langle \pi, \sigma \rangle$ consisting of a bijection $\pi \colon V \to \{1, \ldots, n\}$, defining a total order on V, and a page assignment $\sigma \colon E \to \{1, \ldots, k\}$, partitioning E into k subsets $E_i = \{e \in E \mid \sigma(e) = i\}$ $(i = 1, \ldots, k)$ called *pages* (or *stacks*) such that no two edges $uv, wx \in E$ mapped to the same page $\sigma(uv) = \sigma(wx)$ cross in the following sense. Assume, w.l.o.g., $\pi(u) < \pi(v)$ and $\pi(w) < \pi(x)$ as well as $\pi(u) < \pi(w)$. Then uv and wx *cross* if $\pi(u) < \pi(w) < \pi(v) < \pi(x)$, i.e., their endpoints interleave. The *book thickness* (or *stack number*) of G is the smallest k for which G admits a k-page book embedding. Book embeddings and book thickness of graphs are well-studied topics in graph drawing and graph theory [12,17,24,29]. For instance, it is NP-complete to decide for $k \geq 2$ if the book thickness of a graph is at most k [7,12] and it is known that planar graphs have book thickness at most 4 [30]; this bound has recently been shown tight [6]. More in general, the book thickness of graphs of genus g is $O(\sqrt{g})$ [25] and constant upper bounds are known for some families of non-planar graphs [4,5,18].

© Springer Nature Switzerland AG 2021
H. C. Purchase and I. Rutter (Eds.): GD 2021, LNCS 12868, pp. 242–256, 2021.
https://doi.org/10.1007/978-3-030-92931-2_18

Upward book embeddings (UBEs) are a natural extension of book embeddings to directed acyclic graphs (DAGs) with the additional requirement that the vertex order π respects the directions of all edges, i.e., $\pi(u) < \pi(v)$ for each $uv \in E$ (and hence G must be acyclic). Thus the ordering induced by π is a topological ordering of V. Book embeddings with different constraints on the vertex ordering have also been studied in [2,3,20]. Analogously to book embeddings, the *upward book thickness* (UBT) of a DAG G is defined as the smallest k for which G admits a k-page UBE. The notion of upward book embeddings is similar to upward planar drawings [14,16], i.e., crossing-free drawings, where additionally each directed edge uv must be a y-monotone curve from u to v. Upward book embeddings have been introduced by Heath et al. [22,23]. They showed that graphs with UBT 1 can be recognized in linear time, whereas Binucci et al. [10] proved that deciding the UBT of a graph is generally NP-complete, even for fixed values of $k \geq 3$. On the positive side, deciding if a graph admits a 2-page UBE can be solved in polynomial time for st-graphs of bounded treewidth [10].

Constant upper bounds on the UBT are known for some graph classes: directed trees have UBT 1 [23], unicyclic DAGs, series-parallel DAGs, and N-free upward planar DAGs have UBT 2 [1,15,23,26]. Frati et al. [19] studied UBEs of upward planar triangulations and gave several conditions under which they have constant UBT. Interestingly, upward planarity is a necessary condition to obtain constant UBT, as there is a family of planar, but non-upward planar, DAGs that require $\Omega(n)$ pages in any UBE [21]. Back in 1999, Heath et al. [23] conjectured that the UBT of outerplanar graphs is bounded by a constant, regardless of their upward planarity. Another long-standing open problem [28] is whether upward planar DAGs have constant UBT; in this respect, examples with a lower bound of 5 pages are known [27] and there is no known upper bound better than $O(n)$.

Contributions. In this paper, we contribute to the research on the upward book thickness problem from two different directions. We first report some notable progress towards the conjecture of Heath et al. [23]. We consider subfamilies of upward outerplanar graphs (see Sect. 2 for definitions), namely those whose underlying graph is an internally-triangulated outerpath or a cactus, and those whose biconnected components are st-outerplanar graphs, and provide constant upper bounds on their UBT (Sect. 3). Our proofs are constructive and give rise to polynomial-time book embedding algorithms. We then investigate the complexity of the problem (Sect. 4) and show that for any $k \geq 5$ it remains NP-complete for graphs whose domination number is in $O(k)$. On the positive side, we prove that the upward book thickness problem is fixed-parameter tractable in the vertex cover number. These two results narrow the gap between tractable and intractable parameterizations of the problem. Proofs of statements marked with a $(\star)$ have been sketched or omitted and can be found in [8].

2   Preliminaries

We assume familiarity with basic concepts in graph drawing (see also [13]).

BC-tree. The *BC-tree* of a connected graph G is the incidence graph between the (maximal) biconnected components of G, called *blocks*, and the cut-vertices of G. A block is *trivial* if it consists of a single edge, otherwise it is *non-trivial*.

Outerplanarity. An *outerplanar graph* is a graph that admits an *outerplanar drawing*, i.e., a planar drawing in which all vertices are on the outer face, which defines an *outerplanar embedding*. Unless otherwise specified, we will assume our graphs to have planar or outerplanar embeddings. An outerplanar graph G is *internally triangulated* if it is biconnected and all its inner faces are cycles of length 3. An edge of G is *outer* if it belongs to the outer face of G, and it is *inner* otherwise. A *cactus* is a connected outerplanar graph in which any two simple cycles have at most one vertex in common. Therefore, the blocks of a cactus graph are either single edges (and hence trivial) or cycles. The *weak dual* $\overline{G}$ of a planar graph G is the graph having a node for each inner face of G, and an edge between two nodes if and only if the two corresponding faces share an edge. For an outerplanar graph G, its weak dual $\overline{G}$ is a tree. If $\overline{G}$ is a path, then G is an *outerpath*. A *fan* is an internally-triangulated outerpath whose inner edges all share an end-vertex.

Directed Graphs. A *directed graph* $G = (V, E)$, or *digraph*, is a graph whose edges have an orientation. We assume each edge $e = uv$ of G to be oriented from u to v, and hence denote u and v as the *tail* and *head* of e, respectively. A vertex u of G is a *source* (resp. a *sink*) if it is the tail (resp. the head) of all its incident edges. If u is neither a source nor a sink of G, then it is *internal*. A *DAG* is a digraph that contains no directed cycle. An *st-DAG* is a DAG with a single source s and a single sink t; if needed, we may use different letters to denote s and t. A digraph is *upward (outer)planar*, if it has a (outer)planar drawing such that each edge is a y-monotone curve. Such a drawing (if any) defines an upward (outer)planar embedding. An upward planar digraph G (with an upward planar embedding) is always a DAG and it is *bimodal*, that is, the sets of incoming and outgoing edges at each vertex v of G are contiguous around v (see also [13]). The *underlying graph* of a digraph is the graph obtained by disregarding the edge orientations. An *st-outerplanar graph* (resp. *st-outerpath*) is an *st*-DAG whose underlying graph is outerplanar graph (resp. an outerpath). An *st-fan* is an *st*-DAG whose underlying graph is a fan and whose inner edges have s as an end-vertex.

Lemma 1 ($\star$). *Let G be an upward outerplanar graph and let c be a cut-vertex of G. Then there are at most two blocks of G for which c is internal.*

Basic Operations. Let π and π' be two orderings over vertex sets V and $V' \subseteq V$, respectively. Then π *extends* π' if for any two vertices $u, v \in V'$ with $\pi'(u) < \pi'(v)$, it holds $\pi(u) < \pi(v)$. We may denote an ordering π as a list

$\langle v_1, v_2, \ldots, v_{|V|} \rangle$ and use the *concatenation operator* $\circ$ to define an ordering from other lists, e.g., we may obtain the ordering $\pi = \langle v_1, v_2, v_3, v_4 \rangle$ as $\pi = \pi_1 \circ \pi_2$, where $\pi_1 = \langle v_1, v_2 \rangle$ and $\pi_2 = \langle v_3, v_4 \rangle$. Also, let π be an ordering over V and let $u \in V$. We denote by π_{u^-} and π_{u^+} the two orderings such that $\pi = \pi_{u^-} \circ \langle u \rangle \circ \pi_{u^+}$. Consider two orderings π over V and π' over V' with $V \cap V' = \{u, v\}$, and such that: (i) u and v are consecutive in π, and (ii) u and v are the first and the last vertex of π', respectively. The ordering π^* over $V \cup V'$ obtained by *merging* π and π' is $\pi^* = \pi_{u^-} \circ \pi' \circ \pi_{v^+}$. Note that π^* extends both π and π'.

3   Book Thickness of Outerplanar Graphs

In this section, we study the UBT of three families of upward outerplanar graphs. We begin with internally-triangulated upward outerpaths (Sect. 3.1), which are biconnected and may have multiple sources and sinks. We then continue with families of outerplanar graphs that are not biconnected but whose biconnected components have a simple structure, namely outerplanar graphs whose biconnected components are *st*-DAGs (Sect. 3.2), and cactus graphs (Sect. 3.3).

3.1   Internally-Triangulated Upward Outerpaths

In this subsection, we assume our graphs to be internally triangulated. We will exploit the following definition and lemmas for our constructions.

Definition 1. *Let G be an st-outerpath and uv be an outer edge different from st. An UBE $\langle \pi, \sigma \rangle$ of G is uv-consecutive if the following properties hold: (i) u and v are consecutive in π, (ii) the edges incident to s lie on one page, and (iii) the edges incident to t lie on at most two pages.*

An *st*-outerplanar graph is *one-sided* if the edge st is an outer edge.

Lemma 2. *Let G be a one-sided st-outerplanar graph. Then, G admits a 1-page UBE $\langle \pi, \sigma \rangle$ that is uv-consecutive for each outer edge $uv \neq st$.*

Proof. Consider the path of the outer face of G that encompasses all vertices of G and does not contain the edge st. Let $\langle \pi, \sigma \rangle$ be the 1-page UBE in which π is the ordering defined by such path. One easily verifies that no two edges cross and that the endpoints u, v of each outer edge $uv \neq st$ are consecutive in π.

Lemma 3. *Let G be an st-fan and let $uv \neq st$ be an outer edge of G. Then, G admits a 2-page uv-consecutive UBE.*

Proof. Let $P_\ell = \{s, a_1, \ldots, a_\ell, t\}$ and $P_r = \{s, b_1, \ldots, b_r, t\}$ be the left and right *st*-paths of the outer face of G, respectively. Then the edge uv belongs to either P_ℓ or P_r. We show how to construct an UBE $\langle \pi, \sigma \rangle$ of G that satisfies the requirements of the lemma, when uv belongs to P_ℓ (see Fig. 1); the construction when uv belongs to P_r is symmetric (it suffices to flip the embedding along st).

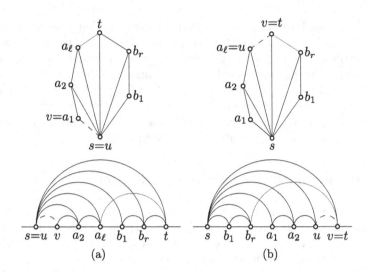

Fig. 1. Cases for the proof of Lemma 3. The edge uv is dashed. In all figures, the drawings are upward, hence the edge orientations are implied. For interpretation of the colors in this and the next figures, the reader is referred to the online coloured version of this article.

Since $uv \neq st$, we have either (a) $u = s$ and $v \neq t$, or (b) $v = t$ and $u \neq s$, or (c) $\{u, v\} \cap \{s, t\} = \emptyset$. In case (a), refer to Fig. 1(a). We set $\pi = \langle s, a_1, \dots, a_\ell, b_1, \dots, b_r, t \rangle$ (that is, we place P_ℓ before P_r), $\sigma(e) = 1$ for each edge $e \neq a_\ell t$, and $\sigma(a_\ell t) = 2$. In case (b), refer to Fig. 1(b). We set $\pi = \langle s, b_1, \dots, b_r, a_1, \dots, a_\ell, t \rangle$ (that is, we place P_ℓ after P_r), $\sigma(e) = 1$ for each edge $e \neq b_r t$, and $\sigma(b_r t) = 2$. In case (c), we can set π and σ as in any of case (a) and (b). In all three cases, all outer edges (including uv) are consecutive, except for one edge e incident to t; also, all edges (including those incident to s) are assigned to the same page, except for e, which is assigned to a second page.

The next definition allows us to split an st-outerpath into two simpler graphs. The *extreme faces* of an st-outerpath G are the two faces that correspond to the vertices of $\overline{G}$ having degree one.

Definition 2. *An st-outerpath G is* primary *if and only if the path forming $\overline{G}$ has one extreme face incident to s.*

Let G' be an st-outerpath and refer to Fig. 2(a). Consider the subgraph F_s of G' induced by s and its neighbors, note that this is an sw-fan. Assuming $F_s \neq G'$, and since G' is an outerpath, one or two edges on the outer face of F_s are separation pairs for G'. In the former case, it follows that G' is primary, since $\overline{G'}$ is a path having one face of F_s as its extreme face. In the latter case, since G' has a single source s and a single sink t, (at least) one of the two separation pairs splits G' into a one-sided uv-outerpath H_1 (for some vertices u, v of F_s) and into a primary st-outerpath G. We will call H_1, the *appendage at uv of G'*.

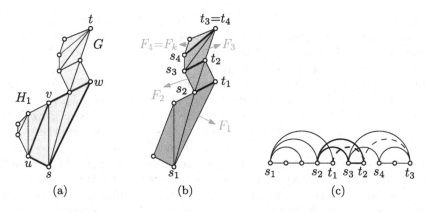

Fig. 2. (a) Decomposing G' into an appendage H_1 (light blue) and a primary st-outerpath G (light gray). (b) An st-fan decomposition of G with differently colored fans, and fat edges e_i. (c) A UBE of G for the proof of Lemma 6.

Let G be an st-outerpath (not necessarily primary). Consider a subgraph F of G that is an xy-fan (for some vertices x, y of G). Let $\langle f_1, \ldots, f_h \rangle$ be the ordered list of faces forming the path $\overline{G}$. Note that F is the subgraph of G formed by a subset of faces that are consecutive in the path $\langle f_1, \ldots, f_h \rangle$. Let f_i be the face of F with the highest index. We say that F is *incrementally maximal* if $i = h$ or $F \cup f_{i+1}$ is not an xy-fan. We state another key definition; refer to Fig. 2(b).

Definition 3. *An st-fan decomposition of an st-outerpath G is a sequence of $s_i t_i$-fans $F_i \subseteq G$, with $i = 1, \ldots, k$, such that: (i) F_i is incrementally maximal; (ii) For any $1 \le i < j \le k$, F_i and F_j do not share any edge if $j > i + 1$, while F_i and F_{i+1} share a single edge, which we denote by e_i; (iii) $s_1 = s$; (iv) the tail of e_i is s_{i+1}; (v) edge $e_i \ne s_i t_i$; and (vi) $\bigcup_{i=1}^k F_i = G$.*

We next show that primary st-outerpaths always have st-fan decompositions.

Lemma 4 ($\star$). *Every primary st-outerpath G admits an st-fan decomposition.*

Proof (Sketch). By Definition 2, one extreme face of $\overline{G}$ is incident to s; we denote such face by f_1, and the other extreme face of $\overline{G}$ by f_h. We construct the st-fan decomposition as follows. We initialize $F_1 = f_1$ and we parse the faces of G in the order defined by $\overline{G}$ from f_1 to f_h. Let F_i be the current $s_i t_i$-fan and let f_j be the last visited face (for some $1 < j < h$). If $F_i \cup f_{j+1}$ is an $s_i t_i$-fan, we set $F_i = F_i \cup f_{j+1}$, otherwise we finalize F_i and initialize $F_{i+1} = f_j$.

Lemma 5 ($\star$). *Let G be a primary st-outerpath and let $F_1, F_2, \ldots, F_k$ be an st-fan decomposition of G. Any two fans F_i and F_{i+1} are such that if F_{i+1} is not one-sided, then $e_i = s_{i+1} t_i$.*

Lemma 6. *Let G be a primary st-outerpath and let $F_1, F_2, \ldots, F_k$ be an st-fan decomposition of G. Also, let $e \ne s_k t_k$ be an outer edge of F_k. Then, G admits a 4-page e-consecutive UBE $\langle \pi, \sigma \rangle$.*

Proof. We construct a 4-page e-consecutive UBE of G by induction on k; see also Fig. 2(c), which shows a UBE of the primary st-outerpath in Fig. 2(b).

Suppose $k = 1$. Then G consists of the single s_1t_1-fan and a 2-page e-consecutive UBE of G exists by Lemma 3.

Suppose now $k > 1$. Let G_i be the subgraph of G induced by $F_1 \cup F_2 \cup \cdots \cup F_i$, for each $1 \leq i \leq k$. Recall that e_i is the edge shared by F_i and F_{i+1} and that the tail of e_i coincides with s_{i+1}. Let $\langle \pi, \sigma \rangle$ be a 4-page e_{k-1}-consecutive UBE of G_{k-1}, which exists by induction since $e_{k-1} \neq s_{k-1}t_{k-1}$ by condition (v) of Definition 3, and distinguish whether F_k is one-sided or not.

If F_k is one-sided, it admits a 1-page e-consecutive UBE $\langle \pi', \sigma' \rangle$ by Lemma 2. In particular, $e \neq e_{k-1}$, since e_{k-1} is not an outer edge of G_k. Also, $e_{k-1} = s_kt_k$, and hence the two vertices shared by π and π' are s_k, t_k, which are consecutive in π and are the first and the last vertex of π'. Then we define a 4-page e-consecutive UBE $\langle \pi^*, \sigma^* \rangle$ of G_k as follows. The ordering π^* is obtained by merging π' and π. Since $e_{k-1} = s_kt_k$ is uncrossed (over all pages of σ), for every edge e of F_k, we set $\sigma^*(e) = \sigma(e_{k-1})$, while for every other edge e we set $\sigma^*(e) = \sigma(e)$.

If F_k is not one-sided, by Lemma 3, F_k admits a 2-page e-consecutive UBE $\langle \pi', \sigma' \rangle$. Then we obtain a 4-page e-consecutive UBE $\langle \pi^*, \sigma^* \rangle$ of G_k as follows. By Lemma 5, it holds $e_{k-1} = s_kt_{k-1}$, and thus s_k and t_{k-1} are the second-to-last and the last vertex in π, respectively; also, s_k is the first vertex in π'. We set $\pi^* = \pi_{t_{k-}} \circ \pi'$. Since $\langle \pi, \sigma \rangle$ is a e_{k-1}-consecutive UBE of G_{k-1}, by Definition 1 we know that the edges incident to t_{k-1} can use up to two different pages. On the other hand, these are the only edges that can be crossed by an edge of F_k assigned to one of these two pages. Therefore, in Lemma 3, we can assume σ' uses the two pages not used by the edges incident to t_{k-1}, and set $\sigma^*(e) = \sigma'(e)$ for every edge e of F_k and $\sigma^*(e) = \sigma(e)$ for every other edge.

Next we show how to reinsert the appendage H_1 (Lemma 8). To this aim, we first provide a more general tool (Lemma 7) that will be useful also in Sect. 3.2.

Lemma 7 ($\star$). *Let $G = (V, E)$ be a primary st-outerpath with a 4-page e-consecutive UBE of G obtained by using Lemma 6, for some outer edge e of G. For $i = 1, \ldots, h$, let $H_i = (V_i, E_i)$ be a one-sided u_iv_i-outerplanar graph such that $E \cap E_i = \{u_i, v_i\}$. Then $G' = G \cup H_1 \cup \cdots \cup H_h$ admits a 4-page e-consecutive UBE, as long as $e \neq e_i$ $(i = 1, \ldots, h)$.*

Lemma 8 ($\star$). *Let $G' = G \cup H_1$ be an st-outerpath, such that G is a primary st-outerpath and H_1 is the appendage at uv of G'. Let $F_1, F_2, \ldots, F_k$ be an st-fan decomposition of G. Let e be an outer edge of G' that belongs to either F_k or H_1, or to F_1 if $H_1 = \emptyset$. Also, if $e \in F_k$, then $e \neq s_kt_k$, otherwise $e \neq s_1t_1$. Then, G' admits an e-consecutive 4-page UBE $\langle \pi, \sigma \rangle$.*

We now define a decomposition of an upward outerpath G; refer to Fig. 3. Let $P \subset G$ be an st-outerpath, then P is the subgraph of G formed by a subset of consecutive faces of $\overline{G} = \langle f_1, \ldots, f_h \rangle$. Let f_j and $f_{j'}$ be the faces of P with the smallest and highest index, respectively. Let F^i be the incrementally maximal

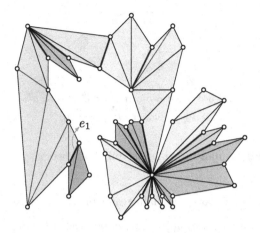

Fig. 3. An st-outerpath decomposition of an upward outerpath; edges e_i are fat.

$s^i t^i$-fan of P_i (assuming $\langle f_j, \ldots, f_{j'} \rangle$ to be the ordered list of faces of $\overline{P_i}$). We say that P is *incrementally maximal* if $i = h$ or if $P \cup f_{i+1}$ is not an st-outerpath or if $P \cup f_{i+1}$ is still an st-outerpath but the edge $s^i t^i$ of F^i is an outer edge of P.

Definition 4. *An st-outerpath decomposition of an upward outerpath G is a sequence of $s_i t_i$-outerpaths $P_i \subseteq G$, with $i = 1, 2, \ldots, m$, such that: (i) P_i is incrementally maximal; (ii) For any $1 \leq i < j \leq m$, P_i and P_j share a single edge if $j = i + 1$, which we denote by e_i, while they do not share any edge otherwise; and (iii) $\bigcup_{i=1}^{m} P_i = G$.*

Lemma 9 ($\star$). *Every upward outerpath admits an st-outerpath decomposition.*

An st-outerpath that is not a single st-fan is called *proper* in the following. Let $P_1, P_2, \ldots, P_m$ be an st-outerpath decomposition of an upward outerpath G, two proper outerpaths P_i and P_j are *consecutive*, if there is no proper outerpath P_a, such that $i < a < j$. We will use the following technical lemmas.

Lemma 10 ($\star$). *Two consecutive proper outerpaths P_i and P_j share either a single vertex v or the edge e_i. In the former case, it holds $j > i + 1$, in the latter case it holds $j = i + 1$.*

Lemma 11 ($\star$). *Let P_i and P_j be two consecutive proper outerpaths that share a single vertex v. Then each P_a, with $i < a < j$, is an $s_a t_a$-fan such that $v = s_a$ if v is the tail of e_i, and $v = t_a$ otherwise.*

Lemma 12 ($\star$). *Let v be a vertex shared by a set $\mathcal{P}$ of $n_\mathcal{P}$ outerpaths. Then at most two outerpaths of $\mathcal{P}$ are such that v is internal for them, and $\mathcal{P}$ contains at most four proper outerpaths.*

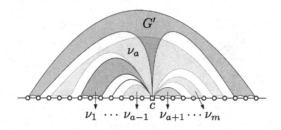

Fig. 4. Illustration for the proof of Theorem 2.

We are now ready to prove the main result of this section.

Theorem 1 (⋆). *Every internally-triangulated upward outerpath G admits a 16-page UBE.*

Proof (Sketch). Let $P_1, P_2, \ldots, P_m$ be an st-outerpath decomposition of G (Lemma 9). Based on Lemma 10, a *bundle* is a maximal set of outerpaths that either share an edge or a single vertex. Let G_b be the graph induced by the first b bundles of G (going from P_1 to P_m). We prove the statement by induction on the number l of bundles of G. In particular, we can prove that G_l admits a $e_{g(l)}$-consecutive 16-page UBE $\langle \pi^l \sigma^l \rangle$, where $g(l)$ is the greatest index such that $P_{g(l)}$ belongs to G_l, and such that each single $s_i t_i$-outerpath uses at most 4 pages. In the inductive case, we distinguish whether the considered bundle contains only two outerpaths that share an edge, or at least three outerpaths that share a vertex. Here, we exploit the crucial properties of Lemmas 8, 11 and 12, which allow us to limit the interaction between different outerpaths in terms of pages.

3.2 Upward Outerplanar Graphs

We now deal with upward outerplanar graphs that may be non-triangulated and may have multiple sources and sinks, but whose blocks are st-DAGs. We begin with the following lemma, which generalizes Lemma 8 in terms of UBT.

Lemma 13 (⋆). *Every biconnected st-outerplanar graph G admits a 4-page UBE.*

Proof (Sketch). By exploiting a technique in [14], we can assume that G is internally triangulated. Let $\langle f_1, \ldots, f_h \rangle$ be a path in $\overline{G}$ whose primal graph $P \subset G$ is a primary st-outerpath. Each outer edge uv of P is shared by P and by a one-sided uv-outerpath. Then a 4-page UBE of G exists by Lemma 7.

We are now ready to show the main result of this subsection.

Theorem 2 (⋆). *Every upward outerplanar graph G whose biconnected components are st-outerplanar graphs admits an 8-page UBE.*

Proof (Sketch). We prove a stronger statement. Let T be a BC-tree of G rooted at an arbitrary block ρ, then G admits an 8-page UBE $\langle \pi, \sigma \rangle$ that has the *page-separation* property: For any block β of T, the edges of β are assigned to at most 4 different pages. We proceed by induction on the number h of cut-vertices in G. If $h = 0$, then G consists of a single block and the statement follows by Lemma 13. Otherwise, let c be a cut-vertex whose children are all leaves. Let $\nu_1, \ldots, \nu_m$ be the $m > 1$ blocks representing the children of c in T, and let μ be the parent block of c. Also, let G' be the maximal subgraph of G that contains μ but does not contain any vertex of $\nu_1, \ldots, \nu_m$ except c, that is, $G = G' \cup \nu_1 \cup \cdots \cup \nu_m$. By induction, G' admits an 8-page UBE $\langle \pi', \sigma' \rangle$ for which the page-separation property holds, as it contains at most $h - 1$ cut-vertices. On the other hand, each ν_i admits a 4-page UBE $\langle \pi^i, \sigma^i \rangle$ by Lemma 13. By Lemma 1, at most two blocks in $\{\mu\} \cup \{\nu_1, \ldots, \nu_m\}$ are such that c is internal.

Let us assume that there are exactly two such blocks and one of these two blocks is μ, as otherwise the proof is just simpler. Also, let ν_a, for some $1 \leq a \leq m$ be the other block for which c is internal. Up to a renaming, we can assume that $\nu_1, \ldots, \nu_{a-1}$ are st-outerplanar graphs with sink c, while $\nu_{a+1}, \ldots, \nu_m$ are st-outerplanar graphs with source c. Refer to Fig. 4. Crucially, we set:

$$\pi = \pi'_{c-} \cup \pi^a_{c-} \cup \pi^1_{c-} \cup \cdots \cup \pi^{a-1}_{c-} \cup \{c\} \cup \pi^{a+1}_{c+} \cup \cdots \cup \pi^m_{c+} \cup \pi^a_{c+} \cup \pi'_{c+}.$$

The page assignment is based on the fact that e and e', such that $e \in \nu_i$ and $e' \in G'$, cross each other only if $i = a$ and in such a case e' is incident to c.

3.3 Upward Cactus Graphs

The first lemma allows us to consider cactus graphs with no trivial blocks.

Lemma 14 ($\star$). *A cactus G' can always be augmented to a cactus G with no trivial blocks and such that the embedding of G' is maintained.*

It is well known that any DAG whose underlying graph is a cycle admits a 2-page UBE [23, Lemma 2.2]. We can show a slightly stronger result, which will prove useful afterwards; see Fig. 5(b).

Lemma 15 ($\star$). *Let G be a DAG whose underlying graph is a cycle and let s be a source (resp. let t be a sink) of G. Then, G admits a 2-page UBE $\langle \pi, \sigma \rangle$ where s is the first vertex (resp. t is the last vertex) in π.*

Using the proof strategy of Theorem 2, we can exploit Lemma 15 to show:

Theorem 3 ($\star$). *Every upward outerplanar cactus G admits a 6-page UBE.*

Proof (Sketch). By Lemma 14, we can assume that all the blocks of G are non-trivial, i.e., correspond to cycles. Also, let T be the BC-tree of G rooted at any block. The theorem can be proved by induction on the number of blocks in T. In fact, we prove the following slightly stronger statement: G admits a 6-page UBE in which the edges of each block lie on at most two pages. The proof crucially relies on Lemma 1 and follows the lines of Theorem 2.

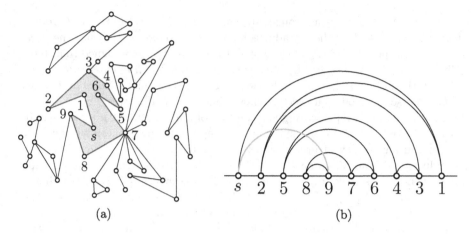

Fig. 5. (a) A cactus G and (b) a 2-page UBE of the red non-trivial block of G.

4 Complexity Results

Recall that the upward book thickness problem is NP-hard for any fixed $k \geq 3$ [10]. This implies that the problem is para-NP-hard, and thus it belongs neither to the FPT class nor to the XP class, when parameterized by its natural parameter (unless P=NP). In this section, we investigate the parameterized complexity of the problem with respect to the domination number and the vertex cover number, showing a lower and an upper bound, respectively.

4.1 Hardness Result for Graphs of Bounded Domination Number

A *domination set* for a graph $G = (V, E)$ is a subset $D \subseteq V$ such that every vertex in $V \setminus D$ has *at least one* neighbor in D. The *domination number* $\gamma(G)$ of G is the number of vertices in a smallest dominating set for G. Given a DAG G such that UBT$(G) \leq k$, one may consider the trivial reduction obtained by considering the DAG G' obtained from G by introducing a new super-source (connected to all the vertices), which has domination number 1 and for which it clearly holds that UBT$(G') \leq k + 1$. However, the other direction of this reduction is not obvious, and indeed for this to work we show a more elaborated construction.

Theorem 4 ($\star$). *Let G be an n-vertex DAG and let k be a positive integer. It is possible to construct in $O(n)$ time an st-DAG G' with $\gamma(G') \in O(k)$ such that* UBT$(G) \leq k$ *if and only if* UBT$(G') = k + 2$.

Proof (Sketch). The proof is based on the construction in Fig. 6. We obtain graph G' by suitably combining G with an auxiliary graph H whose vertices have the same order in any UBE of H, and $\mathrm{UBT}(H) = k + 2$. The key property of G' is that the vertices of G are incident to vertices c and f of H, and that edges incident to each of these vertices must lie in the same page in any UBE of G'.

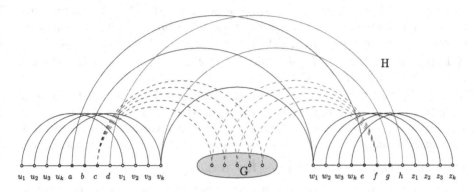

Fig. 6. The graph G' in the reduction of Theorem 4. The edges of the auxiliary graph H are solid. The black edges lie in k pages.

Since testing for the existence of a k-page UBE is NP-hard when $k \geq 3$ [10], Theorem 4 implies that the problem remains NP-hard even for inputs whose domination number is linearly bounded by k. We formalize this in the following.

Theorem 5. *For any fixed $k \geq 5$, deciding whether an st-DAG G is such that $\mathrm{UBT}(G) \leq k$ is NP-hard even if G has domination number at most $O(k)$.*

Theorem 5 immediately implies that the upward book thickness problem parameterized by the domination number is para-NP-hard. On the positive side, we next show that the problem parameterized by the vertex cover number admits a kernel and hence lies in the FPT class.

4.2 FPT Algorithm Parameterized by the Vertex Cover Number

We prove that the upward book thickness problem parameterized by the vertex cover number admits a (super-polynomial) kernel. We build on ideas in [9]. A *vertex cover* of a graph $G = (V, E)$ is a subset $C \subseteq V$ such that each edge in E has at least one incident vertex in C (a vertex cover is in fact a dominating set). The *vertex cover number* of G, denoted by τ, is the size of a minimum vertex cover of G. Deciding whether an n-vertex graph G admits vertex cover of size τ, and if so computing one, can be done in $O(2^\tau + \tau \cdot n)$ time [11]. Let $G = (V, E)$ be an n-vertex DAG with vertex cover number τ. Let $C = \{c_1, c_2, \ldots, c_\tau\}$ be a vertex cover of G such that $|C| = \tau$. The next lemma matches an analogous result in [9].

Lemma 16 ($\star$). *G admits a τ-page UBE that can be computed in $O(\tau \cdot n)$ time.*

For a fixed $k \in \mathbb{N}$, if $k \geq \tau$, then G admits a k-page UBE by Lemma 16. Thus we assume $k < \tau$. Two vertices $u, v \in V \setminus C$ are of the same *type U* if they have the same set of neighbors $U \subseteq C$ and, for every $w \in U$, the edges connecting w to u and w to v have the same orientation. We proceed with the following reduction rule. For each type U, let V_U denote the set of vertices of type U.

R.1: *If there exists a type U such that $|V_U| \geq 2 \cdot k^\tau + 2$, then pick an arbitrary vertex $u \in V_U$ and set $G := G - u$.*

Since there are 2^τ different neighborhoods of size at most τ, and for each of them there are at most 2^τ possible orientations, the type relation yields at most $2^{2\tau}$ distinct types. Therefore assigning a type to each vertex and applying **R.1** exhaustively can be done in $2^{O(\tau)} + \tau \cdot n$ time. We can prove that the rule is safe.

Lemma 17 ($\star$). *The reduction rule **R.1** is safe.*

Proof (Sketch). Let $u \in V_U$, such that $|V_U| \geq 2 \cdot k^\tau + 2$. Suppose that $G - u$ admits a k-page UBE $\langle \pi, \sigma \rangle$. Two vertices $u_1, u_2 \in V_U \setminus \{u\}$ are *page equivalent*, if for each vertex $w \in U$, the edges $u_1 w$ and $u_2 w$ are both assigned to the same page according to σ. By definition of type, each vertex in V_U has degree exactly $|U|$, hence this relation partitions the vertices of V_U into at most $k^{|U|} \leq k^\tau$ sets. Since $|V_U \setminus \{u\}| \geq 2 \cdot k^\tau + 1$, at least three vertices of this set are page equivalent. One can prove that these three vertices are incident to only one vertex in C. Then we can extend π by introducing u right next to any of these three vertices, say u_1, and assign each edge uw incident to u to the same page as $u_1 w$.

Theorem 6 ($\star$). *The upward book thickness problem parameterized by the vertex cover number τ admits a kernel of size $k^{O(\tau)}$.*

Corollary 1 ($\star$). *Let G be an n-vertex graph with vertex cover number τ. For any $k \in \mathbb{N}$, we can decide whether $\mathrm{UBT}(G) \leq k$ in $O(\tau^{\tau^{O(\tau)}} + \tau \cdot n)$ time. Also, within the same time complexity, we can compute a k-page UBE of G, if it exists.*

5 Open Problems

The next questions naturally arise from our research: (i) Is the UBT of upward outerplanar graphs bounded by a constant? (ii) Are there other parameters that are larger than the domination number (and possibly smaller than the vertex cover number) for which the problem is in FPT? (iii) Does the upward book thickness problem parameterized by vertex cover number admit a polynomial kernel?

Acknowledgments. This research was partially supported by MIUR Project "AHeAD" under PRIN 20174LF3T8, by H2020-MSCA-RISE project 734922 – "CONNECT", and by Dipartimento di Ingegneria - Università degli Studi di Perugia grants RICBA19FM and RICBA20EDG.

References

1. Alzohairi, M., Rival, I.: Series-parallel planar ordered sets have pagenumber two. In: North, S. (ed.) GD 1996. LNCS, vol. 1190, pp. 11–24. Springer, Heidelberg (1997). https://doi.org/10.1007/3-540-62495-3_34
2. Angelini, P., Da Lozzo, G., Di Battista, G., Frati, F., Patrignani, M.: 2-level quasi-planarity or how caterpillars climb (SPQR-)trees. In: Marx, D. (ed.) ACM-SIAM Symposium on Discrete Algorithms, (SODA 2021), pp. 2779–2798. SIAM (2021). https://doi.org/10.1137/1.9781611976465.165
3. Angelini, P., Da Lozzo, G., Neuwirth, D.: Advancements on SEFE and partitioned book embedding problems. Theor. Comput. Sci. **575**, 71–89 (2015). https://doi.org/10.1016/j.tcs.2014.11.016
4. Bekos, M.A., Bruckdorfer, T., Kaufmann, M., Raftopoulou, C.N.: The book thickness of 1-planar graphs is constant. Algorithmica **79**(2), 444–465 (2017)
5. Bekos, M.A., Da Lozzo, G., Griesbach, S., Gronemann, M., Montecchiani, F., Raftopoulou, C.N.: Book embeddings of nonplanar graphs with small faces in few pages. In: Cabello, S., Chen, D.Z. (eds.) Symposium on Computational Geometry (SoCG 2020). LIPIcs, vol. 164, pp. 16:1–16:17. Schloss Dagstuhl - Leibniz-Zentrum für Informatik (2020). https://doi.org/10.4230/LIPIcs.SoCG.2020.16
6. Bekos, M.A., Kaufmann, M., Klute, F., Pupyrev, S., Raftopoulou, C.N., Ueckerdt, T.: Four pages are indeed necessary for planar graphs. J. Comput. Geom. **11**(1), 332–353 (2020). https://doi.org/10.20382/jocg.v11i1a12
7. Bernhart, F., Kainen, P.C.: The book thickness of a graph. J. Comb. Theory, Ser. B **27**(3), 320–331 (1979). https://doi.org/10.1016/0095-8956(79)90021-2
8. Bhore, S., Da Lozzo, G., Montecchiani, F., Nöllenburg, M.: On the upward book thickness problem: Combinatorial and complexity results. CoRR abs/2108.12327 (2021). https://arxiv.org/abs/2108.12327
9. Bhore, S., Ganian, R., Montecchiani, F., Nöllenburg, M.: Parameterized algorithms for book embedding problems. J. Graph Algorithms Appl. **24**(4), 603–620 (2020). https://doi.org/10.7155/jgaa.00526
10. Binucci, C., Da Lozzo, G., Di Giacomo, E., Didimo, W., Mchedlidze, T., Patrignani, M.: Upward book embeddings of ST-graphs. In: Barequet, G., Wang, Y. (eds.) SoCG 2019. LIPIcs, vol. 129, pp. 13:1–13:22. Schloss Dagstuhl - Leibniz-Zentrum für Informatik (2019). https://doi.org/10.4230/LIPIcs.SoCG.2019.13
11. Chen, J., Kanj, I.A., Xia, G.: Improved upper bounds for vertex cover. Theor. Comput. Sci. **411**(40–42), 3736–3756 (2010). https://doi.org/10.1016/j.tcs.2010.06.026
12. Chung, F., Leighton, F., Rosenberg, A.: Embedding graphs in books: A layout problem with applications to VLSI design. SIAM J. Alg. Discr. Meth. **8**(1), 33–58 (1987). https://doi.org/10.1137/0608002
13. Di Battista, G., Eades, P., Tamassia, R., Tollis, I.G.: Graph Drawing: Algorithms for the Visualization of Graphs. Prentice-Hall, Upper Saddle River (1999)
14. Di Battista, G., Tamassia, R.: Algorithms for plane representations of acyclic digraphs. Theor. Comput. Sci. **61**, 175–198 (1988). https://doi.org/10.1016/0304-3975(88)90123-5
15. Di Giacomo, E., Didimo, W., Liotta, G., Wismath, S.K.: Book embeddability of series-parallel digraphs. Algorithmica **45**(4), 531–547 (2006)
16. Didimo, W.: Upward graph drawing. In: Kao, M.Y. (ed.) Encyclopedia of Algorithms, pp. 1–5. Springer, Cham (2014). https://doi.org/10.1007/978-3-642-27848-8_653-1

17. Dujmović, V., Wood, D.R.: On linear layouts of graphs. Discrete Math. Theor. Comput. Sci. **6**(2), 339–358 (2004)
18. Dujmovic, V., Wood, D.R.: Graph treewidth and geometric thickness parameters. Discret. Comput. Geom. **37**(4), 641–670 (2007). https://doi.org/10.1007/s00454-007-1318-7
19. Frati, F., Fulek, R., Ruiz-Vargas, A.J.: On the page number of upward planar directed acyclic graphs. J. Graph Algorithms Appl. **17**(3), 221–244 (2013). https://doi.org/10.7155/jgaa.00292
20. Fulek, R., Tóth, C.D.: Atomic embeddability, clustered planarity, and thickenability. In: Chawla, S. (ed.) ACM-SIAM Symposium on Discrete Algorithms (SODA 2020), pp. 2876–2895. SIAM (2020). https://doi.org/10.1137/1.9781611975994.175
21. Heath, L.S., Pemmaraju, S.V.: Stack and queue layouts of posets. SIAM J. Discrete Math. **10**(4), 599–625 (1997)
22. Heath, L.S., Pemmaraju, S.V.: Stack and queue layouts of directed acyclic graphs: Part II. SIAM J. Comput. **28**(5), 1588–1626 (1999)
23. Heath, L.S., Pemmaraju, S.V., Trenk, A.N.: Stack and queue layouts of directed acyclic graphs: Part I. SIAM J. Comput. **28**(4), 1510–1539 (1999). https://doi.org/10.1137/S0097539795280287
24. Kainen, P.C.: Some recent results in topological graph theory. In: Bari, R.A., Harary, F. (eds.) Graphs and Combinatorics, pp. 76–108. Springer, Cham (1974). https://doi.org/10.1007/BFb0066436
25. Malitz, S.M.: Genus g graphs have pagenumber $o(\sqrt{g})$. J. Algorithms **17**(1), 85–109 (1994). https://doi.org/10.1006/jagm.1994.1028
26. Mchedlidze, T., Symvonis, A.: Crossing-free acyclic Hamiltonian path completion for planar st-digraphs. In: Dong, Y., Du, D.-Z., Ibarra, O. (eds.) ISAAC 2009. LNCS, vol. 5878, pp. 882–891. Springer, Heidelberg (2009). https://doi.org/10.1007/978-3-642-10631-6_89
27. Merker, L.: Ordered covering numbers. Masters thesis, Karlsruhe Institute of Technology (2020)
28. Nowakowski, R., Parker, A.: Ordered sets, pagenumbers and planarity. Order **6**(3), 209–218 (1989)
29. Ollmann, L.T.: On the book thicknesses of various graphs. In: Southeastern Conference on Combinatorics, Graph Theory and Computing. Congressus Numerantium, vol. 8, p. 459 (1973)
30. Yannakakis, M.: Embedding planar graphs in four pages. J. Comput. Syst. Sci. **38**(1), 36–67 (1989). https://doi.org/10.1016/0022-0000(89)90032-9

Linear Layouts of Complete Graphs

Stefan Felsner[1], Laura Merker[2(✉)], Torsten Ueckerdt[2], and Pavel Valtr[3]

[1] Institute of Mathematics, Technische Universität Berlin, Berlin, Germany
`felsner@math.tu-berlin.de`
[2] Institute of Theoretical Informatics, Karlsruhe Institute of Technology,
Karlsruhe, Germany
{`laura.merker2,torsten.ueckerdt`}`@kit.edu`
[3] Department of Applied Mathematics, Faculty of Mathematics and Physics,
Charles University, Prague, Czech Republic
`valtr@kam.mff.cuni.cz`

Abstract. A *page* (*queue*) with respect to a vertex ordering of a graph is a set of edges such that no two edges cross (nest), i.e., have their endpoints ordered in an ABAB-pattern (ABBA-pattern). A *union page* (*union queue*) is a vertex-disjoint union of pages (queues). The *union page number* (*union queue number*) of a graph is the smallest k such that there is a vertex ordering and a partition of the edges into k union pages (union queues). The *local page number* (*local queue number*) is the smallest k for which there is a vertex ordering and a partition of the edges into pages (queues) such that each vertex has incident edges in at most k pages (queues).

We present upper and lower bounds on these four parameters for the complete graph K_n on n vertices. In three cases we obtain the exact result up to an additive constant. In particular, the local page number of K_n is $n/3 \pm \mathcal{O}(1)$, while its local and union queue number is $(1-1/\sqrt{2})n \pm \mathcal{O}(1)$. The union page number of K_n is between $n/3 - \mathcal{O}(1)$ and $4n/9 + \mathcal{O}(1)$.

Keywords: Page number · Stack number · Queue number · Local covering numbers · Union covering numbers · Complete graphs

1 Introduction

A *linear layout* of a graph consists of a vertex ordering together with a partition of the edges. For a fixed vertex ordering $\prec$, we say that two independent edges vw and xy with $v \prec w$ and $x \prec y$

- *nest* if $v \prec x \prec y \prec w$ or $x \prec v \prec w \prec y$ and
- *cross* if $v \prec x \prec w \prec y$ or $x \prec v \prec y \prec w$.

A *queue*, respectively a *page* (also called *stack*), is a subset of edges that are pairwise non-nesting, respectively non-crossing. A *queue layout*, respectively a

---

S. Felsner—Partially supported by DFG grant FE 340/13-1.

H. C. Purchase and I. Rutter (Eds.): GD 2021, LNCS 12868, pp. 257–270, 2021.
https://doi.org/10.1007/978-3-030-92931-2_19

book embedding, is a linear layout whose partition of the edge set consists of queues, respectively pages. The *queue number* and *page number* (also called *stack number* or *book thickness*) denote the smallest k such that there is a linear layout consisting of at most k queues, respectively at most k pages. Both queue layouts and book embeddings were intensively investigated in the past decades, where complete graphs are one of the very first considered graph classes [1,9].

Queue layouts and book embeddings model how the edges of a graph can be assigned to and processed by queues, respectively stacks. It was first asked by Heath, Leighton, and Rosenberg [8] whether queues or stacks are more powerful in this context. Recently, Dujmović et al. [4] partly answered this question by presenting a class of graphs with bounded queue number that needs an unbounded number of stacks, showing that stacks are not more powerful than queues for representing graphs. In the classical variant in this question, the total number of necessary queues or stacks serves as a measure for the power of queues and stacks. Local and union variants relax the setting by allowing more queues or stacks, as long as each vertex is touched by a small number of queues, respectively stacks, or if they operate on vertex-disjoint subgraphs.

Local and union variants of queue layouts and book embeddings were recently introduced by the second and third author [13,14]. A linear layout is called k-*local* if each vertex has incident edges in at most k parts of the edge partition. The *local queue number*, respectively *local page number*, is the smallest k such that there is a k-local queue layout, respectively a k-local book embedding for G. Given a fixed vertex ordering, a *union queue* (*union page*) is a vertex-disjoint union of queues (pages). The *union queue number*, respectively *union page number*, then is the smallest k such that there is a linear layout whose edge partition consists of at most k union queues, respectively union pages.

Local and union variants have been considered for numerous graph decomposition parameter, such as boxicity, interval numbers, planar thickness, poset dimension, several arboricities, and many more. Especially in recent years, there has been a lot of interest in these variants in various directions, see e.g. [2,3,6,10–12].

In this paper, we continue the investigation of local and union variants of linear layouts, that is, queue numbers and page numbers. We establish bounds on the respective graph parameters for complete graphs, which interestingly turns out to be non-trivial; in contrast to their classic counterparts. Our results show that the local and union variants of queue numbers and page numbers for n-vertex complete graphs are located strictly between the trivial lower bound of $(n-1)/4$ due to the density [13,14, see also Sect. 2] and $\lfloor n/2 \rfloor$, respectively $\lceil n/2 \rceil$, which is the queue number, respectively page number, of complete graphs [1,9].

Outline. In Sect. 2, we survey the relation between local and union variants of queue layouts and book embeddings. We give a lower bound on the local queue number of complete graphs and sketch how to obtain the same upper bound for the union queue number in Sect. 3 (up to an additive constant). The full proof and an improved upper bound on the local queue number is given in the full version [7]. Section 4 continues with a lower and a matching (up to an additive constant) upper bound on the local page number, while our upper bound on the union page number is proved in the full version [7].

Our Results. Both the local queue number and the union queue number of K_n are linear in n with the leading coefficient being $1 - 1/\sqrt{2} \approx 0.29289$. For the local queue number, the error term $\mathcal{O}(1)$ is small; it is between -0.21 and $+2$.

Theorem 1. *The local queue number and the union queue number of K_n satisfy*

$$\mathrm{qn}_\ell(K_n) = (1 - \frac{1}{\sqrt{2}})n \pm \mathcal{O}(1) \ \text{and} \ \mathrm{qn}_u(K_n) = (1 - \frac{1}{\sqrt{2}})n \pm \mathcal{O}(1).$$

The local page number of K_n is also linear in n with the leading coefficient being $1/3$. In this case, the error term $\mathcal{O}(1)$ is between 0 and 4. The local page number also gives a lower bound on the union page number, whereas the leading coefficient of our upper bound is $4/9$.

Theorem 2. *The local page number and the union page number of K_n satisfy*

$$\frac{1}{3}n \pm \mathcal{O}(1) = \mathrm{pn}_\ell(K_n) \leqslant \mathrm{pn}_u(K_n) \leqslant \frac{4}{9}n + \mathcal{O}(1).$$

2 Preliminaries

In this section, we summarize some known results on local and union linear layouts, which are presented in [13,14] in detail.

First, we have $\mathrm{pn}_\ell(G) \leqslant \mathrm{pn}_u(G) \leqslant \mathrm{pn}(G)$ and $\mathrm{qn}_\ell(G) \leqslant \mathrm{qn}_u(G) \leqslant \mathrm{qn}(G)$, where the gap between the union and global variants can be arbitrarily large. There are graph classes (e.g. k-regular graphs for $k \geqslant 3$) with bounded union queue number and union page number but unbounded queue number and page number. In contrast, the local page number, the local queue number, the union page number, and the union queue number are all tied to the maximum average degree, which is defined by $\mathrm{mad}(G) = \max\{2\,|E(H)|/|V(H)|\colon H \subseteq G, H \neq \emptyset\}$. In particular, we have the following connection between the maximum average degree and the union queue number, respectively the union page number. See also [7] for a proof.

Proposition 3. *Every graph G admits a $(\mathrm{mad}(G)+2)$-union queue layout and a $(\mathrm{mad}(G)+2)$-union book embedding with any vertex ordering.*

In addition, the local queue number and the local page number are lower-bounded by $\mathrm{mad}(G)/4$, which gives a lower bound of $(n-1)/4$ for K_n.

In the following sections, we consider the complete graph K_n with vertex set $V(K_n) = \{v_1, \ldots, v_n\}$. Due to symmetry, we may throughout assume that the vertex ordering of K_n is given by $v_1 \prec \cdots \prec v_n$. Partitioning the edges is the difficult part.

3   Local and Union Queue Numbers

We first establish the lower bound of Theorem 1. As the union queue number is lower-bounded by the local queue number, we only consider the latter. Note that $(9 - 4\sqrt{2})/16 \approx 0.21$.

Lemma 4. *For any n we have* $qn_\ell(K_n) > (1 - \frac{1}{\sqrt{2}})n - \frac{1}{16}(9 - 4\sqrt{2})$.

Proof. Consider a k-local queue layout $\mathcal{Q}$ of K_n. Without loss of generality, each edge is contained in exactly one queue. Moreover, the vertices are ordered $v_1 \prec \cdots \prec v_n$ and the length of an edge $v_i v_j$ is defined as $|i-j|$. Now for any edge $e = v_i v_j$ with $i < j$ consider the queue $Q \in \mathcal{Q}$ containing e. We call e *left-longest* if there is no edge in Q that is longer than e and has the same right endpoint as e, i.e., Q contains no edge $v_{i'} v_j$ with $i' < i$. Similarly, we call $e = v_i v_j \in Q$ *right-shortest* if there is no edge in Q that is shorter than e and has the same left endpoint as e, i.e., Q contains no edge $v_i v_{j'}$ with $i < j' < j$. We have that

(i) every edge of K_n is left-longest or right-shortest (or both).

In fact, if $v_i v_j \in Q$ is of neither type, then Q would contain two edges $v_{i'} v_j$ and $v_i v_{j'}$ with $i' < i < j' < j$, and hence Q would not be a queue.

For each vertex v_i let ℓ_i, respectively r_i, denote the number of left-longest edges whose right endpoint is v_i, respectively the number of right-shortest edges whose left endpoint is v_i. That is,

$$\ell_i = \#\{v_a \in V(K_n) \mid a < i \text{ and } v_a v_i \text{ left-longest}\} \text{ and}$$
$$r_i = \#\{v_b \in V(K_n) \mid i < b \text{ and } v_i v_b \text{ right-shortest}\}.$$

Further let b_i denote the number of queues in $\mathcal{Q}$ with at least one edge whose right endpoint is v_i *and* at least one edge whose left endpoint is v_i. That is,

$$b_i = \#\{Q \in \mathcal{Q} \mid \exists a, b \text{ with } a < i < b \text{ and } v_a v_i, v_i v_b \in Q\}.$$

We can then write the number of queues in $\mathcal{Q}$ containing the vertex v_i in terms of ℓ_i, r_i and b_i. Indeed, if $Q \in \mathcal{Q}$ contains an edge incident to v_i, then it contains a left-longest or a right-shortest or both, i.e., the contribution of Q to $\ell_i + r_i - b_i$ is exactly one.

(ii) Vertex v_i has incident edges in exactly $\ell_i + r_i - b_i$ queues in $\mathcal{Q}$.

As every vertex is in at most k queues, we have $b_i \leq k$ for $i = 1, \ldots, n$. Also every vertex v_i is the right endpoint of at most $i - 1$ edges and thus $b_i \leq i - 1$. Similarly, v_i the left endpoint of at most $n - i$ edges and thus $b_i \leq n - i$. Together,

(iii) for every vertex v_i we have $b_i \leq \min\{i - 1, n - i, k\}$.

Using the above and assuming $k \leqslant n/2$ (we are done otherwise), we calculate

$$kn \geqslant \sum_{i=1}^{n} \#\{Q \in \mathcal{Q} \mid v_i \in V(Q)\} \overset{(ii)}{=} \sum_{i=1}^{n}(\ell_i + r_i - b_i) \overset{(i)}{\geqslant} |E(K_n)| - \sum_{i=1}^{n} b_i$$

$$\overset{(iii)}{\geqslant} \binom{n}{2} - \sum_{i=1}^{k}(i-1) - \sum_{i=n-k+1}^{n}(n-i) - (n-2k)k$$

$$= \binom{n}{2} - 2\binom{k}{2} - (n-2k)k.$$

For $a \geqslant 1$ and $b > 0$ we have

$$\sqrt{a+b} < \sqrt{a} + b/2. \tag{1}$$

Using this we get the desired bound for k as follows.

$$kn \geqslant \binom{n}{2} - 2\binom{k}{2} - (n-2k)k$$

$$\Longleftrightarrow \quad 0 \geqslant k^2 + (1-2n)k + \binom{n}{2}$$

$$\Longrightarrow \quad k \geqslant (n - \tfrac{1}{2}) - \sqrt{(n-\tfrac{1}{2})^2 - \binom{n}{2}} = (n-\tfrac{1}{2}) - \sqrt{\tfrac{1}{2}(n^2 - n + \tfrac{1}{2})}$$

$$= (n-\tfrac{1}{2}) - \sqrt{\tfrac{1}{2}(n-\tfrac{1}{2})^2 + \tfrac{1}{8}} \overset{(1)}{>} (n-\tfrac{1}{2}) - \sqrt{\tfrac{1}{2}(n-\tfrac{1}{2})^2} - \tfrac{1}{16}$$

$$= (1 - \tfrac{1}{\sqrt{2}})(n-\tfrac{1}{2}) - \tfrac{1}{16} = (1 - \tfrac{1}{\sqrt{2}})n - \tfrac{1}{16}(9 - 4\sqrt{2})$$

$\square$

Now let us turn to the upper bound of the union queue number in Theorem 1. We thereby also prove an upper bound for the local queue number. However, we improve on this bound with a different construction in the full version [7].

Lemma 5. *For any $n \geqslant 0$, we have*

$$\mathrm{qn}_\ell(K_n) \leqslant \left\lceil 1 - \frac{1}{\sqrt{2}} \right\rceil n + 11 \text{ and } \mathrm{qn}_u(K_n) \leqslant \left\lceil 1 - \frac{1}{\sqrt{2}} \right\rceil n + 42.$$

We prove that whenever $k \geqslant (1 - 1/\sqrt{2})(n+1)$, there is a $(k+11)$-local queue layout and a $(k+42)$-union queue layout of K_{n+1}. Let $v_1 \prec \cdots \prec v_{n+1}$ be a fixed vertex ordering of K_{n+1}. For ease of presentation, we model the edge set of K_{n+1} as a point set T_n in $\mathbb{Z}^2$ with triangular shape defined by

$$T_n = \{(x,y) \in \mathbb{Z}^2 \mid x + y \leqslant n+1; \ x \geqslant 1; \ y \geqslant 1\}.$$

The elements in T_n correspond to the entries of the adjacency matrix of K_{n+1}. That is, element (x,y) of T_n corresponds to edge $v_{n+2-y}v_x$ in K_{n+1} and conversely edge v_iv_j in K_{n+1} with $i > j$ corresponds to element $(j, n+2-i)$ in T_n.

Two edges $v_i v_j$ with $i > j$ and $v_{i'} v_{j'}$ with $i' > j'$ nest if and only if the corresponding elements $(j, n + 2 - i)$ and $(j', n + 2 - i')$ in T_n are comparable in the strict dominance order of $\mathbb{Z}^2$ (i.e. coordinate-wise strict inequalities of points). To see this, observe that small y-coordinates correspond to a left endpoint having a small index, whereas small x-coordinates correspond to a right endpoint with a large index. Hence, an edge set $Q \subseteq E(K_{n+1})$ forms a queue if and only if the corresponding points in T_n form a weakly monotonically decreasing chain, see Fig. 1.

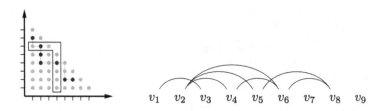

Fig. 1. Left: Triangle T_8 corresponding to K_9 and the hook of vertex 4. The blue (dark) entries represent a queue, the zig-zag (light blue) shows that the edges are non-nesting. Right: Linear layout of the blue queue. (Color figure online)

A vertex v_i of K_{n+1} corresponds to column i and row $n + 2 - i$ in T_n. We call the union of column i and row $n + 2 - i$ the *hook* of vertex v_i. If H is the hook of vertex v_i and Q is a queue corresponding to chain $C \subseteq T_n$, then vertex v_i is contained in queue Q if and only if $H \cap C \neq \emptyset$. For our construction of a $(k + 11)$-local queue assignment of K_{n+1}, we use the equivalent model of covering the triangular point set T_n with monotone chains such that no hook intersects more than $k + 11$ chains.

Analogously to union queues, we call a subset $S \subseteq T_n$ a *union chain* if there is a partition of S into weakly monotonically decreasing chains such that each hook intersects at most one of them. To prove Lemma 5, we partition T_n into $k + 42$ union chains and therefore get a $(k + 42)$-union queue layout for K_{n+1}.

Lemma 6. *For any integer $n \geqslant 0$ and any integer $k \geqslant (1 - 1/\sqrt{2})(n + 1)$, the points of T_n can be partitioned into $k + 42$ union chains. In addition, the points of T_n can be partitioned into weakly monotonically decreasing chains such that each hook intersects at most $k + 11$ chains.*

Proof (sketch). First, we define weakly monotonically decreasing chains that cover T_n such that no hook intersects more than $k + 11$ chains. We then partition these chains into sets of chains that form the basis for our union chains. We assume that n is even and that k is the smallest even integer with $k \geqslant (1 - 1/\sqrt{2})(n + 1)$. To compensate for this assumption, we construct chains such that each hook intersects at most $k + 9$ chains and a partition of T_n into $k + 40$ union chains. We need $3n \geqslant 10k$ for the construction, which is the case for $n \geqslant 294$.

For $n \leqslant 56$, we have $\lfloor (n+1)/2 \rfloor \leqslant \lceil 1 - 1/\sqrt{2} \rceil (n+1) + 11$, so an $\lfloor (n+1)/2 \rfloor$-queue layout of K_{n+1} gives the desired partition of K_{n+1}, respectively T_n, into queues, respectively chains. For any n between 56 and 294, one can check that the desired bounds can be obtained from a queue layout of some $K_{n'}$ with a slightly smaller or greater number of vertices.[1]

We start by defining a family $\mathcal{L}$ of k chains $L_1, \ldots, L_k$, illustrated in Fig. 2. Chain L_i is composed of three blocks. The first block consists of the $2(k-i+1)$ topmost elements in column i of T_n. The second block starts at the lowest element of the first block, continues with a right and down alternation for $2(n - 2(k-1))$ steps, and ends in row i. The last block consists of the $2(k-i+1)$ rightmost elements in row i. Formally, for $i = 1, \ldots, k$ we set

$$L_i = \{(i, y) \in T_n \mid n + 1 - i \geqslant y \geqslant n - 2k + i\}$$
$$\cup \{(x, y) \in T_n \mid x, y \geqslant i \text{ and } n - 2(k-i) \leqslant x + y \leqslant n - 2(k-i) + 1\}$$
$$\cup \{(x, i) \in T_n \mid n - 2k + i \leqslant x \leqslant n + 1 - i\}.$$

The chains of $\mathcal{L}$ cover all points of T_n except for the bottom left triangle T_{n-2k}. The remaining points are covered by chains containing only points of a single column or row. We refer to these chains as *vertical* and *horizontal* chains, respectively. Note that vertical and horizontal chains correspond to stars in K_{n+1}. The resulting layout of T_{n-2k} is illustrated in Fig. 2.

The chains for T_{n-2k} are formally defined and analyzed in the full version [7]. We obtain that each hook intersects at most $k+9$ chains, which yields a $(k+9)$-local queue layout. To obtain a $(k+40)$-union queue layout, we partition the set of all chains into sets of chains $\mathcal{S}_1, \ldots, \mathcal{S}_{k+6}$ with $L_i \in \mathcal{S}_i$ for $i = 1, \ldots, k$.

We first introduce some notions that allow us to transform a set of chains into a union chain. Consider some set $\mathcal{S} = \{S_1, \ldots, S_m\}$ of chains. Note that $\bigcup_{i \in [m]} S_i$ is not necessarily a union chain as two chains may intersect the same hook. We call the set of all hooks that intersect at least two chains of $\mathcal{S}$ the *common hooks* of $\mathcal{S}$. If $\mathcal{S}$ has no common hooks, then $\mathcal{S}$ is already a union chain. Otherwise, we assign each of the common hooks to at most one chain. A point that is contained in some chain $S \in \mathcal{S}$ and in some common hook that is not assigned to S is called a *bad point*. Removing all bad points yields chains that together form a union chain.

We now aim to define $\mathcal{S}_1, \ldots, \mathcal{S}_{k+6}$ such that the resulting bad points can be covered by a constant number of union chains. For this, we associate each vertical

---

[1] Indeed, for each n between 56 and 294 and corresponding k, one of the following two cases applies. First, there is an even $n' \geqslant n$ such that $k' = \lceil (1 - 1/\sqrt{2})(n'+1) \rceil$ is even, $3n' \geqslant 10k'$ holds, and we have $\mathrm{qn}_\ell(K_n) \leqslant \mathrm{qn}_\ell(K_{n'}) \leqslant k' + 9 \leqslant k + 11$, respectively $\mathrm{qn}_\mathrm{u}(K_n) \leqslant \mathrm{qn}_\mathrm{u}(K_{n'}) \leqslant k' + 40 \leqslant k + 42$. Or second, there is an even n' such that $n - 4 \leqslant n' \leqslant n$, $k' = \lceil (1 - 1/\sqrt{2})(n'+1) \rceil$ is even, $3n' \geqslant 10k'$ holds, and the queue layout we obtain for $K_{n'}$ can be augmented to a queue layout of K_n matching the desired bounds. To do so, we can be attach at most two of the additional $n' - n \leqslant 4$ vertices to the left and at most two to the right and use two additional queues to cover the new edges. We then observe that $k' + 2 \leqslant k$, which gives the desired bounds for K_n.

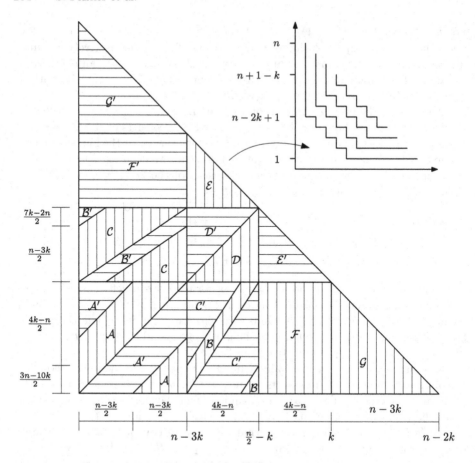

Fig. 2. Chains $L_1, \ldots, L_k$ (top right). The triangle T_{n-2k} is covered by families $\mathcal{A}, \ldots, \mathcal{G}$ and $\mathcal{A}', \ldots, \mathcal{G}'$ of vertical, respectively horizontal, chains (bottom left).

(horizontal) chain C with the interval $I_C \subseteq [k]$ that consists of the y-coordinates (x-coordinates) of the points contained in C. We say two vertical (horizontal) chains *overlap* if the corresponding intervals are not disjoint. Consider the set $\mathcal{S}_i$ for some $i = 1, \ldots, k + 6$. We add vertical and horizontal chains to $\mathcal{S}_i$ such that

(i) chains in $\mathcal{S}_i$ do not overlap,
(ii) the y-coordinates (x-coordinates) of all points in vertical (horizontal) chains in $\mathcal{S}_i$ are smaller than i, and
(iii) there is no vertical (horizontal) chain in $\mathcal{S}_i$ in column (row) i.

We first assume that Conditions (i) to (iii) hold and show that they can indeed be satisfied at the end of the proof. We merge vertical (horizontal) chains of $\mathcal{S}_i$ that are in the same column (row) into a single chain. For some vertical (horizontal) chain C, let H_C denote the hook that contains C. If H_C is a common hook, we assign it to C and say that the bad points in H_C are *caused by* C. Let $G \subseteq K_{n+1}$

denote the graph that is induced by all bad points. For a vertical or horizontal chain $C \subseteq T_{n-2k}$, let $v_C \in V(K_{n+1})$ denote the vertex that is represented by hook H_C. We orient the edges of G such that every edge whose corresponding bad point is caused by chain C is oriented away from v_C. Each hook contains at most four vertical or horizontal chains (see Fig. 2 and recall that each hook either contains vertical chains or horizontal chains). In the full version [7], we show that each chain causes at most four bad points. Thus, the out-degree of every vertex of G is at most 16. By Proposition 3, the graph G can be covered with $\mathrm{mad}(G) + 2 \leqslant 2 \cdot 16 + 2 = 34$ union queues using an arbitrary vertex ordering. Hence, the bad points can be covered by 34 union chains.

Finally, we show how to partition the vertical chains of the presented $(k+9)$-local layout such that Conditions (i) to (iii) are satisfied. Let H denote the interval graph that is given by the intervals that correspond to vertical chains, i.e., $V(H) = \{I_C \subseteq [k] \mid C \in \mathcal{A} \cup \cdots \cup \mathcal{G}\}$ and there is an edge between two vertices if and only if the intervals are not disjoint. A clique of m vertices in H corresponds to a row that intersects m vertical chains. From the analysis in the full version [7], we get that every row $y = 1, \ldots, k$ intersects at most $k - y + 5$ vertical chains. In particular, the clique number of H is at most $k + 4$.

Note that any proper $(k + 6)$-coloring and an arbitrary mapping between color classes and the sets of chains $\mathcal{S}_1, \ldots, \mathcal{S}_{k+6}$ satisfies Condition (i). We next find a coloring with less than $k + 6$ colors that also satisfies Conditions (ii) and (iii). We define an ordering on the vertices of H by decreasing topmost points of the intervals, i.e., $[a, b] \prec [a', b']$ if and only if $b > b'$ or $b = b'$ and $a > a'$. We color the vertices of H greedily with $k + 5$ colors $k + 6, \ldots, 2$. That is, for an interval in column x, we choose the largest color that is not used by any smaller neighbor and that does not equal x. We then define $\mathcal{S}_i$ to contain the vertical chains whose intervals have color i. Since H is an interval graph, the set consisting of a vertex $[a, b] \in V(H)$ and its smaller neighbors induces a clique that corresponds to row b. The vertex $[a, b]$ thus has at most $k - b + 4$ smaller neighbors. There are at least two colors left that are larger than b and that are not already used by a smaller neighbor. Choosing one that does not equal x satisfies Conditions (ii) and (iii).

By symmetry, we color the horizontal chains with colors $k+6, \ldots, 2$ such that Conditions (i) to (iii) hold. We now have a partition of all chains into $k + 6$ sets of chains $\mathcal{S}_1, \ldots, \mathcal{S}_{k+6}$, each forming a union chain when bad points are removed. Together with the 34 union chains for bad points, we get a partition of T_n into $k + 6 + 34 = k + 40$ union chains. □

Lemma 5 provides an upper bound of $(1 - 1/\sqrt{2})n + \mathcal{O}(1)$ both for the union queue number and the local queue number and thus completes the proof of Theorem 1. For the local queue number, however, we obtain an improved upper bound of $\lceil (1 - 1/\sqrt{2})n \rceil + 1$, which is proved in the full version [7].

4 Local and Union Page Numbers

For book embeddings, it is convenient to think of the spine as being circularly closed. The placement of the vertices together with straight-line edges yields a convex drawing of K_n. A page assignment is a partition of the edges into non-crossing subsets, i.e., into outerplanar subdrawings of this drawing of K_n.

First, we analyze the outerplanar subgraphs on each page of a book embedding and thereby show the lower bound of Theorem 2. The proof gives insight into how book embeddings for a matching upper bound on the local or union page number should look like. This bound is also the best lower bound we obtain for the union page number. The upper bound for the union page number is given in the full version [7].

Lemma 7. *For any n we have* $\mathrm{pn}_\ell(K_n) > \dfrac{1}{3}n - 1$.

Proof. Let $\mathcal{P}$ be a page assignment of K_n which minimizes the local page number. We assume that $\mathrm{pn}_\ell(\mathcal{P}) \leqslant n/3$, otherwise we are done. Let k be the average number of vertex-page incidences over all vertices, i.e., $k = \frac{1}{n}\sum_{P\in\mathcal{P}}|V_P|$. We shall show that $k > \frac{1}{3}n - 1$, which in particular proves that $\mathrm{pn}_\ell(\mathcal{P}) > n/3 - 1$. Later we will use that $k \leqslant \mathrm{pn}_\ell(\mathcal{P}) \leqslant n/3$, i.e.,

$$k \leqslant \frac{n}{3}. \tag{2}$$

Note that every edge of K_n belongs to exactly one page of $\mathcal{P}$. Now for each page $P \in \mathcal{P}$ we consider an outerplanar graph O_P consisting of all edges of P and their incident vertices V_P together with the edges of the convex hull C_P of V_P. For each page $P \in \mathcal{P}$ we color the edges of O_P:

- The *black edges* are edges of C_P belonging to P.
- The *red edges* are edges of C_P which do not belong to P.
- The *green edges* are inner edges of O_P which belong to P.

Observe that every edge e of K_n has a color in {black, green} for exactly one page, while e may be red for any number of pages.

For a vertex v and a page P containing v, let the *forward edge* $\mathrm{fwd}_P(v)$ at v be the edge of C_P which leaves v in clockwise direction. Let r_v be the number of pages for which the forward edge of v is red, and b_v be the number of pages for which the forward edge of v is black. As v has exactly one forward edge on each page, v is incident to exactly $r_v + b_v$ pages in $\mathcal{P}$. Hence, denoting $R = \sum_v r_v$ and $B = \sum_v b_v$, we have

$$kn = \sum_{P\in\mathcal{P}}|V_P| = \sum_v(r_v + b_v) = R + B. \tag{3}$$

Now for each page P and each edge $e = uv$ of C_P with u clockwise followed by v, let $\mathrm{len}(e)$ be the distance along K_n when going clockwise from u to v, i.e., if $u = v_i$ and $v = v_j$ and $i < j$ then $\text{len}(e) = j - i$ and $\text{len}(e) = j - i + n$ if $j < i$. Since C_P is a cycle, we have $\sum_{e \in C_P} \text{len}(e) = n$. Thus

$$|\mathcal{P}| \cdot n = \sum_{P \in \mathcal{P}} \left(\sum_{e \in C_P} \text{len}(e) \right) = \sum_{P \in \mathcal{P}} \left(\sum_{v \in V_P} \text{len}(\text{fwd}_P(v)) \right)$$

$$= \sum_{v} \left(\sum_{P: v \in V_P} \text{len}(\text{fwd}_P(v)) \right) \overset{(\diamond)}{\geqslant} \sum_{v} \left(\sum_{\ell=1}^{b_v} \ell \right) \geqslant \sum_{v} \frac{b_v^2}{2} \overset{(*)}{\geqslant} \frac{1}{2n} \left(\sum_{v} b_v \right)^2$$

$$= \frac{1}{2n} B^2 \overset{(3)}{=} \frac{1}{2n}(kn - R)^2 \geqslant \frac{1}{2n}(k^2 n^2 - 2knR) = \frac{k^2}{2} n - kR \overset{(2)}{\geqslant} \frac{k^2}{2} n - \frac{n}{3} R.$$

For $(\diamond)$ ignore red forward edges at v and use that the black forward edges at v are pairwise distinct and for $(*)$ use the Cauchy-Schwarz inequality with the vectors $(b_{v_1}, \ldots, b_{v_n})$ and $(1, \ldots, 1)$.

Dividing both sides of the above by n we get

$$|\mathcal{P}| \geqslant \frac{k^2}{2} - \frac{R}{3}. \tag{4}$$

Now consider the green edges in K_n. Since O_P is outerplanar and the green edges of O_P are the inner edges there are at most $|V_P| - 3$ green edges on page P. Therefore, we have

$$\text{\#green edges} \leqslant \sum_{P \in \mathcal{P}} (|V(P)| - 3) = kn - 3|\mathcal{P}| \overset{(4)}{\leqslant} kn - \frac{3k^2}{2} + R \text{ and} \tag{5}$$

$$\text{\#green edges} = |E(K_n)| - \text{\#black edges} = \binom{n}{2} - B \overset{(3)}{=} \binom{n}{2} - kn + R. \tag{6}$$

Combining (5) and (6) we conclude:

$$kn - \frac{3k^2}{2} + R \geqslant \binom{n}{2} - kn + R$$

$$\iff \qquad 0 \geqslant \frac{3k^2}{2} - 2kn + \binom{n}{2}$$

$$\iff \qquad 0 \geqslant k^2 - \frac{4n}{3}k + \frac{n(n-1)}{3}$$

$$\implies \qquad k \geqslant \frac{2n}{3} - \sqrt{\left(\frac{2n}{3}\right)^2 - \frac{n(n-1)}{3}}$$

$$\implies \qquad k \geqslant \frac{2n}{3} - \sqrt{\frac{n^2}{9} + \frac{n}{3}} > \frac{2n}{3} - \sqrt{\frac{(n+3)^2}{9}} = \frac{n}{3} - 1$$

Thus we have $k > \frac{n}{3} - 1$, as desired. $\qquad\qquad\qquad\qquad\qquad\qquad\qquad\qquad\qquad \square$

Note that for the lower bound to be tight there can be no red edges, i.e., each page contains the edges of the convex hull of its vertices. This is the case in the construction for the upper bound for Theorem 2, which is given next.

Lemma 8. *For any n, we have* $\text{pn}_\ell(K_n) \leqslant \frac{1}{3}n + 4.$

Proof. We shall show that if $n = 18k - 3$ for some positive integer k, then $\text{pn}_\ell(K_n) \leqslant n/3 = 6k - 1$. For n of the form $n = 18k - 3 + i$ with $i < 18$ we get a page assignment with locality $6k - 1 + i$ by adding the stars of the i additional vertices on an extra page each. A second option is to use a page assignment of $K_{18(k+1)-3}$ and remove $18 - i$ vertices, this yields a page assignment with locality $6(k + 1) - 1 = 6k + 5$. By taking the better of these two choices we achieve a locality of at most $n/3 + 4$, as desired.

From now on we assume that $n = 18k - 3$, i.e., $k = (n + 3)/18$. We define the length of an edge as the shorter distance between its two endpoints along the cyclic ordering. The length of edge e is denoted $\text{len}(e)$. As $n = 18k - 3$ is odd, there are exactly $(n - 1)/2 = 9k - 2$ different lengths, each realized by exactly n edges. For each vertex v of K_n we define a set of k pages, each containing v, and together covering exactly one edge of each length $1, \ldots, 9k - 2$. These pages each contain an outerplanar graph and are denoted by $O_v(t)$, where $t = 0, \ldots, k - 1$. The page $O_v(0)$ contains seven edges (and five vertices) while for $t > 0$ page $O_v(t)$ has nine edges (and six vertices). In total this makes the needed $9(k - 1) + 7 = 9k - 2 = (n - 1)/2$ edge lengths.

Recall that the vertices of K_n are $v_1, \ldots, v_n$ in this cyclic ordering. Below we describe the pages corresponding to $v_1 = v_{n+1}$. For ease of notation, let $O(t) = O_{v_1}(t)$. For $t = 0, \ldots, k - 1$, we define the vertices $r_1(t), \ldots, r_6(t)$ of $O(t)$:

$$r_1(t) = v_1 = v_{18k-2} \qquad r_2(t) = v_{2k-2t} \qquad r_3(t) = v_{5k+1}$$

$$r_4(t) = v_{8k-t} \qquad r_5(t) = v_{8k+t} \qquad r_6(t) = v_{13k+2t}$$

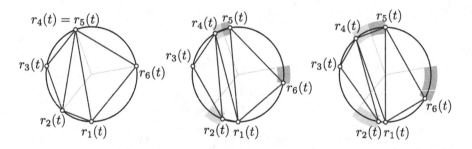

Fig. 3. $O(t)$ for $t = 0$ (left), $t \approx k/2$ (middle) and $t = k - 1$ (right)

We refer to Fig. 3 for an illustration. Note that for $t = 0$ we have $r_4(t) = r_5(t)$ and for all $t \leqslant k$ the vertices $r_1(t), \ldots, r_6(t)$ appear in the order of their indices in the cyclic ordering of K_n. The edges of $O(t)$ are the cycle edges $e_{12}(t) = r_1(t)r_2(t)$, $e_{23}(t) = r_2(t)r_3(t)$, $e_{34}(t) = r_3(t)r_4(t)$, $e_{45}(t) = r_4(t)r_5(t)$, $e_{56}(t) = r_5(t)r_6(t)$, $e_{61}(t) = r_6(t)r_1(t)$, except for $e_{45}(0)$ which would be a loop, and the inner edges $e_{14}(t) = r_1(t)r_4(t)$, $e_{24}(t) = r_2(t)r_4(t)$, $e_{15}(t) = r_1(t)r_5(t)$, again for

$t = 0$ there is an exception, we disregard $e_{15}(0)$ because it equals $e_{14}(0)$. Note that $O(t)$ is indeed outerplanar. We claim that for every length ℓ in the interval $[1, 9k - 2]$ there is an edge of length ℓ in some $O(t)$.

$$\begin{aligned}
\ell \text{ odd},\ \ell \in [1, 2k-1] \quad &: \quad \text{len}(e_{12}(t)) = (2k - 2t) - 1 = 2k - 2t - 1 \\
\ell \text{ even},\ \ell \in [1, 2k-1] \quad &: \quad \text{len}(e_{45}(t)) = (8k + t) - (8k - t) = 2t \\
\ell \in [2k, 3k-1] \quad &: \quad \text{len}(e_{34}(t)) = (8k - t) - (5k + 1) = 3k - t - 1 \\
\ell \text{ odd},\ \ell \in [3k, 5k-1] \quad &: \quad \text{len}(e_{23}(t)) = (5k + 1) - (2k - 2t) = 3k + 2t + 1 \\
\ell \text{ even},\ \ell \in [3k, 5k-1] \quad &: \quad \text{len}(e_{61}(t)) = (18k - 2) - (13k + 2t) = 5k - 2t - 2 \\
\ell \in [5k, 6k-1] \quad &: \quad \text{len}(e_{56}(t)) = (13k + 2t) - (8k + t) = 5k + t \\
\ell \in [6k, 7k-1] \quad &: \quad \text{len}(e_{24}(t)) = (8k - t) - (2k - 2t) = 6k + t \\
\ell \in [7k, 8k-1] \quad &: \quad \text{len}(e_{14}(t)) = (8k - t) - 1 = 8k - t - 1 \\
\ell \in [8k, 9k-2] \quad &: \quad \text{len}(e_{15}(t)) = (8k + t) - 1 = 8k + t - 1
\end{aligned}$$

For a vertex v_i and some t we obtain the page $O_{v_i}(t)$ from $O(t)$ by a rotation which maps v_1 to v_i. Hence, for each v_i we get a collection $\mathcal{P}_i = \{O_{v_i}(t) \mid t = 0, \ldots, k-1\}$ of pages. We claim that $\mathcal{P} = \bigcup_v \mathcal{P}_v$ covers all the edges of K_n. Consider an arbitrary edge $v_a v_b$, we assume that the arc from v_a to v_b is the shorter arc, i.e., the length of the arc is $\text{len}(v_a v_b) = \ell$. From the analysis above we know that there is a unique t and a unique edge $e_{ij}(t) \in O(t)$ with $\text{len}(e_{ij}(t)) = \ell$. There is a rotation which maps $r_i(t)$ to v_a and consequently also $r_j(t)$ to v_b. If this rotation maps $v_1 = r_1(t)$ to v_c, then $O_{v_c}(t)$ contains the edge $v_a v_b$. Hence, $\mathcal{P}$ is a covering of the edges of K_n with outerplanar graphs.

In $\mathcal{P}_1$ there are $6(k-1) + 5 = 6k - 1$ vertex-page incidences, hence the total number of vertex-page incidences in $\mathcal{P}$ is $n(6k - 1) = n^2/3$. Due to symmetry, each of the n vertices is incident to exactly $n/3$ pages. This proves that $\text{pn}_\ell(K_n) \leqslant n/3$ whenever n is of the form $n = 18k - 3$ for some positive integer k. $\qquad\square$

5 Conclusions

We have shown bounds on the local page number, the local queue number, and the union queue number of complete graphs that are tight up to a constant additive term. However, there remains a gap between the lower bound of $n/3 - \mathcal{O}(1)$ and the upper bound of $4n/9 + \mathcal{O}(1)$ on the union page number of K_n.

Question 9. What is the union page number of complete graphs?

Comparing queues and stacks, we find that both in the local and in the union setting, queues are more powerful than stacks for representing complete graphs as both the local and the union queue number is smaller than the respective variant of the page number.

Finally, we point out complete bipartite graphs as another dense graph class. Heath and Rosenberg [9] proved $\text{qn}(K_{m,n}) = \lceil m/2 \rceil$, where $m \leqslant n$. For the page number, it is known that $\mathrm{pn}(K_{m,n}) = m$ if $n \geqslant m^2 - m + 1$ [1], $\mathrm{pn}(K_{n,n}) \leqslant \lfloor 2n/3 \rfloor + 1$, $\mathrm{pn}(K_{\lfloor n^2/4 \rfloor, n}) \leqslant n - 1$ [5], and in general $\mathrm{pn}(K_{m,n}) \leqslant \lceil (m + 2n)/4 \rceil$ [15]. In light of the unclear situation for the page number, we ask for the local and union variants of queue number and page number of complete bipartite graphs.

Acknowledgments. The first, third and fourth author would like to thank the organizers and all participants of the Seventh Annual Workshop on Geometry and Graphs in Barbados, where part of this research was carried out.

References

1. Bernhart, F., Kainen, P.C.: The book thickness of a graph. J. Comb. Theory Ser. B **27**(3), 320–331 (1979). https://doi.org/10.1016/0095-8956(79)90021-2
2. Bläsius, T., Stumpf, P., Ueckerdt, T.: Local and union boxicity. Discrete Math. **341**(5), 1307–1315 (2018). https://doi.org/10.1016/j.disc.2018.02.003
3. Damásdi, G., et al.: On covering numbers, Young diagrams, and the local dimension of posets. SIAM J. Discrete Math. **35**(2), 915–927 (2021). https://doi.org/10.1137/20M1313684
4. Dujmović, V., Eppstein, D., Hickingbotham, R., Morin, P., Wood, D.R.: Stack-number is not bounded by queue-number. arXiv:2011.04195 (2020)
5. Enomoto, H., Nakamigawa, T., Ota, K.: On the pagenumber of complete bipartite graphs. J. Comb. Theory Ser. B **71**(1), 111–120 (1997). https://doi.org/10.1006/jctb.1997.1773
6. Esperet, L., Lichev, L.: Local boxicity. arXiv:2012.04569 (2020)
7. Felsner, S., Merker, L., Ueckerdt, T., Valtr, P.: Linear layouts of complete graphs. arXiv:2108.05112 (2021)
8. Heath, L.S., Leighton, F., Rosenberg, A.: Comparing queues and stacks as machines for laying out graphs. SIAM J. Discrete Math. **5**(3), 398–412 (1992). https://doi.org/10.1137/0405031
9. Heath, L.S., Rosenberg, A.: Laying out graphs using queues. SIAM J. Comput. **21**(5), 927–958 (1992). https://doi.org/10.1137/0221055
10. Kim, J., et al.: On difference graphs and the local dimension of posets. Eur. J. Comb. **86**, 103074 (2020). https://doi.org/10.1016/j.ejc.2019.103074
11. Knauer, K., Ueckerdt, T.: Three ways to cover a graph. Discrete Math. **339**(2), 745–758 (2016). https://doi.org/10.1016/j.disc.2015.10.023
12. Majumder, A., Mathew, R.: Local boxicity and maximum degree. arXiv:1810.02963 (2021)
13. Merker, L., Ueckerdt, T.: Local and union page numbers. In: Archambault, D., Tóth, C.D. (eds.) GD 2019. LNCS, vol. 11904, pp. 447–459. Springer, Cham (2019). https://doi.org/10.1007/978-3-030-35802-0_34
14. Merker, L., Ueckerdt, T.: The local queue number of graphs with bounded treewidth. In: GD 2020. LNCS, vol. 12590, pp. 26–39. Springer, Cham (2020). https://doi.org/10.1007/978-3-030-68766-3_3
15. Muder, D.J., Weaver, M.L., West, D.B.: Pagenumber of complete bipartite graphs. J. Graph Theory **12**(4), 469–489 (1988). https://doi.org/10.1002/jgt.3190120403

On the Queue Number of Planar Graphs

Michael A. Bekos[1]([✉])[iD], Martin Gronemann[2][iD],
and Chrysanthi N. Raftopoulou[3][iD]

[1] Department of Computer Science, University of Tübingen, Tübingen, Germany
bekos@informatik.uni-tuebingen.de
[2] Theoretical Computer Science, Osnabrück University, Osnabrück, Germany
martin.gronemann@uni-osnabrueck.de
[3] School of Applied Mathematics and Physical Sciences, NTUA, Athens, Greece
crisraft@mail.ntua.gr

Abstract. A k-queue layout is a special type of a linear layout, in which the linear order avoids $(k+1)$-rainbows, i.e., $k+1$ independent edges that pairwise form a nested pair. The optimization goal is to determine the *queue number* of a graph, i.e., the minimum value of k for which a k-queue layout is feasible. Recently, Dujmović et al. [13] showed that the queue number of planar graphs is at most 49, thus settling in the positive a long-standing conjecture by Heath, Leighton and Rosenberg. To achieve this breakthrough result, their approach involves three different techniques: (i) an algorithm to obtain straight-line drawings of outerplanar graphs, in which the y-distance of any two adjacent vertices is 1 or 2, (ii) an algorithm to obtain 5-queue layouts of planar 3-trees, and (iii) a decomposition of a planar graph into so-called tripods. In this work, we push further each of these techniques to obtain the first non-trivial improvement on the upper bound from 49 to 42.

Keywords: Queue layouts · Planar graphs · Queue number

1 Introduction

Linear layouts of graphs have a long tradition of study in different contexts, including graph theory and graph drawing, as they form a framework for defining different graph-theoretic parameters with several applications; see, e.g., [10]. Here, we seek to find a total order of the vertices of a graph that reaches a certain optimization goal [3,8,19]. In this work, we focus on a well-studied type of linear layouts, called *queue layout* [13,17,18,21], in which the goal is to minimize the size of the largest *rainbow*, i.e., a set of independent edges that are pairwise nested. Equivalently, the problem asks for a linear order of the vertices and a partition of the edges into a minimum number of queues (called *queue number*), such that no two independent edges in the same queue are nested [18]; see Fig. 1b.

The work of C. Raftopoulou is funded by the National Technical University of Athens research program ΠΕΒΕ 2020.

© Springer Nature Switzerland AG 2021
H. C. Purchase and I. Rutter (Eds.): GD 2021, LNCS 12868, pp. 271–284, 2021.
https://doi.org/10.1007/978-3-030-92931-2_20

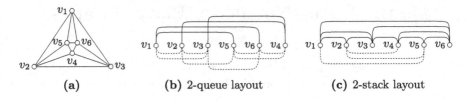

Fig. 1. (a) The octahedron graph and (b)–(c) different linear layouts of it.

Queue layouts of graphs were introduced by Heath and Rosenberg [18] in 1992 as the "dual problem" of *stack layouts* (widely known also as *book embeddings*), in which the edges must be partitioned into a minimum number of stacks (called *stack number*), such that no two edges in the same stack cross [7]; see Fig. 1c. Since their introduction, queue layouts of graphs have been a fruitful subject of intense research with several important milestones over the years [1, 2, 4, 14, 17, 21, 22].

The most intriguing problem in this research field is undoubtedly the problem of specifying the queue number of planar graphs. This problem dates back to a conjecture by Heath, Leighton and Rosenberg, who in 1992 conjectured that the queue number of planar graphs is bounded [17]. Notably, despite the different efforts [2, 9, 11], this conjecture remained unanswered for more than two decades. That only changed in 2019 with a breakthrough result of Dujmović et al. [12], who managed to settle in the positive the conjecture, as they showed that the queue number of planar graphs is at most 49. The best-known corresponding lower bound is 4 due to Alam et al. [1].

The gap between the currently best-known lower and upper bounds is rather large, which implies that the exact queue number of planar graphs is, up to the point of writing, still unknown. Note that this is in contrast with the maximum stack number of planar graphs, which was recently shown to be exactly 4 [6, 23]. Also, the existing gap in the bounds on the queue number of planar graphs gives the intuition that it is unlikely that the upper bound of 49 by Dujmović et al. [12] is tight, even though in the last two years no improvement appeared in the literature.

Our Contribution. We verify the aforementioned intuition by reducing the upper bound on the queue number of planar graphs from 49 to 42 (see Theorem 1 in Sect. 4). To achieve this, we present improvements to each of the following three main techniques (outlined in Sect. 2) involved in the approach by Dujmović et al. [12]: (i) an algorithm to obtain straight-line drawings of outerplanar graphs, in which the y-distance of any two adjacent vertices is 1 or 2 [15], (ii) an algorithm to obtain 5-queue layouts of planar 3-trees [1], and (iii) a decomposition of a planar graph into so-called tripods [12]; see Sect. 3.

Preliminaries. A *vertex order* $\prec$ of a simple undirected graph G is a total order of its vertices, such that for any two vertices u and v of G, $u \prec v$ if and only if u precedes v in the order. Let F be a set of $k \geq 2$ independent edges (u_i, v_i) of G, that is, $F = \{(u_i, v_i); \; i = 1, \ldots, k\}$. If $u_1 \prec \ldots \prec u_k \prec v_k \prec \ldots \prec v_1$, then we say that the edges of F form a k-*rainbow*, while if $u_1 \prec v_1 \prec \ldots \prec u_k \prec v_k$, then the edges of F form a k-*necklace*. The edges of F form a k-*twist*, if $u_1 \prec \ldots \prec$

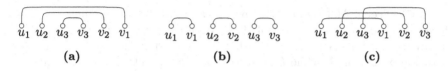

(a)   (b)   (c)

Fig. 2. Illustration of (a) a 3-rainbow, (b) a 3-necklace, and (c) a 3-twist.

$u_k \prec v_1 \prec \ldots \prec v_k$; see Fig. 2. Two independent edges that form a 2-rainbow (2-necklace, 2-twist) are referred to as *nested* (*disjoint, crossing,* respectively).

A *k-queue layout* of a graph consists of a vertex order $\prec$ of G and a partition of the edge-set of G into k sets of pairwise non-nested edges, called *queues*. A preliminary result by Heath and Rosenberg [18] states that a graph admits a k-queue layout if and only if it admits a vertex order in which no $(k+1)$-rainbow is formed. The *queue number* of a graph G, denoted by qn(G), is the minimum k, such that G admits a k-queue layout. Accordingly, the queue number of a class of graphs is the maximum queue number over all its members.

2 Sketch of the Involved Techniques

Here, we outline the main aspects of the three algorithms mentioned in Sect. 1.

Outerplanar Graphs. The main ingredient of the algorithm by Dujmović, Pór and Wood [15] is an algorithm to obtain a straight-line drawing $\Gamma(G)$ of a maximal outerplane graph G whose output can be transformed into a 2-queue layout of G. The recursive construction of $\Gamma(G)$ maintains the following invariant properties:

(O.1) The cycle delimiting the outerface consists of two strictly x-monotone paths, referred to as *upper* and *lower envelopes,* respectively.
(O.2) The y-coordinates of the endvertices of each edge differ by either one (*span-1* edge) or two (*span-2* edge).

To maintain (O.1) and (O.2), Dujmović et al. adopt an approach in which at each recursive step a vertex of degree 2 is added to the already constructed drawing; see Fig. 3 (for details refer to [5]). Drawing $\Gamma(G)$ is transformed to a 2-queue layout of G as follows: (i) for any two vertices u and v of G, $u \prec v$ if and only if either $y(u) > y(v)$, or $y(u) = y(v)$ and $x(u) < x(v)$ in $\Gamma(G)$, (ii) edge (u, v) is assigned to the first (second) queue if it has span 1 (2, respectively) in $\Gamma(G)$.

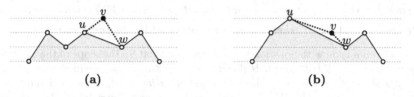

(a)   (b)

Fig. 3. Introducing a degree-2 vertex v along the upper envelope, when its two neighbors u and w are connected with (a) a span-1 edge, and (b) a span-2 edge.

Planar 3-Trees. Alam et al. [1] adopt a *peeling-into-levels* approach [16] to produce a 5-queue layout of a maximal plane 3-tree H. Initially, the vertices of H are partitioned into *levels* $L_0, \ldots, L_\lambda$ with $\lambda \geq 1$, such that L_0-vertices are incident to the outer face of H, while L_{i+1}-vertices are in the outer face of the subgraph of H obtained by the removal of all vertices in $L_0, \ldots, L_i$. The edges of H are partitioned into *level* and *binding*, depending on whether their endpoints are on the same or on consecutive levels. As each connected component of the subgraph H_i of H induced by the edges of level L_i is an internally triangulated outerplane graph, it is embeddable in two queues. This implies that each connected component c of H_{i+1} (which is outerplane) lies in the interior of a triangular face of H_i, therefore there are exactly three vertices of H_i that are connected to c. The constructed 5-queue layout of H satisfies the following invariant properties:

(T.1) The linear order $\prec_H$ is such that all vertices of level L_j precede all vertices of level L_{j+1} for every $j = 0, \ldots, \lambda - 1$;

(T.2) Vertices of each connected component of level L_j appear consecutively in $\prec_H$ for every $j = 0, \ldots, \lambda$;

(T.3) Level edges of each of the levels $L_0, \ldots, L_\lambda$ are assigned to two queues denoted by Q_0 and Q_1;

(T.4) For every $j = 0, \ldots, \lambda - 1$, the binding edges between L_j and L_{j+1} are assigned to three queues Q_2, Q_3 and Q_4 as follows. For each connected component c of H_{j+1}, let x, y and z be its three neighbors in H_j so that $x \prec_H y \prec_H z$. Then, the binding edges between L_j and L_{j+1} incident to c are assigned to Q_2, Q_3 and Q_4 if they lead to x, y and z, respectively.

General Planar Graphs. Central in the algorithm by Dujmović et al. [13] is the notion of H-partition,[1] defined as follows. Given a graph G, an H-*partition* of G is a partition of the vertices of G into sets A_x with $x \in V(H)$, called *bags*, such that for each edge (u, v) of G with $u \in A_x$ and $v \in A_y$ either $x = y$ holds or (x, y) is an edge of H. In the former case, (u, v) is called *intra-bag* edge, while in the latter case *inter-bag*. A *BFS-layering* of G is a partition $\mathcal{L} = (V_0, V_1, \ldots)$ of its vertices according to their distance from a specific vertex r of G, i.e., it is a special type of H-partition, where H is a path and each bag V_i corresponds to a *layer*. In this regard, an intra-bag edge is called *intra-layer*, while an inter-bag edge is called *inter-layer*.[2] An H-partition has *layered-width* ℓ with respect to a BFS-layering $\mathcal{L}$ if each bag of H has at most ℓ vertices on each layer of $\mathcal{L}$.

Lemma 1 (Dujmović et al. [13]). *For all graphs G and H, if H admits a k-queue layout and G has an H-partition of layered-width ℓ with respect to some layering $\mathcal{L} = (V_0, V_1, \ldots)$ of G, then G admits a $(3\ell k + \lfloor \frac{3}{2}\ell \rfloor)$-queue layout using vertex order $\overrightarrow{V_0}, \overrightarrow{V_1}, \ldots$, where $\overrightarrow{V_i}$ is some order of V_i. In particular,*

---

[1] To avoid confusion with notation used earlier, note that, in the scope of the algorithm by Dujmović et al. [13], graph H denotes a plane 3-tree, as we will shortly see.

[2] Dujmović et al. [13] refer to the intra- and inter-*layer* edges as intra- and inter-*level* edges, respectively. We adopt the terms intra- and inter-layer edges to avoid confusion with the different type of leveling used in the algorithm of Alam et al. [1].

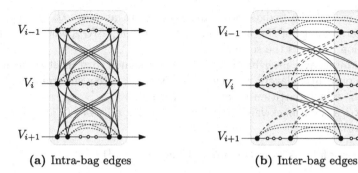

(a) Intra-bag edges   **(b)** Inter-bag edges

Fig. 4. Illustration of (a) Intra-bag edges; the intra-layer ones are dotted, while the inter-layer ones are solid, and (b) inter-bag edges; the intra-layer ones are dotted, while the inter-layer ones are solid (forward) and dashed (backward).

$$\mathrm{qn}(G) \le 3\ell\,\mathrm{qn}(H) + \lfloor \tfrac{3}{2}\ell \rfloor. \tag{1}$$

In the proof of Lemma 1, the order of the vertices of G on each layer of $\mathcal{L}$ is defined as follows. Let $x_1, \dots, x_h$ be the vertices of H as they appear in a k-queue layout $QL(H)$ of H and let $A_{x_1}, \dots, A_{x_h}$ be the corresponding bags of the H-partition. Then, the order $\vec{V_i}$ of each layer V_i with $i \ge 0$ is:

$$\vec{V_i} = A_{x_1} \cap V_i, A_{x_2} \cap V_i, \dots, A_{x_h} \cap V_i$$

where each subset $A_{x_j} \cap V_i$ is ordered arbitrarily. This gives the total order $\prec_G$ for the vertices of G. The edge-to-queue assignment, which completes the construction of $QL(G)$, exploits the following two properties; see Fig. 4:

(P.1) Two intra-bag edges nest in $\prec_G$, only if they belong to the same bag.
(P.2) Two inter-layer edges nest in $\prec_G$, only if their endpoints belong to the same pair of consecutive layers of $\mathcal{L}$.

For the edge-to-queue assignment, the edges of G are classified into four categories given by the bags of the H-partition and the layers of $\mathcal{L}$. We start with edges whose endpoints belong to the same bag (i.e., intra-bag edges); see Fig. 4a.

(E.1) Intra-layer intra-bag edges of G are assigned to at most $\lfloor \tfrac{\ell}{2} \rfloor$ queues, as the queue number of K_ℓ is $\lfloor \tfrac{\ell}{2} \rfloor$ [16].
(E.2) Inter-layer intra-bag edges of G are assigned to at most ℓ queues, as the queue number of $K_{\ell,\ell}$ is ℓ, when all vertices of the first bipartition precede those of the second.

The remaining edges of G connect vertices of different bags (i.e., inter-bag edges); see Fig. 4b. We further partition the inter-layer inter-bag edges into two categories. Let (u, v) be an inter-layer inter-bag edge with $u \in A_x \cap V_i$ and $v \in A_y \cap V_{i+1}$, for some $i \ge 0$. Then (u, v) is *forward*, if $x \prec_H y$ holds in $QL(H)$; otherwise, it is *backward*. For all inter-bag edges, in total, $3\ell k$ queues suffice (see [13, Lemma 9]).

(E.3) Intra-layer inter-bag edges of G are assigned to at most ℓk queues; on each layer, an edge of H corresponds to a subgraph of $K_{\ell,\ell}$, where the first bipartition precedes the second.

(E.4) Forward inter-layer inter-bag edges of G are assigned to at most ℓk queues; for two consecutive layers, an edge of H corresponds to a subgraph of $K_{\ell,\ell}$, where the first bipartition precedes the second.

(E.5) Symmetrically all backward inter-layer inter-bag edges of G are assigned to at most ℓk queues.

The next property follows from the proof of Lemma 9 in [13]:

(P.3) For $1 \leq i \leq r$, let (u_i, v_i) be an edge of G, such that $u_i \prec_G v_i$, $u_i \in A_{x_i}$ and $v_i \in A_{y_i}$. If all these r edges belong to one of (E.3)–(E.5) and form an r-rainbow in $\prec_G$, while edges $(x_1, y_1), \ldots, (x_r, y_r)$ of H are assigned to the same queue in $QL(H)$, then $r \leq \ell$ and either $u_1, \ldots, u_r$ or $v_1, \ldots, v_r$ belong to the same bag of the H-partition of G.

If G is maximal plane, few more ingredients are needed to apply Lemma 1. A *vertical path* of G in a BFS-layering $\mathcal{L}$ is a path $P = v_0, \ldots, v_k$ of G consisting only of edges of the BFS-tree of $\mathcal{L}$ and such that if v_0 belongs to V_i in $\mathcal{L}$, then v_j belongs to V_{i+j}, with $j = 1, \ldots, k$. Further, we say that v_0 and v_k are the *first* and *last* vertices of P. A *tripod* of G consists of up to three pairwise vertex-disjoint vertical paths in $\mathcal{L}$ whose last vertices form a clique of size at most 3 in G. We refer to this clique as the *base* of the tripod. Dujmović et al. [13] showed that for any BFS-layering $\mathcal{L}$, G admits an H-partition with the following properties:

(P.4) H is a planar 3-tree and thus $QL(H)$ is a k-queue layout with $k \leq 5$ [1].

(P.5) Its layered-width ℓ is at most 3, since each bag induces a *tripod* in G, whose base is a triangular face of G, if it is a 3-clique.

Properties (P.4) and (P.5) along with Eq. (1) imply that the queue number of planar graphs is at most $3 \cdot 3 \cdot 5 + \lfloor \frac{3}{2} \cdot 3 \rfloor = 49$.

3   Refinements of the Involved Techniques

In this section, we present refinements of the algorithms outlined in Sect. 2.

Outerplanar Graphs. We modify the algorithm by Dujmović et al. [15] to guarantee two additional properties (stated in Lemma 2) of the outerplanar drawing. To this end, besides Invariants (O.1) and (O.2), we maintain a third one:

(O.3) The lower envelope consists of a single edge.

To achieve this, we use the fact that a biconnected maximal outerplane graph with at least four vertices contains at least two non-adjacent degree-2 vertices, x and y. We fix one of them, say x, and we do not remove x at any recursive step in the approach by Dujmović et al. [15]. Hence, x will be drawn at the base of the recursion, such that the bottom envelope consists of one edge incident to x that lies on the outer face of G. Subsequent vertices are always added along the upper envelope, which guarantees the invariant (see [5] for a formal proof).

Let $\langle u, v, w \rangle$ be a face of drawing $\Gamma(G)$ such that $y(u) - y(w) = 2$ and $y(u) - y(v) = y(v) - y(w) = 1$. We refer to vertices u, v and w as the *top*, *middle* and *bottom* vertex of the face, respectively.[3] Further, we say that face $\langle u, v, w \rangle$ is a *bottom*, *side* and *top triangle* for vertices u, v and w, respectively.

Lemma 2. *Let $\Gamma(G)$ be an outerplanar drawing satisfying Invariants (O.1)–(O.3) of a biconnected maximal outerplane graph G. Then, each vertex of G is*

(a) the top vertex of at most two triangular faces of $\Gamma(G)$ and
(b) the side vertex of at most two triangular faces of $\Gamma(G)$.

Proof. For (a), consider a vertex u of G. If u is the top vertex of a face, then u is incident to a span-2 edge (u, v) with $y(u) > y(v)$. By Invariant (O.3), u is a successor of v in the recursive approach by Dujmović et al. [15], i.e., when u is placed in $\Gamma(G)$, vertex v belongs to the upper envelope. We now claim that u cannot be incident to two edges (u, v) and (u, v') with the properties mentioned above; this claim implies the lemma. Assuming the contrary, by Invariant (O.2), when u is placed in $\Gamma(G)$ at most one edge incident to u has span 2. So, at most one of (u, v) and (u, v') is drawn when u is placed in $\Gamma(G)$, which implies that at least one of v and v', say v', is a successor of u. Thus, $y(u) < y(v')$ holds; a contradiction. The proof of (b) is deferred to [5]. □

Planar 3-Trees. To maintain Invariant (T.3), Alam et al. [1] use the algorithm by Dujmović et al. [15] to assign the level edges of $L_0, \ldots, L_\lambda$ of the input plane 3-tree H to two queues $\mathcal{Q}_0$ and $\mathcal{Q}_1$, since on each level these edges induce a (not necessarily connected) outerplane graph. Unlike in the original algorithm, in our approach we adopt the above modification for the algorithm by Dujmović et al. [15]. As Invariants (O.1) and (O.2) are preserved, queues $\mathcal{Q}_0$ and $\mathcal{Q}_1$ suffice.

To maintain Invariant (T.4), Alam et al. [1] adopt the following assignment scheme for the binding edges between L_j and L_{j+1} to queues $\mathcal{Q}_2$, $\mathcal{Q}_3$ and $\mathcal{Q}_4$, for each $j = 0, \ldots, \lambda - 1$. Consider a binding edge (u, v) with $u \in L_j$ and $v \in L_{j+1}$. Then, vertex u belongs to a connected component C_u of the subgraph H_j of H induced by the level-L_j vertices, while vertex v belongs to a connected component C_v of H_{j+1}. Further, C_u is outerplane and its 2-queue layout has been computed by the algorithm by Dujmović et al. [15], while C_v resides in the interior of a triangular face $\mathcal{T}_v$ of C_u in the embedding of H, such that u is on the boundary of $\mathcal{T}_v$. Edge (u, v) is assigned to $\mathcal{Q}_2$, $\mathcal{Q}_3$ or $\mathcal{Q}_4$ if and only if u is the top, middle or bottom vertex of $\mathcal{T}_v$, respectively [1]. For a vertex $u \in L_j$ with $j = 0, \ldots, \lambda - 1$, we denote by $N_2(u)$, $N_3(u)$ and $N_4(u)$ the neighbors of u in L_{j+1} such that the edges connecting them to u are assigned to queue $\mathcal{Q}_2$, $\mathcal{Q}_3$ and $\mathcal{Q}_4$, respectively.

---

[3] Alam et al. [1] refer to the middle vertex of a triangular face in $\Gamma(G)$ as its *anchor*.

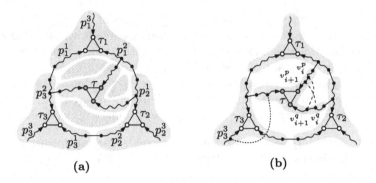

Fig. 5. (a) Tripod τ with parents τ_1, τ_2 and τ_3. (b) Illustration for (P.6) and (P.8).

Lemma 3. *For $j = 0, \ldots, \lambda - 1$, let $u \in L_j$ be a vertex of H in our modification of the peeling-into-levels approach by Alam et al. [1]. Then, the vertices of $N_2(u)$ precede those of $N_3(u)$, and the vertices of $N_2(u)$ (or of $N_3(u)$) belong to at most two connected components of H_{j+1} (residing within distinct faces of H_j).*

Proof. The first part of the lemma is proven in [1]. For the second part, consider a binding edge (u, v) with $v \in N_2(u)$; a similar argument applies when $v \in N_3(u)$. Thus, u is the top vertex of the triangular face $\mathcal{T}_v$ of H_j, in which the connected component C_v of H_{j+1} that contains v resides. Since by Lemma 2(a) vertex u can be the top vertex of at most two triangular faces of H_j, there exist at most two connected components of H_{j+1}, to which vertex v can belong. $\square$

General Planar Graphs. Dujmović et al. [13] recursively compute the bags (i.e., the tripods) of the H-partition. Each newly discovered tripod τ is adjacent to at most three tripods τ_1, τ_2 and τ_3 already discovered. We say that τ_1, τ_2 and τ_3 are the *parents* of τ; see Fig. 5a. Also, each of the three vertical paths of τ is connected to only one of its parents via an edge of the BFS-tree of $\mathcal{L}$ (black in Fig. 5a). This property gives rise to at most three sub-instances (gray in Fig. 5a), which are processed recursively to compute the final tripod decomposition. For $i \in \{1, 2, 3\}$, let p_i^1, p_i^2 and p_i^3 be the three vertical paths of τ_i (if any). Up to renaming, assume that τ lies in the cycle bounded by (parts of) p_1^1, p_1^2, p_2^1, p_2^2, p_3^1 and p_3^2 as in Fig. 5a. The next properties follow by planarity and the BFS-layering:

(P.6) There is no edge connecting a vertex of τ to a vertex of p_i^3 for $i = 1, 2, 3$; see the dotted edge in Fig. 5b.

(P.7) Let v_i^p be the vertex of vertical path p of τ on layer V_i of $\mathcal{L}$. For two vertical paths p and q of τ, edge (v_i^p, v_j^q) belongs to G only if $|i - j| \leq 1$.

(P.8) For vertical paths p and q of τ, at most one of the edges (v_i^p, v_{i+1}^q) and (v_{i+1}^p, v_i^q) exists in G; see the dashed edges of Fig. 5b.

Note that (P.6)–(P.8) hold even if τ has less than three parents, or if the cycle bounding the region of τ does not contain two vertical paths of each parent tripod.

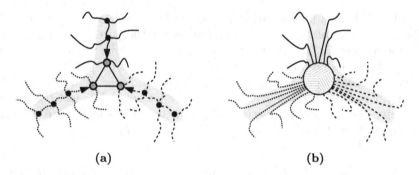

Fig. 6. (a) A tripod τ in G where the edges incident to its three vertical paths are drawn dotted, and (b) the result after contracting τ to v_τ.

In the original algorithm by Dujmović et al. [13], each vertex v_τ in H corresponds to a tripod τ in G, and an edge $(v_\tau, v_{\tau'})$ exists in H, if τ is a parent of τ' in G, or vice versa. Also, H is a connected partial planar 3-tree, which is arbitrarily augmented to a maximal planar 3-tree H' (to compute a 5-queue layout of it). Here, we adopt a particular augmentation to guarantee an additional property for the graph H' (see Lemma 4). Similarly to the original approach, we contract the vertices of each tripod τ of G to a single vertex v_τ. However, in our modification, we keep self-loops that occur when an edge of G has both endpoints in τ (unless this edge belongs to one of its vertical paths), as well as, parallel edges that occur when two vertices of τ have a common neighbor not in τ. Since G is planar, this contraction ensures the next two properties; see Fig. 6:

(P.9) The edges around each contracted vertex v_τ appear in the same clockwise cyclic order as they appear in a clockwise traversal along τ in G.
(P.10) The edges having at least one endpoint on the same vertical path of τ appear consecutively around v_τ; see Fig. 6b.

To guarantee simplicity, we focus on *homotopic* self-loops and pairs of parallel edges, which contain no vertex either in the interior or in the exterior region that they define. We remove such self-loops and keep one copy of such parallel edges. Then, we subdivide each self-loop twice, and for each edge with multiplicity $m > 1$, we subdivide all but one of its copies. In this way, each vertex v_τ corresponding to a tripod τ in G always lies in the interior of a separating 3-cycle C that contains all the vertices corresponding to its parent tripods on its boundary (see [5] for details). Since subdivision vertices are of degree 2, the result is a simple (possibly not maximal) planar 3-tree, which is a supergraph of H. To derive H', we augment it to maximal by adding edges, while maintaining its embedding [20].

Lemma 4. *Let v_τ and v_{τ_p} be two vertices of H' that correspond to a tripod τ and to a parent tripod τ_p of τ in G. If L_i and L_j are the levels of v_τ and v_{τ_p} in the peeling-into-levels approach for H', then $i \geq j$.*

Proof. Let C be the inclusion-minimal separating 3-cycle of H' containing v_τ in its interior and all vertices that correspond to the parent tripods of τ on its boundary. Let L_k, L_l and L_m be the levels of the three vertices of C, with $k \le l \le m$, in the peeling-into-levels approach for H'. As τ_p is a parent of τ, $j \in \{k, l, m\}$ holds. Since C is a 3-cycle and since each edge in the peeling-into-level approach is either level or binding, $m \le k + 1$ holds. The fact that v_τ lies in the interior of C and is connected to each of the vertices in H' corresponding to its parent tripods in G, implies that v_τ is on level L_{k+1}, i.e., $i = k + 1$. So, $j \le m \le k + 1 = i$ holds.

As in the original algorithm by Dujmović et al. [13], we compute a 5-queue layout $QL(H')$ of H'. However, we use our modification of the algorithm by Alam et al. [1] described earlier. Denote by $x_1, \ldots, x_h$ the vertices of the subgraph H of H' as they appear in $QL(H')$ (i.e., we ignore subdivision vertices introduced when augmenting H to H') and by $\mathcal{Q}_0, \ldots, \mathcal{Q}_4$ the queues of $QL(H')$ as described in Invariants (T.1)–(T.4). To compute the linear layout $QL(G)$ of G, we use Lemma 1, which orders the vertices of each layer V_i, with $i \ge 0$, of $\mathcal{L}$ as:

$$\overrightarrow{V_i} = A_{x_1} \cap V_i, A_{x_2} \cap V_i, \ldots, A_{x_h} \cap V_i,$$

where $A_{x_1}, \ldots, A_{x_h}$ are the bags (i.e., the tripods) of the H-partition of G. Unlike in the original algorithm, we do not order the vertices in each subset $A_{x_j} \cap V_i$, with $j \in \{1, \ldots, h\}$, arbitrarily. Instead, we carefully choose their order as follows. Let τ be the tripod of bag A_{x_j}. Then, $A_{x_j} \cap V_i$ contains at most one vertex of each vertical path of τ. We will order the three vertical paths of τ, which defines the order of the (at most three) vertices of $A_{x_j} \cap V_i$ for every $i \ge 0$.

Let L_l be the level of v_τ in the peeling-into-levels of H', with $0 \le l < \lambda$. By Lemma 3, there are at most four connected components c_s^1, c_s^2, c_t^1 and c_t^2 of the subgraph H'_{l+1} of H' induced by the vertices of L_{l+1}, such that the edges connecting v_τ to vertices of c_s^1 and c_s^2 (c_t^1 and c_t^2) belong to $\mathcal{Q}_2$ ($\mathcal{Q}_3$, respectively). Let c be one of c_s^1, c_s^2, c_t^1 and c_t^2; c may contain vertices that correspond to tripods in G (i.e., not to subdivisions introduced while augmenting H to H'). We refer to the union of these vertices of G as the *tripod-vertices* of c. By Invariant (T.1), v_τ precedes the vertices of c_s^1, c_s^2, c_t^1 and c_t^2 in $QL(H')$. Also by (T.2), we may assume that the vertices of c_s^1 (c_t^1) precede those of c_s^2 (c_t^2, respectively). Additionally, Lemma 3 ensures that the vertices of c_s^2 precede those of c_t^1.

Since an edge $(v_\tau, v_{\tau'})$ exists in H, if τ is a parent of τ', or vice versa, by Lemma 4, for each vertex $v_{\tau'}$ of H that is a neighbor of v_τ in one of c_s^1, c_s^2, c_t^1 and c_t^2, it follows that τ is a parent of τ'. By Property (P.6), there is a vertical path of τ in G, say p (q), such that no tripod-vertex of c_s^2 (c_t^2, respectively) is adjacent to it in G. Note that p and q might be the same vertical path of τ.

We now describe the order of the three vertical paths of τ. We only specify the first one; the other two can be arbitrarily ordered: (i) if the tripod-vertices of c_s^1 and c_s^2 are connected to all three vertical paths of τ in G, then p is the first vertical path of τ; (ii) if the tripod-vertices of c_t^1 and c_t^2 are connected to all

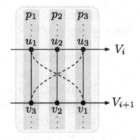

Fig. 7. Illustration for the proof of Lemma 5.

three vertical paths of τ in G, then q is the first vertical path of τ; (iii) otherwise, any vertical path of τ can be first. We prove in [5] that Cases (i) and (ii) cannot apply simultaneously. Next, we state two important implications of the described choice for the first vertical path of τ.

(P.11) Under our assumption that all vertices of c_s^1 precede those of c_s^2, if tripod-vertices of c_s^1 and c_s^2 are connected to all three vertical paths of τ, then tripod-vertices of c_s^2 are not connected to the first vertical path of τ.

(P.12) Also, under our assumption that all vertices of c_t^1 precede those of c_t^2, if tripod-vertices of c_t^1 and c_t^2 are connected to all three vertical paths of τ, then tripod-vertices of c_t^2 are not connected to the first vertical path of τ.

4   Reducing the Bound

For intra-bag inter-layer edges (E.2), the original algorithm by Dujmović et al. [13] uses three queues, since $\ell = 3$. We prove that no three intra-bag inter-layer edges form a 3-rainbow, implying that the upper bound on the queue number of planar graphs can be improved from 49 to 48.

Lemma 5. *In the queue layout computed by our modification of the algorithm by Dujmović et al. [13], no three intra-bag inter-layer edges of G form a 3-rainbow.*

Proof. Assume to the contrary that there exist three such edges (u_1, v_1), (u_2, v_2) and (u_3, v_3) forming a 3-rainbow in $QL(G)$ so that $u_1 \prec_G u_2 \prec_G u_3 \prec_G v_3 \prec_G v_2 \prec_G v_1$. By (P.1) these edges belong to the same bag A of the H-partition, while by (P.2) their endpoints belong to two consecutive layers V_i and V_{i+1} of $\mathcal{L}$. Due to the order, $u_1, u_2, u_3 \in V_i$ and $v_1, v_2, v_3 \in V_{i+1}$. The order of $A \cap V_i$ and $A \cap V_{i+1}$ is $u_1 \prec_G u_2 \prec_G u_3$ and $v_3 \prec_G v_2 \prec_G v_1$; see Fig. 7. Let p_1, p_2 and p_3 be the first, second and third vertical paths of tripod τ forming A. Then, $(u_1, v_3) \in p_1$, $(u_2, v_2) \in p_2$ and $(u_3, v_1) \in p_3$. However, (u_1, v_1) and (u_3, v_3) contradict (P.8). ∎

For inter-bag edges (E.3)–(E.5), the algorithm by Dujmović et al. [13] uses $3 \cdot 15$ queues, since $k = 5$ and $\ell = 3$. We exploit (P.3) to prove that $3 \cdot 13$ queues suffice. This further improves the upper bound on the queue number of planar graphs from 48 to 42 (see [5] for omitted details).

Lemma 6. *In the queue layout computed by our modification of the algorithm by Dujmović et al. [13], the inter-bag edges of G do not form a 40-rainbow.*

Proof sketch. We partition the inter-bag edges into 15 sets $E_j^i = \{(u,v) \in (\text{E.j}) : u \in A_x, v \in A_y, (x,y) \in \mathcal{Q}_i\}$, where $i \in \{0, \ldots, 4\}$ and $j \in \{3, 4, 5\}$. We prove that the edges of E_j^i do not form a 3-rainbow, when $i \in \{2, 3\}$ and $j \in \{3, 4, 5\}$. Hence, we save one queue for each of the six sets. For a contradiction, assume that (u_1, v_1), (u_2, v_2) and (u_3, v_3) form a 3-rainbow in E_3^2; similarly we argue for the other sets. By (P.3) and (T.4), u_1, u_2 and u_3 belong to the same bag of the H-partition; in particular, to the first, second and third vertical paths of the tripod τ of this bag. By Lemma 4 followed by Lemma 3, v_1 v_2 and v_3 are tripod-vertices of exactly two connected components c_s^1 and c_s^2 of H'. In fact, if the vertices of c_s^1 precede those of c_s^2 in $QL(H')$, then v_1 is a tripod-vertex of c_s^2. Further, the tripod-vertices of c_s^1 and c_s^2 are connected to all three vertical paths of τ in G, and (P.11) implies that tripod-vertices of c_s^2, are not adjacent to the first vertical path of τ. Hence, (u_1, v_1) cannot exist in G; a contradiction.

We are now ready to state the main theorem of this section.

Theorem 1. *Every planar graph has queue number at most 42.*

5 Conclusions

We believe that our approach has the potential to further reduce the upper bound by at least 3 (i.e., from 42 to 39). However, more elaborate arguments that exploit deeper the planarity of the graph are required, and several details need to be worked out. Still the gap with the lower bound of 4 remains large and needs to be further reduced. In this regard, determining the exact queue number of planar 3-trees becomes critical, since an improvement of the current upper bound of 5 (to meet the lower bound of 4) directly implies a corresponding improvement on the upper bound on the queue number of general planar graphs. On the other hand, to obtain a better understanding of the general open problem, it is also reasonable to further examine subclasses of planar graphs, such as bipartite planar graphs or planar graphs with bounded degree (e.g., max-degree 3).

Acknowledgments. The authors would like to thank the anonymous referees for useful comments and suggestions.

References

1. Alam, J.M., Bekos, M.A., Gronemann, M., Kaufmann, M., Pupyrev, S.: Queue layouts of planar 3-trees. Algorithmica **82**(9), 2564–2585 (2020). https://doi.org/10.1007/s00453-020-00697-4

2. Bannister, M.J., Devanny, W.E., Dujmovic, V., Eppstein, D., Wood, D.R.: Track layouts, layered path decompositions, and leveled planarity. Algorithmica **81**(4), 1561–1583 (2019). https://doi.org/10.1007/s00453-018-0487-5

3. Barth, D., Pellegrini, F., Raspaud, A., Roman, J.: On bandwidth, cutwidth, and quotient graphs. RAIRO Theor. Inform. Appl. **29**(6), 487–508 (1995). https://doi.org/10.1051/ita/1995290604871

4. Bekos, M.A., et al.: Planar graphs of bounded degree have bounded queue number. SIAM J. Comput. **48**(5), 1487–1502 (2019). https://doi.org/10.1137/19M125340X

5. Bekos, M.A., Gronemann, M., Raftopoulou, C.N.: On the queue number of planar graphs. CoRR arXiv:2106.08003 (2021)

6. Bekos, M.A., Kaufmann, M., Klute, F., Pupyrev, S., Raftopoulou, C.N., Ueckerdt, T.: Four pages are indeed necessary for planar graphs. J. Comput. Geom. **11**(1), 332–353 (2020). https://journals.carleton.ca/jocg/index.php/jocg/article/view/504

7. Bernhart, F., Kainen, P.C.: The book thickness of a graph. J. Comb. Theory Ser. B **27**(3), 320–331 (1979). https://doi.org/10.1016/0095-8956(79)90021-2

8. Chinn, P.Z., Chvatalova, J., Dewdney, A.K., Gibbs, N.E.: The bandwidth problem for graphs and matrices - a survey. J. Graph Theory **6**(3), 223–254 (1982). https://doi.org/10.1002/jgt.3190060302

9. Di Battista, G., Frati, F., Pach, J.: On the queue number of planar graphs. SIAM J. Comput. **42**(6), 2243–2285 (2013). https://doi.org/10.1137/130908051

10. Díaz, J., Petit, J., Serna, M.J.: A survey of graph layout problems. ACM Comput. Surv. **34**(3), 313–356 (2002). https://doi.org/10.1145/568522.568523

11. Dujmović, V., Frati, F.: Stack and queue layouts via layered separators. J. Graph Algorithms Appl. **22**(1), 89–99 (2018). https://doi.org/10.7155/jgaa.00454

12. Dujmović, V., Joret, G., Micek, P., Morin, P., Ueckerdt, T., Wood, D.R.: Planar graphs have bounded queue-number. In: Zuckerman, D. (ed.) FOCS, pp. 862–875. IEEE Computer Society (2019). https://doi.org/10.1109/FOCS.2019.00056

13. Dujmovic, V., Joret, G., Micek, P., Morin, P., Ueckerdt, T., Wood, D.R.: Planar graphs have bounded queue-number. J. ACM **67**(4), 22:1–22:38 (2020)

14. Dujmović, V., Morin, P., Wood, D.R.: Layout of graphs with bounded tree-width. SIAM J. Comput. **34**(3), 553–579 (2005). https://doi.org/10.1137/S0097539702416141

15. Dujmović, V., Pór, A., Wood, D.R.: Track layouts of graphs. Discrete Math. Theor. Comput. Sci. **6**(2), 497–522 (2004). http://dmtcs.episciences.org/315

16. Heath, L.S.: Embedding planar graphs in seven pages. In: FOCS, pp. 74–83. IEEE Computer Society (1984). https://doi.org/10.1109/SFCS.1984.715903

17. Heath, L.S., Leighton, F.T., Rosenberg, A.L.: Comparing queues and stacks as mechanisms for laying out graphs. SIAM J. Discrete Math. **5**(3), 398–412 (1992). https://doi.org/10.1137/0405031

18. Heath, L.S., Rosenberg, A.L.: Laying out graphs using queues. SIAM J. Comput. **21**(5), 927–958 (1992). https://doi.org/10.1137/0221055

19. Horton, S.B., Parker, R.G., Borie, R.B.: On minimum cuts and the linear arrangement problem. Discrete Appl. Math. **103**(1–3), 127–139 (2000). https://doi.org/10.1016/S0166-218X(00)00173-6

20. Kratochvíl, J., Vaner, M.: A note on planar partial 3-trees. CoRR arXiv:1210.8113 (2012)

21. Wiechert, V.: On the queue-number of graphs with bounded tree-width. Electr. J. Comb. **24**(1), P1.65 (2017). http://www.combinatorics.org/ojs/index.php/eljc/article/view/v24i1p65

22. Wood, D.R.: Queue layouts, tree-width, and three-dimensional graph drawing. In: Agrawal, M., Seth, A. (eds.) FSTTCS 2002. LNCS, vol. 2556, pp. 348–359. Springer, Heidelberg (2002). https://doi.org/10.1007/3-540-36206-1_31
23. Yannakakis, M.: Embedding planar graphs in four pages. J. Comput. Syst. Sci. **38**(1), 36–67 (1989). https://doi.org/10.1016/0022-0000(89)90032-9

Contact and Visibility Representations

Optimal-Area Visibility Representations
of Outer-1-Plane Graphs

Therese Biedl[1(✉)] [ID], Giuseppe Liotta[2] [ID], Jayson Lynch[1] [ID],
and Fabrizio Montecchiani[2] [ID]

[1] David R. Cheriton School of Computer Science, University of Waterloo,
Waterloo, Canada
{biedl,jayson.lynch}@uwaterloo.ca
[2] Department of Engineering, University of Perugia, Perugia, Italy
{giuseppe.liotta,fabrizio.montecchiani}@unipg.it

Abstract. This paper studies optimal-area visibility representations of
n-vertex outer-1-plane graphs, i.e. graphs with a given embedding where
all vertices are on the boundary of the outer face and each edge is
crossed at most once. We show that any graph of this family admits
an embedding-preserving visibility representation whose area is $O(n^{1.5})$
and prove that this area bound is worst-case optimal. We also show
that $O(n^{1.48})$ area can be achieved if we represent the vertices as L-
shaped orthogonal polygons or if we do not respect the embedding but
still have at most one crossing per edge. We also extend the study to
other representation models and, among other results, construct asymptotically optimal $O(n\,pw(G))$ area bar-1-visibility representations, where
$pw(G) \in O(\log n)$ is the pathwidth of the outer-1-planar graph G.

Keywords: Visibility representations · Outer-1-plane graphs ·
Optimal area

1 Introduction

Visibility representations are one of the oldest topics studied in graph drawing:
Otten and van Wijk showed in 1978 that every planar graph has a visibility
representation [38]. A *rectangle visibility representation* consists of an assignment
of disjoint axis-parallel boxes to vertices, and axis-parallel segments to edges in
such a way that edge-segments end at the vertex boxes of their endpoints and do
not intersect any other vertex boxes. (They can hence be viewed as lines-of-sight,
though not every line-of-sight needs to give rise to an edge.)

Vertex-boxes are permitted to be degenerated into a segment or a point
(in our pictures we thicken them slightly for readability). The construction by

---

Work of TB supported by NSERC; FRN RGPIN-2020-03958. Work of GL supported
by MIUR, grant 20174LF3T8 "AHeAD: efficient Algorithms for HArnessing networked
Data". Work of FM supported by Dipartimento di Ingegneria, Università degli Studi
di Perugia, grant RICBA19FM.

H. C. Purchase and I. Rutter (Eds.): GD 2021, LNCS 12868, pp. 287–303, 2021.
https://doi.org/10.1007/978-3-030-92931-2_21

Otten and van Wijk is also *uni-directional* (all edges are vertical) and all vertices are *bars* (horizontal segments or points). Multiple other papers studied uni-directional bar-visibility representations and showed that these exist if and only if the graph is planar [24,39,41,42].

Unless otherwise specified, we assume throughout this paper that any visibility representation Γ (as well as the generalizations we list below) are on an *integer grid*. This means that all corners of vertex polygons, as well as all *attachment points* (places where edge-segments end at vertex polygons) have integer coordinates. The *height [width]* of Γ is the number of grid rows [columns] that intersect Γ. The *area* of Γ is its width times its height. Any visibility representation can be assumed to have area $O(n^2)$ (see also Observation 1). Efforts have been made to obtain small constants factors [28,32].

In this paper, we focus on *bi-directional* rectangle visibility representations, i.e., both horizontally and vertically drawn edges are allowed. For brevity we drop 'bi-directional' and 'rectangle' from now on. Recognizing graphs that have a visibility representation is NP-hard [40]. Planar graphs have visibility representations where the area is at most n^2, and $\Omega(n^2)$ area is sometimes required [30]. For special graph classes, $o(n^2)$ area can be achieved, such as $O(n \cdot pw(G))$ area for outer-planar graphs [5] (here $pw(G)$ denotes the *pathwidth* of G, defined later), and $O(n^{1.5})$ area for series-parallel graphs [4]. The latter two results do not give embedding-preserving drawings (defined below).

Variations of Visibility Representations. For graphs that do not have visibility representations (or where the area-requirements are larger than desired), other models have been introduced that are similar but more general. One option is to increase the dimension, see e.g. [1,2,14]. We will not do this here, and instead allow more complex shapes for vertices or edges. Define an *orthogonal polygon [polyline]* to be a polygon [polygonal line] whose segments are horizontal or vertical. We use OP as convenient shortcut for 'orthogonal polygon'. All variations that we study below are what we call *OP-∞-orthogonal drawings*.[1] Such a drawing is an assignment of disjoint orthogonal polygons $P(\cdot)$ to vertices and orthogonal poly-lines to edges such that the poly-line of edge (u,v) connects $P(u)$ and $P(v)$. Edges can intersect each other, and they are specifically *allowed* to intersect arbitrarily many vertex-polygons (hence the "∞"), but no two edge-segments are allowed to overlap each other. The *vertex complexity* is the maximum number of reflex corners in a vertex-polygon, and the *bend complexity* is the maximum number of bends in an edge-poly-line.

One variation that has been studied is *bar-(k,j)-visibility representation*, where vertices are bars, edges are vertical line segments, edges may intersect up to k bars that are not their endpoints, and any vertex-bar is intersected by at most j edges that do not end there. Bar-(k,∞)-visibility representations were introduced by Dean et al. [19], and testing whether a graph has one is textsfNP-hard [17]. All 1-planar graphs have a bar-$(1,1)$-visibility representation [15,26].

---

[1] We do not propose actually drawing graphs in this model (its readability would not be good), but it is convenient as a name for "all drawing models that we study here".

In this paper, we will use *bar-1-visibility representation* as a convenient shortcut for "unidirectional bar-$(1,1)$-visibility representation".

Another variation is *OP visibility representation*, where edges must be horizontal or vertical segments that do not intersect vertices except at their endpoints. OP visibility representations were introduced by Di Giacomo et al. [22] and they exist for all 1-planar graphs. There are further studies, considering the vertex complexity that may be required in such drawings [16,22,27,36,37].

Finally, there are *orthogonal box-drawings*, where vertices must be boxes and edges do not intersect vertices except at their endpoints. We will not review the (vast) literature on orthogonal box-drawings (see e.g. [10,13] and the references therein), but they exist for all graphs.

All OP-∞-orthogonal drawings can be assumed to have area $O(n^2)$ (assuming constant complexity and $O(n)$ edges), see also Observation 1. We are not aware of any prior work that tries to reduce the area to $o(n^2)$ for specific graph classes.

Drawing Outer-1-planar Graphs. An *outer-1-planar graph* (first defined by Eggleton [25]) is a graph that has a drawing Γ in the plane such that all vertices are on the infinite region of Γ and every edge has at most one crossing. We will not review the (extensive) literature on their superclass of 1-planar graphs here; see e.g. [35] or [23,34] for even more related graph classes. Outer-1-planar graphs can be recognized in linear time [3,33]. All outer-1-planar graphs are planar [3], and so can be drawn in $O(n^2)$ area, albeit not embedding-preserving.

Very little is know about drawing outer-1-planar graphs in area $o(n^2)$. Auer et al. [3] claimed to construct planar visibility representations of area $O(n \log n)$, but this turns out to be incorrect [7] since some outer-1-planar graphs require $\Omega(n^2)$ area in planar drawings. Outer-1-planar graphs do have orthogonal box-drawings with bend complexity 2 in $O(n \log n)$ area [7].

Our Results. We study visibility representations (and variants) of outer-1-planar graphs, especially drawings that preserve the given outer-1-planar embedding. Table 1 gives an overview of all results that we achieve. As our main result, we give tight upper and lower bounds on the area of embedding-preserving visibility representations (Sect. 3 and 4): It is $\Theta(n^{1.5})$. We find it especially interesting that the lower bound is neither $\Theta(n \log n)$ nor $\Theta(n^2)$ (the most common area lower bounds in graph drawing results). Also, a tight area bound is not known for embedding-preserving visibility representations of outerplanar graphs.

We also show in Sect. 5 that the $\Omega(n^{1.5})$ area bound can be undercut if we relax the drawing-model slightly, and show that area $O(n^{1.48})$ can be achieved in three other drawing models. Finally we give further area-optimal results in other drawing models in Sect. 6. To this end, we generalize a well-known lower bound using the pathwidth to *all* OP-∞-orthogonal drawings, and also develop an area lower bound for the planar visibility representations of outer-1-planar graphs based on the number of crossings in an outer-1-planar embedding. Then we give constructions that show that these can be matched asymptotically. We conclude in Sect. 7 with open problems.

For space reasons we only sketch the proofs of most theorems; a $(\star)$ symbol indicates that further details can be found in [12].

Table 1. Upper and lower bound on the area achieved in various drawing styles in this paper. The column title e-p stands for embedding-preserving, $pw(G)$ denotes the pathwidth of G, $\chi(G)$ denotes the number of crossings in the 1-planar embedding.

Drawing-style	e-p	Lower bound	Upper bound
Visibility representation	✓	$\Omega(n^{1.5})$ [Theorem 1]	$O(n^{1.5})$ [Theorem 3]
Complexity-1 OP vis.repr.	✓	$\Omega(n\,pw(G))$ [Theorem 6]	$O(n^{1.48})$ [Theorem 4]
1-bend orth. box-drawing	✓	$\Omega(n\,pw(G))$ [Theorem 6]	$O(n^{1.48})$ [Theorem 4]
Visibility representation	✗	$\Omega(n\,pw(G))$ [Theorem 6]	$O(n^{1.48})$ [Theorem 4]
Bar visibility representation	✓	$\Omega(n^2)$ [Theorem 2]	$O(n^2)$ [Theorem 5]
Bar-1-visibility representation	✗	$\Omega(n\,pw(G))$ [Theorem 6]	$O(n\,pw(G))$ [Theorem 8]
Planar visibility representation	✗	$\Omega(n(pw(G)+\chi(G)))$ [Theorem 6 & 7]	$O(n(pw(G)+\chi(G)))$ [Theorem 8]

2 Preliminaries

We assume familiarity with standard graph drawing terminology [21]. Throughout the paper, n and m denotes the number of vertices and edges.

A planar drawing of a graph subdivides the plane into topologically connected regions, called *faces*. The unbounded region is called the *outer-face*. An *embedding* $\mathcal{E}(G)$ of a graph G is an equivalence class of drawings whose *planarizations* (i.e., planar drawings obtained after replacing crossing points by dummy vertices) define the same set of circuits that bound faces. An *outer-1-planar drawing* is a drawing with at most one crossing per edge and all vertices on the outer-face. An *outer-1-planar graph* is a graph admitting an outer-1-planar drawing. An *outer-1-plane graph* G is a graph with a given outer-1-planar embedding $\mathcal{E}(G)$. We use $\chi(G)$ for the number of crossings in $\mathcal{E}(G)$. An outer-1-plane graph G is *plane-maximal* if it is not possible to add any uncrossed edge without losing outer-1-planarity or simplicity. The *planar skeleton* of an outer-1-plane graph G, denoted by $\overline{G}$, is the graph induced by its uncrossed edges. If G is plane-maximal, then $\overline{G}$ is a 2-connected graph whose interior faces have degree 3 or 4 [20]. Let $\overline{G}^*$ be the weak dual of $\overline{G}$ and call it the *inner tree* of G. Since $\overline{G}$ is outer-plane, $\overline{G}^*$ is a tree (as the name suggests), and since each face of $\overline{G}$ has degree 3 or 4, every vertex of $\overline{G}^*$ has degree at most 4. An *outer-1-path* P is an outer-1-plane graph whose inner tree $\overline{P}^*$ is a path.

Consider a graph G with a fixed embedding $\mathcal{E}(G)$. An OP-∞-orthogonal drawing is *embedding-preserving* if (1) walking around each vertex-polygon we encounter the incident edges in the same cyclic order as in $\mathcal{E}(G)$, and (2) no edge crosses a vertex, and the planarization of the OP-∞-orthogonal drawing has the same set of faces as $\mathcal{E}(G)$. Note that bar-1-visibility representations by definition violate (2), but we call them embedding-preserving if (1) holds.

Our results will only consider the smaller dimension of the drawing (up to rotation the height), because the other dimension does not matter (much):

Observation 1. *Let Γ be an OP-∞-orthogonal drawing with constant vertex and bend complexity. Then we may assume that the width and height is $O(n+m)$.*

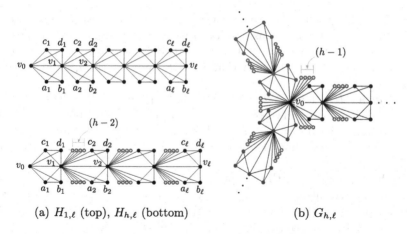

(a) $H_{1,\ell}$ (top), $H_{h,\ell}$ (bottom)   (b) $G_{h,\ell}$

Fig. 1. The graph for Lemma 1. For interpretation of the colors in this and the next figures, the reader is referred to the online coloured version of this article.

Observation 2. *Let Γ be an OP-∞-orthogonal drawing with constant vertex complexity in a $W \times H$-grid. Then $\max\{W, H\} \in \Omega(\text{maximum degree of } G)$.*

Observation 1 holds because we can delete empty rows and columns (and was mentioned for visibility representations in [6]); Observation 2 holds since some vertex-polygons must have sufficient width or height for its incident edges. ($\star$)

Remark. In consequence of Observation 2, if we know a lower bound $f(G)$ on the width and height of a drawing, then (after adding degree-1 vertices to achieve maximum degree $\Theta(n)$) we know that any drawing of the resulting graph G' has (up to rotation) width $\Omega(n)$ and height $\Omega(f(G))$, so area $\Omega(n\,f(G))$. This is assuming G' is within the same graph class and $f(G') \in \Theta(f(G))$; both hold when we apply this below.

3   Lower Bound on the Height

In this section, we show that embedding-preserving visibility representations must have height $\Omega(\sqrt{n})$ for some outer-1-plane graphs. A crucial ingredient is a lemma that studies the case where the height of vertex-boxes is restricted.

Lemma 1. *For any $h, \ell > 0$ there exists an outer-1-plane graph $G_{h,\ell}$ with $O(h \cdot \ell)$ vertices such that any embedding-preserving visibility representation in which each vertex-box intersects at most h rows, has width and height $\Omega(\ell)$.*

Proof. To build graph $G_{h,\ell}$, we first need a graph $H_{1,\ell}$ with $5\ell + 1$ vertices, depicted in Fig. 1(a) (top). This graph consists of a path $v_0, \ldots, v_\ell$ such that at each edge (v_{i-1}, v_i) (for $1 \le i \le \ell$) there are two attached K_4's $\{v_{i-1}, a_i, b_i, v_i\}$ and $\{v_{i-1}, c_i, d_i, v_i\}$ (drawn such that (v_{i-1}, b_i) crosses (v_i, a_i) and (v_{i-1}, d_i) crosses (v_i, c_i)).

Next define $H_{h,\ell}$ for $h \ge 2, \ell \ge 1$ by taking $H_{1,\ell}$ and adding $2(\ell-1)(h-2)$ vertices of degree 1 (we call these *leaves*), as shown in Fig. 1(a) (bottom). Namely, at each vertex v_i for $0 < i < \ell$, we add $h - 2$ leaves between b_i and a_{i+1} (in the order around v_i), and another $h - 2$ leaves between c_{i+1} and d_i. Clearly graph $H_{h,\ell}$ is outer-1-planar.

Graph $G_{h,\ell}$ consists of three copies of $H_{h,\ell}$, with the three vertices v_0 combined into one, see also Fig. 1(b). Furthermore, add $h - 1$ leaves at v_0 between any two copies, i.e., between c_1 of one copy and a_1 of the next copy. Graph $G_{h,\ell}$ has $n = 15\ell + 1 + 6(\ell-1)(h-2) + 3(h-1) \in \Theta(\ell h)$ vertices.

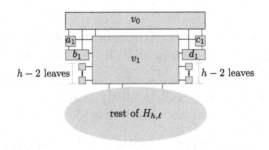

Fig. 2. Illustration for the proof of Lemma 1.

One can argue $(\star)$ that inside any embedding-preserving visibility representation Γ of $G_{h,\ell}$, there exists a copy of $H_{h,\ell}$ whose drawing Γ_H satisfies (up to symmetry) the premise of the following claim.

Claim. Let Γ_H be an embedding-preserving visibility representation of $H_{h,\ell}$ such that all edges at box $P(v_0)$ go downward, with edge (v_0, a_1) leftmost among them. Assume that all boxes of Γ_H intersect at most h rows. Then Γ_H uses at least $\ell + 1$ rows and $P(v_0)$ has width at least $4\ell + 1$.

Proof. We proceed by induction on ℓ. In the base case ($\ell = 1$) we have five vertical downward edges at v_0; this means that the height is at least 2 and $P(v_0)$ must have width at least 5 as required.

Now assume $\ell \ge 2$ and study the five downward edges from v_0 to a_1, b_1, v_1, c_1, d_1, see also Fig. 2. The vertical edges (v_0, b_1) and (v_0, d_1) are crossed by edges (a_1, v_1) and (v_1, c_1), which means that the latter two edges must be horizontal. Since (v_0, a_1) is leftmost, and the embedding is preserved, edge (a_1, v_1) attaches on the left side of $P(v_1)$ while (v_1, c_1) attaches on the right side.

The counter-clockwise order of edges at v_1 contains $h - 1$ edges (to b_1 and leaves) between a_1 and a_2. Since $P(v_1)$ intersects at most h rows, and (v_1, a_1) attaches on its left side, therefore (v_1, a_2) can *not* attach on its left side. Likewise (v_1, c_2) can *not* attach on the right side of $P(v_1)$. To preserve the embedding, therefore the edges from v_1 to the rest of $H_{h,\ell}$ must be drawn downward from $P(v_1)$, with (v_1, a_2) leftmost. Also observe that $H_{h,\ell} \setminus \{v_0, a_1, b_1, c_1, d_1\}$ contains a copy of $H_{h,\ell-1}$. Applying induction, there are at least ℓ rows below $P(v_1)$, and $P(v_1)$ has width at least $4\ell - 3$. Adding at least one row for $P(v_0)$, and observing that $P(v_0)$ must be at least four units wider than $P(v_1)$ proves the claim.   □

So the claim holds, and Γ_H (and with it Γ) has width and height $\Omega(\ell)$.   □
As a consequence, we obtain two lower bound for visibility representation.

Theorem 1. *For any N there is an n-vertex outer-1-plane graph with $n \geq N$ such that any embedding-preserving visibility representation has area $\Omega(n^{1.5})$.*

Theorem 2. *For any N there is an n-vertex outer-1-plane graph with $n \geq N$ such that any embedding-preserving bar visibility representation has area $\Omega(n^2)$.*

Roughly speaking, Theorem 1 uses $G_{\sqrt{N},\sqrt{N}}$, with leaves added to have maximum degree $\Theta(n)$, while Theorem 2 uses $G_{1,N}$. ($\star$). The bounds then hold by Lemma 1 and Observation 2.

4   Optimal Area Drawings

In this section we show how to compute an embedding-preserving visibility representation of area $O(n^{1.5})$ which is tight by Theorem 1. By Observation 1 it suffices to construct a drawing of height $O(\sqrt{n})$.

Our construction is quite lengthy, so we mostly sketch it here via figures. We assume that G is maximal-planar and a *reference-edge* (s,t) on the outer-face of G is fixed, and first choose a path π in dual tree $\overline{G}^*$ (rooted at the face incident to (s,t)). Let $F \in \Theta(n)$ be the size of $\overline{G}^*$ (hence the number of inner faces of $\overline{G}$). As shown by Chan for binary trees [18] and generalized by us to arbitrary trees [11], π can be chosen such that $\alpha^p + \beta^p \leq (1 - \delta)F^p$, where α [β] is the maximum size of a left [right] subtree of π, $p = 0.48$ and $\delta > 0$ is a constant. Define a recursive function $h(F) = \max_{\alpha,\beta}\{h(\alpha) + h(\beta)\} + O(\sqrt{F})$ (with appropriate constants and base cases). Here the maximum is over all choices of α, β that satisfy the inequality. We construct a drawing of height $h(F)$.

So we first discuss how to draw the outer-1-path P_π whose inner dual is π, plus all its *hanging subgraphs* (i.e., maximum subgraphs in $G \setminus P_\pi$). To do so, first create a visibility representation of P_π on 5 rows such that edges with attached hanging subgraphs are drawn horizontally in the top or bottom row (see Fig. 3). Assume that each hanging subgraph H has a $TC_{\sigma,\tau}$-*drawing* (for $\{\sigma, \tau\} = \{1, 2\}$), i.e., a drawing where the endpoints of the reference-edge occupy the *T*op *C*orners and have height σ and τ. Then we can easily merge all hanging subgraphs, after expanding some boxes of P_π one row outward. The resulting drawing has height $h(\alpha)+h(\beta)+O(1)$ since all hanging subgraphs below [above] correspond to left [right] subtrees of π.[2]

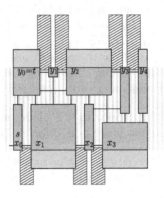

Fig. 3. Drawing an outer-1-path (dark gray) and merging hanging subgraphs (blue striped) after expanding some vertex-boxes (light gray). (Color figure online)

Alas, this path-drawing is not a $TC_{\sigma,\tau}$-drawing as required for the recursion. So we change the approach and first draw a larger subgraph that includes P_π. First extract the *cap* C_1, consisting of all neighbours of s and t, see Fig. 4. This is an outer-1-path, so we can use a path-drawing and get a $TC_{\sigma,\tau}$-drawing of the cap (see the corresponding part in Fig. 5). The part of P_π not in C_1 could be drawn as for outer-1-paths, but instead we first extract another cap C_2 at the edge common to C_1 and the rest of P_π. We draw C_2 as a path and place it (after suitable expansion of the vertices of C_1) below the drawing of C_1. This repeats k times for some parameter k of our choice ($k = 2$ in the example). Then we draw the rest of P_π (which we call *handle*) as a path.

A major difficulty is combining the drawing of the caps with the handle-drawing. Let (x_i, y_j) be the edge common to caps and handle. It is not too difficult to change the boxes of x_i and y_j to combine the two boxes that represented them in the two drawings, see also Fig. 5. The main challenge is that for the two hanging subgraphs incident to y_j, there is no suitable place to merge a $TC_{\sigma,\tau}$-drawing. To resolve this, we split these hanging subgraphs further, and can then merge all their parts after adding $d(y_j)$ more rows, where $d(y_j)$ is the number of edges that y_j has in these subgraphs (Fig. 6).

So the goal is to choose the parameter k such that $d(y_j)$ is small, because we need $d(y_j)$ additional rows beyond the $h(\alpha) + h(\beta)$ that we budget for hanging subgraphs. Each extra cap also requires $O(1)$ additional rows but changes which vertex will take on the role of y_j. Crucially, the vertices $Y_1, \dots, Y_k$ that take on the role of y_j have disjoint edge sets that count for $d(\cdot)$. Since there are $O(n)$ edges

---

[2] Readers familiar with LR-drawings [11,18,31] may notice the similarity of constructing the path-drawing with the (rotated) LR-drawing of π, except that we draw the outer-1-planar graph rather than its dual tree.

in total, there exists a $k \in O(\sqrt{n})$ such that $d(Y_k) \in O(\sqrt{n})$. With this choice of k, the recursive formula for the height hence becomes $h(\alpha) + h(\beta) + O(\sqrt{n})$, which by $F \in \Theta(n)$ and $\alpha^p + \beta^p \leq (1 - \delta)n^p$ resolves to $O(\sqrt{n})$. $(\star)$

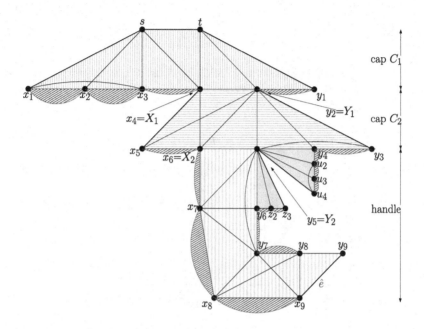

Fig. 4. Our running example.

Theorem 3. *Every n-vertex outer-1-plane graph has an embedding-preserving visibility representation of area $O(n^{1.5})$, which is worst-case optimal.*

5 Breaking the $\sqrt{n}$-barrier

We know that the height-bound of Theorem 3 is asymptotically tight due to Theorem 1. But the lower bound only holds for embedding-preserving visibility representations—can we get better height-bounds if we relax this restriction?

Theorem 4. *Any outer-1-planar graph G has*

- *An embedding-preserving OPVR of complexity 1, and*
- *An embedding-preserving 1-bend orthogonal box-drawing, and*
- *A visibility representation that is not necessarily embedding-preserving and has at most one crossing per edge,*

and the drawings have area $O(n^{1.48})$.

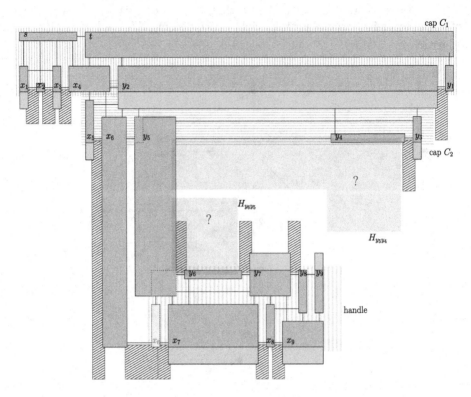

Fig. 5. Drawing the complete example except for two hanging subgraphs $H_{y_j y_{j+1}}$ and $H_{y_{j-1} y_j}$. Here $j = 5$.

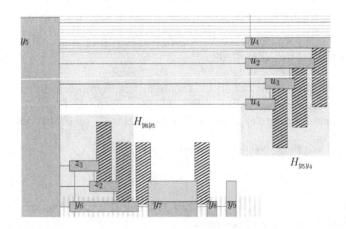

Fig. 6. Closeup on breaking up and merging $H_{y_j y_{j+1}}$ and $H_{y_j y_{j-1}}$.

We again give the proof mostly in figures. $(\star)$ We assume as in Sect. 4 that the graph is planar-maximal, a reference-edge (s,t) is given, and we construct a $TC_{\sigma,\tau}$-drawing for any given $\{\sigma,\tau\} = \{1,2\}$. We use $k = 1$, i.e., we draw one cap and use the rest of P_π as handle. Recall that the main difficulty in Sect. 4 was that two hanging subgraphs could not be merged using $TC_{\sigma,\tau}$-drawings since no suitable space was available.

If we change the drawing model (using a Γ-shape or a box in the cap-drawing for y_j) then one of these hanging subgraphs can use a $TC_{\sigma,\tau}$-drawing, and all edges can still be drawn, perhaps after adding a bend or changing the embedding. See Fig. 7. The other hanging subgraph uses a new drawing-type (i.e., different restrictions on shapes and locations of the endpoints of the reference-edge). It is not obvious that this exists, but we can show that it can be constructed by adding two rows. With this, the recursion for the height-function becomes $h(\alpha) + h(\beta) + O(1)$, which resolves to $O(n^{0.48})$ [18].

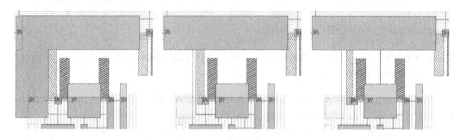

Fig. 7. Inserting the two remaining hanging subgraphs when permitting ortho-polygons, or bends in edges, or after changing the embedding.

6   Optimum-height Drawings in Other Drawing Models

In this section we give drawings whose height (and area) is also optimal, but they are in a different drawing model (hence different lower bounds apply).

6.1   Embedding-preserving Bar Visibility Representations

We proved in Theorem 2 that any embedding-preserving bar-visibility-representation has height $\Omega(n)$ for some outer-1-plane graphs. A fairly straight-forward greedy-construction shows that we can match this. The main difficulty is showing that such a drawing exists as all; the area-bound then follows from Observation 1). $(\star)$

Theorem 5. *Any outer-1-planar graph G has an embedding-preserving bar-visibility representation of area $O(n^2)$ which is worst-case optimal.*

6.2   More Lower Bounds

We now prove other lower bounds on the height that depend on the *pathwidth* $pw(G)$ and the number of crossings $\chi(G)$ of the outer-1-plane graph.

We recall that a *path decomposition* of a graph G consists of a collection $B_1, \ldots, B_\xi$ of vertex-sets ("bags") such that every vertex belongs to a consecutive set of bags, and for every edge (u, v) at least one bag contains both u and v. The *width* of such a path decomposition is $\max_i\{|B_i| - 1\}$, and the *pathwidth* $pw(G)$ is the minimum width of a path decomposition of G. Any outer-1-planar graph has pathwidth $O(\log n)$, since it has treewidth 3 [3].

For planar drawings, the width and height of a drawing is lower-bounded by the pathwidth of the graph [29].

Less is known for non-planar drawings. It follows from the proof of Corollary 3 in [9] that any bar-1-visibility representation of graph G has height at least $pw(G) + 1$. Roughly speaking, we can extract a path decomposition of G by scanning left-to-right with a vertical line and attaching a new bag whenever the set of intersected vertices changes. We use the same proof-idea here to show a lower bound for *all* OP-∞-drawings.   ($\star$)

Theorem 6. *Any OP-∞-drawing of a graph G (not necessarily outer-1-planar) has height and width $\Omega(pw(G))$.*

By the remark after Observation 2 hence some outer-1-planar graphs require area $\Omega(n\,pw(G))$ in all OP-∞-orthogonal drawings.

If we specifically look at drawings that have no crossings, then we can also create a lower bound based on the number of crossings. This is easily obtained by modifying the lower-bound example from [7].   ($\star$)

Theorem 7. *For any k and $n \geq 4k$, there exists an outer-1-plane graph with n vertices and k crossings that requires at least $2k$ height and width in any planar drawing.*

In particular, the lower bound on the height of a planar OP-∞-orthogonal drawing of G is $\Omega(\max\{pw(G), \chi(G)\})$, which is the same as $\Omega(pw(G) + \chi(G))$.

6.3   More Constructions

We now turn towards creating bar representations that prove that Theorems 6 and 7 are tight.

Theorem 8. *Every outer-1-plane graph G has a planar bar visibility representation of area $O((pw(G)+\chi(G))n)$ and a bar-1-visibility representation of area $O(n{\cdot}pw(G))$.*

We can again only sketch the proof ($\star$). We first draw the planar skeleton of some outer-1-path and the hanging subgraphs much as was done for outer-planar graphs in [5]. Based on the pathwidth (or actually the closely related parameter *rooted pathwidth*), extract a root-to-leaf path π in the dual tree $\overline{G^*}$

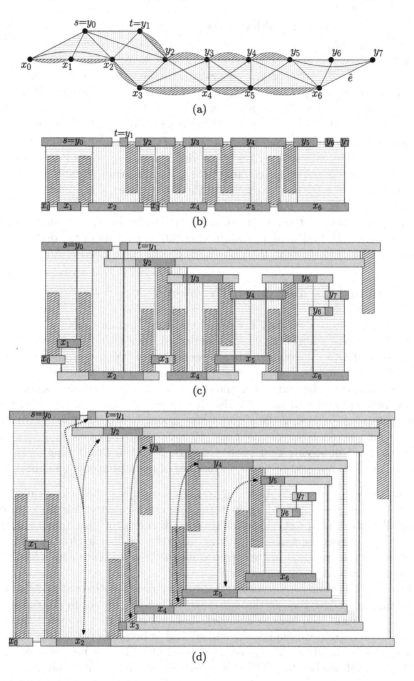

Fig. 8. (a) Example-graph. (b) Drawing its skeleton and merging hanging subgraphs inward. Some vertex-boxes are artificially wide to match (c). (c) The bar-1-visibility representation obtained by moving some bars and sometimes traversing bars. (d) The planar bar-visibility representation obtained by moving some bars and extending them rightwards. Arrows indicate vertices that get moved outward beyond their neighbour on the right.

such that the rooted pathwidth of all subtrees is smaller. Expand P_π by adding all neighbours of s to get P_π^+. Create a bar visibility representation of $\overline{P_\pi^+}$ on three rows. See Fig. 8(b). Now merge hanging subgraphs "inward", i.e., inside the faces of $\overline{P_\pi}$. They hence share rows and the height is only $O(1)$ more than the one of the subgraphs and works out to $O(pw(G))$. For the merging we need $TC_{1,1}$-drawings, but with our placement of (s,t) this can easily be achieved.

However, we have not yet drawn the crossings in P_π^+. One of each pair of crossing edges can be realized inside a face of $\overline{P_\pi^+}$. For bar-1-visibility representations, we realize the other edges by moving vertex-bars inward or outward by one unit (plus some special handling near s and t). After suitable lengthening of bars the other edge in a crossing can then be realized, sometimes by traversing a bar. See Fig. 8(c). For planar drawings, we move bars outward sufficiently far (proportionally to the number of crossings on the right) such that they can be extended rightward without intersecting other elements of the drawing. The other edge in a crossing can then be drawn on the right. See Fig. 8(d).

7   Conclusions and Open Problems

In this paper, we studied visibility representations of outer-1-planar graphs. We showed that if the embedding must be respected, then $\Omega(n^{1.5})$ area is sometimes required, and $O(n^{1.5})$ area can always be achieved. We also studied numerous other drawing models, showing that $o(n^{1.5})$ area can be achieved as soon as we allow bends in the vertices or the edges or can change the embedding. We also achieve optimal area for bar-1-visibility representations and planar visibility representations. Following the steps of our proofs, it is clear that the drawings can be constructed in polynomial time; with more care when handing subgraph-drawings (and observing that path π can be found in linear time [8,11]) the run-time can be reduced to linear. A number of open problems remain:

- Our drawings of height $O(n^{0.48})$ were based on the idea of so-called LR-drawings of trees [18], which in turn were crucial ingredients for obtaining small embedding-preserving straight-line drawings of outer-planar graphs. With a different approach, Frati et al. [31] achieved height $O(n^\varepsilon)$ for drawing outer-planar graphs. Can we achieve height $O(n^\varepsilon)$ (hence area $O(n^{1+\varepsilon})$ in some of our constructions as well?
- Our bar-1-visibility representations do not preserve the embedding, both because the edges that go through some vertex-bar are not in the right place in the rotation, and because we merge hanging subgraphs inward. What area can we achieve if we require the embedding to be preserved?
- We achieved height $O(n^{0.48})$ in complexity-1 OPVRs. It is not hard to achieve the optimal height $O(pw(G))$ if we allow higher complexity (complexity 4 is enough; we leave the details to the reader). What is the status for complexity 2 or 3, can we achieve height $o(n^{0.48})$?

Finally, are there other significant subclasses of 1-planar graphs for which we can achieve $o(n^2)$-area drawings, either straight-line or visibility representations?

References

1. Angelini, P., Bekos, M.A., Kaufmann, M., Montecchiani, F.: On 3D visibility representations of graphs with few crossings per edge. Theor. Comput. Sci. **784**, 11–20 (2019)
2. Arleo, A., et al.: Visibility representations of boxes in 2.5 dimensions. Comput. Geom. **72**, 19–33 (2018)
3. Auer, C., et al.: Outer 1-planar graphs. Algorithmica **74**(4), 1293–1320 (2016)
4. Biedl, T.: Small drawings of outerplanar graphs, series-parallel graphs, and other planar graphs. Discrete Comput. Geom. **45**(1), 141–160 (2011)
5. Biedl, T.: A 4-approximation for the height of drawing 2-connected outer-planar graphs. In: Erlebach, T., Persiano, G. (eds.) WAOA 2012. LNCS, vol. 7846, pp. 272–285. Springer, Heidelberg (2013). https://doi.org/10.1007/978-3-642-38016-7_22
6. Biedl, T.: Height-preserving transformations of planar graph drawings. In: Duncan, C., Symvonis, A. (eds.) GD 2014. LNCS, vol. 8871, pp. 380–391. Springer, Heidelberg (2014). https://doi.org/10.1007/978-3-662-45803-7_32
7. Biedl, T.: Drawing outer-1-planar graphs revisited. In: Auber, D., Valtr, P. (eds.) GD. LNCS, vol. 12590, pp. 526–527. Springer (2020), CoRR 2009.07106
8. Biedl, T.: Horton-Strahler number, rooted pathwidth and upward drawings of tree, accepted pending revisions at Information Processing Letters. CoRR 1506.02096 (2021)
9. Biedl, T., Chaplick, S., Kaufmann, M., Montecchiani, F., Nöllenburg, M., Raftopoulou, C.: On layered fan-planar graph drawings. In: Esparza, J., Král', D. (eds.) MFCS 2020. LIPIcs, vol. 170, pp. 14:1–14:13. LZI (2020)
10. Biedl, T.C., Kaufmann, M.: Area-efficient static and incremental graph drawings. In: Burkard, R., Woeginger, G. (eds.) ESA 1997. LNCS, vol. 1284, pp. 37–52. Springer, Heidelberg (1997). https://doi.org/10.1007/3-540-63397-9_4
11. Biedl, T., Liotta, G., Lynch, J., Montecchiani, F.: Generalized LR-drawings of trees. In: Canadian Conference on Computational Geometry (CCCG), pp. 78–88 (2021)
12. Biedl, T.C., Liotta, G., Lynch, J., Montecchiani, F.: Optimal-area visibility representations of outer-1-plane graphs. CoRR abs/2108.11768 (2021)
13. Bläsius, T., Brückner, G., Rutter, I.: Complexity of higher-degree orthogonal graph embedding in the kandinsky model. In: Schulz, A.S., Wagner, D. (eds.) ESA 2014. LNCS, vol. 8737, pp. 161–172. Springer, Heidelberg (2014). https://doi.org/10.1007/978-3-662-44777-2_14
14. Bose, P., et al.: A visibility representation for graphs in three dimensions. J. Graph Algorithms Appl. **2**(3), 1–16 (1998)
15. Brandenburg, F.J.: 1-visibility representations of 1-planar graphs. J. Graph Algorithms Appl. **18**(3), 421–438 (2014)
16. Brandenburg, F.J.: T-shape visibility representations of 1-planar graphs. Comput. Geom. **69**, 16–30 (2018)
17. Brandenburg, F.J., Heinsohn, N., Kaufmann, M., Neuwirth, D.: On bar $(1,j)$-visibility graphs. In: Rahman, M.S., Tomita, E. (eds.) WALCOM 2015. LNCS, vol. 8973, pp. 246–257. Springer, Cham (2015). https://doi.org/10.1007/978-3-319-15612-5_22
18. Chan, T.M.: A near-linear area bound for drawing binary trees. Algorithmica **34**(1), 1–13 (2002)

19. Dean, A.M., Evans, W.S., Gethner, E., Laison, J.D., Safari, M.A., Trotter, W.T.: Bar k-visibility graphs. J. Graph Algorithms Appl. **11**(1), 45–59 (2007)
20. Dehkordi, H.R., Eades, P.: Every outer-1-plane graph has a right angle crossing drawing. Int. J. Comput. Geom. Appl. **22**(6), 543–558 (2012)
21. Di Battista, G., Eades, P., Tamassia, R., Tollis, I.G.: Graph Drawing: Algorithms for the Visualization of Graphs. Prentice-Hall, Hoboken (1999)
22. Di Giacomo, E., et al.: Ortho-polygon visibility representations of embedded graphs. Algorithmica **80**(8), 2345–2383 (2018)
23. Didimo, W., Liotta, G., Montecchiani, F.: A survey on graph drawing beyond planarity. ACM Comput. Surv. **52**(1), 4:1–4:37 (2019)
24. Duchet, P., Hamidoune, Y.O., Vergnas, M.L., Meyniel, H.: Representing a planar graph by vertical lines joining different levels. Discret. Math. **46**(3), 319–321 (1983)
25. Eggleton, R.: Rectilinear drawings of graphs. Util. Math. **29**, 149–172 (1986)
26. Evans, W.S., Kaufmann, M., Lenhart, W., Mchedlidze, T., Wismath, S.K.: Bar 1-visibility graphs vs. other nearly planar graphs. J. Graph Algorithms Appl. **18**(5), 721–739 (2014)
27. Evans, W.S., Liotta, G., Montecchiani, F.: Simultaneous visibility representations of plane st-graphs using L-shapes. Theor. Comput. Sci. **645**, 100–111 (2016)
28. Fan, J.H., Lin, C.C., Lu, H.I., Yen, H.C.: Width-optimal visibility representations of plane graphs. In: Tokuyama, T. (ed.) ISAAC, pp. 160–171. Springer, LNCS (2007)
29. Felsner, S., Liotta, G., Wismath, S.: Straight-line drawings on restricted integer grids in two and three dimensions. J. Graph Algorithms Appl. **7**(4), 335–362 (2003)
30. Fößmeier, U., Kant, G., Kaufmann, M.: 2-Visibility drawings of planar graphs. In: North, S. (ed.) GD 1996. LNCS, vol. 1190, pp. 155–168. Springer, Heidelberg (1997). https://doi.org/10.1007/3-540-62495-3_45
31. Frati, F., Patrignani, M., Roselli, V.: LR-drawings of ordered rooted binary trees and near-linear area drawings of outerplanar graphs. J. Comput. Syst. Sci. **107**, 28–53 (2020)
32. He, X., Zhang, H.: Nearly optimal visibility representations of plane graphs. SIAM J. Discrete Math. **22**(4), 1364–1380 (2008)
33. Hong, S., Eades, P., Katoh, N., Liotta, G., Schweitzer, P., Suzuki, Y.: A linear-time algorithm for testing outer-1-planarity. Algorithmica **72**(4), 1033–1054 (2015)
34. Hong, S., Tokuyama, T. (eds.): Beyond Planar Graphs. Springer, Singapore (2020). https://doi.org/10.1007/978-981-15-6533-5
35. Kobourov, S., Liotta, G., Montecchiani, F.: An annotated bibliography on 1-planarity. Comput. Sci. Rev. **25**, 49–67 (2017)
36. Liotta, G., Montecchiani, F.: L-visibility drawings of IC-planar graphs. Inf. Process. Lett. **116**(3), 217–222 (2016)
37. Liotta, G., Montecchiani, F., Tappini, A.: Ortho-polygon visibility representations of 3-connected 1-plane graphs. Theor. Comput. Sci. **863**, 40–52 (2021)
38. Otten, R., van Wijk, J.: Graph representations in interactive layout design. In: IEEE ISCS, pp. 914–918 (1978)
39. Rosenstiehl, P., Tarjan, R.E.: Rectilinear planar layouts and bipolar orientation of planar graphs. Discrete Comput. Geom. **1**, 343–353 (1986)
40. Shermer, T.: Block visibility representations III: external visibility and complexity. In: CCCG. International Informatics Series, vol. 5, pp. 234–239. Carleton University Press (1996)

41. Tamassia, R., Tollis, I.G.: A unified approach to visibility representations of planar graphs. Discrete Comput. Geom. **1**(4), 321–341 (1986). https://doi.org/10.1007/BF02187705
42. Wismath, S.: Characterizing bar line-of-sight graphs. In: Snoeyink, J. (ed.) SoCG, pp. 147–152. ACM (1985)

Unit Disk Representations of Embedded Trees, Outerplanar and Multi-legged Graphs

Sujoy Bhore[1] , Maarten Löffler[2], Soeren Nickel[3(✉)] ,
and Martin Nöllenburg[3]

[1] Indian Institute of Science Education and Research, Bhopal, India
sujoy@iiserb.ac.in
[2] Department of Computing and Information Sciences,
Utrecht University, Utrecht, The Netherlands
m.loffler@uu.nl
[3] Algorithms and Complexity Group, TU Wien, Vienna, Austria
{soeren.nickel,noellenburg}@ac.tuwien.ac.at

Abstract. A unit disk intersection representation (UDR) of a graph G
represents each vertex of G as a unit disk in the plane, such that two
disks intersect if and only if their vertices are adjacent in G. A UDR with
interior-disjoint disks is called a unit disk contact representation (UDC).
We prove that it is NP-hard to decide if an outerplanar graph or an
embedded tree admits a UDR. We further provide a linear-time decidable
characterization of caterpillar graphs that admit a UDR. Finally we show
that it can be decided in linear time if a lobster graph admits a *weak*
UDC, which permits intersections between disks of non-adjacent vertices.

1 Introduction

The representation of graphs as contacts or intersections of disks has been a
major topic of investigation in geometric graph theory and graph drawing. The
famous circle packing theorem states that every planar graph has a contact
representation by touching disks (of various size) and vice versa [14]. Since then,
a large body of research has been devoted to the representation of planar graphs
as contacts or intersections of various kinds of geometric objects [5,6,9,10]. In
this paper, we are interested in unit-radius disks. A set of unit disks in $\mathbb{R}^2$ is a
unit disk intersection representation (UDR) of a graph $G = (V, E)$, if there is a
bijection between V and the set of unit disks such that two disks intersect if and
only if they are adjacent in G. *Unit disk graphs* are graphs that admit a UDR.
Unit disk contact graphs (also known as *penny graphs*) are the subfamily of unit
disk graphs that have a UDR with interior-disjoint disks, which is thus called
a *unit disk contact representation* (UDC). This can be relaxed to *weak* UDCs,
which permit contact between non-adjacent disks; see Fig. 1 for examples.

The recognition problem, where the objective is to decide whether a given
graph admits a UDR, has a rich history [2,3,11,12]. Breu and Kirkpatrick [4]

All figures are available in a colored variant in the online version.

© Springer Nature Switzerland AG 2021
H. C. Purchase and I. Rutter (Eds.): GD 2021, LNCS 12868, pp. 304–317, 2021.
https://doi.org/10.1007/978-3-030-92931-2_22

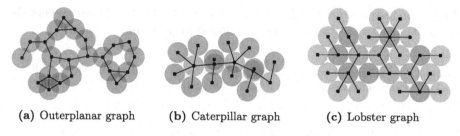

(a) Outerplanar graph   (b) Caterpillar graph   (c) Lobster graph

Fig. 1. Specific contact and intersection graphs of unit disks. In a UDR (a–b) disks overlap is allowed, and contact of disks implies an edge between their vertices. In a weak UDC disks are interior disjoint, but contact between non-adjacent disks is allowed. The disks of backbone vertices are colored grey (b–c).

proved that it is NP-hard to decide whether a graph G admits a UDR or a UDC, even for planar graphs. Klemz et al. [13] showed that recognizing outerplanar unit disk contact graphs is already NP-hard, but it is decidable in linear time for caterpillars, i.e., trees whose internal vertices form a path (see Fig. 1b).

Recognition with a fixed embedding is an important variant of the recognition problem. Given a plane graph G, the objective in this problem is to decide whether G admits a UDC in the plane that preserves the cyclic order of the neighbors at each vertex. Some recent works investigated the recognition problem of UDCs with/without fixed embedding, and narrowed down the precise boundary between hardness and tractability; see [2,7,8,13]. A remaining open question is to settle the complexity of recognizing non-embedded trees that admit a UDC. Some of these works focused on *weak* UDCs, where disks of non-adjacent vertices may touch. In this model, the recognition of non-embedded trees that admit a weak UDC is NP-hard [8]. We summarize the results on weak UDCs in Table 1.

While several results of the past years have shed light into the recognition complexity gap for UDCs, not much is known in this regard for the more general class of UDRs since the NP-hardness for planar graphs from 1998 [4].

Our Contribution. We investigate the unit disk graph recognition problem for subclasses of planar graphs. We show that recognizing unit disk graphs remains NP-hard for non-embedded outerplanar graphs (see Fig. 1a) – strengthening the previous hardness result for planar graphs [4] – and for embedded trees (Sect. 3). This line of research aims to extend earlier investigations [2,7,8] of UDCs to the UDR model and builds in particular on the work of Bowen et al. [2].

On the positive side, we provide a linear-time algorithm to recognize caterpillar graphs (see Fig. 1b) that admit a UDR (Sect. 4). In Sect. 5, we return to the problem of recognizing unit disk contact graphs and extend the tractability boundary for non-embedded graphs. While it was known that a weak UDC for caterpillar graphs can be constructed in linear time (if one exists), but the same recognition problem is NP-hard for trees [8], we prove that we can decide in linear time if a lobster graph admits a weak UDC on the triangular grid, where a *lobster* is a tree whose internal vertices form a caterpillar (see Fig. 1c).

Table 1 summarizes our results and remaining open problems. Proofs and details of results marked with a star ($\star$) can be found in the complete version [1].

Table 1. State of the art, our results, and open problems on unit disk graph recognition. Upward arrows indicate, that a result follows from the one below.

Graph class	Weak UDC		UDR	
	Non-embedded	Embedded	Non-embedded	Embedded
Planar	↑ NP-hard	↑ NP-hard	NP-hard [4]	↑ NP-hard
Outerplanar	↑ NP-hard	↑ NP-hard	NP-hard (Thm. 1)	↑ NP-hard
Trees	NP-hard [8]	↑ NP-hard	Open	NP-hard (Thm. 2)
Lobsters	$O(n)$ (Thm. 4)	↑ NP-hard	Open	Open
Caterpillars	$O(n)$ [8]	NP-hard [7]	$O(n)$ (Thm. 3)	Open

2   Preliminaries

A graph $G = (V, E)$ with $V = \{v_1, \ldots, v_n\}$ is a unit disk graph if there exists a set of closed unit disks $\mathcal{D} = \{d_1, \ldots, d_n\}$ and a bijective mapping $d\colon V \to \mathcal{D}$, s.t., $d(v_i) = d_i$ and $v_i v_j \in E$ if and only if d_i and d_j intersect. We call $\mathcal{D}$ a *unit disk intersection representation* (UDR) of G. If all disks in $\mathcal{D}$ are pairwise interior disjoint we also call $\mathcal{D}$ a *unit disk contact representation* (UDC) of G. A graph is a *unit disk contact graph* if it admits a UDC. A *weak* UDC permits contact between two disks d_i and d_j, even if $v_i v_j \notin E$.[1]

A *caterpillar* graph G is a tree, which yields a path, when all leaves are removed. We call this path, the backbone B_G of G. Similarly a *lobster* graph G' is a tree, which yields a caterpillar graph G'', when all leaves are removed. The backbone of G' is the backbone of G''. For each vertex v of the backbone, we denote the set of vertices that are reachable from v on a path that does not include any other backbone vertex, as the *descendants* of v.

The set of disks $\mathcal{D}$ induces an embedding $\mathcal{E}_\mathcal{D}(G)$ of G by placing every vertex v at the center of $d(v)$ and linking neighboring vertices by straight-line edges. We will therefore also use v as the center of $d(v)$. We say that a UDR $\mathcal{D}$ preserves the embedding of an embedded graph G with embedding $\mathcal{E}(G)$ if $\mathcal{E}_\mathcal{D}(G) = \mathcal{E}(G)$. Let a, b, c be three points in $\mathbb{R}^2$. We use $\angle abc$ to denote the clockwise angle defined between the segments ab and bc. We define the angle $\angle d_i d_j d_k$ as the clockwise angle $\angle v_i v_j v_k$ in $\mathcal{E}_\mathcal{D}(G)$.

---

[1] Note that weak UDRs, in contrast, are not interesting, since a complete graph K_n can easily be represented as a UDR and therefore every graph admits a weak UDR.

3   NP-Hardness Results

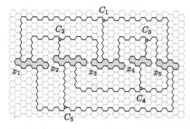

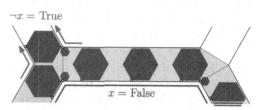

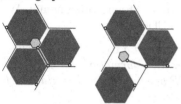

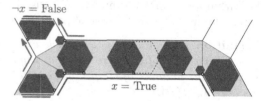

(a) Auxiliary hexagonal grid structure with embedded incidence graph

(b) False variable x. The connected wire on the top left is in a true state and therefore corresponds to the negated literal $\neg x$.

(c) Unsatisfied state (left) and one possible satisfied state (right) of the clause gadget

(d) True variable x. The wire is in a false state. The dotted hexagon indicates the overlap, if hexagons would have inconsistent states.

Fig. 2. Auxiliary structure details used by Bowen et al. [2]. All Figures are recreations/adaptions from their paper. The incidence graph is embedded on a hexagonal grid (a). The edges are short corridors in which the blue hexagons are fitted, hinged at white vertices. Hexagons in the variable cycle (red line, grey backdrop) have two states (b) and (d). The clause gadget (c) requires one hexagon, which does not enter the junction. (Color figure online)

In this section, we prove that recognizing unit disk graphs remains NP-hard for non-embedded outerplanar graphs and for embedded trees. Our proofs apply the generic machinery of Bowen et al. [2] to decide realizability of polygonal linkages, which requires to construct gadgets that can model hexagons and rhombi in a stable way. First we sketch the reduction of Bowen et al. (full details in Appendices A.1–A.3 of the full version [1]), then we describe our constructions of the required stable structures with outerplanar (Sect. 3.1) and embedded tree gadgets (Sect. 3.2).

Bowen et al. [2] proved that recognizing unit disk contact graphs is NP-hard for embedded trees, via a reduction from planar 3-SAT, which uses an auxiliary construction formulated as a realization of a polygonal linkage. A polygonal linkage is a set of polygons, which are realizable if they can be placed in the plane, s.t., predefined sets of points on the boundary of these polygons are identified. Bowen et al. define a set of hexagons in a hexagonal tiling (Fig. 2a) with small gaps between them, which form a hexagonal grid, in which a representation of the incidence graph of the planar 3-SAT instance is fitted. Smaller hexagons are fitted into cycles in this grid, s.t., they admit only two different realizations, see Fig. 2b, and determine the state of neighboring small hexagons, see Fig. 2d. The cycles represent variables in a true or false state. The states of the cycles are transmitted via chains of smaller hexagons into the gaps. The vertex, where three such chains meet, contains a small hexagon on a thin connection, which can only be realized if at least one transmitted state is *true*, see Fig. 2c. The polygonal linkage is realizable if and only if the planar 3-SAT instance is satisfiable. For a detailed description, we refer to Appendix A of the full version [1] and Bowen et al. [2]. The building blocks of this reduction are hexagons of variable sizes and short segments. Bowen et al. [2] approximate the hexagons by creating graphs, whose UDCs must be within a constant (asymmetric) Hausdorff distance[2] of a hexagon and the segments similarly by providing graphs, whose UDCs must be within a constant (asymmetric) Hausdorff distance of long thin rhombi. We extend their notion of λ-stable approximations to UDRs.

Definition 1. *A graph G is a λ-stable approximation of a region P in the plane if, in every UDR of G, there exists a congruence transformation $f : \mathbb{R}^2 \to \mathbb{R}^2$ such that the union U of all unit disks in the UDR has an asymmetric Hausdorff distance $d_H(f(P), U) \leq \lambda$.*

3.1 Non-embedded Outerplanar Graphs

We prove that recognition of unit disk graphs is hard for non-embedded outerplanar graphs by providing outerplanar graphs G_k^H and G_k^R, which are 7-stable approximations of a hexagon of side length $2k-1$ and a rhombus of width $2k+6$, respectively. Then, the NP-hardness follows immediately from the construction of Bowen et al. [2] sketched above. To obtain the 7-stable approximations we present two graphs, which enforce local bends in one direction of at least π and $\frac{4\pi}{3}$ in any UDR.

A *ladder* L_k (see Figs. 3a and 3c) is a chain of pairwise connected vertices v_i and v_i', for $i = 1, \ldots, k$, also called the *outer* and *inner* vertices of L_k, respectively. Additionally, so-called *extension neighbors*, which are connected to only one outer vertex each, are added on one side of the ladder, s.t., the outer vertices have alternating degrees of four and five. Since the ladder consists of a chain of C_4's, these neighbors are forced to be placed all on the outside in an outerplanar embedding. The minimal height of such a ladder is $2\sqrt{3} + 2 - \varepsilon$, which is the height of the smallest bounding box of a tight packing of three rows of unit disks minus an (arbitrarily) small constant $\varepsilon > 0$.

The permission of overlap between adjacent disks allows for a placement of one extension neighbor almost on top of its adjacent outer vertex, which leads to an ever so slight inwards bend and, more importantly, any outwards bend is impossible. In order to enforce an inward bend of at least $\frac{4\pi}{3}$, we connect

---

[2] Recall that the asymmetric Hausdorff distance from a set A to a set B in a metric space with metric d is defined as $d_H(A, B) = \sup_{a \in A} \inf_{b \in B} d(a, b)$.

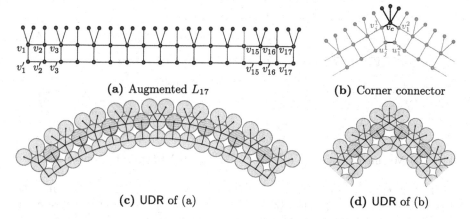

(a) Augmented L_{17} **(b)** Corner connector

(c) UDR of (a)   **(d)** UDR of (b)

Fig. 3. Components for creating λ-stable approximations G_k^H and G_k^R. Ladder L_{17} (a) and its UDR (c), as well as a corner connector (b) and its UDR (d). The corner connector connects two ladders (lighter colored parts) and its actual parts have a darker color in (b). The bends in the UDRs are required but exaggerated. (Color figure online)

the last inner vertex u_j^1 of one ladder with the first inner vertex u_1^2 of a second ladder and the last and first outer vertex v_j^1/v_1^2 of the first and second ladder, respectively, both with a vertex v_c, which has three attached extension neighbors, see Fig. 3b. This construction is called a *corner connector*. Since it is impossible to place the disk of any extension neighbor of v_c inside the 5-cycle $d(v_j^1), d(u_j^1), d(u_1^2), d(v_1^2), d(v_c)$, without overlapping at least two disks in the UDR, all extension neighbors are still forced to the outside and therefore $\angle v_j^1 v_c v_1^2 > \frac{4\pi}{3}$, see Fig. 3d.

Placing two ladders opposite each other and connecting them on one end with three corner connectors as shown in Fig. 4, yields a 7-stable approximation G_k^R of a thin rhombus. The following lemma is analogue to Lemma 10 in [2].

Lemma 1 ($\star$). *For every positive integer k the outerplanar graph G_k^R in Fig. 4 is a 7-stable approximation of a rhombus of width $2k+6$ and height $6\sqrt{3}+2$.*

Proof (Sketch). The graph G_k^R is made up of components, which make any bend to the outside impossible, see Fig. 4a. Since two ladders need to be overlap free, the minimum height of the ladders guarantees that part of the boundary of the union over all disks in the UDR of G_k^R lies above and below the line $\overline{c_0 l_0}$, see Fig. 4b. The largest vertical distance of the rhombus to this line is $3\sqrt{3}+1 \approx 6.196 < 7$.

Lemma 2 ($\star$). *For every integer $k = 6n+4, n \in \mathbb{N}$, the outerplanar graph G_k^H in Fig. 5a is a 7-stable approximation of a regular hexagon of side length $2k-1$.*

Proof (Sketch). The 7-stable approximation G_k^H of a hexagon of side length $2k-1$ uses again ladders and corner connectors to trace the outline of the

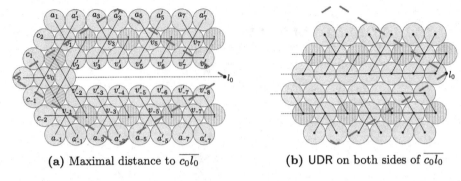

(a) Maximal distance to $\overline{c_0 l_0}$ (b) UDR on both sides of $\overline{c_0 l_0}$

Fig. 4. A 7-stable approximation G_7^R of a rhombus superimposed on its UDR. The maximal distance of any point of the UDR to $\overline{c_0 l_0}$ is smaller or equal 7 (a) and at all points of $\overline{c_0 l_0}$ a part of the UDR lies above and below $\overline{c_0 l_0}$ (b). Both UDRs require a small inward bend to be valid and hatched disks indicate almost overlapping placement of a red disk on a blue disk, with a small shift to the outer side. Both inward bends and outward shifts are omitted. (Color figure online)

hexagon. Then the inside of this construction is filled with a set of ladders until no additional ladders can be added, see Fig. 5a. Outwards bends are impossible by construction of the ladders and corner connectors and inwards bends are strongly limited since the interior of the hexagon is almost completely filled with ladders. Since the amount of compression in a ladder is very limited, this leaves only a constant amount of space on the inside of a UDR of G_k^H.

Theorem 1. *Recognizing unit disk graphs is* NP-*hard for outerplanar graphs.*

Proof. This result follows from Lemmas 1 and 2 by using the polygonal linkage reduction of Bowen et al. [2] (see also Appendix A of the full version [1]). Note that we can emulate hinges exactly as in the original reduction.

3.2 Embedded Trees

By slightly adapting the construction of the outerplanar graphs of Sect. 3.1, we can prove that recognizing unit disk graphs is NP-hard for embedded trees. The crucial observation is that we used the outerplanarity of G_k^R and G_k^H exclusively to be able to build a tree-like structure out of chains of 4- and 5-cycles. We used this to force the placement of leaf disks to a specific side of these chains in any UDR of G_k^R and G_k^H. As we are concerned with embedded trees in this section, we can omit the inner vertices of the ladder, as the given embedding puts the leaves on the desired side of the chains. This results in a tree. We call a ladder without the inner vertices a *chain*.

We can now use a very similar construction idea as for G_k^R and G_k^H above. We need to augment both gadgets with an additional chain in order to retain the property, that no parts of these gadgets can be folded onto themselves. The resulting trees T_k^R and T_k^H are shown in Figs. 5b and 6.

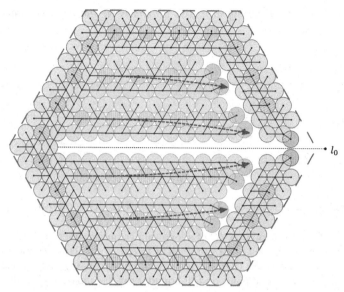

(a) Combination of the ladder and corner connectors into 7-stable approximation G_k^H of a hexagon superimposed on its UDR.

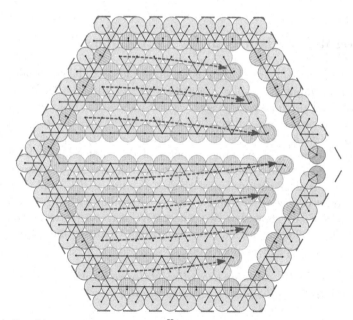

(b) A 7-stable approximation tree T_k^H of a hexagon superimposed on its UDR.

Fig. 5. 7-stable hexagon approximations. Hatched disks indicate almost overlapping placement of disks. Necessary infinitesimal bends are omitted. The bend directions of the inner components are indicated with gray arrows. The approximated regular hexagon is indicated by the dashed green outlines. (Color figure online)

Theorem 2 (⋆). *Recognizing unit disk graphs is* NP-*hard for embedded trees.*

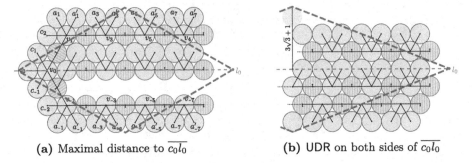

(a) Maximal distance to $\overline{c_0 l_0}$ **(b)** UDR on both sides of $\overline{c_0 l_0}$

Fig. 6. A 7-stable approximation T_k^R of a long thin rhombus superimposed on its UDR. The maximal distance of any point of the UDR to $\overline{c_0 l_0}$ is smaller or equal 7 (a) and at all points of $\overline{c_0 l_0}$ a part of the UDR lies above and below $\overline{c_0 l_0}$ (b). In both cases at any point along $\overline{c_0 l_0}$ at least one point on the boundary of the union of all disks in a UDR of T_k^R lies on or above $\overline{c_0 l_0}$ and on or below $\overline{c_0 l_0}$.

4 Recognition Algorithm for Caterpillars

We propose a linear-time algorithm using similar ideas to Klemz et al. [13], that recognizes if an input caterpillar graph $G = (V, E)$ admits a UDR or not; it is constructive and provides a representation if one exists. However, we need to address several new issues as we show that a larger class of graphs admits a UDR compared to a UDC. Clearly, if G contains a vertex of degree at least 6, then due to the unit disk packing property, it does not admit a UDR. Hence, every realizable caterpillar must have maximum degree $\Delta \leq 5$. Moreover, it is easy to observe that all caterpillars with $\Delta \leq 4$ admit a UDC (and thus a UDR), as also noted by Klemz et al. [13]. Not every caterpillar with $\Delta = 5$, however, is realizable as UDR. We show that two consecutive degree-5 vertices on B_G cannot be realized. The following lemma gives a sufficient condition for a "No" instance to be used in the recognition algorithm.

Lemma 3 (⋆). *If B_G contains two adjacent degree 5 vertices u, v, then it does not admit a unit disk intersection representation.*

4.1 The Algorithm

As a preprocessing step we augment all backbone vertices of degree 3 or lower with additional degree-1 neighbors, s.t., they have degree 4. Consider a chain $v_1, \ldots, v_n$ of backbone vertices. Now assume all vertices are of degree 4 or lower. We place them on a horizontal line. For each $1 \leq i \leq n$ at disk $d(v_i)$, we place its leaf neighbor disks $d(v_i^t), d(v_i^b)$ first at the top and then at the bottom of $d(v_i)$, see Fig. 7a, s.t., the clockwise angle $\angle v_i^t v_i v_i^b = \frac{4\pi}{3} - 2i\varepsilon$. The rotational ε offset avoids adjacencies between the leaf disks. While these offsets can add up, we can choose ε small enough for every finite caterpillar, s.t., this is negligible.

Now we assume that not all vertices are of degree 4 or lower. To keep the entire construction of the backbone x-monotone, whenever we encounter a degree 5 vertex u after a degree four vertex v_k, we place $d(u^{t'})$ of its additional leaf $u^{t'}$ alternatingly on the top or the bottom side with a $\frac{\pi}{3} + \varepsilon$ rotational offset to $d(u^t)$ (or $d(u^b)$). We will assume that we placed the disk at the top. Therefore $\angle u^{t'} u u^b \leq \pi - (2k+1)\varepsilon$, i.e., an almost horizontal connection, see Fig. 7b.

If the next vertex x has also degree five, then due to Lemma 3 we know that the sequence is not realizable. Otherwise, we place $d(x)$, s.t., it is touching $d(u)$ with a $\frac{\pi}{3} + \varepsilon$ rotational offset to $d(x)$, see Fig. 7b. We place $d(x^b)$ at the planned position relative to $d(x)$ at the bottom, i.e., with a $\frac{\pi}{3} + (k+2)\varepsilon$ counterclockwise offset relative to the x-axis, however, we place $d(x^t)$ almost exactly on top of $d(x)$ with a very small shift of $\frac{\varepsilon}{Cn}$ orthogonal to $\overline{ux}$, for some large constant C. This prevents touching of $d(u)$ and $d(x^t)$, without creating an adjacency between $d(u^{t'})$ and $d(x^t)$.

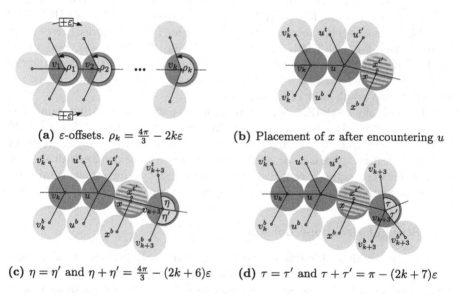

(a) ε-offsets. $\rho_k = \frac{4\pi}{3} - 2k\varepsilon$ (b) Placement of x after encountering u

(c) $\eta = \eta'$ and $\eta + \eta' = \frac{4\pi}{3} - (2k+6)\varepsilon$ (d) $\tau = \tau'$ and $\tau + \tau' = \pi - (2k+7)\varepsilon$

Fig. 7. Chains of degree-4 vertices are placed in a dense packing formation with small offsets (a). A degree-5 vertex places an additional leaf on one side (b). The next vertex v_{k+3} can again be placed with the desired angle of just over $\frac{2\pi}{3}$ between two neighbors (c). Placement of v_{k+3} is possible if its degree is 5 (d). Note that, the rotational offset angles are exaggerated, for better readability.

From this point onwards, we consider the direction of $\overline{ux}$ to be the direction in which we extend the backbone of the caterpillar. Any following disk $d(v_{k+3})$

can be placed again in the new extension direction touching $d(x)$. Its leaf disks $d(v_{k+3}^t)$ and $d(v_{k+3}^b)$ can be placed in their planned positions, i.e. with a clockwise or counterclockwise offset of $\frac{\pi}{3} + (k+3)\varepsilon$ relative to the new extension direction, respectively, which results in a clockwise angle $\angle u_t u u_b \leq \frac{4\pi}{3} - (2k+6)\varepsilon$. Note that v_{k+3} can have a degree of four (Fig. 7c) or five (Fig. 7d) and that at this point, if v_{k+3} has degree five, we can immediately repeat this procedure.

As a postprocessing step, we remove all degree-1 vertices that were added in the preprocessing step. Then from the above description of the algorithm and the correctness analysis in Appendix B.2 of the full version [1] we obtain the following theorem.

Theorem 3. *Let $G = (V, E)$ be a caterpillar graph. G admits a UDR if and only if G does not contain any two adjacent degree-5 vertices in the backbone path B_G of G. This property can be tested in linear time and if a UDR exists then it can be constructed in linear time.*

5   Weak **UDC**s of Lobsters on the Triangular Grid

We have shown that recognition of UDRs is NP-hard for outerplanar graphs and linear-time solvable for caterpillars, which mirrors the results for UDCs and weak UDCs; it leaves the recognition complexity for (non-embedded) trees as an open question for both UDRs and UDCs. For weak UDCs, however, recognition has been proven NP-hard for trees [8]. In order to investigate the complexity of weak UDCs further, we zoom in on the gap between trees and caterpillars and investigate the graph class of lobsters.

The *spine* of a weak UDC of a lobster G is the polyline defined by connecting the centers of all disks belonging to the vertices of B_G in order. A weak UDC is *straight*, if its spine is a straight line segment. Similarly, a weak UDC is x- or y-monotone, if its spine is x- or y-monotone. Since we consider weak UDCs with contacts between non-adjacent disks permitted, we focus our attention on weak UDCs placed on a triangular grid (similarly to previous work on weak UDCs [8]).

5.1   Straight Backbone Lobsters

Since any caterpillar G, admits a weak UDC if and only if it admits a straight weak UDC [8] we investigate lobster graphs, which admit a straight weak UDC. These are not all lobsters, since any simple lobster graph containing a non-backbone vertex of degree 6 only admits a non-straight weak UDC. We observe that already for this restricted subclass, a greedy placement scheme similar to Cleve's approach [8] for caterpillars is not possible; again shown by an example.

We specify two lobster graphs G and G', see Fig. 8. It can be checked via exhaustive enumeration that G admits 18 different weak UDCs, while G' admits only 12. The subgraphs induced by their first three backbone vertices are identical, however the realization of the descendants of v_3 (highlighted in red) is unique for both graphs (up to symmetry) and dependent on the structure of the

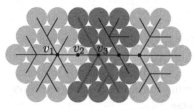 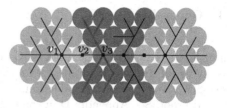

(a) G has 18 possible weak UDCs   **(b)** G' has 12 possible weak UDCs

Fig. 8. The subgraphs of G and G' induced by their first three backbone vertices are equal, however, depending on the following vertices a different realization of the neighbors of v_3 is necessary.

graphs beyond this point. We can therefore not simply scan over the backbone in a greedy manner and fix all positions for the disks of descendants of a backbone vertex and then continue on to the next. It is, however, still possible to do this in linear time with dynamic programming. The requirements for this are actually less strict, as it is already sufficient to have a strictly x-monotone rather than a straight backbone, which we show in the next section.

5.2   Monotone Weak UDCs

If we can guarantee that a lobster can be realized as a strictly x-monotone weak UDC, we can compute such a weak UDC with a linear-time dynamic programming algorithm. The dynamic program uses the following three observations.

Observation 1 ($\star$). *The number of possible placements of a backbone vertex v_i and its descendants is constant for a fixed position of v_i.*

Observation 2 ($\star$). *For a fixed grid position, the number of backbone vertices of a strictly x-monotone weak UDC, who can occupy this position by themselves or with a descendant is constant. Moreover, the distance in graph between the first and the last such vertex is constant.*

Observation 3 ($\star$). *Let C be a sufficiently large constant, and let A, B be two weak UDC, whose last C backbone vertices have placed themselves and their descendants in such a manner, that the pattern of occupied grid positions and the placement of the last backbone vertex is equivalent up to translation and rotation. Then any extension graph C, which can be appended to A, s.t., the combined graph admits a weak UDC can also be appended to B, s.t., their combined graph admits a weak UDC and vice versa.*

With these three claims we obtain the following Lemma.

Lemma 4 ($\star$). *Using dynamic programming it can be checked in linear time if a lobster graph admits an x-monotone weak UDC on the triangular grid.*

5.3   General Lobsters

The algorithm sketched in the previous section recognizes lobster graphs, which admit a strictly x-monotone weak UDC in linear time. Now we set out to prove that every lobster which admits a weak UDC also admits a strictly x-monotone weak UDC. We prove this by induction. The induction step is done as a computer-assisted proof. See Appendix D in the full version [1].

Lemma 5. *Every lobster graph, which admits a weak UDC on the triangular grid, also admits an x-monotone weak UDC on the triangular grid.*

Proof. We use induction on the length of the backbone. The base cases are backbones of length one, two or three. The spine of any realization is at most a polyline consisting of two segments and can therefore always be rotated to be x-monotone. The induction hypothesis is, that any lobster graph, with a backbone of length k admits an x-monotone weak UDC. In the induction step we need to show that any extension to a graph G' with a backbone of length $k + 1$, can be realized as a weak UDC if and only if it can be realized in an x-monotone way. The extensions are done by appending a single new backbone vertex v_{k+1}, whose descendants are specified as a sorted list of the degrees of its direct neighbors. Since the total degree of every vertex is at most 6, the set Γ of options for v_{k+1} is constant (Observation 1). Let Θ be the set of possible combinations of already occupied grid positions where v_{k+1} is placed such that the spine remains x-monotone. Let Δ_6 and Δ_3 be the sets of possible placements of disks of descendants of v_{k+1}, when v_{k+1} is placed at one of six (in the unrestricted case) or one of three (in the strictly x-monotone case) positions. Therefore we can enumerate all triples $(\gamma \in \Gamma, \delta_3 \in \Delta_3, \theta \in \Theta)$ and $(\gamma \in \Gamma, \delta_6 \in \Delta_6, \theta \in \Theta)$ and check if they are realizable. By exhaustive enumeration,[3] we have found that for every possible $(\gamma, \delta_6, \theta)$, which is realizable, we can find a suitable $(\gamma, \delta_3, \theta)$, which is realizable, too. This concludes the induction step.

From Lemmas 4 and 5, we conclude the following theorem.

Theorem 4. *It can be decided linear time if a lobster graph admits a weak UDC on the triangular grid.*

6   Conclusions

We have investigated the existing complexity gap for the recognition problem of UDRs and weak UDCs. In addition to the open problems for various graph classes in different settings (recall Table 1 in Sect. 1) there are two main open questions. First, we have investigated weak UDCs of lobsters on the triangular grid, however, it is not entirely clear if every lobster, which admits a weak UDC, also does so on the grid. Second, it seems reasonable to assume that our enumeration approach can

---

[3] The cases were reduced, by considering symmetry and infeasibility beforehand. Enumeration was done in the form of a computer-assisted proof. Details are explained in Appendix D of the full version [1].

be extended to graph classes beyond lobsters, which admit a weak UDC at least on the triangular grid. In fact, we conjecture that every class of trees, in which each vertex has bounded distance to a central backbone in extension of caterpillars and lobsters, can be recognized in polynomial time by such an approach.

Acknowledgements. We thank Jonas Cleve and Man-Kwun Chiu for fruitful discussions about the project during their research visits in Vienna.

References

1. Bhore, S., Löffler, M., Nickel, S., Nöllenburg, M.: Unit disk representations of embedded trees, outerplanar and multi-legged graphs. CoRR abs/2103.08416 (2021). https://arxiv.org/abs/2103.08416
2. Bowen, C., Durocher, S., Löffler, M., Rounds, A., Schulz, A., Tóth, C.D.: Realization of simply connected polygonal linkages and recognition of unit disk contact trees. In: Di Giacomo, E., Lubiw, A. (eds.) GD 2015. LNCS, vol. 9411, pp. 447–459. Springer, Cham (2015). https://doi.org/10.1007/978-3-319-27261-0_37
3. Breu, H., Kirkpatrick, D.G.: On the complexity of recognizing intersection and touching graphs of disks. In: Brandenburg, F.J. (ed.) GD 1995. LNCS, vol. 1027, pp. 88–98. Springer, Heidelberg (1996). https://doi.org/10.1007/BFb0021793
4. Breu, H., Kirkpatrick, D.G.: Unit disk graph recognition is NP-hard. Comput. Geom. Theory Appl. 9(1–2), 3–24 (1998). https://doi.org/10.1016/S0925-7721(97)00014-X
5. Chalopin, J., Gonçalves, D., Ochem, P.: Planar graphs have 1-string representations. Discrete Comput. Geom. 43(3), 626–647 (2009). https://doi.org/10.1007/s00454-009-9196-9
6. Chaplick, S., Ueckerdt, T.: Planar graphs as VPG-graphs. In: Didimo, W., Patrignani, M. (eds.) GD 2012. LNCS, vol. 7704, pp. 174–186. Springer, Heidelberg (2013). https://doi.org/10.1007/978-3-642-36763-2_16
7. Chiu, M., Cleve, J., Nöllenburg, M.: Recognizing embedded caterpillars with weak unit disk contact representations is NP-hard. CoRR abs/2010.01881 (2020)
8. Cleve, J.: Weak unit disk contact representations for graphs without embedding. CoRR abs/2010.01886 (2020)
9. Felsner, S.: Rectangle and square representations of planar graphs. In: Pach, J. (ed.) Thirty Essays on Geometric Graph Theory, pp. 213–248. Springer (2013). https://doi.org/10.1007/978-1-4614-0110-0_12
10. Gonçalves, D., Isenmann, L., Pennarun, C.: Planar graphs as L-intersection or L-contact graphs. In: Discrete Algorithms (SODA 2018), pp. 172–184. SIAM (2018). https://doi.org/10.1137/1.9781611975031.12
11. Hliněný, P.: Classes and recognition of curve contact graphs. J. Comb. Theor. Ser. B 74(1), 87–103 (1998). https://doi.org/10.1006/jctb.1998.1846
12. Hliněný, P., Kratochvíl, J.: Representing graphs by disks and balls (a survey of recognition-complexity results). Discrete Math. 229(1), 101–124 (2001). https://doi.org/10.1016/S0012-365X(00)00204-1
13. Klemz, B., Nöllenburg, M., Prutkin, R.: Recognizing weighted disk contact graphs. In: Di Giacomo, E., Lubiw, A. (eds.) GD 2015. LNCS, vol. 9411, pp. 433–446. Springer, Cham (2015). https://doi.org/10.1007/978-3-319-27261-0_36
14. Koebe, P.: Kontaktprobleme der konformen abbildung. Ber. Sächs. Akad. Wiss. Leipzig, Math. Phys. Klasse 88, 141–164 (1936)

Layered Area-Proportional Rectangle Contact Representations

Martin Nöllenburg🆔, Anaïs Villedieu$^{(\boxtimes)}$🆔, and Jules Wulms🆔

Algorithms and Complexity Group, TU Wien, Vienna, Austria
{noellenburg,avilledieu,jwulms}@ac.tuwien.ac.at

Abstract. We investigate two optimization problems on area-proportional rectangle contact representations for layered, embedded planar graphs. The vertices are represented as interior-disjoint unit-height rectangles of prescribed widths, grouped in one row per layer, and each edge is ideally realized as a rectangle contact of positive length. Such rectangle contact representations find applications in semantic word or tag cloud visualizations, where a collection of words is displayed such that pairs of semantically related words are close to each other. In this paper, we want to maximize the number of realized rectangle contacts or minimize the overall area of the rectangle contact representation, while avoiding any false adjacencies. We present a network flow model for area minimization, a linear-time algorithm for contact maximization of two-layer graphs, and an ILP model for maximizing contacts of k-layer graphs.

Keywords: Contact graphs · Layered planar graphs · Semantic word clouds

1 Introduction

Contact representations of planar graphs are a well-studied topic in graph theory, graph drawing, and computational geometry [7,8,10]. Vertices are represented by geometric objects, e.g., disks or polygons, and two objects touch if and only if they are connected by an edge. They find many applications, for instance in VLSI design [19], cartograms [13], or semantic word clouds [1,18].

Word or tag clouds are popular visualizations that summarize textual information in an aesthetically pleasing way. They show the main themes of a text by displaying the most important keywords obtained from text analysis and scale the word size to their frequency in the text. Word clouds became widespread after the first automated generation tool "Wordle" was published in 2009 [15].

Word clouds with their different font sizes and words packed without semantic context, such as the one shown in Fig. 1, have also received some criticism as their audience sometimes fails at understanding the underlying data (while

---

We acknowledge funding by the Austrian Science Fund (FWF) under grant P31119. Colored versions of figures can be found in the online version of this paper.

H. C. Purchase and I. Rutter (Eds.): GD 2021, LNCS 12868, pp. 318–326, 2021.
https://doi.org/10.1007/978-3-030-92931-2_23

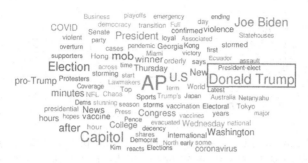

Fig. 1. Word cloud generated from the apnews.com frontpage by worditout.com on the day of the certification of Joe Biden's victory in the 2020 US elections.

enjoying their playful nature) [9]. For example, neighboring words that are not semantically related can be misleading (see marked words in Fig. 1). As a way to improve readability, *semantic* word clouds have been introduced [1,6,18]. In semantic word clouds, an underlying edge-weighted graph indicates the semantic relatedness of two words, whose positions are chosen such that semantically related words are next to each other while unrelated words are kept far apart.

Classic word clouds are often generated using forced-based approaches, alongside with a spiral placement heuristic [15–17] that allows for a very compact final layout. This method is powerful even when the rough position of a word is dictated by an underlying map [4,12]. Semantic word clouds on the other hand have been approached with many different techniques, e.g., force directed [6], seam-carving [18], and multidimensional scaling [2]. The problem has also been studied from a theoretical point of view, where an edge of the semantic word graph is realized if the bounding boxes of two related words properly touch; the realized edge weight is gained as profit. Then the semantic word cloud problem can be phrased as the optimization problem to maximize the total profit. Barth et al. [1] and later Bekos et al. [3] gave several hardness and approximation results for this problem (and some variations) on certain graph classes. The underlying geometric problem also has links to more general contact graph representation problems, like rectangular layouts [5] or cartograms [11].

In most of the literature about layered graphs, vertices are assigned to rows without a predefined left-to-right order, yet this has interesting properties in the context of word clouds. For instance, layered rectangle contact representations are compact, assuming a good assignment they have an even distribution of words and our eye naturally understands words grouped in rows or tables. In this paper we study row-based contact graphs of unit-height but arbitrary-width rectangles, which may represent the bounding boxes of words with fixed font size.

Problem Description. As input we take a layered graph $G = (V, E)$ on L layers, with an arbitrary number of vertices per layer. Each vertex $v_{i,j} \in V$ is indexed by its layer $i \in [0, L - 1]$ and its position j within the layer: $v_{i,j}$ is the jth vertex on the ith layer. The edge set E consists of edges connecting each vertex $v_{i,j}$ to its neighbors $v_{i,j-1}$ and $v_{i,j+1}$ on the same row (if they exist),

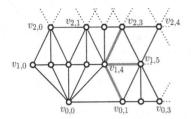
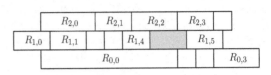

Fig. 2. Partial drawing of a graph G, along with a representation $\mathcal{R}$ of the visible vertices of G. Fat red edges are not realized, due to the gray gap in $\mathcal{R}$. (Color figure online)

and connections between adjacent rows form an internally triangulated graph. We associate each vertex with an axis-aligned unit-height rectangle $R_{i,j}$ with width $w_{i,j}$, and y-coordinate i. We want to compute its x-position $x_{i,j}$ given by the x-coordinate of its bottom left corner such that the rectangles do not overlap except on their boundaries (see Fig. 2). Leaving whitespace between two rectangles on the same layer is allowed and forms a *gap*. Such a layout $\mathcal{R}$ is called a *representation* of G. An edge $(u,v) \in E$ is *realized* in a representation $\mathcal{R}$ if rectangles R_u and R_v, representing vertices u and v, intersect along their boundaries for a positive length $\varepsilon > 0$, which we denote by $(R_u, R_v) \in \mathcal{R}$. If R_u and R_v are horizontally adjacent we call the contact a *horizontal contact*.

Otherwise, the intersection is located along a horizontal boundary if R_u and R_v are on adjacent layers; these are called *vertical contacts*. Contacts between rectangles whose vertices are not adjacent in G are *false adjacencies*. Such adjacencies can mislead a user to infer a link between unrelated words, invalidating the representation. Within this model we study two problem variations, *area minimization* and *contact maximization*.

For the area minimization problem the goal is to produce a representation $\mathcal{R}$ that minimizes the total width of the gaps in $\mathcal{R}$. The contact maximization problem asks to maximize the number of adjacencies realized in $\mathcal{R}$, as specified by edge set E. For both optimization criteria, false adjacencies are forbidden: otherwise a trivial gap-less representation would always be a solution to the area minimization problem and in the case of contact maximization, false adjacencies may reduce the number of lost contacts with respect to a valid optimal solution as Fig. 3 shows. We say a representation is *valid* if it has no false adjacencies.

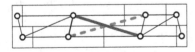
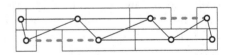

Fig. 3. Allowing false adjacencies (fat/red) could reduce lost contacts (dashed/blue). (Color figure online)

2   Area Minimization

To solve the area minimization problem, we construct a flow network $N = (G' = (V', E'); l; c; b; cost)$ for a given vertex-weighted layered graph $G = (V, E)$, with edge capacity lower bound $l: E' \to \mathbb{R}_0^+$, edge capacity $c: E' \to \mathbb{R}_0^+$, vertex production/consumption $b: V' \to \mathbb{R}$ and cost function $cost: E' \to \mathbb{R}_0^+$. Each unit of cost will represent a unit length gap and each unit of flow on an edge will represent a unit length contact. To build the network we create two vertices v^a and v^b for each rectangle, that respectively receive the flow from the lower layer and output flow to the upper layer, and one for each potential gap, located between each sequential pair of rectangles in the same layer. Every edge e that ends on a gap vertex has $cost(e) = 1$. We also add an edge e between v^a and v^b for each $R_{i,j}$ with $l(e) = c(e) = w_{i,j}$ and no cost to ensure that rectangle nodes receive exactly as much flow as they are wide.

The intuition behind the network is that it represents a stack of layers consisting of rectangles and gaps, with a maximum width of $w_{\max} \cdot K$, K being the maximum number of rectangles per layer, and $w_{\max}$ the width of the widest rectangle. To facilitate this flow on all layers, there are buffer vertices on both sides of each layer. Each rectangle is as wide as the amount of flow its vertices v^a and v^b receive, and has contacts with its upper and lower neighbors as wide as the flow on the edges representing these contacts. Every vertex has edges to the layer above as far as its rectangle is allowed to have contacts: a rectangle $R_{i,j}$ that has only one upper neighbor, will have an edge to that neighbor, and to the gaps on that neighbor's right and left side. Any further edge would be to another rectangle with which $R_{i,j}$ should not share a contact, and such edges would hence result in false adjacencies. We picture the stack bottom-up, meaning that the flow comes in at the bottom layer and exits from the top layer. A gap block $g_{i,j}$ will reach as far left and right as its left and right neighbors in the same row: if rectangle $R_{i,j}$ lies directly left of $g_{i,j}$, then the furthest left upward neighbor of $R_{i,j}$ is the furthest left upward neighbor of $g_{i,j}$. If $g_{i,j}$ could reach even further, then it would essentially push $R_{i,j}$ into a false adjacency. The exact construction is detailed in the full paper [14] and sketched in Fig. 4.

Theorem 1. *Given a graph $G = (V, E)$, the cost of a minimum-cost flow f in N equals the minimum total gap length of any valid representation of G. An area-minimal representation of G is constructed from f in polynomial time.*

Proof. Given any graph $G = (V, E)$, the associated network N has production equal to its consumption $\sum_{v \in V'} b(v) = b(s) + b(t) = 0$. The source produces $b(s) = w_{max} \cdot K$ flow, which is available to every vertex on layer 1 (except v^b vertices whose incoming neighbor is v^a on the same layer). Any vertex $v_{1,j}^a$ must receive $w_{1,j}$ units of flow, as its only edge towards $v_{1,j}^b$ has capacity constraint $c = l = w_{1,j}$. Because $w_{1,j} \le w_{max}$ and since there are at most K vertices $v_{1,j}^a$ on layer 1, the capacities can be satisfied. Any edge in E' goes from layer i to $i + 1$, except for $(v_{i,j}^a, v_{i,j}^b)$, but the exact amount of flow that comes from layer

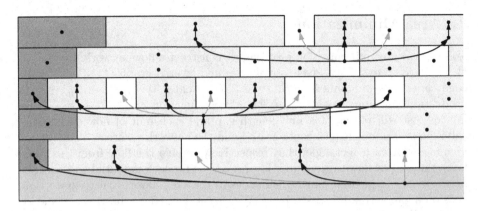

Fig. 4. Parts of a flow network: outgoing edges from the source (orange, bottom), a rectangle (green) and a gap (white); buffer rectangles (red, left); gray edges cost 1. (Color figure online)

$i - 1$ into $v^a_{i,j}$ will go through $v^b_{i,j}$ to layer $i + 1$. Hence for the same reason as in layer 1, there is enough flow to satisfy the edge capacity constraints, while excess flow is routed through (0-cost) buffers or (1-cost) gaps.

Since in this network only flow that goes into a gap vertex has a (non-zero) cost, flow into gap vertices, and therefore also the total gap width, is minimized. The minimum cost procedure finds this optimal flow f.

We construct a minimum-area representation $\mathcal{R}$ by placing rectangles row by row: we leave buffers and gaps equal to the flow routed through the corresponding vertices, and align all rows on the left. The total width of each row, including buffers and gaps, is $w_{max} \cdot K$. Since the flow through each rectangle is constant, the area of the buffers is maximized, to minimize the area occupied by gaps. □

While this method minimizes the area occupied by the drawing it will not always lead to the representation with the minimum bounding box. To minimize the size of the bounding box, we propose to limit the amount of flow outgoing from the source node and incoming into the sink node. If the chosen bound is too small then the flow network will not be realizable. We can thus perform a binary search between the width of the longest layer $W_{\max}$ as a lower bound and $w_{\max} \cdot K$ as an upper bound. Assuming input widths have integer values, this method would add a $\mathcal{O}(\log(w_{\max} \cdot K - W_{\max}))$ factor to the flow runtime.

3   Maximization of Realized Contacts

In this section we propose algorithms that maximize the number of realized contacts. We start with a linear-time algorithm for $L = 2$, followed by an integer linear programming model for $L > 2$. The complexity for $L > 2$ remains open.

3.1   Linear-Time Algorithm for $L = 2$

In this section we describe an algorithm $\mathcal{A}$ for the case where the input has 2 layers. On a 2-layer graph, a vertex either has one neighbor in the adjacent layer, or more than one. If a vertex has one neighbor we call it a *T-vertex* and if there are more neighbors, it is called a *fan*. A *block* is a maximal sequence of consecutive rectangles in a layer i, for which each horizontal contact is realized. A block from the jth until the lth vertex of row i is the sequence $(R_{i,j}, \ldots, R_{i,l})$, where for each $k \in [j, l-1]$ holds that $(R_{i,k}, R_{i,k+1}) \in \mathcal{R}$.

In a given 2-layer graph, there will always be a layer that starts on a fan, while the opposite layer starts on a T-vertex. Assume without loss of generality that $R_{0,0}$ is a fan, otherwise swap the two rows for the duration of the algorithm. Algorithm $\mathcal{A}$ first places $R_{0,0}$, followed by all its neighbors on the adjacent layer, from left to right, ending with $R_{1,j}$. Rectangle $R_{1,j}$ is again a fan, and the process of placing all opposite-row neighbors, left to right, is repeated for $R_{1,j}$ and every consecutive fan, as they are encountered. We call this placement ordering $\prec$.

When we add a rectangle R_i (fan or T-vertex), we always first attempt to add it next to its horizontal predecessor, if possible (no false adjacency). Though, if the horizontal predecessor is too far left, we place R_i in the leftmost allowed position. Let R_0 be the first rectangle in $\prec$, which is placed on position x_0 by $\mathcal{A}$. Algorithm $\mathcal{A}$ then proceeds by adding R_1, representing a T-vertex in the opposite row. Rectangle R_1, with width w_1, is placed leftmost, on coordinate $x_0 + \varepsilon - w_1$. We then proceed to add all rectangles corresponding to other T-vertices of R_0 one by one, such that all horizontal contacts are realized. Once a T-vertex R_i cannot reach R_0, we store the amount of contacts currently realized by R_0 as well as its position x_0, and slide R_0 rightward, to the leftmost position x'_0 that allows a contact of ε with R_i. Note that, since we placed R_1 in the leftmost position that allowed a contact of width ε with R_0, we lose at least one contact by moving R_0 rightward. If placing R_0 at x'_0 ties the amount of contacts of x_0, then we set $x_0 := x'_0$. If x'_0 is strictly worse, then the representation is reset to having R_0 at x_0. From that point on, every time we add a new rectangle, we attempt this shift of the fan and update the position when we find a tie or when we realize more contacts. We repeat this operation for each rectangle, following the order $\prec$, always shifting the last encountered fan.

However, once we consider a fan R_f that is not R_0, any sliding operation will be attempted on the block containing R_f, rather than just R_f. As before, we always shift the block to the leftmost position that realizes the contact between R_f and the newly placed rectangle. We remember position x_f that realizes most contacts, and favor the newest position on a tie. In case moving the block containing R_f leads to strictly less contacts, we also try to move only R_f instead. This starts a new block containing just R_f. Below we sketch the proof for Theorem 2, the complete proof can be found in the full paper version [14].

Theorem 2. *Algorithm $\mathcal{A}$ computes a contact maximal valid representation with contacts of length at least ε for a given 2-layer graph G in linear time.*

324   M. Nöllenburg et al.

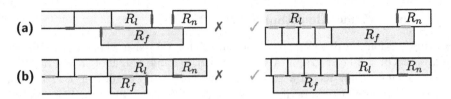

Fig. 5. Two configurations where R_l is necessarily a fan (blue) or a T-vertex (yellow). (a) If R_n can achieve only a vertical contact, R_f is a fan. (b) If R_n can achieve only a horizontal contact and is a fan, R_f is a T-vertex. (Color figure online)

Proof (Sketch). We show that during algorithm $\mathcal{A}$, the invariant holds that a representation of the first n rectangles in $\prec$ maximizes the number of contacts.

We assume that the invariant holds after $\mathcal{A}$ placed $n-1$ rectangles, such that the current representation $\mathcal{R}^*$ is a contact maximal representation of the first $n-1$ rectangles in $\prec$, and realizes k contacts. Algorithm $\mathcal{A}$ now adds the next rectangle R_n. We show that the new representation is contact maximal.

We first prove that the maximum number of contacts that the new representation can realize is $k+2$, since R_n can achieve at most one vertical and one horizontal contact. If these contacts happen naturally, when placing R_n leftmost, then the invariant trivially holds. We therefore prove via a case distinction that in all other cases $k+1$ adjacencies are optimal. We distinguish between the contact that is achieved by R_n, either vertical or horizontal. A sole vertical contact with fan (necessarily, forced by the placement ordering) R_f is achieved only if the horizontal predecessor R_l of R_n is a fan, as shown in Fig. 5a. If R_n is a T-vertex, a single horizontal contact can arise only if R_f is not moved to R_n, as can be seen in Fig. 6. However, when R_n is a fan, this requires R_l to be a T-vertex, see Fig. 5b. The invariant will therefore still be true after $\mathcal{A}$ added all rectangles, producing a contact maximal representation.

Algorithm $\mathcal{A}$ considers each rectangle either once, or its degree many times, when a fan is shifted. As the input graph G is planar, $\mathcal{A}$ runs in linear time. $\square$

3.2   ILP

To solve the contact maximization problem on $L > 2$ layers we propose an ILP formulation, which intuitively works as follows. We create a binary contact variable $c(e)$ for each edge e in the input graph. If a contact is not realized, we set $c(e) = 1$ to satisfy the position constraints, otherwise we can set $c(e) = 0$. To handle false adjacencies we add for each rectangle a constraint on the first false contact that happens from the right and left on the row above, if they exist. We use hard constraints on the rectangle coordinates to prevent the false adjacencies. The objective is to minimize the sum over all contact variables, under all these constraints, to maximize the number of realized contacts in a solution. Additional details can be found in the full paper [14].

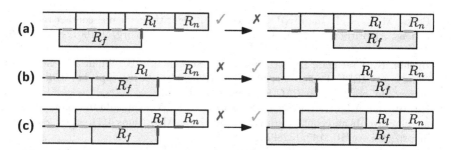

Fig. 6. Three configurations where T-vertex R_n does not realize vertical contacts with R_f initially. We move R_f and either **(a)** reset if the number of contacts is strictly worse, or save when we find **(b)** a tie, or **(c)** an increase in contacts.

$$\text{minimize} \sum_{(v,v')\in E} c(v,v') \tag{1}$$

The following inequalities ensure that there is no overlap between rectangles on the same layer (2), and check whether the horizontal contact is realized (3).

$$x_{i,j} + w_{i,j} \le x_{i,j+1} \qquad\qquad \forall(v_{i,j},v_{i,j+1}) \in E \tag{2}$$

$$x_{i,j+1} \le x_{i,j} + w_{i,j} + c(v_{i,j},v_{i,j+1})M \qquad\qquad \forall(v_{i,j},v_{i,j+1}) \in E \tag{3}$$

The next inequalities verify that the contacts between rectangle $R_{i,j}$ and all of its neighbors on layer $i+1$ are realized.

$$x_{i+1,j'} \le x_{i,j} + w_{i,j} - \varepsilon + c(v_{i,j},v_{i+1,j'})M \qquad \forall e(v_{i,j},v_{i+1,j'}) \in E \tag{4}$$

$$x_{i,j} \le x_{i+1,j'} + w_{i+1,j'} - \varepsilon + c(v_{i,j},v_{i+1,j'})M \qquad \forall e(v_{i,j},v_{i+1,j'}) \in E \tag{5}$$

Finally, we model false adjacencies using pairs $(v_{i,j},v_{i+1,j'})$ stored in sets F_L (resp. F_R) that indicate the index of the first block in row $i+1$ that is left (resp. right) of a neighbor of $R_{i,j}$, but is not itself a neighbor of $R_{i,j}$.

$$x_{i+1,j'} + w_{i+1,j'} \le x_{i,j} \qquad\qquad \forall(v_{i,j},v_{i+1,j'}) \in F_L \tag{6}$$

$$x_{i,j} + w_{i,j} \le x_{i+1,j'} \qquad\qquad \forall(v_{i,j},v_{i+1,j'}) \in F_R \tag{7}$$

References

1. Barth, L.: Semantic word cloud representations: hardness and approximation algorithms. In: Pardo, A., Viola, A. (eds.) LATIN 2014. LNCS, vol. 8392, pp. 514–525. Springer, Heidelberg (2014). https://doi.org/10.1007/978-3-642-54423-1_45
2. Barth, L., Kobourov, S.G., Pupyrev, S.: Experimental comparison of semantic word clouds. In: Gudmundsson, J., Katajainen, J. (eds.) SEA 2014. LNCS, vol. 8504, pp. 247–258. Springer, Cham (2014). https://doi.org/10.1007/978-3-319-07959-2_21

3. Bekos, M.A., et al.: Improved approximation algorithms for box contact representations. Algorithmica **77**(3), 902–920 (2016). https://doi.org/10.1007/s00453-016-0121-3
4. Buchin, K., Creemers, D., Lazzarotto, A., Speckmann, B., Wulms, J.: Geo word clouds. In: IEEE Pacific Visualization (PacificVis 2016), pp. 144–151 (2016). https://doi.org/10.1109/PACIFICVIS.2016.7465262
5. Buchsbaum, A.L., Gansner, E.R., Procopiuc, C.M., Venkatasubramanian, S.: Rectangular layouts and contact graphs. ACM Trans. Algorithms **4**(1), 8:1–8:28 (2008). https://doi.org/10.1145/1328911.1328919
6. Cui, W., Wu, Y., Liu, S., Wei, F., Zhou, M.X., Qu, H.: Context preserving dynamic word cloud visualization. In: IEEE Pacific Visualization (PacificVis 2010), pp. 121–128 (2010). https://doi.org/10.1109/PACIFICVIS.2010.5429600
7. Felsner, S.: Rectangle and square representations of planar graphs. In: Pach, J. (ed.) Thirty Essays on Geometric Graph Theory, pp. 213–248. Springer (2013). https://doi.org/10.1007/978-1-4614-0110-0_12
8. de Fraysseix, H., de Mendez, P.O., Rosenstiehl, P.: On triangle contact graphs. Comb. Probab. Comput. **3**, 233–246 (1994). https://doi.org/10.1017/S0963548300001139
9. Hearst, M.A., Pedersen, E., Patil, L., Lee, E., Laskowski, P., Franconeri, S.: An evaluation of semantically grouped word cloud designs. IEEE Trans. Vis. Comput. Graph. **26**(9), 2748–2761 (2020). https://doi.org/10.1109/TVCG.2019.2904683
10. Koebe, P.: Kontaktprobleme der konformen abbildung. Ber. Sächs. Akad. Wiss. Leipzig, Math. Phys. Klasse **88**, 141–164 (1936)
11. van Kreveld, M., Speckmann, B.: On rectangular cartograms. Comput. Geom. **37**(3), 175–187 (2007). https://doi.org/10.1016/j.comgeo.2006.06.002
12. Li, C., Dong, X., Yuan, X.: Metro-wordle: an interactive visualization for urban text distributions based on wordle. Vis. Inf. **2**(1), 50–59 (2018). https://doi.org/10.1016/j.visinf.2018.04.006
13. Nusrat, S., Kobourov, S.: The state of the art in cartograms. Comput. Graph. Forum **35**(3), 619–642 (2016). https://doi.org/10.1111/cgf.12932
14. Nöllenburg, M., Villedieu, A., Wulms, J.: Layered area-proportional rectangle contact representations. CoRR abs/2108.10711 (2021)
15. Viegas, F.B., Wattenberg, M., Feinberg, J.: Participatory visualization with wordle. IEEE Trans. Vis. Comput. Graph. **15**(6), 1137–1144 (2009). https://doi.org/10.1109/TVCG.2009.171
16. Wang, Y., et al.: Edwordle: consistency-preserving word cloud editing. IEEE Trans. Vis. Comput. Graph. **24**(1), 647–656 (2018). https://doi.org/10.1109/TVCG.2017.2745859
17. Wang, Y., et al.: Shapewordle: tailoring wordles using shape-aware archimedean spirals. IEEE Trans. Vis. Comput. Graph. **26**(1), 991–1000 (2020). https://doi.org/10.1109/TVCG.2019.2934783
18. Wu, Y., Provan, T., Wei, F., Liu, S., Ma, K.L.: Semantic-preserving word clouds by seam carving. Comput. Graph. Forum **30**(3), 741–750 (2011). https://doi.org/10.1111/j.1467-8659.2011.01923.x
19. Yeap, K.H., Sarrafzadeh, M.: Floor-planning by graph dualization: 2-concave rectilinear modules. SIAM J. Comput. **22**(3), 500–526 (1993). https://doi.org/10.1137/0222035

Geometric Aspects in Graph Drawing

Arrangements of Orthogonal Circles
with Many Intersections

Sarah Carmesin$^{(\boxtimes)}$(iD) and André Schulz$^{(\boxtimes)}$(iD)

FernUniversität in Hagen, Universitätsstraße 47, 58097 Hagen, Germany
sarah.carmesin@studium.fernuni-hagen.de, andre.schulz@fernuni-hagen.de

Abstract. An arrangement of circles in which circles intersect only in angles of $\pi/2$ is called an *arrangement of orthogonal circles*. We show that in the case that no two circles are nested, the intersection graph of such an arrangement is planar. The same result holds for arrangement of circles that intersect in an angle of at most $\pi/2$.

For the general case we prove that the maximal number of edges in an intersection graph of an arrangement of orthogonal circles lies in between $4n - O\left(\sqrt{n}\right)$ and $\left(4 + \frac{5}{11}\right)n$, for n being the number of circles. Based on the lower bound we can also improve the bound for the number of triangles in arrangements of orthogonal circles to $(3+5/9)n - O\left(\sqrt{n}\right)$.

Keywords: Circle arrangements · Orthogonal intersection · Intersection graphs · Planar graphs

1 Introduction

A collection of n circles in the plane, is called an *arrangement of orthogonal circles* if any two intersecting circles intersect orthogonally. Here, we call an intersection orthogonal, if the tangents at the intersection point form an angle of $\pi/2$. By definition circles cannot touch in an arrangement of orthogonal circles.

A natural object that arises from an arrangement of orthogonal circles is its intersection graph. A graph G is a *(geometric) intersection graph* if its vertices can be realized by a set of geometric objects, such that two objects intersect if and only if their corresponding vertices form an edge in G. Thus, for an arrangement of orthogonal circles $\mathcal{A}$ we define its intersection graph $G(\mathcal{A})$ as the graph, whose vertices correspond to the circles in $\mathcal{A}$ and two vertices are adjacent, if and only if the associated circles intersect in $\mathcal{A}$. The graph $G(\mathcal{A})$ is called an *orthogonal circle intersection graph*.

Arrangements of orthogonal circles and their intersection graphs were recently introduced by Chaplick et al. [5]. Here it was shown that the intersection graph of n circles contains at most $7n$ edges. Furthermore, it is NP-hard to test whether a graph is an orthogonal unit circle intersection graph. Chaplick et al. also provide bounds for the maximal number of digonal, triangular and quadrilateral cells in arrangements of orthogonal circles.

H. C. Purchase and I. Rutter (Eds.): GD 2021, LNCS 12868, pp. 329–342, 2021.
https://doi.org/10.1007/978-3-030-92931-2_24

Related Work. General (non-orthogonal) arrangements of circles or disks have been studied extensively before. Giving a complete overview over the results in this field is out of scope for this article. We will hence only mention a few selected results. For the special case where all circles have the same radius the intersection graphs are known as *unit disk graphs*. For general arrangements of circles or balls the recognition problems for the corresponding intersection graphs are usually hard (for example for unit disk graphs [3]). We refer the reader to the survey of Hlinený and Kratochvíl [11] for more information. Other work focused on bounding the number of small faces in arrangements of circles [1] or about the circleability of topologically described arrangements [10,13].

Note that we can have general circle arrangements in which all circles pairwise intersect. Thus, the density of the intersection graph can be $\Theta(n^2)$, although many graphs are not intersection graphs of circle arrangements [16] (for example every graph containing $K_{3,3}$ as a subgraph [11]). Hence, asking for the maximum density for intersection graphs in this setting is not an interesting question.

If the circles are allowed to only intersect pairwise in one point, then the intersection graph is called a *contact graph* and the corresponding arrangement is a circle packing. Due to the famous Andreev–Koebe–Thurston circle packing theorem [2,15] the disk contact graphs coincide with the planar graphs. One direction of the circle packing theorem is obvious, a planar straight-line drawing of the contact graph can be derived by placing the vertices at the disk centers. A related result is due to Alon et al. [1]. A *lune* is a digonal cell in an arrangement of circles. If we restrict the intersection graph of the (general) circle arrangement to intersections that are formed by lunes (we call this the *lune-graph*) then also in this setting we can obtain a planar straight-line drawing by placing the vertices at the circle centers.

Every arrangement of orthogonal circles with the same radius can be turned into a unit circle packing by shrinking the circle size by a factor of $\sqrt{2}/2$, but there are unit disk contact graphs that are not intersection graphs of an arrangement of orthogonal circles [5].

A well established quality criteria for drawing graphs is to avoid crossings. However, crossings with large angles are considered less problematic [12]. For this reason graphs that can be drawn with right-angle crossings (known as RAC-drawings) are considered an interesting class from a graph drawing perspective. It was shown that graphs that have straight-line RAC-drawings have at most $4n - 10$ edges [7], for n being the number of vertices. Recently this approach was carried over to drawings with circular arcs that can intersect in right angles only. Chaplick et al. showed that graphs that have circular arc RAC-drawings can have at most $14n - 12$ edges and there are such graphs with $4.5n - O(\sqrt{n})$ edges [6].

Orthogonal circle arrangements can also be seen as circular arc drawings (of 4-regular graphs) with perfect angular resolution. Such drawings are known as Lombardi drawings and have been studied deeply [8,9,14].

Results. We prove bounds for the maximal number of edges in an intersection graph of an arrangement of n orthogonal circles. We show an upper bound of

$\left(4 + \frac{5}{11}\right) n$ and present a lower bound of $4n - O\left(\sqrt{n}\right)$. As a crucial intermediate result we show that in the case of arrangements without nested circles, the intersection graph is planar. In particular, (in a similar vein to disk contact graphs and lune graphs) we obtain a planar straight-line drawing by placing the vertices at the centers of the corresponding circles. As an immediate consequence we get that for arrangements of nonnested orthogonal circles the intersection graph has at most $3n - 6$ edges. We can refine the analysis to improve this bound to $3n - 8$. This bound is tight, since we can show a matching lower bound. Our lower bound constructions can be slightly modified to also improve the bounds for the maximal number of triangular cells in arrangements of orthogonal circles to $(3 + 5/9)n - O\left(\sqrt{n}\right)$.

Organization. We first prove in Sect. 2 that the orthogonal circle intersection graphs are planar in the nonnested case. In Sect. 3 we extend our ideas to general circle arrangements and prove the upper bound. In Sect. 4 we discuss lower bound constructions.

2   Bounds for Nonnested Arrangements

For an arrangement of orthogonal circles we call the straight-line drawing of its intersection graph that is obtained by placing the vertices on the corresponding circle centers the *embedded intersection graph*. Figure 1 depicts such an arrangement and its embedded intersection graph. In this section we prove that the embedded intersection graph is noncrossing.

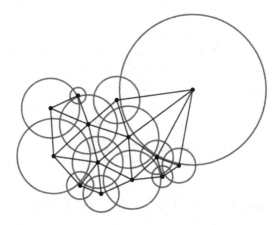

Fig. 1. A nonnested circle arrangement and its embedded intersection graph.

We start with properties of arrangements of two or three nonnested orthogonal circles. The first observation is a simple application of Pythagoras' theorem.

Observation 1. *Let A and B be two circles with centers C_A and C_B and radii r_A and r_B, respectively. Then A and B are orthogonal if and only if $|C_A C_B|^2 = r_A^2 + r_B^2$.*

Lemma 1. *In an arrangement of nonnested orthogonal circles, the center of a circle A is not contained inside a circle other than A.*

Proof. Let A and B be two nonnested circles with centers C_A and C_B and radii r_A and r_B, respectively. Assume that C_A lies inside B. Obviously, A and B intersect, since otherwise the circles are nested. Since A and B intersect orthogonally it holds that $|C_A C_B|^2 = r_A^2 + r_B^2$. Further, since C_A is in B, we have $|C_A C_B| < r_B$ and thus $|C_A C_B|^2 < r_B^2$. We get that $r_A^2 + r_B^2 < r_B^2$, which is a contradiction for $r_A, r_B \in \mathbb{R}$.

Lemma 2. *In an arrangement of nonnested orthogonal circles, for every pair of circles A and B and every point p on A it holds that B intersects the line segment $C_A p$ in at most one point.*

Proof. Let A and B be two circles with centers C_A and C_B and radii r_A and r_B, respectively. Assume that there exist a point p on A such that B intersects $C_A p$ twice. We call these intersection points q and s with $|C_A q| < |C_A s| < r_A$ and denote the midpoint between q and s with t. By Lemma 1 C_B lies outside of A. So, for the circle B to have a point inside of A, B has to intersect A in some point u (see Fig. 2). Since the circles intersect orthogonally, $C_A u C_B$ is a right triangle. Further, since qs is a chord of the circle B, the triangle sqC_B is isosceles and its height is $C_B t$. Thus, we have a right angle at t between $C_A p$ and $C_B t$ and $C_A t C_B$ is a right triangle.

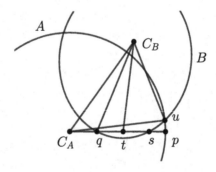

Fig. 2. Illustration of the construction in the proof of Lemma 2.

Since the right triangles $C_A t C_B$ and $C_A u C_B$ share the same hypotenuse $C_A C_B$ we get by Thales' theorem that t and u have to be on the circle with diameter $C_A C_B$. We know that $|C_B u| = r_B$ and $|C_B t| < |C_B q| = r_B$. Hence, t lies closer to C_B than u. It follows that u is closer to C_A than t, thus $C_A t > C_A u = r_A$. This implies that t is outside of A, which is a contradiction.

Lemma 3. *In an arrangement of nonnested orthogonal circles, for every intersecting pair of circles A and B there is no third circle that intersects the line segment between the centers of A and B.*

Proof. Let A, B and D be three circles with centers C_A, C_B and C_D and radii r_A, r_B and r_D, respectively. The circles A and B intersect. Assume for a contradiction that the circle D intersects the line between C_A and C_B.

If the circle D intersects the line segment between C_A and C_B just once, either C_A or C_B would be inside D, a contradiction to Lemma 1. Thus, D has to intersect the line segment $C_A C_B$ twice. We denote these intersection points by q and s with $|C_A q| < |C_A s| < |C_A C_B|$ and midpoint between q and s by t. Due to Lemma 2 q and s cannot lie in the same circle, so one lies in A and the other in B. Thus, D intersects both A and B. By Lemma 1 the center C_D of D has to be outside of the circles A and B. Thus, for the circle D to have a point in the inside of the circles A and B, the circle D has to intersect the circles A (in some point u_A) and B (in some point u_B). The situation is depicted in Fig. 3.

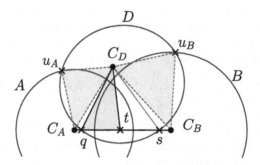

Fig. 3. Illustration of the construction in the proof of Lemma 3.

Since all circles intersect orthogonally we have right angles at u_A between $C_A u_A$ and $C_D u_A$ and at u_B between $C_B u_B$ and $C_D u_B$. Also, since sq is a chord of the circle D, the triangle sqC_D is isosceles and its height is $C_D t$. Thus, we have a right angle at t between $C_A C_B$ and $C_D t$. This gives us five right triangles $C_A C_D u_A$ and $C_B C_D u_B$ (red, dashed), $qC_D t$, $C_B C_D t$ and $C_A C_D t$. We obtain

$$|C_A C_D|^2 = r_A^2 + r_D^2, \quad |C_B C_D|^2 = r_B^2 + r_D^2, \quad |C_B C_D|^2 = |C_B t|^2 + |C_D t|^2$$

$$|C_A C_D|^2 = |C_A t|^2 + |C_D t|^2, \quad r_D^2 = \left(\frac{|qs|}{2}\right)^2 + |C_D t|^2.$$

Combining these equations we get

$$|C_A t|^2 = |C_A t|^2 + |C_D t|^2 - |C_D t|^2 = |C_A C_D|^2 - |C_D t|^2 = r_A^2 + r_D^2 - |C_D t|^2$$

$$= r_A^2 + \left(\frac{|qs|}{2}\right)^2.$$

It follows that $|C_A t| > r_A$. By a symmetric argument we see also that $|C_B t| > r_B$. We get $|C_A t| + |C_B t| > r_A + r_B$, which is a contradiction.

We can now combine our observations to prove the following result.

Theorem 1. *The embedded intersection graph of an arrangement of nonnested orthogonal circles is noncrossing.*

Proof. Suppose for contradiction four circles A, B, D, E with centers C_A, C_B, C_D and C_E that are arranged in such way that their embedded intersection graph has two edges $C_A C_B$ and $C_D C_E$ that cross in the point h. This means we have two pairs of intersecting circles A, B and D, E. Note that $C_A C_B$ is contained in the union of A and B. Hence, h has to lie in at least one of the circles A or B. By the same reasoning h also has to lie in at least one of the circles D or E. Without loss of generality we can assume that h lies in D. By Lemma 1 the circle D cannot enclose $C_A C_B$ completely, thus it has to intersect the line segment $C_A C_B$. This, however, contradicts Lemma 3.

By Theorem 1 the intersection graph is planar and we can further show that the boundary face of the embedded intersection graph is at least a pentagon if we have five or more circles (the proof can be found in the full version [4]). Applying Euler's formula yields the following result.

Corollary 1. *The intersection graph of an arrangement of nonnested orthogonal circles has at most $3n - 8$ edges for $n \geq 5$.*

In Sect. 4 we show that the bound of $3n - 8$ in Corollary 1 is tight. In the full version [4] we show that Theorem 1 also holds when all circles intersect in a (not necessarily identical) angle of at most $\pi/2$.

3   Bounds for General Orthogonal Arrangements

In this section we prove an upper bound of $\left(4 + \frac{5}{11}\right) n$ edges for intersection graphs of orthogonal circle arrangements with nested circles. We first discuss the general approach and introduce necessary terminology before continuing with details and proofs.

For every circle C in an arrangement $\mathcal{A}$ we define its *depth* $t(C)$ as the maximum cardinality of a set of pairwise nested circles in $\mathcal{A}$ that are properly contained in C. A circle with depth 0, i.e., it contains no circles properly, is referred to as *shallow* otherwise as *deep*.

As a first step we show that in every arrangement we can find a circle with depth at most 1 that is orthogonal to at most seven deep circles (Lemma 9). We select one circle with this property and name it the *red circle*. We then look at the circles properly contained in the red circle. We call these circles *black circles*; see Fig. 4. The key observation is that we can delete the set of black circles from the arrangement and by doing so we only lose few edges from the intersection graph, i.e. at most $(4 + 5/11) \cdot n$ for n black circles. To obtain this bound we distinguish between intersections between the black circles and intersections between a black circle and a circle that intersects a black and the red circle (such circles are called *green circles*). To make our analysis work we have to partition the black circles further. If a black circle center lies on the boundary of the embedded intersection graph induced by the vertices of the black circles we call the corresponding circle *boundary black circle*, otherwise *inner black circle*.

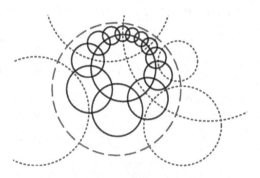

Fig. 4. Illustration of the red (dashed), black and green (dotted) circles. This arrangement has only one inner black circle. (Color figure online)

We color edges in the intersection graph according to the color of the corresponding circles as follows: An intersection between a black and a green circle yields a *green edge* and an intersection between two black circles yields a *black edge*. If there are n black circles and b of those are boundary black circles, then we have at most $3n - b - 3$ black edges as a consequence of Euler's formula and Theorem 1. We will prove that each black circle can be orthogonal to at most two green circles (Lemma 5). However, the inner black circles can only be intersected by green circles with depth at least 1 (Lemma 11). We can chose the red circle so that there are at most seven deep green circles. The intersection graph of these seven circles has at most eight edges (Observation 2 and Lemma 12). We exploit this fact to show that only eight inner circles can be orthogonal to two green circles (Lemma 13). As a final observation we show that if there are at most 11 black circles in the red circle, there are at most 3 inner black circles that intersect two of the green circles. We can then combine our findings to prove that we can always find a set of n black circles that intersects at most $(4 + 5/11) n$ circles.

We are now continuing with the proofs and details. We begin by stating a few properties of arrangements of orthogonal circles. The first lemma was proven by Chaplick et al. [5].

Lemma 4 ([5]). *No orthogonal circle intersection graph contains a K_4 or an induced C_4.*

Lemma 5. *Let A and B be two nested circles. There are at most two circles that intersect both A and B orthogonally.*

Proof. Suppose that there are two nested circles A and B (B lies inside A) that both intersect at least three circles D, E and F orthogonally. Consider the intersection graph of A, B, D, E and F. If the circles D, E and F are pairwise orthogonal to each other the vertices corresponding of A, D, E and F form a K_4, a contradiction due to Lemma 4. However, if two of the circles D, E and F are not orthogonal to each other their corresponding vertices together with B and A induce a C_4, which yields a contradiction due to Lemma 4.

Thus, at most two circles that intersect both A and B orthogonally.

Lemma 6. *If a circle C intersects the circles A and B orthogonally, then one of the following holds: (i) A and B do not intersect, or (ii) A and B are orthogonal and C contains precisely one of the intersection points of A and B.*

Proof. We prove that if (i) does not hold, then (ii) holds. So assume A and B intersect. We apply a Möbius transformation that maps A to a straight line. Note that such a transformation is conformal and thus maintains the angles; see Fig. 5. The centers of B and C will then have to lie on A. Clearly, if C contains both points of $A \cap B$ then it also has to contain B, but since B intersects C, we have a contradiction. Also, if C does not contain any point of $A \cap B$, then it has to be either contained in B or is to the left or right of B along A, but since B intersects C, we have again a contradiction.

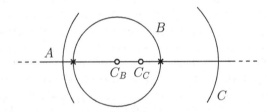

Fig. 5. Situation in the proof of Lemma 6 when C contains $A \cap B$. C_B (C_C) is the center of $B(C)$.

Lemma 7. *In an arrangement of orthogonal circles let A and B be two circles that intersect. All circles that are orthogonal to A and B that contain the same intersection point of A and B are nested.*

Proof. Assume that there are two nonnested circles C and D that both contain the same intersection point u of A and B. Since C and D contain u but are not nested, they must intersect each other. Both also intersect A and B. This means the intersection graph of the four circles is a K_4. This contradicts Lemma 4.

The following lemma is again taken from Chaplick et al. [5, Lemma 5]. The "Moreover"-part is not explicitly written down, but it is apparent from the construction given in its proof.

Lemma 8 ([5]). *Every arrangement of orthogonal circles has a circle that is orthogonal to at most seven other circles. Moreover, this circle is a shallow circle.*

We can deduce a similar lemma for deep circles.

Lemma 9. *Every arrangement of orthogonal circles with nested circles has a circle C with depth $t(C) = 1$ that is orthogonal to at most seven other circles with depth at least 1.*

Proof. Let $\mathcal{A}$ be an arrangement of orthogonal circles. By deleting all shallow circles we obtain the arrangement $\mathcal{A}'$. According to Lemma 8 we can find a shallow circle C in $\mathcal{A}'$ that is orthogonal to at most seven other circles. Since C is shallow in $\mathcal{A}'$ it has depth $t(C) = 1$ in the arrangement $\mathcal{A}$.

In the following we select any circle that meets the requirements of Lemma 9 and refer to it as the red circle. We remind the reader that we call the circles contained in the red circle the black circles.

Lemma 10. *The set of black circles S_B inside a red circle C corresponds to a vertex set V_B in the intersection graph incident to no more than $4n_B + i - 3$ edges, for $n_B = |S_B|$ and i being the number of inner black circles in S_B, that each are orthogonal to two circles not in S_B.*

Proof. Let C be the red circle. We count the edges incident to V_B. Edges with two endpoints in V_B are black edges, edges with one endpoint in V_B are green edges. We denote the number of boundary black circles by b. According to Theorem 1 the intersection graph of the arrangement restricted to the S_B is planar. Moreover this planar graph has b vertices on its outer face. Thus, by Euler's formula we have at most than $3n_B - b - 3$ black edges.

We now count the green edges. Every circle $D \notin S_B$ that intersects a circle in S_B has to intersect C as well. According to Lemma 5, each of the n_B black circles is orthogonal to at most two green circles. By our assumption $n_B - b - i$ black inner circles intersect at most one green circle. Thus, we have at most $2n_B - (n_B - b - i) = n_B + b + i$ green edges. Adding the $3n_B - b - 3$ black edges yields the upper bound of $4n_B + i - 3$ as stated in the lemma.

Lemma 11. *Every green circle intersecting an inner black circle is a deep circle.*

Proof. Let D be the red circle and S_D be the set of black circles. Suppose for a contradiction that there is a shallow green circle E with center C_E that intersects an inner circle $F \in S_D$ with center C_F. Note that in this case E also has to intersect D. By Lemma 1 C_E, is therefore outside of D; see Fig. 6.

Let $\mathcal{A}$ be the arrangement consisting of the circles in S_D and E. All circles in S_D and the circle E have depth 0 so the arrangement $\mathcal{A}$ is nonnested. According to Theorem 1 the embedded intersection graph $G(\mathcal{A})$ is noncrossing.

Let $\mathcal{A}'$ be the arrangement consisting only of the circles in S_D. Again, all the circles are shallow, thus the arrangement is nonnested and the intersection graph $G(\mathcal{A}')$ noncrossing.

Since C_E is outside D it lies in the outer face of $G(\mathcal{A}')$. On the other hand F is an inner circle, so its corresponding vertex is not on the boundary of $G(\mathcal{A}')$. The straight-line edge between C_E and C_F must intersect an edge on the boundary of the outer face of $G(\mathcal{A}')$. This yields a crossing and thus a contradiction.

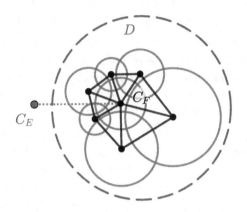

Fig. 6. Illustration of the arrangement in Lemma 11.

By Lemma 8 a red circle C intersects at most 7 deep circles. We now take a look at the possible intersections of the seven deep circles. We start with the following observation.

Observation 2. *Let I_C be the set of deep circles that intersect a red circle. The intersection graph of I_C has*

- *No induced C_4, according to Lemma 4 and*
- *No induced C_3, since every circle in I_C is orthogonal to the red circle and according to Lemma 4 there is no K_4 in the intersection graph of the arrangement consisting of I_C and C.*

By a case distinction we can limit the graphs that fulfil the constraints listed in Observation 2. The proof is given in the full version [4].

Lemma 12. *Every graph G with at most seven vertices without an induced C_3 or C_4 has at most 8 edges.*

We can now bound the number of intersection points of the circles in I_C.

Lemma 13. *Let C be the red circle and let I_C be the set of deep circles intersecting C. The arrangement of circles in I_C has at most sixteen intersection points of which eight are inside of C.*

Proof. According to Lemma 12 the intersection graph of I_C has at most eight edges. Hence, there are eight pairs of intersection points in the arrangement consisting of the circles in I_C. Due to Lemma 6 for every pair exactly one intersection point is inside of C. Thus, at most eight intersection points of circles in I_C are inside the circle C.

Lemma 14. *In the intersection graph of every arrangement of orthogonal circles we can find a nonempty subset V_C that is incident to at most $4n + 5$ edges, where $n = |V_C|$.*

Proof. Let $\mathcal{A}$ be an arrangement of orthogonal circles. According to Lemma 9 we can a find a red circle C with depth $t(C) = 1$ that is orthogonal to at most seven deep circles. We denote the black circles by S_C and set $n = |S_C|$. Further let V_C denote the vertex set corresponding to S_C.

We now prove that there are at most 8 inner black circles in S_C that are orthogonal to two circles not in S_C. According to Lemma 11 the inner black circles can only be intersected by deep green circles. If a black circle intersects two green circles, then the green circles have to intersect, otherwise the intersection graph of the black, the two green and the red circle would induce a C_4. According to Lemma 6 a black circle that intersects two green circles contains their intersection point. Lemma 7 states that all circles containing the same intersection point must be nested. Since the black circles are not nested, only one black circle contains a given intersection point. By Lemma 13 the seven deep green circles have at most eight intersection points inside C. Thus, at most eight inner black circles are orthogonal to two deep green circles. We now apply Lemma 10 with $i_2 \le 8$ to obtain that V_C is incident to at most $4n - 3 + i_2 = 4n + 5$ edges.

Our goal is to apply the last lemma for bounding the density of the intersection graph. If we can repeatedly take out vertex sets of size k with ck incident edges (for a constant c), then the density of the graph is no more than cn, for n being the number of vertices. Unfortunately, because of the additive constant Lemma 14 is too weak if the subsets are small. Hence, we analyse small sets separately to get a better bound. The analysis including the proofs can be found in the full version [4]. It culminates in the following statement.

Lemma 15. *In the intersection graph of an arrangement of orthogonal circles we can find a subset V_C of n vertices that has at most $\left(4 + \frac{5}{11}\right) n$ edges.*

Theorem 2. *The intersection graph of an arrangement of n orthogonal circles has at most $\left(4 + \frac{5}{11}\right) n$ edges.*

Proof. Assume there exist arrangements with n orthogonal circles, whose intersection graphs have more than $\left(4 + \frac{5}{11}\right) n$ edges. Consider a smallest such arrangement $\mathcal{A}$ in terms of numbers of circles and its intersection graph $G(\mathcal{A}) = (V, E)$. By Lemma 15 there exists a subset $S \subset V$ of n' vertices that is incident to at most $\left(4 + \frac{5}{11}\right) n'$ edges. We take out S and all incident edges. The new graph has $(n - n')$ vertices and more than $\left(4 + \frac{5}{11}\right) (n - n')$ edges. This contradicts the assumption that $\mathcal{A}$ is minimal.

.

4 Lower Bounds

Fig. 7. The arrangement $\mathcal{B}_{3,15}$. Hub circles are drawn with thick, satellite circles with thin lines. Corresponding satellite and hub circles have the same color.

In this section we discuss lower constructions. Our ideas are based on the arrangement $\mathcal{B}_{x,a}$, parametrized by two integers $a \geq 5$ and $x \geq 1$, which is constructed as follows. We start with arranging a circles with the same radius in such a way that their centers lie on a circle and two neighboring circles intersect. We call these circles the *satellite circles*. We add another circle (called *hub circle*) to this arrangement such that it intersects every satellite circle orthogonally. We name this arrangement a *wheel of circles*. An arrangement $\mathcal{B}_{x,a}$ is then constructed by "nesting" x wheels of circles with a satellite circles each inside each other such that each satellite circle of one wheel intersects two satellite circles of the next wheel and two satellite circles of the previous wheel (Fig. 7). The details of this construction (including the proof that the arrangement is orthogonal) can be found in the full version [4].

Lemma 16. *The intersection graph of $\mathcal{B}_{x,a}$ has $x \cdot (a+1)$ vertices and $4xa - 2a$ edges.*

Proof. The arrangement consists of x wheel of circles, each having a satellite circles and one hub circles. Thus, the intersection graph has $x \cdot (a+1)$ vertices. Every vertex corresponding to a hub circle has clearly degree a. Further, every vertex corresponding to a satellite circle has degree 7, except those corresponding to a satellite circle on the inner or outermost wheel of circles, which have degree 5. So the sum of the vertex degrees is $\sum_{v \in V(G_{x,a})} \deg(v) = ax + 7a(x-2) + 5a \cdot 2 = 8xa - 4a$. This number equals twice the number of edges, and therefore the intersection graph has $4xa - 2a$ edges.

Lemma 17. *For every n there is an arrangement of orthogonal circles, whose intersection graph has n vertices and $4n - O(\sqrt{n})$ edges.*

The proof of the lemma is obtained by counting the edges in the arrangement $\mathcal{B}_{x,a}$ with x and a being $\Theta(\sqrt{n})$. Details are given in the full version [4].

We now give a lower bound for nonnested orthogonal circles based on the construction shown in Fig. 8. The proof is given in the full version [4].

Lemma 18. *For every* $n \geq 6$ *for which* $n \bmod 5 = 1$ *the arrangement* $\mathcal{B}_{((n-1)/5),5}$ *with only the innermost hub circle is nonnested and its n-vertex intersection graph has $3n - 8$ edges.*

Fig. 8. Detail of the arrangement used to prove the lower bound in the nonnested case in Lemma 18.

Chaplick et al. [5] investigated the maximal number of triangular cells in an orthogonal circle arrangement. They proved an upper bound of $4n$ and gave a lower bound of $2n$ triangular cells, which they later improved to $3n - 3$. This bound can be improved by taking the arrangement $\mathcal{B}_{x,a}$ and place a small (orthogonal) circle around every intersection point. This implies the following lemma which proof is given in the full version [4].

Lemma 19. *For infinitely many values of n there is an arrangement of n orthogonal circles with $\left(3 + \frac{5}{9}\right) n - O\left(\sqrt{n}\right)$ triangular cells.*

References

1. Alon, N., Last, H., Pinchasi, R., Sharir, M.: On the complexity of arrangements of circles in the plane. Discrete Comput. Geom. **26**(4), 465–492 (2001). https://doi.org/10.1007/s00454-001-0043-x
2. Andreev, E.M.: Convex polyhedra in Lobačevskiĭ spaces. Mat. Sb. (N.S.) **81**(123)(3), 445–478 (1970). https://doi.org/10.1070/SM1970v010n03ABEH001677

3. Breu, H., Kirkpatrick, D.G.: Unit disk graph recognition is NP-hard. Comput. Geom. **9**(1–2), 3–24 (1998). https://doi.org/10.1016/S0925-7721(97)00014-X

4. Carmesin, S., Schulz, A.: Arrangements of orthogonal circles with many intersections (2021). https://arxiv.org/abs/2106.03557v2

5. Chaplick, S., Förster, H., Kryven, M., Wolff, A.: On arrangements of orthogonal circles. In: Archambault, D., Tóth, C.D. (eds.) GD 2019. LNCS, vol. 11904, pp. 216–229. Springer, Cham (2019). https://doi.org/10.1007/978-3-030-35802-0_17

6. Chaplick, S., Förster, H., Kryven, M., Wolff, A.: Drawing graphs with circular arcs and right-angle crossings. In: Albers, S. (ed.) 17th Scandinavian Symposium and Workshops on Algorithm Theory (SWAT 2020). Leibniz International Proceedings in Informatics (LIPIcs), vol. 162, pp. 21:1–21:14. Schloss Dagstuhl-Leibniz-Zentrum für Informatik, Dagstuhl, Germany (2020). https://doi.org/10.4230/LIPIcs.SWAT.2020.21

7. Didimo, W., Eades, P., Liotta, G.: Drawing graphs with right angle crossings. Theor. Comput. Sci. **412**(39), 5156–5166 (2011). https://doi.org/10.1016/j.tcs.2011.05.025

8. Duncan, C.A., Eppstein, D., Goodrich, M.T., Kobourov, S.G., Nöllenburg, M.: Lombardi drawings of graphs. J. Graph Algorithms Appl. **16**(1), 85–108 (2012). https://doi.org/10.7155/jgaa.00251

9. Eppstein, D.: A Möbius-invariant power diagram and its applications to soap bubbles and planar lombardi drawing. Discrete Comput. Geom. **52**(3), 515–550 (2014). https://doi.org/10.1007/s00454-014-9627-0

10. Felsner, S., Scheucher, M.: Arrangements of pseudocircles: on circularizability. Discrete Comput. Geom. **64**(3), 776–813 (2019). https://doi.org/10.1007/s00454-019-00077-y

11. Hlinený, P., Kratochvíl, J.: Representing graphs by disks and balls (a survey of recognition-complexity results). Discret. Math. **229**(1–3), 101–124 (2001). https://doi.org/10.1016/S0012-365X(00)00204-1

12. Huang, W., Hong, S., Eades, P.: Effects of crossing angles. In: IEEE VGTC Pacific Visualization Symposium 2008, PacificVis 2008, Kyoto, Japan, 5–7, March 2008, pp. 41–46. IEEE Computer Society (2008). https://doi.org/10.1109/PACIFICVIS.2008.4475457

13. Kang, R.J., Müller, T.: Arrangements of pseudocircles and circles. Discrete Comput. Geom. **51**(4), 896–925 (2014). https://doi.org/10.1007/s00454-014-9583-8

14. Kindermann, P., Kobourov, S.G., Löffler, M., Nöllenburg, M., Schulz, A., Vogtenhuber, B.: Lombardi drawings of knots and links. J. Comput. Geom. **10**(1), 444–476 (2019)

15. Koebe, P.: Kontaktprobleme der konformen abbildung. Berichte über die Verhandlungen der Sächsischen Akad. der Wissen. zu Leipzig. Math. Phys. Klasse **88**, 141–164 (1936)

16. McDiarmid, C., Müller, T.: The number of disk graphs. Eur. J. Comb. **35**, 413–431 (2014). https://doi.org/10.1016/j.ejc.2013.06.037

Limitations on Realistic Hyperbolic Graph Drawing

David Eppstein[(✉)]

Computer Science Department, University of California, Irvine,
Irvine, CA 92697, USA
eppstein@uci.edu

Abstract. We show that several types of graph drawing in the hyperbolic plane require features of the drawing to be separated from each other by sub-constant distances, distances so small that they can be accurately approximated by Euclidean distance. Therefore, for these types of drawing, hyperbolic geometry provides no benefit over Euclidean graph drawing.

Keywords: Hyperbolic graph drawing · Realistic graph drawing · Vertex-edge resolution · Vertex-vertex resolution · Angular resolution

1 Introduction

Although most graph drawing algorithms place vertices and edges in the Euclidean plane, several past works instead use a different geometry, the hyperbolic plane. Beginning in the 1990s, researchers proposed hyperbolic graph drawings to combine focus and context: the fisheye-like view provided by the Poincaré disk visualization of the hyperbolic plane allows parts of the drawing to be shown in an expanded view, with the rest compressed into the margins of the Poincaré disk but remaining entirely visible [22]. This line of research also includes similar techniques using three-dimensional hyperbolic geometry [26–28,35]. The hyperbolic plane has also been used for greedy graph drawings, with the property that a path to any vertex can be found by always moving to a neighboring vertex that is closer to the eventual destination. Unlike the Euclidean plane, the hyperbolic plane allows such drawings for any graph [4,15,19]. Hyperbolic geometry was central to our construction of Lombardi drawings for graphs of maximum degree three [14] and for Halin graphs [11]. We have also developed algorithms for finding a good choice of initial views in hyperbolic visualizations [3], and used spring embedding techniques to find high-quality hyperbolic graph drawings [20]. Other investigations of hyperbolic graph drawing include the use of circle packings to construct hyperbolic drawings [25], interactive systems using hyperbolic drawing [13,38], hyperbolic drawing of power-law graphs [5], hyperbolic Euler diagrams [34], distance distortion of hyperbolic embeddings [7,32,36], and hyperbolic multidimensional scaling [31,37].

© Springer Nature Switzerland AG 2021
H. C. Purchase and I. Rutter (Eds.): GD 2021, LNCS 12868, pp. 343–357, 2021.
https://doi.org/10.1007/978-3-030-92931-2_25

In this paper, we investigate hyperbolic geometry from the point of view of *realistic graph drawing*. This type of analysis, previously applied to Euclidean graph drawing [1,10], treats the vertices and edges of a graph drawing as having nonzero radius or thickness, rather than being idealized mathematical points and curves, so that they can be seen by a reader of the drawing. This required thickness has been formulated mathematically in several related ways, including placing constraints on the vertex-vertex resolution (the minimum distance between center points of vertices) or vertex-edge resolution (the minimum distance of the center point of any vertex from an edge that it is not an endpoint of). In ink-based *bold graph drawing* methods all features of the drawing must have visible parts that are not covered by other features, so that the graph may be unambiguously determined from its drawing [21,29]. These parameters are also related to *angular resolution*, the sharpest angle between two edges incident at the same vertex, as edges forming sharp angles need high length to be visibly separated from each other [16,18,23].

Any Euclidean drawing can be scaled to achieve constant vertex-vertex or vertex-edge resolution, so these parameters are typically compared against the area of a bounding box of the drawing. A drawing style is considered to be good when it achieves polynomial area, and bad when the area is exponential [12]. But in hyperbolic graph drawing, there is an absolute length scale, and it is not possible to rescale a drawing without changing its shape. This hyperbolic length scale is essential to focus+context applications of hyperbolic visualizations, as it controls the sizes of objects near the center of the visualization, relative to the overall view. In greedy drawings, constant vertex separation in this absolute length scale is necessary, because drawings with smaller vertex distances would be approximately Euclidean, constraining greedy drawings to have bounded vertex degree. More generally, parts of a hyperbolic graph drawing with features significantly smaller than the unit of absolute length would be approximately Euclidean, failing to take advantage of any differences between hyperbolic and Euclidean geometry. Therefore, we will define a realistic hyperbolic drawing to be one in which the resolution parameters of the realistic graph drawing model, such as vertex-vertex resolution, vertex-edge resolution, or line thickness, are at least constant in absolute length. We consider realistic drawing to be impossible when these parameters are forced by the constraints of the drawing to be $o(1)$.

Our work shows that this realistic model of hyperbolic graph drawing is severely limited, providing a partial explanation of the failure of hyperbolic approaches to focus+context to come into wider use. In particular, we prove:

- Every straight-line crossing-free drawing of an n-vertex maximal planar graph in the hyperbolic plane has vertex-edge resolution $O(1/\sqrt{n})$ (Theorem 1). Moreover, there exist n-vertex planar graphs (the well-known nested triangle graphs) for which every straight-line crossing-free drawing has vertex-edge resolution $O(1/n)$ (Theorem 2). Both bounds are tight.
- Although planar graphs have hyperbolic drawings with high vertex-vertex resolution, these drawings have exponentially small angular resolution for all

maximal planar graphs (Theorem 3). This differs from Euclidean drawings, whose angular resolution is bounded by a function of the degree [23].

- Simple structure is not enough to avoid these problems: some series-parallel graphs of bounded bandwidth require polynomially small vertex-edge resolution and either small vertex-vertex resolution or exponentially small angular resolution (Theorem 4). Grid graphs also obey similar bounds (Theorem 5).
- Beyond planar graph drawing, every n-vertex graph has a Euclidean drawing with unit vertex-vertex resolution and angular resolution $\Theta(1/n)$. However, we prove that for hyperbolic drawings with unit vertex-vertex resolution, some graphs require angular resolution $O(1/n^2)$ (Theorem 6) and their bold drawings require edge width $O(1/n)$ (Theorem 7).

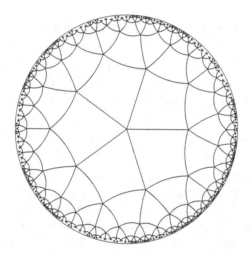

Fig. 1. A tessellation of the Poincaré disk model of the hyperbolic plane by squares. Arbitrarily large subgraphs of this graph have constant vertex-edge resolution, unlike the maximal planar graphs. Although the internal faces of this tessellation can be triangulated, it is impossible to add edges that make the outer face triangular without violating planarity.

2   Vertex-Edge Resolution

In this section, we examine the vertex-edge resolution (the minimum hyperbolic distance between a vertex and an unrelated edge) of graphs, drawn in the hyperbolic plane with (hyperbolically) straight edges. There exist planar graphs that can be drawn in this way with at least constant vertex-edge resolution (Fig. 1). However, as we show, this is not the case for any maximal planar graph.

Lemma 1. *Every hyperbolic triangle has area at most π.*

Proof. This follows from the well-known formula for the area of a hyperbolic triangle as $\pi - \sum \theta_i$, where θ_i are the internal angles of the triangle.

Lemma 2. *In a planar straight-line hyperbolic drawing of an n-vertex maximal planar graph, at least one face has area $\leq \frac{\pi}{2n-3}$.*

Proof. By Lemma 1 the exterior face has area $\leq \pi$. The remaining $2n - 3$ faces partition this area into disjoint subsets, one of which must be $\leq \pi/(2n - 3)$.

Lemma 3 (tangent rule for hyperbolic right triangles). *If a hyperbolic right triangle has legs of length x and y, the angle θ opposite x satisfies*

$$\tan \theta = \frac{\tanh x}{\sinh y}.$$

Proof. See [24, Corollary 32.13, p. 431].

Lemma 4. *A hyperbolic right triangle with leg lengths $x \leq y \leq 1$ has area $\Theta(xy)$.*

Proof. This follows by expressing the area as π minus the sum of angles, expressing these angles in terms of x and y according to Lemma 3, replacing these expressions by their power expansions, and omitting lower-order terms; see [24, p. 434].

Lemma 5. *Let h be the height (minimum distance from any vertex to the opposite edge) of a hyperbolic triangle T. Then T has area $\Omega\big(\min(1, h^2)\big)$.*

Proof. Let height h be achieved by a line segment from vertex v to the opposite side S. In order avoid smaller height at another vertex, S extends for distance at least $h/2$ on either side of this segment. The result follows by applying Lemma 4 to the right triangles formed by the segment of length h and the perpendicular segments of S of length $h/2$. They are disjoint and lie entirely within T, so their total area lower-bounds that of T.

Theorem 1. *Every straight-line planar hyperbolic drawing of an n-vertex maximal planar graph has vertex-edge resolution $O(1/\sqrt{n})$.*

Proof. By Lemma 2, at least one face has area $O(1/n)$. By Lemma 5, the height of this face is $O(1/\sqrt{n})$. Therefore, this face has a vertex and non-incident edge that are at distance $O(1/\sqrt{n})$ from each other.

Theorem 1 is tight: some n-vertex maximal planar graphs have vertex-edge resolution $O(1/\sqrt{n})$. For instance, obtain G from a square grid graph by triangulating each square and adding three surrounding vertices to form an outer face, shrink the drawing to have unit radius, and draw it within a unit disk of the Klein model of the hyperbolic plane (which preserves straight line drawings) giving a hyperbolic drawing whose vertex-edge resolution is the scale factor, $O(1/\sqrt{n})$.

To strengthen Theorem 1 for some graphs, we use a maximal planar version of the *nested triangles graph* (Fig. 2), formed from $n/3$ nested triangles by adding edges between consecutive triangles to make the graph maximal planar.

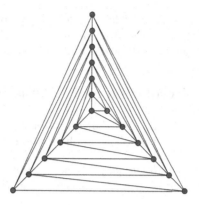

Fig. 2. A maximal planar version of the nested triangles graph, for which any planar straight-line hyperbolic drawing has vertex-edge resolution $O(1/n)$.

Theorem 2. *If the maximal planar nested triangles graph on n vertices is given a straight-line planar drawing in the hyperbolic plane, then the drawing must have vertex-edge resolution $O(1/n)$.*

Proof. Among the $n/3$ nested triangles of the graph, at least $n/6$ must be drawn as nested regardless of the choice of outer face. In any drawing, consider the middle triangle T of these $n/6$ triangles. We distinguish two cases.

– If the hyperbolic diameter of triangle T is ≥ 1, then each of the $n/12$ rings of six faces that surround it in the drawn-nested subset of $n/6$ triangles must, to completely surround T, include at least two triangular faces of diameter $\Omega(1)$. Because these $n/6$ high-diameter triangular faces are all disjoint, and all lie within the $\leq \pi$ area of the outer face of the drawing (Lemma 1), one of them must have area $\leq 6\pi/n$. By Lemma 4, this triangular face, with diameter $\Omega(1)$ and area $O(1/n)$, must have height $O(1/n)$.
– If the hyperbolic diameter of triangle T is ≤ 1, then T surrounds a drawing of a nested triangles graph consisting of $n/4$ vertices in $n/12$ nested triangles, all drawn within a region of the hyperbolic plane of diameter ≤ 1. Within this region, hyperbolic distances can be approximated to within a constant factor by Euclidean distances. It is known that Euclidean straight-line drawings of the nested triangles graph must have two vertices whose distance is $O(1/n)$ times the diameter [9], and the same follows for the hyperbolic drawing within T. If these two close-together vertices are adjacent on a single face of the drawing, then that face must have height $O(1/n)$, and otherwise they are separated by an edge and their distance from that edge is $O(1/n)$.

As both cases give a vertex-edge pair at distance $O(1/n)$, the result follows.

Again, this result is tight. One may draw any planar graph in the hyperbolic plane with vertex-edge resolution $\Omega(1/n)$, in an essentially non-hyperbolic way, by first constructing a straight-line drawing in a Euclidean grid of size $O(n) \times$

$O(n)$ [6,17,33] and then using the Klein model of hyperbolic geometry to map this drawing to a straight-line grid drawing within a subset of the hyperbolic plane of diameter $O(1)$, with constant distortion of distances.

3   Vertex-Vertex Resolution and Angular Resolution

It is not possible to upper-bound only the vertex-vertex resolution of hyperbolic drawings of planar graphs, as the following observation shows.

Observation 6. *For every planar graph G, and every distance d, there is a planar hyperbolic drawing of G with vertex-vertex resolution $\geq d$.*

Proof. We may assume without loss of generality that G is maximal planar, and use a de Fraysseix–Pach–Pollack [17] style graph drawing algorithm (without horizontal shifting) in which vertices are placed into the drawing one-by-one in a canonical ordering, starting from two adjacent vertices on the outer face of the eventual drawing, so that when each vertex is added to the drawing it is adjacent to a consecutive subsequence of vertices on the outer face of the current drawing.

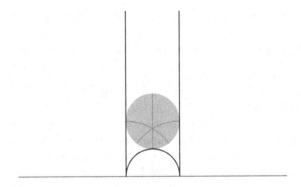

Fig. 3. An ideal hyperbolic triangle (black) in the upper halfplane model of the hyperbolic plane (above the horizontal line), with an inscribed circle of radius $\ln\sqrt{3}$ (shaded). The hyperbolic line segments within the circle meet at its center (the incenter of the triangle), which is somewhat below its Euclidean center.

In the upper halfplane model of the hyperbolic plane, place the first two vertices on two arbitrary points with distinct x-coordinates. Place each subsequent vertex above the midpoint of the x-interval spanned by its earlier neighbors, so that at each stage the upper boundary of the drawing is an x-monotone piecewise-linear curve. Choose the vertical position of each vertex, both high enough that it is visible (along a hyperbolic line segment) to all of its previously-placed neighbors, and high enough that it is at distance at least d from all previously placed vertices; these two requirements do not interfere with each other.

Nevertheless, when the vertex-vertex resolution is large, it may force the drawing to be bad in other ways, as the remainder of this section shows.

Lemma 7. *Every hyperbolic triangle has an inscribed circle touching all three of its sides. Its radius (the inradius of the triangle) is at most* $\ln \sqrt{3} \approx 0.5493$.

Proof. See for instance [30, Theorem 13.4, p. 103]. The limiting case of inradius $= \ln \sqrt{3}$ occurs for *ideal triangles* (Fig. 3), with all vertices at infinity.

Lemma 8. *Let v be a vertex of a hyperbolic triangle T, at distance d from the incenter of T. Then the angle of T at v is upper bounded by an exponentially small function of d.*

Proof. Assume without loss of generality that $d > 2$ because otherwise the upper bound of the lemma is $O(1)$, trivially valid for all angles.

Let V be the subset of T between the inscribed circle C and vertex v. Replace the sides of T by two asymptotic lines, tangent to the inscribed circle of T at points wider than the tangent points of T to the same circle, and expand C if necessary until the radius of the expanded circle $\hat{C}$ is exactly $\ln \sqrt{3}$, enclosing V in the corresponding subset $\hat{V}$ of an ideal triangle $\hat{T}$ having the same incenter, with v equally far from the two sides of $\hat{T}$. By choosing the point at infinity where the two sides of $\hat{T}$ meet to be the point at vertical infinity of an upper halfplane model of the hyperbolic plane, $\hat{V}$ can be made to be the region two vertical lines and above a circle $\hat{C}$ seen in the upper center of Fig. 3.

In the upper halfplane model, the infinitesimal unit of length ds is given by

$$(ds)^2 = \frac{(dx)^2 + (dy)^2}{y^2},$$

where x and y are Cartesian coordinates. This distance is locally Euclidean, with a scale factor inversely proportional to height, so two points on the vertical lines of Fig. 3 at Euclidean distance y above the blue line are at hyperbolic distance $O(1/y)$ from each other. Integrating this scaled distance over a vertical line segment, the distance from (x, y_1) to (x, y_2) (with $y_1 < y_2$) simplifies to $\ln(y_2/y_1)$. Therefore, for point v to be at hyperbolic distance d above the center of $\hat{C}$ in this model, its y-coordinate is exponentially larger than the y-coordinate of the center of C. At that height, the scale factor of hyperbolic distance is so small that the two vertical sides of $\hat{T}$ are exponentially close. The points along the two sides of T at unit distance from v are at a height corresponding to hyperbolic distance $\geq d - 1$ above the center of $\hat{C}$, and are also exponentially close in hyperbolic distance, because they are sandwiched between the two sides of $\hat{T}$ that are exponentially close at that height.

The angle of T at v equals the angle formed at v by these two exponentially-close points at unit distance from v, so it is exponentially small.

Theorem 3. *For every constant c, and every n-vertex maximal planar graph G, every hyperbolic planar straight line drawing of G that has vertex-vertex resolution $\geq c$ also has angular resolution that is exponentially small in n, with the base of the exponential depending on c.*

Proof. Let T be the outer face of a drawing of G, and partition T into four regions: its inscribed circle C, and the three regions between this circle and the three vertices of T. The inscribed circle has bounded diameter by Lemma 7 and can contain only $O(1)$ points of minimum separation c, so there must be a vertex v of the outer face and a region V between C and v that contains $\Omega(n)$ vertices of the drawing. Partition V into regions of bounded diameter by classifying the points of V according to their distance from the center of C, rounded to an integer; each of these regions can contain only $O(1)$ points of minimum separation c, so there must be a vertex w within a region at distance $\Omega(d)$ from the center of C. Vertex v itself must be even farther away, in order for T to enclose w. Therefore, by Lemma 8, the angle of T at v is exponentially small.

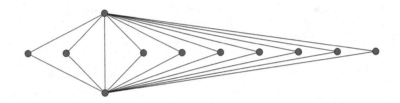

Fig. 4. $K_{1,1,9}$

4 Special Classes of Planar Graphs

4.1 Series-Parallel Graphs

The proofs of Theorem 1 and of Theorem 3 depend only on the property of maximal planar graphs that they have a triangle (the outer triangle of their planar embedding) that contains a linear number of other vertices. Therefore, it can be adapted to other graphs with analogous properties.

Theorem 4. *For every n, there exists an n-vertex series-parallel graph G of bounded bandwidth with the following properties:*

- *Every hyperbolic planar straight line drawing of G has vertex-edge resolution $O(1/\sqrt{n})$.*
- *For every c, every hyperbolic planar straight line drawing of G that has vertex-vertex resolution $\geq c$ also has angular resolution that is exponentially small in n, with the base of the exponential depending on c.*

Proof. Let $G = K_{1,1,n-2}$ (Fig. 4). Up to permutation of the vertices all its planar straight-line drawings consist of a nested triangles on one side of the edge between the two singleton sets of the tripartition, for $0 \leq a \leq n - 2$, and of $n - 2 - a$ nested triangles on the other side. Therefore, if T is the outermost triangle on the side with the larger number of triangles, T contains $\Omega(n)$ other vertices of G.

The bound on vertex-edge resolution follows the same lines as the proof of Theorem 1: T is partitioned into $\Omega(n)$ faces (one triangle and the rest quadrilaterals), so one of these faces has area $O(1/n)$. If this low-area face is a triangle, the bounds on triangle height used in the proof of Theorem 1 show that it already has vertex-edge resolution $O(1/\sqrt{n})$. If it is a quadrilateral, add a diagonal, subdividing it into two triangles of area $O(1/n)$ and height $O(1/\sqrt{n})$. If the heights of both triangles are defined by the distances of a vertex to the added diagonal, then each of these vertices also has distance $O(1/\sqrt{n})$ to a non-adjacent side of the quadrilateral. If one or both of the triangles has a height that does not involve the added diagonal, then we get vertex-edge resolution $O(1/\sqrt{n})$ directly.

The result on vertex-vertex resolution and angular resolution follows directly by applying the argument from the proof of Theorem 3 to T.

4.2   Grid Graphs

The $n \times n$ grid graphs, in their standard Euclidean drawing, are particularly well-behaved: all edges have the same length, and the vertex-vertex, vertex-edge, and angular resolutions are all constant. Unlike the maximal planar graphs, they do not have cycles of bounded length containing many vertices. This does not prevent problems when drawing them hyperbolically:

Theorem 5. *Every hyperbolic planar straight line drawing of an $n \times n$ grid has vertex-edge resolution $O(1/\sqrt{n})$. All such drawings that have vertex-vertex resolution $\Omega(1)$ have angular resolution exponentially small in n.*

Proof. If the grid is drawn with the standard outer face, a polygon with $4(n-1)$ sides, this polygon can be triangulated into $O(n)$ triangles. By Lemma 1 the bounded faces of the drawing cover an area of $O(n)$, and one of the $(n-1)^2$ grid quadrilaterals has area $O(1/n)$. For a different outer face, the area is even smaller. The argument from a quadrilateral of small area to small vertex-edge resolution is the same as for Theorem 4.

Now suppose we have a drawing with vertex-vertex resolution $\Omega(1)$, and triangulate its boundary polygon. Let v be any non-boundary vertex of the grid. Define two polygonal curves to the boundary from v, that step to the nearest points on two different sides of the triangle containing v in the boundary triangulation, then repeatedly cross successive triangles on shortest crossing segments until reaching the boundary. The dual graph of the boundary triangulation (like any triangulation of any simple polygon) is a tree, and these curves follow a path in this dual tree, so they must terminate at the boundary, crossing each triangle at most once. We say that v is captured by triangle T if one of these curves crosses T, v is within distance $\leq \gamma n$ of the incenter of T (for a constant of proportionality γ to be determined later) or within unit distance of a vertex of T, and v is not captured by any triangle crossed earlier by the curve.

If vertex v is captured by triangle T, then before the curve for v reaches T it can only cross segments of other triangles that are exponentially short (as a function of n), by the same reasoning as in Lemma 8. Therefore, the grid vertices captured by T are exponentially close to T. Partitioning the region close to T into subsets of bounded diameter in the same way as in the proof of Theorem 3 shows that, if T captures k_T grid vertices, then one must be at distance $\Omega(k)$ from the incenter and more than unit distance from all vertices of T. However, captured vertices are defined as being within distance $\leq \gamma n$ of the incenter or unit distance of a vertex of T, so $k_T = O(\gamma n)$. The boundary triangulation has $O(n)$ triangles, so the total number of captured vertices over all triangles is $O(\gamma n^2)$. If we choose γ to be a sufficiently small constant (depending on the vertex-vertex resolution), this number will be less than the number $(n-2)^2$ of interior vertices of the grid, and at least one interior vertex v will remain uncaptured.

Each of the $O(n)$ steps in the two curves from v to the boundary has exponentially-small length, so both paths are exponentially short. Therefore, v is sandwiched between two exponentially-close edges of the boundary, and moreover is at least unit distance from the edge endpoints. Because v is an interior vertex of the grid, it has degree four, and has at least two edges extending in at least one of the two directions approximately parallel to the boundary edges that sandwich v. These two edges must form an exponentially small angle at v.

5   Nonplanar Drawings

Beyond planar graph drawing, any n-vertex graph has a Euclidean drawing with unit vertex-vertex resolution and angular resolution $\Theta(1/n)$, with vertices on a unit regular n-gon. The analogous placement in hyperbolic geometry has angular resolution $\Theta(1/n^2)$. In this section, we prove that this is optimal: every hyperbolic drawing of K_n with unit vertex-vertex resolution has angular resolution $O(1/n^2)$.

A key ingredient is the relation between hyperbolic circle area and perimeter:

Lemma 9. *Every hyperbolic circle of perimeter $p > 1$ has area $p + o(p)$.*

Proof. This follows immediately from the formulas for the area $4\pi \sinh^2(r/2)$ and perimeter $2\pi \sinh r$ of a hyperbolic circle of radius r [2]. In the limit as r becomes large, the ratio of these two formulas converges to 1. ∎

We also need the following counterintuitive property of hyperbolic geometry, which in this respect is very different from Euclidean geometry:

Lemma 10. *Divide the hyperbolic plane into four quadrants by two perpendicular lines. Then every line from a point in one quadrant to a point in the opposite quadrant passes within distance $\ln(1 + \sqrt{2}) \approx 0.8814$ of the crossing.*

Proof. Choose an upper halfplane model of the hyperbolic plane that represents the two perpendicular lines as congruent semicircles, and the two opposite quadrants crossed by any given line as the two congruent regions to the left and right of their crossing (Fig. 5). The convex hull of the two quadrants is bounded by two more hyperbolic lines, asymptotic to the crossing lines (the boundaries of the light yellow region in the figure). All lines from one quadrant to another remain within the hull, crossing the red circle in the figure.

The figure may be given Cartesian coordinates in which the semicircles representing the two perpendicular lines have (Euclidean) centers at the points $(-1, 0)$ and $(1, 0)$ and cross at the point $(0, 1)$, the hyperbolic center of the red circle. With these coordinates, these semicircles have Euclidean radius $\sqrt{2}$. The top point of the red circle is at $(0, 1 + \sqrt{2})$, and the result follows from the formula $\ln(y_2/y_1)$ for the hyperbolic length of a vertical line segment.

Theorem 6. *For every constant c, every hyperbolic drawing of the complete graph K_n with vertex-vertex resolution $\geq c$ has angular resolution $O(1/n^2)$.*

Proof. By ignoring one vertex if necessary, assume without loss of generality that n is even. Consider any drawing with vertex-vertex resolution $\geq c$, find a hyperbolic line splitting the vertices of the drawing into equal subsets, and (by applying the intermediate value theorem to the partitions by perpendicular lines) find a second perpendicular line splitting the vertices into two equal subsets. Let x be the crossing point of these two lines. Among the four quadrants formed by these two lines, opposite quadrants necessarily have equal numbers of points, so two opposite quadrants Q and Q' each contain at least $n/4$ vertices.

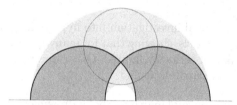

Fig. 5. Illustration for Lemma 10. Two perpendicular lines (shown as semicircles in an upper halfplane model above the horizontal line) determine two opposite quadrants (dark shading) whose convex hull (light shading) is bounded by two lines asymptotic to the two perpendicular lines. All lines from the left quadrant to the right quadrant remain within the hull and pass through the circle shown, of radius $\ln(1 + \sqrt{2})$. (Color figure online)

Within Q, because of the vertex spacing, not all vertices can be in a quarter-circle centered at r of area $o(n)$. Therefore, by Lemma 9, for some vertex v in Q, a circle of radius xv has perimeter $\Omega(n)$. If we center such a circle at v, then the circle of radius $\ln(1 + \sqrt{2})$ centered at x (the red circle in Fig. 5) spans only a constant number of units of the perimeter, an $O(1/n)$ fraction of the total perimeter, so the angle spanned by the red circle as viewed from v is $O(1/n)$.

Within this $O(1/n)$ angle as viewed from v, at least $n/4$ vertices in Q' are also visible, by Lemma 10. By the pigeonhole principle, some two of these vertices in Q' must be within an angle of $O(1/n^2)$ of each other as viewed from v.

Following van Kreveld [21] we define a *bold* hyperbolic drawing to draw vertices as hyperbolic disks of a given radius and edges as thickened hyperbolic line segments of width less than this radius, with all edges and all vertices having part of their boundary visible.

Theorem 7. *Any bold hyperbolic drawing of K_n has edge width $O(1/n)$.*

Proof. Let the edge width of a drawing be $w < 1$. As in the proof of Theorem 6, find two perpendicular lines partitioning the centers of the vertices into quadrants, with two opposite quadrants Q and Q' containing at least $n/4$ vertex centers each; let C be the circle of radius $\ln 1 + \sqrt{2}$ centered at the crossing point, and let D be the diameter of C halfway between Q and Q'. Each vertex in Q has $\geq n/4$ edges to Q', all crossing D, among which $O(1/w)$ can have an exposed portion of boundary between the vertex and D, because each edge whose boundary is not entirely covered by edges from the same vertex covers a segment of D of length at least w. Similarly, each vertex in Q' has $O(1/w)$ edges with exposed boundaries. Unless w is $O(1/n)$, the total number of edges with exposed boundaries either near their endpoint in Q or near their endpoint in Q' will be less than the $(n/4)^2$ number of edges from Q to Q', and at least one edge will be totally covered.

6   Conclusions

We have performed an initial investigation into hyperbolic graph drawing under realistic graph drawing models, showing that for many variations of these models, drawings are impossible or seriously limited. Other questions in this area, which we leave open for future research, include:

- Which planar graphs have planar hyperbolic drawings with bounded vertex-edge resolution? These include all trees, and all outerplanar graphs (using a placement of vertices on a large regular polygon), but not all planar graphs and not even all bounded-bandwidth series-parallel graphs. Are there other natural classes of planar graphs that always have such drawings?
- For Euclidean planar drawings, edge-edge resolution is not usually studied separately, because it is essentially the same as vertex-edge resolution. However, in the hyperbolic plane, edges may approach each other closely even when all vertex-edge pairs are well separated. Does this cause differences between hyperbolic vertex-edge resolution and edge-edge resolution?
- RAC graphs (graphs drawn with right-angle crossings) are motivated by realistic graph drawing: crossings with high angles are easier to understand than sharp crossing angles [8]. However their definition strongly depends on geometry: edges that cross at right angles in the Euclidean plane have bipartite intersection graphs such as cycles of four edges. In the hyperbolic plane, odd cycles of right-angle-crossing edges are possible; however, 4-cycles are not possible. How do these differences affect the hyperbolic RAC graphs?

Our results should not be interpreted as shutting off research on hyperbolic graph drawing, which remains important in applications such as greedy routing where realistic drawing assumptions do not fit the problem, and as a building block for Euclidean drawing methods such as Lombardi drawing.

References

1. Barequet, G., Goodrich, M.T., Riley, C.: Drawing planar graphs with large vertices and thick edges. J. Graph Algorithms Appl. **8**, 3–20 (2004). https://doi.org/10.7155/jgaa.00078
2. Bennett, A.G.: Hyperbolic geometry. J. Online Math. Appl. **31**, 2 (2001). https://www.maa.org/press/periodicals/loci/joma/hyperbolic-geometry-introduction
3. Bern, M., Eppstein, D.: Optimal Möbius transformations for information visualization and meshing. In: Dehne, F., Sack, J.-R., Tamassia, R. (eds.) WADS 2001. LNCS, vol. 2125, pp. 14–25. Springer, Heidelberg (2001). https://doi.org/10.1007/3-540-44634-6_3
4. Bläsius, T., Friedrich, T., Katzmann, M., Krohmer, A.: Hyperbolic embeddings for near-optimal greedy routing. ACM J. Exp. Algorithmics **25**(A1.3), 1–18 (2020). https://doi.org/10.1145/3381751
5. Bläsius, T., Friedrich, T., Krohmer, A., Laue, S.: Efficient embedding of scale-free graphs in the hyperbolic plane. IEEE/ACM Trans. Netw. **26**(2), 920–933 (2018). https://doi.org/10.1109/TNET.2018.2810186
6. Brandenburg, F.J.: Drawing planar graphs on $\frac{8}{9}n^2$ area. In: Proceedings of the International Conference on Topological and Geometric Graph Theory. Electronic Notes in Discrete Mathematics, vol. 31, pp. 37–40 (2008). https://doi.org/10.1016/j.endm.2008.06.005
7. Coudert, D., Ducoffe, G.: A simple approach for lower-bounding the distortion in any Hyperbolic embedding. In: Drmota, M., Kang, M., Krattenthaler, C., Nešetřil, J. (eds.) Proceedings of European Conference on Combinatorics, Graph Theory and Applications (EUROCOMB 2017). Electronic Notes in Discrete Mathematics, vol. 61, pp. 293–299 (2017). https://doi.org/10.1016/j.endm.2017.06.051
8. Didimo, W., Eades, P., Liotta, G.: Drawing graphs with right angle crossings. Theor. Comput. Sci. **412**(39), 5156–5166 (2011)
9. Dolev, D., Leighton, F.T., Trickey, H.: Planar embedding of planar graphs. Adv. Comput. Res. **2**, 147–161 (1984). https://noodle.cs.huji.ac.il/~dolev/pubs/planar-embed.pdf
10. Duncan, C.A., Efrat, A., Kobourov, S.G., Wenk, C.: Drawing with fat edges. Int. J. Found. Comput. Sci. **17**(5), 1143–1164 (2006). https://doi.org/10.1142/S0129054106004315
11. Duncan, C.A., Eppstein, D., Goodrich, M.T., Kobourov, S.G., Nöllenburg, M.: Lombardi drawings of graphs. J. Graph Algorithms Appl. **16**(1), 85–108 (2012). https://doi.org/10.7155/jgaa.00251
12. Duncan, C.A., Eppstein, D., Goodrich, M.T., Kobourov, S.G., Nöllenburg, M.: Drawing trees with perfect angular resolution and polynomial area. Discret. Comput. Geom. **49**(2), 157–182 (2012). https://doi.org/10.1007/s00454-012-9472-y
13. Eklund, P.W., Roberts, N., Green, S.: OntoRama: browsing RDF ontologies using a hyperbolic-style browser. In: Proceedings of the 1st International Symposium on Cyber Worlds (CW 2002), pp. 405–411. IEEE Computer Society (2002). https://doi.org/10.1109/CW.2002.1180907

14. Eppstein, D.: A Möbius-invariant power diagram and its applications to soap bubbles and planar Lombardi drawing. Discret. Comput. Geom. **52**(3), 515–550 (2014). https://doi.org/10.1007/s00454-014-9627-0
15. Eppstein, D., Goodrich, M.T.: Succinct greedy geometric routing using hyperbolic geometry. IEEE Trans. Comput. **60**(11), 1571–1580 (2011). https://doi.org/10.1109/TC.2010.257
16. Formann, M., et al.: Drawing graphs in the plane with high resolution. SIAM J. Comput. **22**(5), 1035–1052 (1993). https://doi.org/10.1137/0222063
17. de Fraysseix, H., Pach, J., Pollack, R.: How to draw a planar graph on a grid. Combinatorica **10**(1), 41–51 (1990). https://doi.org/10.1007/BF02122694
18. Garg, A., Tamassia, R.: Planar drawings and angular resolution: algorithms and bounds. In: van Leeuwen, J. (ed.) ESA 1994. LNCS, vol. 855, pp. 12–23. Springer, Heidelberg (1994). https://doi.org/10.1007/BFb0049393
19. Kleinberg, R.: Geographic routing using hyperbolic space. In: Proceedings of the 26th IEEE International Conference on Computer Communications (INFOCOM 2007), pp. 1902–1909. IEEE (2007). https://doi.org/10.1109/INFCOM.2007.221
20. Kobourov, S.G., Wampler, K.: Non-Euclidean spring embedders. IEEE Trans. Vis. Comput. Graph. **11**(6), 757–767 (2005). https://doi.org/10.1109/TVCG.2005.103
21. van Kreveld, M.J.: Bold graph drawings. Comput. Geom. **44**(9), 499–506 (2011). https://doi.org/10.1016/j.comgeo.2011.06.002
22. Lamping, J., Rao, R.: The hyperbolic browser: a focus + context technique for visualizing large hierarchies. J. Vis. Lang. Comput. **7**(1), 33–55 (1996). https://doi.org/10.1006/jvlc.1996.0003
23. Malitz, S., Papakostas, A.: On the angular resolution of planar graphs. SIAM J. Discret. Math. **7**(2), 172–183 (1994). https://doi.org/10.1137/S0895480193242931
24. Martin, G.E.: The Foundations of Geometry and the Non-Euclidean Plane. Undergraduate Texts in Mathematics. Springer, Heidelberg (1982). https://doi.org/10.1007/978-1-4612-5725-7
25. Mohar, B.: Drawing graphs in the hyperbolic plane. In: Kratochvíyl, J. (ed.) GD 1999. LNCS, vol. 1731, pp. 127–136. Springer, Heidelberg (1999). https://doi.org/10.1007/3-540-46648-7_13
26. Munzner, T.: H3: laying out large directed graphs in 3D hyperbolic space. In: Proceedings of the 1997 IEEE Symposium on Information Visualization (InfoVis 1997), pp. 2–10. IEEE Computer Society (1997). https://doi.org/10.1109/INFVIS.1997.636718
27. Munzner, T.: Exploring large graphs in 3D hyperbolic space. IEEE Comput. Graph. Appl. **18**(4), 18–23 (1998). https://doi.org/10.1109/38.689657
28. Munzner, T., Burchard, P.: Visualizing the structure of the world wide web in 3D hyperbolic space. In: Nadeau, D.R., Moreland, J.L. (eds.) Proceedings of the 1995 Symposium on Virtual Reality Modeling Language (VRML 1995), pp. 33–38. ACM (1995). https://doi.org/10.1145/217306.217311
29. Pach, J.: Every graph admits an unambiguous bold drawing. J. Graph Algorithms Appl. **19**(1), 299–312 (2015). https://doi.org/10.7155/jgaa.00359
30. Petrunin, A.: Euclidean Plane and its Relatives; a Minimalist Introduction, 3rd edn. CreateSpace (December 2020). https://arxiv.org/abs/1302.1630v18
31. Sala, F., De Sa, C., Gu, A., Ré, C.: Representation tradeoffs for hyperbolic embeddings. In: Dy, J.G., Krause, A. (eds.) Proceedings of the 35th International Conference on Machine Learning (ICML 2018). Proceedings of Machine Learning Research, vol. 80, pp. 4457–4466. ML Research Press (2018). https://proceedings.mlr.press/v80/sala18a.html

32. Sarkar, R.: Low distortion delaunay embedding of trees in hyperbolic plane. In: van Kreveld, M., Speckmann, B. (eds.) GD 2011. LNCS, vol. 7034, pp. 355–366. Springer, Heidelberg (2012). https://doi.org/10.1007/978-3-642-25878-7_34
33. Schnyder, W.: Embedding planar graphs on the grid. In: Proceedings of the 1st ACM-SIAM Symposium on Discrete Algorithms (SODA 1990), pp. 138–148 (1990). https://dl.acm.org/citation.cfm?id=320176.320191
34. Suzuki, R., Takahama, R., Onoda, S.: Hyperbolic disk embeddings for directed acyclic graphs. In: Chaudhuri, K., Salakhutdinov, R. (eds.) Proceedings of the 36th International Conference on Machine Learning. Proceedings of Machine Learning Research, vol. 97, pp. 6066–6075. ML Research Press (2019). https://proceedings.mlr.press/v97/suzuki19a.html
35. Türker, U.C., Balcisoy, S.: A visualisation technique for large temporal social network datasets in hyperbolic space. J. Vis. Lang. Comput. 25(3), 227–242 (2014). https://doi.org/10.1016/j.jvlc.2013.10.008
36. Verbeek, K., Suri, S.: Metric embedding, hyperbolic space, and social networks. Comput. Geom. 59, 1–12 (2016). https://doi.org/10.1016/j.comgeo.2016.08.003
37. Walter, J.A.: H-MDS: a new approach for interactive visualization with multidimensional scaling in the hyperbolic space. Inf. Syst. 29(4), 273–292 (2004). https://doi.org/10.1016/j.is.2003.10.002
38. Walter, J.A., Ritter, H.J.: On interactive visualization of high-dimensional data using the hyperbolic plane. In: Proceedings of the 8th ACM International Conference on Knowledge Discovery and Data Mining (SIGKDD 2002), pp. 123–132. ACM (2002). https://doi.org/10.1145/775047.775065

Embedding Ray Intersection Graphs and Global Curve Simplification

Mees van de Kerkhof[1]($\boxtimes$), Irina Kostitsyna[2], and Maarten Löffler[1]

[1] Department of Computing and Information Sciences, Utrecht University,
Utrecht, The Netherlands
{m.vandekerkhof,m.loffler}@uu.nl
[2] Department of Mathematics and Computer Science, TU Eindhoven,
Eindhoven, The Netherlands
i.kostitsyna@tue.nl

Abstract. We prove that circle graphs (intersection graphs of circle chords) can be embedded as intersection graphs of rays in the plane with polynomial-size bit complexity.

We use this embedding to show that the global curve simplification problem for the directed Hausdorff distance is NP-hard. In this problem, we are given a polygonal curve P and the goal is to find a second polygonal curve P' such that the directed Hausdorff distance from P' to P is at most a given constant, and the complexity of P' is as small as possible.

1 Introduction

Problems in the area of graph drawing often find application in complexity theory by providing a basis for NP-hardness proofs for geometric problems. In this paper, we study an application of embedding *circle graphs* (intersection graphs of chords of a circle) as *ray graphs* (intersection graphs of half-lines) to the analysis of the complexity of *global curve simplification*. In particular, we prove (refer to Sect. 2 for precise problem definitions):

- All circle graphs are ray graphs that have a representation as a set of intersecting rays described by coordinates that have a polynomial number of bits (Theorem 1);
- HAMILTONIAN PATH is NP-hard on such ray graphs (Corollary 1);
- DIRECTED CURVE SIMPLIFICATION is NP-hard (Theorem 4).

1.1 Global Curve Simplification

Curve simplification is a long-studied problem in computational geometry and has applications in many related disciplines, such as graphics, and geographical information science (GIS). Given a polygonal curve P with n vertices, the goal is to find another polygonal curve P' with a smaller number of vertices such that P' is sufficiently similar to P. Methods proposed for this problem famously

H. C. Purchase and I. Rutter (Eds.): GD 2021, LNCS 12868, pp. 358–371, 2021.
https://doi.org/10.1007/978-3-030-92931-2_26

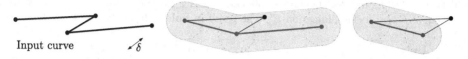

Fig. 1. For a target Hausdorff distance δ, the boldened red curve (middle) is a global simplification of the input curve (left), but it is not a local simplification, since the first shortcut does not closely represent its corresponding curve section (right). (Color figure online)

include a simple heuristic scheme by Douglas and Peucker [12], and a more involved classical algorithm by Imai and Iri [18]; both are frequently implemented and cited. Since then, numerous further results on curve simplification, often in specific settings or under additional constraints, have been obtained [1–5,9,10, 14,17].

Recently, the distinction was made between *global* simplification, when a bound on a distance measure must be satisfied between P and P', and *local* simplification when a bound on a distance measure must be satisfied between each edge of P' and its corresponding section of P [20]. A local simplification is also a global simplification, but the reverse is not necessarily true, see Fig. 1.

Agarwal *et al.* [2] were first to consider the idea of global simplification under the *Fréchet distance*. They introduce what they call a *weak simplification*: a model in which the vertices of the simplification are not restricted to be a subset of the input vertices, but can lie anywhere in the ambient space. Kostitsyna *et al.* [22] present a polynomial-time algorithm for this model but for the *Hausdorff distance*; in particular, the directed Hausdorff distance from the simplification curve to input curve. Van Kreveld *et al.* [24] consider a different setting in which the output vertices should be a subsequence of the input, and they also consider the Hausdorff distance. They give a polynomial-time algorithm for the directed Hausdorff distance from the simplification curve to input curve, but they show the problem is NP-hard for the directed Hausdorff distance in the opposite direction, and also for the symmetric (undirected) Hausdorff distance. Van de Kerkhof *et al.* [20] prove that the hardness result for the unrestricted Hausdorff distance can be extended to the non-restricted case as well; in addition, they introduce an intermediate *curve-restricted* model where the vertices of the simplified curve should lie on the input curve. Surprisingly, the problem is hard under this model for all three variants of the Hausdorff distance. Table 1 summarizes the state of the art for global curve simplification under the Hausdorff distance.

1.2   Embeddings of Geometric Intersection Graphs

Geometric intersection graphs have long been studied due to their wide range of applications, and lie on the interface between computational geometry, graph theory, and graph drawing [26]. The graph classes corresponding to intersections

Table 1. Results for global curve simplification under the Hausdorff distance between the curve P and its simplification P'. The result in **bold** is from this work.

Distance	Vertex-restricted $(\mathcal{V})$	Curve-restricted $(\mathcal{C})$	Non-restricted $(\mathcal{N})$
$\overrightarrow{\mathsf{H}}(P,P')$	NP-hard [24]	NP-hard [20]	**NP-hard**
$\overrightarrow{\mathsf{H}}(P',P)$	$O(n^4)$ [24] $O(n^2\text{polylog } n)$ [20]	NP-hard [20]	poly(n) [22]
$\mathsf{H}(P,P')$	NP-hard [24]	NP-hard [20]	NP-hard [20]

of geometric shapes form a natural hierarchy that links to the complexity of those shapes: more complex shapes allow to represent more graphs. Arguably the most restricted class in this family are the *unit interval graphs* [19], and the most general class of intersection graphs of connected shapes in $\mathbb{R}^2$ are the *string graphs* [13]. Between these two, a hierarchy of classes exist; part of it is illustrated in Fig. 2.

Hartmann *et al.* [16] introduce *grid intersection graphs* where the shapes are aligned to an orthogonal grid; Mustata [27] gives an overview of the state of the art and also discusses the complexity of computational problems on such classes. Cardinal *et al.* [7] prove several relations between segment intersection graphs and ray intersection graphs; in particular they introduce *downward ray graphs*: intersection graphs of rays that all point into a common half-plane. Their main result is that *recognition* of several classes is complete for the existential theory of the reals. *Circle graphs* are intersection graphs of chords of a circle; equivalently, they may be defined as interval graphs where there is an edge between two intervals on a line when they intersect but are not nested [15]. Circle graphs are known to be contained in 1-string graphs [8]. We are not aware of any published statements of stricter containment; we show in this paper that they are in fact contained in the downward ray graphs.

When utilizing graph embedding algorithms in hardness proofs, one important issue is the representation of the embedding. The class of graphs which can be represented as intersection graphs of a given set of shapes is not necessarily the same as the class of graphs which can be represented as intersection graphs of shapes which each can be represented with coordinates of bounded complexity. For instance, McDiarmid and Müller [25] show that not all realizable unit disk graphs can be realized with coordinates of logarithmic complexity, and the same is true for *segment* graphs [23].

2 Preliminaries, Overview and Challenges

2.1 Polygonal Curves and the Hausdorff Distance

A *polygonal curve* (also called a *polyline*) $P = \{p_1, p_2, \ldots, p_n\}$ is defined by an ordered sequence of n vertices. We can treat P as a continuous map $P : [1, n] \to \mathbb{R}^d$ that maps real values in the interval $[1 \ldots n]$ to points on the polyline by linearly interpolating between the vertices, which allows us to visualize a polyline as $n - 1$ line segments linked one after the other. We will refer to these segments as the polyline's *links*. For integer i, $P(i)$ will return the vertex p_i. Points on P's i'th link are parametrized as $P(i + \lambda) = (1 - \lambda)p_i + \lambda p_{i+1}$. The *directed Hausdorff distance* from curve P to curve Q, which have n and m vertices respectively, is given by $\overrightarrow{H}(P,Q) = \max_{i \in [1...n]} \min_{j \in [1...m]} \|P(i) - Q(j)\|$. I.e. it is equal to the Euclidean distance from the point on P furthest from Q to the point on Q closest to that point. The *(undirected) Hausdorff distance* is the maximum over both directions, i.e. $H(P,Q) = \max\{\overrightarrow{H}(P,Q), \overrightarrow{H}(Q,P)\}$.

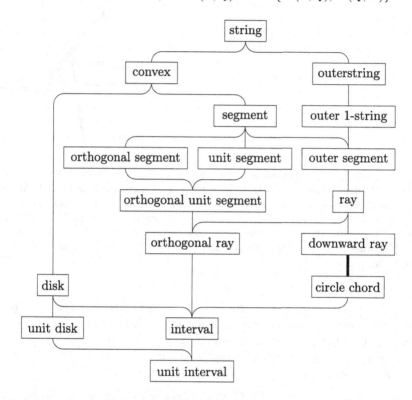

Fig. 2. Intersection graph classes and their inclusion relations. The thickened edge indicates our contribution. In order to keep the figure readable many classes and refinements have been omitted; for an extensive overview we refer the reader to e.g. [6,7,27] or to graphclasses.org.

2.2 Problem

The problem we wish to tackle (and which we will prove NP-hard) is:

Problem 1. DIRECTED CURVE SIMPLIFICATION. Given a polyline P, integer k and a value δ, find another polyline P' such that the directed Hausdorff distance from P to P' is at most δ and the number of links in P' is at most k.

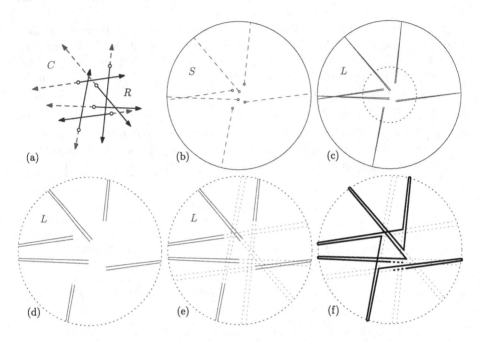

Fig. 3. The idea for a small example (which does not admit a Hamiltonian cycle). (a) A set of rays R (blue) whose intersection graph is G, and the complement C (red, dashed). (b) Zooming out until we can draw a circle that contains all intersections among rays in C. (c) Replacing each ray in C by a *needle*. (d) Zooming back in. (e) The extensions of the needles (blue, dotted) correspond to the original rays. (f) A polygon covering all needles must correspond to a Hamiltonian cycle in G (here, there is no solution). (Color figure online)

Note that van de Kerkhof *et al.* [20] call this problem the *non-restricted global curve simplification problem* to distinguish it from other variants; in the remainder of the present paper we use the shorter name for convenience. We will find that the key difficulty in solving DIRECTED CURVE SIMPLIFICATION lies in the following similar problem:

Problem 2. SEGMENT POLYLINE COVER. Given a set L of line segments in the plane and integer k, find a polyline P such that every segment in L is *covered* by P (contained in at least one segment of P), and P has at most k links.

2.3 Proof Idea

Our approach is to show that SEGMENT POLYLINE COVER is hard by a reduction from HAMILTONIAN PATH on ray intersection graphs. Specifically, we use the following idea.

Observation 1. *Let G be a ray intersection graph with n vertices. There exists a set L of $2n$ segments such that G has a Hamiltonian cycle if and only if there is a polygon covering L with $2n$ vertices.*

We can use Observation 1 to prove SEGMENT POLYLINE COVER is NP-hard and then reduce SEGMENT POLYLINE COVER to DIRECTED CURVE SIMPLIFICATION, proving it NP-hard as well.

We sketch the proof of Observation 1 here; the rest of the paper is devoted to making it precise. The high level proof idea is illustrated in Fig. 3. Let R be a set of rays in $\mathbb{R}^2$, and G its intersection graph. The *complement* of a ray r is the ray with the same origin and the same supporting line as r which points in the opposite direction. Let C be the complement of R. We cut the rays in C to a set of segments S in such a way that C and S have the same intersection graph. Then we replace each segment $s \in S$ by a *needle*: a pair of segments both very close to s that share one endpoint (different from the corresponding ray's origin). Let L be the resulting set of $2n$ segments. Now, any polygon with $2n$ segments covering L must use the two edges of one needle consecutively (since, by construction, the extension of these segments does not intersect the supporting line of any other segment), and it can connect an edge from one needle to an edge of another needle exactly when the corresponding original rays in R intersect.

2.4   Challenges

Though the idea is conceptually simple, there are several difficulties in turning Observation 1 into a proof that DIRECTED CURVE SIMPLIFICATION is NP-hard.

– The simple idea above is phrased in terms of a HAMILTONIAN CYCLE and covering segments by a polygon; for our proof we need to use a polyline. We need to be careful in how to handle the endpoints.
– We need to establish that HAMILTONIAN PATH is indeed NP-hard on ray intersection graphs.
– We need to know how to embed a ray intersection graph as an actual set of rays with limited bit complexity.
– We need to model the input to DIRECTED CURVE SIMPLIFICATION as an instance of SEGMENT POLYLINE COVER. Specifically, the complement of a set of rays is not necessarily connected; but the input to DIRECTED CURVE SIMPLIFICATION must be connected.
– The SEGMENT POLYLINE COVER problem closely resembles DIRECTED CURVE SIMPLIFICATION for $\delta = 0$; to extend it to the case $\delta > 0$ we (again) need to carefully consider the complexity of the embedding.

Most of these challenges can be overcome, as we show in the remainder of this paper. However, since the problem of recognizing if a graph can be embedded as a set of intersecting rays is complete for the existential theory of the reals [7], we know that there are ray intersection graphs that cannot be embedded by a set of rays with subexponential bit complexity, unless $\mathbf{NP} = \exists\mathbb{R}$. In this paper, we work around this problem by considering a smaller class of graphs, and allowing a superpolynomial grid for our embeddings, which we show is sufficient for the proof of Theorem 4.

3   Hamiltonian Cycles in Ray Intersection Graphs

3.1   Embedding Circle Graphs as Ray Graphs

We will show that each circle graph can be embedded as a ray intersection graph. To show this, we construct a set of n points that lie on a convex, increasing curve such that all chords connecting a pair of points can be extended to a ray to the right, and none of these rays will intersect below the curve. This requires the curve to grow very fast. We use the points $(x, x!)$ for $x \in [1..n]$, where $x! = \Pi_{i=1}^{x} i$ is the factorial function. Indeed, these points have the following property.

Lemma 1. *Let $a, b, c, d \in [1..n]$ be four numbers such that $a < b$ and $c < d$. Let A be the ray starting at $(a, a!)$ and containing $(b, b!)$, and let B be the ray starting at $(b, b!)$ such that $B \subset A$. Similarly, let C be the ray starting at $(c, c!)$ and containing $(d, d!)$, and let D be the ray starting at $(d, d!)$ such that $D \subset C$. Then B and D do not intersect; hence, A and C intersect if and only if $A \backslash B$ and $C \backslash D$ intersect.*

Proof. Since every ray is drawn between two points on the curve of the function $x!$, we know that it intersects this curve only at these points. The distance between y-coordinates of successive points keeps rapidly increasing as x increases, but the distance between x-coordinates of successive points is constant. Thus the slope of a ray r_1 whose intersection points with the curve lie to the right of those of ray r_2 will be greater than the slope of r_2. Without loss of generality we assume $a < c$. There are three possible cases, see Fig. 4:

- $c < b < d$: Here it is clear that A and C will intersect at an x-coordinate somewhere between c and b, and so B and D will not intersect.
- $b < c$: Here we can easily see that B and D do not intersect, as B starts below D and has a lower slope.
- $d < b$: Whereas the first two cases only require the curve to be convex and increasing, this case also requires the function to grow quick enough: Since D starts to the left of B it could possibly intersect B if its slope was higher. We will now show, however, that the factorial function grows quick enough so that this cannot happen. For a fixed b, the lowest slope that B can have is when $a = 1$. The highest slope that D can have occurs when $c = b - 2$ and $d = b - 1$. The slope of B is equal to the slope of A, which would be $\frac{b! - 1!}{b - 1} = b \cdot (b-2)! - \frac{1}{b-1}$. The slope of D (and C) in this scenario would be $\frac{(b-1)! - (b-2)!}{1} = (b-2) \cdot (b-2)!$. We can see that the slope of B is higher than D if $2 \cdot (b-2)! \geq \frac{1}{b-1}$ which obviously holds for all $b > 2$. So since B starts above D and has higher slope, B and D will not intersect.

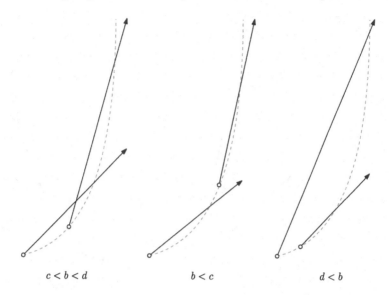

$$c < b < d \qquad\qquad b < c \qquad\qquad d < b$$

Fig. 4. The three cases for two rays. a and b are the x-coordinates of the points where the first ray intersects with the curve $y = x!$, and c and d are those values for the second ray. No matter the case, the rays will not intersect below the curve.

Once we have constructed these points we can "unroll" any circle graph by picking one chord endpoint on the circle to be the first point and then traversing the circle in clockwise order and assigning each chord endpoint we encounter the next point of our set. See Fig. 5 for a sketch. Because the y-coordinate for a point will not grow bigger than $O(n^n)$ we can represent the points using polynomial bit complexity. At this point, we have shown that circle graphs are contained in ray graphs. In fact, our construction gives a bit more:

Theorem 1. *The class of circle graphs is contained in the class of ray intersection graphs. Furthermore, every circle graph can be embedded as the intersection graph of a set of rays such that:*

- *every ray is grounded on a common curve* (grounded ray graph [7]);
- *every ray points towards the upper right quadrant* (downward ray graph [7]);
- *every ray is described by a point and a vector with polynomial bit complexity.*

3.2   Hamiltonian Paths and Cycles

Next, we show that Hamiltonian Path problem is NP-hard on ray graphs, and in particular, on ray graphs with polynomial bit complexity.

We reduce from the HAMILTONIAN PATH problem on circle graphs. We make use of the proof from Damaschke [11]. He shows that Hamiltonian cycle is NP-hard on circle graphs, by reducing from Hamiltonian Cycle in cubic bipartite graphs. He also claims that there is an easy adaptation that shows the Hamiltonian path problem is also NP-hard for circle graphs. We will start by making

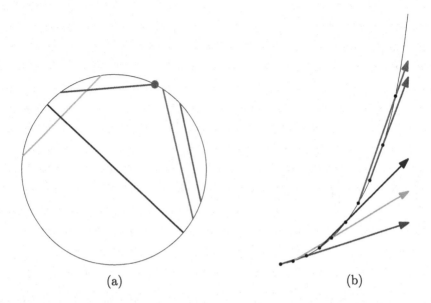

(a)   (b)

Fig. 5. (a) A circle graph with colors assigned to the chords. The chosen starting point is marked with a red dot. (b) Unrolled version of (a), by assigning chord endpoints in counterclockwise order to points on the convex curve they can be extended into rays without intersecting. (Color figure online)

this adaptation explicit: We construct an instance of the circle graph problem as described in [11], but then we replace one of the X-chords with two parallel chords close to where the X-chord was, so that they both intersect the same chords that were intersected by the X-chord. For both of the new chords we then add one new chord that only intersects that chord and no others. Now we know that the circle graph will have a Hamiltonian path if and only if the bipartite graph has a Hamiltonian cycle. From Theorem 1 we now immediately have:

Corollary 1. *Hamiltonian Path is NP-hard on intersection graphs of rays that have a polynomial bit complexity.*

4   Connected Segment Polyline Cover

Next, we introduce the CONNECTED SEGMENT POLYLINE COVER problem, and show that it is NP-hard by a reduction from Hamiltonian Path problem on circle graphs through the construction outlined above.

Problem 3. CONNECTED SEGMENT POLYLINE COVER. Given a set L of n line segments whose union is connected, and an integer k, decide if there exists a polyline of k links that fully covers all segments in L.

We start by embedding the circle graph as a ray intersection graph in the manner outlined above. Then, we compute all intersection points between supporting lines of the rays. One of these intersection points will have the lowest y-coordinate. We will then choose a value that is lower than this lowest y-coordinate, which we will denote as y_ℓ. For each ray r, let p_r be its starting point. Let $\bar{r}$ be r's complement: the part of the supporting line that is not covered by r. Let $\tilde{r}$ be the part of $\bar{r}$ that has $y \geq y_\ell$. Now we construct a *needle* for each ray's complement: Two line segments that share one endpoint at the point where $\bar{r}$ has y-coordinate y_ℓ. The other endpoint for both segments lies very close to p_r. The endpoints are on opposite sides of the ray starting point so we get a wedge-like shape that runs nearly parallel to $\tilde{r}$. In addition to these $2n$ segments, which we will refer to as *needle segments*, we create three more segments which we will refer to as the *leading segments*: We create one horizontal segment we call s_h with y-coordinate between y_ℓ and the lowest intersection point between ray supporting lines, that starts far to the right of the needle segments and ends to the left of them, intersecting all of the needles. Attached to s_h is a large vertical segment we call s_v, running up to a point above the highest starting point of a ray. Attached to that is another horizontal segment we call s_t, this one being short and ending to the left of any ray starting point. See Fig. 6 for a sketch.

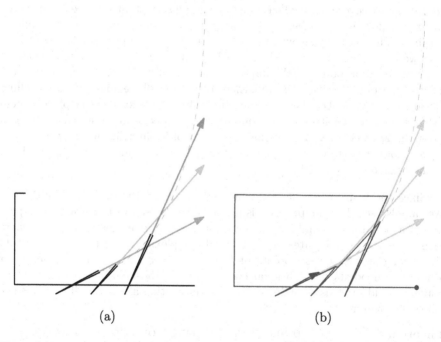

(a)   (b)

Fig. 6. (a) Sketch of a reduction of a circle graph with three chords. Segments shown in black. (b) Polyline of $2n + 3$ links covering the constructed segments, corresponding to a Hamiltonian Path traversing the rays, starting with the ray with the largest slope and ending with the ray with the smallest slope.

Now we have $2n + 3$ segments in total, where n is the number of chords in the original graph.

Lemma 2. *We can cover all segments using a polyline of $2n + 3$ links if and only if the circle graph has a Hamiltonian Path.*

Proof. To see why this is true, consider that since none of the segments are collinear and no three segments intersect in the same point, a suitable polyline must fully cover one segment with each link. For a polyline to be able to bend from fully covering one segment to fully covering another, either the segments must have a shared endpoint, or the supporting lines of the segments must intersect in a point not contained in either segment. Segment s_h intersects all needle segments in their interior and is parallel to s_t, so we know that a suitable polyline must start[1] by covering s_h, and it must bend at the common endpoint with s_v and then fully cover s_v. All of the intersection points between s_v and the supporting lines of needle segments lie below the endpoint it shares with s_h, so to be able to cover s_v with the second link the polyline must next connect to s_t, meaning it bends at the shared endpoint of s_v and s_t. The third link is horizontal, covering s_t. Since the supporting line of s_t intersects all of the rays, the polyline can bend to any needle segment for its next link.

Since we have covered our additional segments s_h, s_v, and s_t, the rest of the $2n$ links must cover one needle segment each. Observe that the needle segments all extend downward to below the lowest intersection point between supporting lines. This means that when a link covers a needle segment when travelling downward, the next link must then travel upward on the other half of the needle, as all intersection points with supporting lines of other segments lie in the segment's interior. When the next link then covers a needle segment when travelling upward, the only places the polyline can viably bend next are near places where the ray associated with the previous needle intersects another ray. So we can cover the $2n$ needle segments using a polyline of $2n$ links if and only if there is a Hamiltonian Path in the ray intersection graph and thus a Hamiltonian Path in the original circle graph.

Since the transformation is polynomial, we know the problem is NP-hard. We can also see that the problem is in NP, since for any instance we can expect that if a polyline of k links exists covering a set L, one must also exist where each vertex has coordinates of polynomial complexity, since the vertices could all lie on the intersection points of the supporting lines of the segments, or otherwise on points with rational coordinates on those supporting lines. This polyline could serve as a certificate for the verification algorithm. This gives the following theorem:

Theorem 2. CONNECTED SEGMENT POLYLINE COVER *is NP-complete.*

---

[1] A suitable polyline could also end with s_h, but we will define the polyline to be in this direction for ease of notation.

5   Directed Curve Simplification

Finally, we reduce CONNECTED SEGMENT POLYLINE COVER to DIRECTED CURVE SIMPLIFICATION. As a problem instance, we are given a set L of n non-collinear line segments in the plane whose union is connected. We construct an input polyline of polynomial size that completely covers the set of segments and no other points. We could do this, for example, by treating the segment endpoints and intersection points as vertices of a graph connected by edges, and have our polyline be the path of a breadth-first search through the graph. We set δ to 0. Now we know that, since a simplification must cover the union of L, any simplification of our input polyline that has n links must cover each segment in L completely with one link. This means such a simplification would be a solution to our instance of CONNECTED SEGMENT POLYLINE COVER. Since the reduction is polynomial in size, we know that this variant of the GCS problem is NP-hard, and using a similar argument to the one for CONNECTED SEGMENT POLYLINE COVER it is easy to see that it is in NP as well.

Theorem 3. DIRECTED CURVE SIMPLIFICATION, *restricted to instances where* $\delta = 0$, *is NP-complete.*

5.1   Non-zero δ

We can also extend this reduction to non-zero δ by picking $\delta > 0$ but still small enough such that it would not change the combinatorial structure of the space the polyline can lie in, so each link of the polyline must still correspond to exactly one segment in L. For the segments we have constructed in the earlier reduction, we will show that setting $\delta < \frac{3}{4n!}$ will guarantee the structure of the space will not change. So a simplified polyline of $2n + 3$ links with $0 < \delta < \frac{3}{4n!}$ will only exist if and only if it also exists for $\delta = 0$.

For space reasons, we only sketch the ideas of the proof here; details can be found in the full version [21].

The main idea is to choose δ sufficiently small so that there are no additional intersections between extensions of segments that are not supposed to intersect. When $\delta = 0$, the space the simplified polyline can occupy is exactly the input segments, but for non-zero δ, the polyline does not have to exactly cover the original segment. If we center two circles with radius δ on the endpoints of a segment, the two inner tangents of these circles will form the bounding lines of a cone that covers all possible polyline links that are able to "cover" a segment. We will call the part of the cone that is within δ of the segment the *tip* of the cone, and the rest of the cone the *tail* of the cone. For $\delta = 0$, the supporting line for the segment forms a degenerate cone of width 0. To preserve the combinatorial structure, fattening the cones cannot introduce intersections between cone tails, as these correspond to two segments' supporting lines intersecting in the exterior of the segments. To simplify the algebra, we consider a slightly larger cone, between the lines connecting points created by going 2δ to the left and right of the original endpoints of the needle segments. It is easy to see that these larger cones contain the true cones. We reach the bound on δ given above by case distinction of different configurations of potentially intersecting cones, and taking the minimum.

Since we can have a small enough δ of polynomial bit complexity, this means the DIRECTED CURVE SIMPLIFICTION problem is NP-hard in general, as for larger values of δ the construction could be scaled up.

If the general problem is in NP is hard to say, since our approach for showing this for the previous problems does not extend, and it might be possible that inputs exist where the only possible simplifications of k links have vertex coordinates of exponential bit complexity. This remains an open problem.

Theorem 4. DIRECTED CURVE SIMPLIFICATION *is NP-hard.*

6   Conclusion

We have shown that DIRECTED CURVE SIMPLIFICATION is NP-hard, which completes the results in Table 1 and completely settles the complexity of global curve simplification under the Hausdorff distance.

As the main tool in our reduction, we have shown that every circle graph can be embedded as an intersection graph of rays with coordinates of polynomial complexity. It is still an open question if it is possible to embed every circle graph as rays with coordinates of logarithmic complexity. Whether DIRECTED CURVE SIMPLIFICATION is in NP is another open problem that remains.

Acknowledgements. The authors would like to thank Birgit Vogtenhuber, Tillman Miltzow, and anonymous reviewers for valuable discussions and suggestions. Mees van de Kerkhof and Maarten Löffler are (partially) supported by the Dutch Research Council under grant 628.011.00. Maarten Löffler is supported by the Dutch Research Council under grant 614.001.50.

References

1. Abam, M.A., de Berg, M., Hachenberger, P., Zarei, A.: Streaming algorithms for line simplification. Discret. Comput. Geom. **43**(3), 497–515 (2010)
2. Agarwal, P.K., Har-Peled, S., Mustafa, N., Wang, Y.: Near linear time approximation algorithm for curve simplification. Algorithmica **42**(3–4), 203–219 (2005)
3. Barequet, G., Chen, D.Z., Daescu, O., Goodrich, M.T., Snoeyink, J.: Efficiently approximating polygonal paths in three and higher dimensions. Algorithmica **33**(2), 150–167 (2002)
4. de Berg, M., van Kreveld, M., Schirra, S.: Topologically correct subdivision simplification using the bandwidth criterion. Cartogr. Geogr. Inf. Syst. **25**(4), 243–257 (1998)
5. Buzer, L.: Optimal simplification of polygonal chain for rendering. In: Proceedings of 23rd Annual ACM Symposium on Computational Geometry (SoCG), pp. 168–174 (2007)
6. Cabello, S., Jejčič, M.: Refining the hierarchies of classes of geometric intersection graphs. Electron. J. Combin. **24**(1), 1–33 (2017)

7. Cardinal, J., Felsner, S., Miltzow, T., Tompkins, C., Vogtenhuber, B.: Intersection graphs of rays and grounded segments. J. Graph Algorithms Appl. **22**(2), 273–295 (2018)
8. Chalopin, J., Gonçalves, D., Ochem, P.: Planar graphs are in 1-string. In: Proceedings of the Eighteenth Annual ACM-SIAM Symposium on Discrete Algorithms, SODA 2007, pp. 609–617 (2007)
9. Chan, S., Chin, F.: Approximation of polygonal curves with minimum number of line segments or minimum error. Int. J. Comput. Geom. Appl. **6**(1), 59–77 (1996)
10. Chen, D.Z., Daescu, O., Hershberger, J., Kogge, P.M., Mi, N., Snoeyink, J.: Polygonal path simplification with angle constraints. Comput. Geom. **32**(3), 173–187 (2005)
11. Damaschke, P.: The Hamiltonian circuit problem for circle graphs is NP-complete. Inf. Process. Lett. **32**(1), 1–2 (1989)
12. Douglas, D.H., Peucker, T.K.: Algorithms for the reduction of the number of points required to represent a digitized line or its caricature. Cartographica **10**(2), 112–122 (1973)
13. Ehrlich, G., Even, S., Tarjan, R.E.: Intersection graphs of curves in the plane. J. Comb. Theory. Ser. B **21**, 8–20 (1976)
14. Godau, M.: A natural metric for curves - computing the distance for polygonal chains and approximation algorithms. In: Proceedings of the 8th Annual Symposium on Theoretical Aspects of Computer Science (STACS), pp. 127–136 (1991)
15. Golumbic, M.C.: Algorithmic graph theory and perfect graphs (2004)
16. Hartman, I.A., Newman, I., Ziv, R.: On grid intersection graphs. Discret. Math. **87**, 41–52 (1991)
17. Imai, H., Iri, M.: An optimal algorithm for approximating a piecewise linear function. J. Inf. Process. **9**(3), 159–162 (1986)
18. Imai, H., Iri, M.: Polygonal approximations of a curve - formulations and algorithms. In: Computational Morphology: A Computational Geometric Approach to the Analysis of Form (1988)
19. Jinjiang, Y., Sanming, Z.: Optimal labelling of unit interval graphs. Appl. Math. **10**(3), 337–344 (1995)
20. van de Kerkhof, M., Kostitsyna, I., Löffler, M., Mirzanezhad, M., Wenk, C.: Global curve simplification. In: Proceedings of the 27th European Symposium on Algorithms (ESA) (2019)
21. van de Kerkhof, M., Kostitsyna, I., Löffler, M.: Embedding ray intersection graphs and global curve simplification (2021). https://arxiv.org/abs/2109.00042
22. Kostitsyna, I., Löffler, M., Polishchuk, V., Staals, F.: On the complexity of minimum-link path problems. J. Comput. Geom. **8**(2), 80–108 (2017)
23. Kratochvíl, J., Matoušek, J.: Intersection graphs of segments. J. Comb. Theory, Ser. B **62**(2), 289–315 (1994)
24. van Kreveld, M., Löffler, M., Wiratma, L.: On optimal polyline simplification using the Hausdorff and Fréchet distance. In: Proceedings of the 34th International Symposium on Computational Geometry (SoCG), vol. 56, pp. 1–14 (2018)
25. McDiarmid, C., Müller, T.: Integer realizations of disk and segment graphs. J. Comb. Theory, Ser. B **103**(1), 114–143 (2013)
26. McKee, T.A., McMorris, F.: Topics in Intersection Graph Theory. Society for Industrial and Applied Mathematics, USA (1999)
27. Mustata, I.: On subclasses of grid intersection graphs. Doctoral thesis, Technische Universität Berlin, Berlin (2014)

AI Applications

Deep Neural Network for DrawiNg Networks, $(DNN)^2$

Loann Giovannangeli$^{(\boxtimes)}$[ID], Frederic Lalanne[ID], David Auber[ID],
Romain Giot[ID], and Romain Bourqui[ID]

LaBRI, UMR CNRS 5800, University Bordeaux, 33405 Talence, France
{loann.giovannangeli,frederic.lalanne,david.auber,
romain.giot,romain.bourqui}@u-bordeaux.fr

Abstract. By leveraging recent progress of stochastic gradient descent methods, several works have shown that graphs could be efficiently laid out through the optimization of a tailored objective function. In the meantime, Deep Learning (DL) techniques achieved great performances in many applications. We demonstrate that it is possible to use DL techniques to learn a graph-to-layout sequence of operations thanks to a graph-related objective function. In this paper, we present a novel graph drawing framework called $(DNN)^2$: Deep Neural Network for DrawiNg Networks. Our method uses Graph Convolution Networks to learn a model. Learning is achieved by optimizing a graph topology related loss function that evaluates $(DNN)^2$ generated layouts during training. Once trained, the $(DNN)^2$ model is able to quickly lay any input graph out. We experiment $(DNN)^2$ and statistically compare it to optimization-based and regular graph layout algorithms. The results show that $(DNN)^2$ performs well and are encouraging as the Deep Learning approach to Graph Drawing is novel and many leads for future works are identified.

Keywords: Graph drawing · Deep Learning · Graph Convolutions

1 Introduction

Optimization-based (OPT) and Deep Learning (DL) methods are gaining increasing interest in the information visualization field [30,34]. From the very design of visualizations to their evaluations, such techniques have shown to perform well and present benefits over standard methods. These advances motivated the exploration of these techniques adaptation to graph drawing. Some studies [1,17,35] used OPT approaches to optimize an objective function for a single graph with Stochastic Gradient Descent (SGD) and obtained good results; Zheng et al. [35] even outperformed some state-of-the-art layout algorithms. On the other hand, if DL techniques have been applied on graph and graph drawing related problems (e.g., evaluate aesthetic metrics) [9,11,20], to the best of our knowledge, only one study made use of this technique to *draw* graphs, *Deep-Drawing* [31]. Their framework leverages DL techniques to learn a model to reproduce layouts (i.e., ground truths) by optimizing a Procrustes-based cost

© Springer Nature Switzerland AG 2021
H. C. Purchase and I. Rutter (Eds.): GD 2021, LNCS 12868, pp. 375–390, 2021.
https://doi.org/10.1007/978-3-030-92931-2_27

function that compares the produced layout to the ground truth one. A major flaw of optimizing such a cost function by opposition to a graph topology related function is that the model is trained to optimize a similarity to a ground truth graph layout (that can be suboptimal) rather than the emphasis of the topology. This paper presents Deep Neural Networks for DrawiNg Networks, $(DNN)^2$, a graph layout framework relying on *unsupervised* Deep Learning. It proposes to adapt well-proven Convolutional Neural Network architecture to graph context using Graph Convolutions [5,16]. To the best of our knowledge, it is the first Deep Neural Network (DNN) architecture trained to lay generic graphs out by optimizing a graph-drawing related cost function. We propose an experimentation of $(DNN)^2$ where we use *ResNet* [14] architecture as a basis to optimize the Kruiger *et al.* [17] adaptation of the *Kullback-Leibler* divergence. In DL, as a model performance and its capability to generalize to unseen data are often incompatible, we also study the benefits of pre-training $(DNN)^2$. Finally, we statistically compare $(DNN)^2$ with state-of-the-art methods on aesthetic metrics and find that it competes with them. By efficiently learning a bounded sequence of operations that lays generic graphs out, $(DNN)^2$ experimentation suggests that graph drawing can be modeled as a mathematical function.

The remainder of the paper is organized as follows. Section 2 presents related works on OPT and DL methods in graphs context. Section 3 introduces $(DNN)^2$ and its key concepts while Sect. 4 presents the results of its experimental evaluation. Section 5 discusses the visual aspect of $(DNN)^2$ layouts and its limitations. Conclusions and leads for future works are presented in Sect. 6.

2 Related Works

First, we define the conventional notations used in this paper. Let $G(V, E)$ be a graph: V is its set of nodes $\{v_i\}, i \in [1, N], N = |V|$ and $E \subseteq V \times V$ its set of edges. Graphs are considered simple and connected. Let nodes positions be encoded in a vector $X \in \mathbb{R}^{N \times 2}$ where X_i is the 2D position of node v_i, $||X_i - X_j||$ relates to the Euclidean distance between points X_i and X_j.

Optimization-based (OPT) and Deep Learning (DL) techniques applications to Graph Drawing are gaining popularity and have been applied to a variety of graph and graph drawing related problems. For instance, Kwon *et al.* [19] used Machine Learning techniques to approximate a graph layout and its aesthetic metrics at the same time. Haleem *et al.* [11] also proposed to predict aesthetic metrics using a DL model. Several studies [10,22,27] used OPT to compute a feature vector embedding of a graph nodes. Kwon and Ma [20] proposed a Deep encoder-decoder to learn smooth transitions between different layouts of a graph.

Recently, OPT and DL techniques were proposed to lay graphs out and did compete with state-of-the-art layout algorithms. Kruiger *et al.* [17] proposed to optimize the Kullback-Leibler divergence by gradient descent. Kullback-Leibler divergence is a measure of dissimilarity between two probabilities distribution P and Q which was used to visualize data [13,28] and is defined as: $D_{KL} = \sum_i P(i) \log \frac{P(i)}{Q(i)}$. The proposed optimization framework, *tsNET*, showed to perform well, although its execution time is extremely high (*i.e.*, several seconds for graphs with $N < 100$). The authors proposed an improved variant of their method for which nodes positions are initialized with PivotMDS rather than randomly. This variant showed to be more efficient in terms of aesthetic metrics and converged faster on larger graphs. S_GD^2 [35] relies on the optimization of *stress* by stochastic gradient descent (SGD). Stress is modeled by a set of constraints between nodes that are relaxed by iteratively moving pairs of nodes. GD^2 [1] also leveraged SGD to optimize a set of aesthetic metrics whose combination can be tuned by associating a weight to each metric.

On the other hand, GraphTSNE [21] learned a shallow Neural Network made of Graph Convolutions to predict a graph layout. The key idea of their work is to train a model for each graph to draw, the train dataset being the graph nodes themselves. Even if their model cannot be described as *deep*, their work confirms that a t-SNE based loss can be optimized by Graph Convolutions networks. *Deep-Drawing* [31] is the first method to train a DNN to compute graph layouts. It aims to mimic a *target* algorithm given as ground truth and can be seen as a fast approximation of its *target*. This was also studied by Espadoto *et al.* [7] and both studies raised several limitations to this approach. First, it requires to run the *target* algorithm thousands of times to generate labeled training data. Due to model convergence issue, the labeled data generation should be manually supervised and the model cannot reproduce results of a non-deterministic algorithm either. Second, as the model learns to mimic an algorithm, it cannot produce better results than its *target* baseline and it also learns its defects. Finally, as the function optimized by the model is not related only to its input data, it does not learn features from its input but rather from its combination input–target algorithm. Hence, it is unclear how well it can generalize to unseen data for which no *target* result was ever provided. As opposed to DeepDrawing, $(DNN)^2$ training is *unsupervised* (*i.e.*, no groundtruth layout is provided) and generated graph layouts are evaluated according to a graph topology related cost function based on t-SNE.

3 $(DNN)^2$ Framework Design

3.1 $(DNN)^2$ Architecture

A ResNet-Like Basis. The design of $(DNN)^2$ architecture leverages Convolutional Neural Networks (CNNs) by adapting them to a graph context with Graph Convolutions [5,16]. The architecture reproduces *ResNet* [14], a CNN designed to classify images and reaching a high accuracy on the ImageNet challenge [6]. It is composed of *residual blocks* that contain *shortcuts* which enable the model to work on several levels of abstraction. It is made of 52 spectral *Graph Convolutions* (see Sect. 3.1) organized in 16 residual blocks as in the ResNet architecture (see Fig. 1). In addition, three node-wise fully connected layers with shared weights are added after the last convolution, the final layer being the model output. To handle graphs of varying sizes, the model inputs are fixed to an arbitrary size N_{max} and are padded with *fictive nodes* to fit this size. After each residual

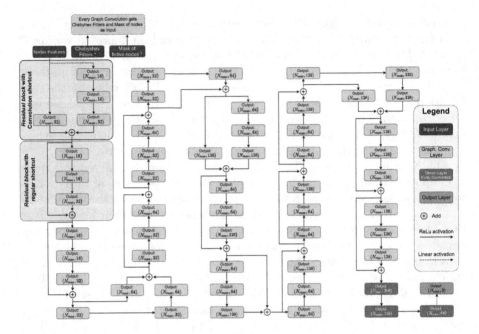

Fig. 1. $(DNN)^2$ architecture based on ResNet50 [14]. Some details have been omitted: (i) Chebyshev filters input is provided up to order 4 to all Graph Convolution layers except the 9 last layers (up to order 2); (ii) features vectors (*i.e.*, convolutions outputs) are normalized after every convolution; (iii) normalized features vectors are applied a mask encoding the *real-fictive* nodes information; and (iv) only the first two residual blocks are emphasized out of the 16 blocks.

block, its resulting features tensor is multiplied with a mask of real-fictive nodes $Mask \in \mathbb{1}^{N_{max}}$ where $Mask_i = 0$ if v_i is a fictive node, 1 otherwise. Padding the model inputs to match the expected shape could create a bias during the training: if fictive nodes are always padded at the same position in the tensors, some trainable weights will mostly see irrelevant features of *fictive nodes* and be underfitted. To avoid this bias, the padded model inputs are randomly permuted.

Spectral Graph Convolutions. Abbreviated Graph Convolutions, they were defined by Kipf and Welling [16] to operate on a graph *signal* encoded as a features vector for every node. The convolution kernel size K is defined to convolve a node with its K-hop neighborhood. The graph topology is provided through the graph spectrum (*i.e.*, eigendecomposition of the normalized Laplacian matrix) [3], approximated with Chebyshev polynomials [12]. Graph Convolutions are formally defined as a function of a signal x:

$$g_\theta \star x = U g_\theta U^T x \tag{1}$$

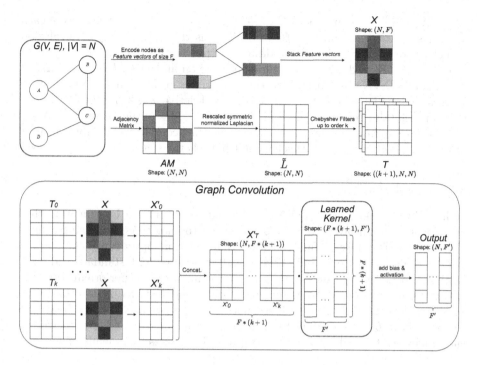

Fig. 2. Graph Convolutional layer diagram. It takes two tensors as input: (i) a feature vector to convolve (X); and (ii) a data structure that encodes the graph topology (T). X can refer to a node features vector at any step of the training.

where U is the matrix of eigenvectors of the symmetric normalized Laplacian matrix L so that $L = U \Lambda U^T$ where Λ are the eigenvalues of L. As the evaluation of Eq. 1 and the eigendecomposition of L are expensive, the operation can be approximated [12] using the Chebyshev polynomials $T_k(x)$ up to order K:

$$g_{\theta'} \star x \approx \sum_{k=0}^{K} \theta'_k T_k \left(\tilde{L} \right) x \qquad (2)$$

where $\tilde{L}$ is the rescaled symmetric normalized Laplacian $\tilde{L} = \frac{2}{\lambda_{max}} L - I_N$, λ_{max} being the highest eigenvalue in Λ and I_N the identity matrix of size N. $\theta' \in \mathbb{R}^K$ is a vector of Chebyshev coefficients and $T_k(x)$ is the Chebyshev polynomial defined as $T_0(x) = 1$, $T_1(x) = x$ and $T_k(x) = 2xT_{k-1}(x) - T_{k-2}(x), \forall k \geq 2$ and that costs $\mathcal{O}(K|E|)$ to be computed up to order K [12]. The Graph Convolution computation in this paper (illustrated in Fig. 2) can be formally defined as:

$$Z = \prod_{k=0}^{K} T_k \left(\tilde{L} \right) \cdot X \cdot \Theta \qquad (3)$$

where $X \in \mathbb{R}^{N \times F}$ is the nodes features vectors (*i.e.*, graph signal) where each node has F features, and $\Theta \in \mathbb{R}^{(F*(K+1)) \times F'}$ is the learned graph convolution kernel where F' is the size of the desired output feature vector for every node. The symbol $\|$ is used as a *concatenate* operator on all the $T_k(\tilde{L}) \cdot X$ tensors.

Finally, $(DNN)^2$ is fed with three tensors: the graph signal (nodes feature vectors, defined later in Sect. 3.3), a mask of real-fictive nodes and the Chebyshev polynomials (also referred to as *Chebyshev filters*). Its output is set to a $N_{max} \times 2$ tensor of nodes positions in the plane. The time complexity of a forward pass in the model is $\mathcal{O}(N_{max})$ as this constant bounds the tensors size.

3.2   Loss Function

Unlike *DeepDrawing* [31], $(DNN)^2$ is trained to optimize a loss function that captures the graph layout quality based on its topology. As optimizing a function for a whole dataset is fundamentally different from optimizing it for specific graphs, the loss function should have already been used with standard and OPT methods to lay graphs out so that we can compare their performances. This mainly let us with two possible functions: *stress* and *Kullback-Leibler(KL) minimization* (see Sect. 2). If we believe both can be optimized by $(DNN)^2$, we selected the *KL minimization* from Kruiger *et al.* [17] as it adapted better to the framework throughout experimentations. The loss is then defined as:

$$C = \lambda_{KL}C_{KL} + \frac{\lambda_c}{2N} \sum_i \|X_i\|^2 - \frac{\lambda_r}{2N^2} \sum_{i,j \in V, i \neq j} \log\left(\|X_i - X_j\| + \epsilon_r\right) \quad (4)$$

where C_{KL} is the main topology-related cost term based on the Kullback-Leibler divergence proposed by Kruiger *et al.* [17]. The second and third terms are respectively a *compression* that minimizes the scale of the drawing and a *repulsion* that counter-balances the compression. $(\lambda_{KL}, \lambda_c, \lambda_r)$ are weights used to tune the loss function during the optimization. $\epsilon_r = \frac{1}{20}$ is a regularization constant.

Kruiger *et al.* [17] defined two *stages* for their *tsNET* algorithm. In the first stage, the three λ factors are set to $(\lambda_{KL} = 1, \lambda_c = 1.2, \lambda_r = 0)$ while in the second stage, they are switched to $(1, 0.01, 0.6)$. They also proposed a variant called *tsNET** with two differences: nodes positions are initialized with PivotMDS [2] and the first stage lambda factors are $(1, 0.1, 0)$. In this paper, $(DNN)^2$ extends both *tsNET* variants and is compared to their implementation[1].

3.3   Graph Signal: Initial *Nodes Features*

The graph signal is defined as a features vector for every node. Some methods already exist to extract a graph signal [10,22,27]. As standard layout algorithms achieved to lay graphs out only using their topology [8,15,35], we assume it can be sufficient to feed the model with this information encoded through Chebyshev

---

[1] https://github.com/HanKruiger/tsNET, consulted on February 2021.

Table 1. Random and Rome graphs datasets properties.

	Graphs distribution			Dataset size		
	|V|	|E|	Degree	Train	Validation	Test
Random graphs	[2, 128]	[1, 6502]	[1, 118]	127 000	25 400	–
Rome graphs	[10, 107]	[9, 158]	[1, 13]	8000	1600	1931

filters. *Nodes features* are then represented by a tensor $F \in \mathbb{R}^{N \times 2}$ with nodes *id* to help the model differentiate them and a *random metric* to reduce overfitting.

With this *nodes features* tensor, it can be expected that adding meaningful features should help the model achieving better layouts. We experimented additional features by adding PivotMDS 2D positions such as in $tsNET^*$ variant [17], raising its size to $F \in \mathbb{R}^{N \times 4}$. This tensor is then transformed throughout the model successive Graph Convolution and Dense layers as presented in Fig. 1.

As all the nodes features are not necessarily of the same order of magnitude, they are normalized to give them the same importance.

4 Experimentation and Statistical Comparison

4.1 Datasets

Two datasets were considered for this experimentation (see Table 1), both being split for Deep Learning validation purposes (*i.e.*, *hold out* validation). In our terminology, *train* and *validation* sets are used during training to feed the model and evaluate it. *Test* set is used to benchmark models on unseen data.

Random Graphs. Used to pretrain $(DNN)^2$, its *train* set was generated to sample 1000 random graphs for each graph size between 2 and N_{max}. It is noteworthy that by generating 1000 instances of each graph size, the model will see many isomorphic graphs (mainly of small size). It means the model could overfit on small graphs, but this kind of overfitting could be beneficial for it. Since graphs can be decomposed into subgraphs of smaller size, the model capability to layout a small graph g can help it laying out a larger graph G having g as a subgraph. The *validation* set was generated with 200 instances per graph size.

Rome Graphs. Rome is a dataset of undirected graphs provided by the Graph Drawing symposium[2] made of 11 534 graphs, 3 of them being excluded as they are disconnected. The set was randomly split as presented in Table 1 and the layout methods of this experiment will be evaluated on the Rome *test* set.

---

[2] http://www.graphdrawing.org/data.html, consulted on February 2021.

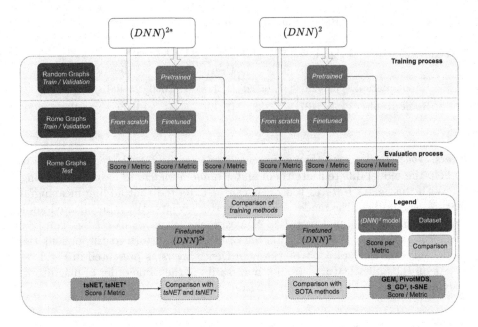

Fig. 3. $(DNN)^2$ training and evaluation pipeline. Six models are initially trained and compared. Then, the best selected models for the two loss variants are compared to $tsNET$, $tsNET^*$ and state-of-the-art layout methods.

4.2   Training

In this experiment, N_{max} was set to 128 to be slightly larger than the biggest graph in the Rome dataset. Transformed features tensors sizes were defined experimentally and are described in Fig. 1. Chebyshev filters were computed up to order 4 for all the Graph Convolution layers except the nine last ones which were only provided up to order 2. Therefore, more weight is given to direct neighborhood which minimizes overdraws that are critical for the drawing quality.

Since we want to compare our DL approach to the original $tsNET$ and $tsNET^*$ algorithms, an instance of $(DNN)^2$ is trained for each of them. We refer to these two variants as $(DNN)^2$ and $(DNN)^{2*}$. Similarly to $tsNET$, the models were trained in two stages. First, to optimize the loss C (see Eq. 4) with their respective $tsNET$ lambda weights (see Sect. 3.2). Second, the optimizer is reset and models are trained to optimize C with *second stage* lambda weights.

$(DNN)^2$ variants were trained with three methods to be evaluated on Rome graphs: (i) *pretraining* on Random graphs, (ii) *finetuning* (after the *pretraining*) on Rome graphs, (iii) training *from scratch* on Rome graphs. The goal is to verify if pretraining the model on a large set of random graphs improves its performances, and whether training on a specific dataset leads to better results than on random graphs. There are six $(DNN)^2$ instances in total (see Fig. 3).

Table 2. Quality metrics used in our benchmark and references to their definition. * represents metrics inverted to allow a *lower is better* reading for all of them.

Metric	Reference
*Aspect ratio**	As defined in [1]
*Angular resolution**	As defined in [1]
Edge crossings number	Well-known aesthetic metric [1, 24]
Cluster overlap	Autocorrelation metric in [32] with MCL clustering [29]
*Neighborhood preservation**	As defined in [17]
Stress	Well-known aesthetic metric [1], normalized by N

The nodes features are rescaled in $[0; 1]$ based on the *train* set. The random permutation of the model inputs (see Sect. 3.1) is fixed for the *test* set graphs so that every model is evaluated on the same permuted graphs.

4.3 Metrics and Comparison Procedure

Designing quality metrics for assessing a graph drawing quality that corroborates how well human subjects understand the drawing is a challenging question [23, 24, 26, 33]. We use a set of common metrics to assess $(DNN)^2$ efficiency and statistically compare it to state-of-the-art methods. Following the recommendations of Purchase [25], some metrics (marked with *) were inverted so that all metrics can be read as *lower is better* (see table Table 2). In addition, we measured Execution times of each algorithm in milliseconds (ms).

In the next, the efficiency of different graph drawing techniques are statistically compared on the presented metrics. To assess which method performs significantly better, a Kruskal-Wallis test [18] first verifies whether the differences of performances between all the compared methods on a given metric are significant or not. If so, a post-hoc Conover test [4] is applied to verify which pairs of methods are performing significantly different on that metric. For both tests, the acceptance threshold is set to $\alpha = 0.05$ and all Kruskal-Wallis tests passed.

4.4 Training Methods Evaluation

This section compares the 6 variants of $(DNN)^2$ to determine which training method is the most beneficial for the model. Execution times are not studied here and Fig. 4 presents other metrics averages and standard deviations on the Rome test set for each $(DNN)^2$ instance. A red bar indicates that the corresponding model performance is significantly different to *all* others. An arc between two blue bars indicates that the difference of their performance is statistically significant.

Pretrained instances perform significantly worse than others on all metrics but aspect ratio where they lead by a fair margin. *From scratch* instances never perform the best on any metric. Overall, *finetuned* $(DNN)^2$ lead to better scores with most metrics. It could be expected as it is well known in the Image Processing community that initialize weights to *pretrained* values tends to speed up the training process and to lead to better performances including generalization to unseen data. The idea is that it is easier for the model to learn to solve a specific task if it already knows high-level features. As *finetuned* models results are best, pretraining effectively learned the model such features that helped it to finetune.

In the next, only *finetuned* instances of $(DNN)^2$ are compared to state-of-the-art methods since they perform better on the graph drawing task.

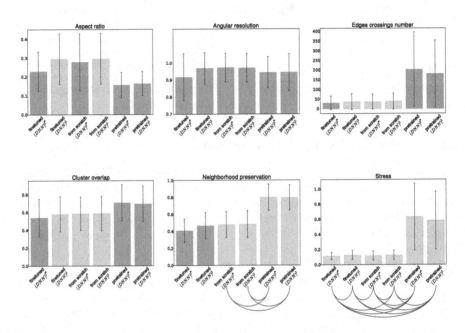

Fig. 4. Comparison of $(DNN)^2$ instances on the Rome test set. A red bar indicates that the corresponding model performance is significantly different to *all* others. An arc between two blue bars indicates pairwise significance. (Color figure online)

4.5   Comparison with *tsNET*

In this section, we study how the $(DNN)^2$ adaptation of *tsNET* loss performs compared to the original Optimization-based implementation of *tsNET* as both use the same cost function. The results are presented in Table 3.

$(DNN)^2$ has significantly lower scores than *tsNET* on all metrics but Stress and Execution time. Though it is significant, the difference on Edge Crossings Number is small. The Execution time difference is heavily in favor of $(DNN)^2$: 20.4 ms as opposed to 6541 ms for *tsNET*. The trends are about the same between $(DNN)^{2*}$ and *tsNET**, but their scores are closer and the differences in Angular Resolution and Edge Crossings Number are not significantly different anymore. $(DNN)^{2*}$ is also better on Stress and strongly better on Execution time.

Table 3. $(DNN)^2$ and $(DNN)^{2*}$ pair comparison with their respective $tsNET$ algorithm on the Rome test set. ° (resp. $^+$) indicates a significant difference with $(DNN)^2$ (resp. $(DNN)^{2*}$). The best *significant* results are bold (*i.e.*, no bold value when the difference is *not* significant).

	Aspect ratio	Angular res.	Cross. number	Cluster overlap	Neighb. preserv.	Stress	Exec. time
$(DNN)^2$ °	0.3 ± 0.136	0.963 ± 0.101	34.9 ± 40.5	0.55 ± 0.207	0.451 ± 0.159	**0.117 ± 0.041**	20.4 ± 10.4
$tsNET$	0.191 ± 0.091 °	0.885 ± 0.155 °	27.7 ± 31.8 °	0.489 ± 0.229 °	0.407 ± 0.1°	0.144 ± 0.155 °	6541 ± 5068 °
$(DNN)^{2*}$ +	0.229 ± 0.104	0.905 ± 0.15	30.0 ± 35.7	0.507 ± 0.214	0.397 ± 0.14	**0.111 ± 0.042**	24.8 ± 8.61
$tsNET^*$	0.206 ± 0.10	**0.872 ± 0.181** +	27.1 ± 32.0	**0.49 ± 0.218** +	**0.386 ± 0.115** +	0.124 ± 0.049 +	5836 ± 5933 +

Table 4. $(DNN)^2$ and $(DNN)^{2*}$ pair comparisons with selected state-of-the-art algorithms. ° (resp. $^+$) indicates a significant difference with $(DNN)^2$ (resp. $(DNN)^{2*}$). The best *significant* result(s) for each metric is(are) bold (*i.e.*, several bold values when the differences between the *best* algorithms is *not* significant).

	Aspect ratio	Angular res.	Cross. number	Cluster overlap	Neighb. preserv.	Stress	Exec. time
$(DNN)^2$ °	0.294 ± 0.134 +	0.969 ± 0.092 +	36.3 ± 39.9 +	0.58 ± 0.197 +	0.468 ± 0.154 +	0.128 ± 0.06 +	21.0 ± 10.3 +
$(DNN)^{2*}$ +	0.229 ± 0.105 °	0.917 ± 0.138 °	30.6 ± 34.8 °	0.541 ± 0.206 °	0.409 ± 0.136 °	0.115 ± 0.046 °	25.1 ± 8.36 °
t-SNE	0.276 ± 0.158 °+	0.97 ± 0.038 °+	69.1 ± 48.9 °+	0.598 ± 0.252 °+	0.584 ± 0.097 °+	0.56 ± 0.771 °+	166 ± 71.7 °+
PivotMDS	0.298 ± 0.125 +	0.978 ± 0.088 °+	38.7 ± 43.6 +	0.623 ± 0.202 °+	0.49 ± 0.17 °+	0.104 ± 0.035 °+	**0.546 ± 0.478** °+
GEM	0.573 ± 0.197 °+	0.972 ± 0.034 °+	54.4 ± 61.2 °+	0.722 ± 0.162 °+	0.617 ± 0.123 °+	0.24 ± 0.062 °+	5.22 ± 3.83 °+
S_GD^2	0.263 ± 0.123 °+	**0.812 ± 0.208** °+	32.2 ± 36.6 °	0.583 ± 0.204 +	0.439 ± 0.181 °+	**0.066 ± 0.027** °+	1.13 ± 0.91 °+

It is noteworthy that $tsNET$ and $tsNET^*$ suffers from a significantly high Execution time standard deviation, meaning that the methods hardly converge on some graphs. In addition, 450 out of 1931 (*i.e.*, 23%) graphs were excluded from the test set as the $tsNET^*$ implementation would not complete on these.

We can conclude that the $(DNN)^2$ implementation that adapts $tsNET$ to a Deep Learning approach is faster but does not lead to better drawings according to most of the metrics. Although quality metrics differences are significant, they remain small and should undeniably be alleviated by future works.

$tsNET$ is designed to optimize a specific input graph at a time whereas, with the Deep Learning approach, we aim at optimizing the model and not the drawing of a single graph. If the DL training process is computationally expensive, the resulting model should be capable of computing the layout without any further need for optimization. In fact, if a DL model learns well to lay graphs out by optimizing a generic cost function, it suggests that there exists a bounded sequence of operations that efficiently projects a graph in a 2D space.

4.6   Comparison with State-of-the-Art Layout Algorithms

This section studies how $(DNN)^2$ performs compared to selected layout algorithms from the literature: t-SNE [28], since we leverage the Kullback-Leibler divergence, PivotMDS [2], a deterministic Multidimensional Scaling used by $(DNN)^{2*}$ and $tsNET^*$, GEM [8], a well-established force-directed technique and S_GD^2 [35], a *stress* Optimization-based approach with SGD. The methods are compared on the Rome test set and the results are reported in Table 4.

$(DNN)^2$ scores are slightly different from Table 3 since all test graphs are taken into account here.

$(DNN)^{2*}$ performs better than $(DNN)^2$ as all aesthetic metrics are significantly in its favor. $(DNN)^{2*}$ is slower due to the extra processing of PivotMDS it requires. This outcome was expected in view of $tsNET$ variants comparisons in [17].

$(DNN)^2$ is better than GEM on all quality metrics; and is significantly better than t-SNE on all metrics but Aspect ratio. It performs better than PivotMDS on Angular resolution, Cluster overlap and Neighborhood preservation, but is outperformed on Aspect ratio and Stress, while the difference is not significant on Edge Crossings Number. Finally, S_GD^2 performs significantly better than $(DNN)^2$ on all metrics but Cluster overlap.

Overall, $(DNN)^{2*}$ is significantly better on Aspect ratio, Cluster overlap and Neighborhood preservation than all the other considered methods. It is also the best in Edge Crossings Number with S_GD^2. While it was observed to be better than $tsNET^*$ on Stress, it is here outperformed by PivotMDS and S_GD^2.

As for Execution time, we can see that both $(DNN)^2$ variants are slower than other methods except t-SNE. However, they are less sensible to graph size variations: $(DNN)^2$ variants execution time standard deviations are 33% and 47% of their average, while they range between 43% and 87% for other methods. It is important to note that a forward pass time in $(DNN)^2$ is almost constant and only takes 1.4 ms (*i.e.*, 6% of its total execution time), the remaining time being used to pre-process data for the model inputs.

Although $(DNN)^2$ is not the best performing variant, its results indicate that a Deep Learning framework, without any knowledge of what is a graph layout, can learn a sequence of operations that lays graphs out. $(DNN)^{2*}$ leveraged its PivotMDS input and drawn better layouts according to the quality metrics. Its performances make it a good trade-off between $tsNET^*$ and S_GD^2. The latter performed surprisingly well, while GEM underperformed in this evaluation.

5   Discussion

5.1   Visual Evaluation

Graph layout examples of $(DNN)^2$ are presented in Fig. 5 alongside $tsNET^*$ and S_GD^2 ones. S_GD^2 drawings being all pleasing and only a few defects away from being perfect, we can use them to see how the layouts should look like. For $(DNN)^2$, the dodecahedron and the grid graph structures can be observed but are severely distorted. It seems that *topologically equivalent* nodes (*i.e.*, nodes that can be mapped to each other by an automorphism) are grouped together. Both drawings are therefore folded, which also emphasizes their symmetry. From what we experienced, this behavior might be caused by the *compression* of the first stage training (see Sect. 3.2) and too similar *nodes features* and Chebyshev filters. Another explanation might be that the model had seen such small *patterns* more often during the training stage and somehow overfitted on them. On the two Rome graphs, the model has successfully laid the graphs structures out, but

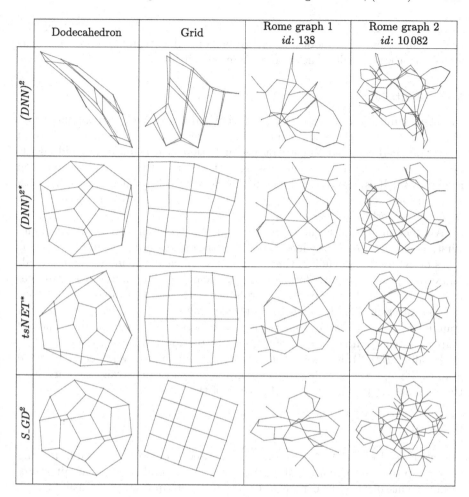

	Dodecahedron	Grid	Rome graph 1 id: 138	Rome graph 2 id: 10 082

Fig. 5. Layout examples for $(DNN)^2$, $(DNN)^{2*}$, $tsNET^*$ and S_GD^2.

its tendency to group *topologically equivalent* nodes leads to unbalanced edge lengths, edge crossings and overplots. On the other hand, $(DNN)^{2*}$layouts are visually more pleasing. The dodecahedron structure can clearly be identified. Despite a lack of regularity, the grid layout is also acceptable. The two Rome graph layouts demonstrate $(DNN)^{2*}$good performances. The model was able to separate topologically equivalent nodes, though they could have been repulsed a little more. $tsNET^*$ also produces nodes overlaps where neighborhoods are similar, as it can be observed in the top right of its *Rome graph 1* layout and on the right side of *Rome graph 2*.

Overall $(DNN)^2$and in particular $(DNN)^{2*}$performed well even compared to OPT methods. The latter optimizing their cost function for a specific graph at a time, we could expect them to provide better results than a DL approach.

Nevertheless, $(DNN)^2$ results acts as a proof-of-concept showing that we can learn unsupervised DL models to lay graphs out. It is therefore encouraging as we believe there is still a large room for improvement.

5.2 Limitations

The main limitation of $(DNN)^2$ is the technical need to set a maximum graph size N_{max} so that the architecture tensors size is static. Setting N_{max} to an arbitrarily high number might not be an option either since it would significantly increase the data pre-processing cost which is the most expensive step in $(DNN)^2$. It is also not certain such model would learn if it is only fed with *small* graphs (*i.e.*, with $N \ll N_{max}$), as each graph convolution kernel weight would be underfitted.

Another limitation is the resources required to obtain a well-trained model. First, if the use of the trained model is straightforward, the model *training* relies on many design choices that can only be efficiently made through a trials and errors process by an informed expert. Second, the computational resources required to train the model can be prohibitive. If Deep Learning-designed computers can easily handle small to mid-scale training (*e.g.*, *finetuning*), heavier training (*e.g.*, *pretraining*) can require to generate hundreds of thousands of graphs, which required the use of a Big Data platform in this experiment.

Finally, $(DNN)^2$ has not been tested on disconnected, weighted or directed graphs. Though the handling of these graph properties is straightforward with this framework, it is not part of the scope of this study.

6 Conclusion

We introduced $(DNN)^2$, a Deep Learning based framework for graph drawing. $(DNN)^2$ proposes to adapt well-established Deep Neural Network architectures in image classification to compute the layout of an input graph by using Graph Convolutions. To the best of our knowledge, it is the first DL model trained to lay graphs out by directly optimizing a graph topology related cost function.

We provided an experimentation of the framework and compared its performances to graph drawing algorithms from the literature. The experiment showed that $(DNN)^2$ performs well compared to these algorithms despite some Deep Learning related limitations. The results highly suggest that Deep Learning is a promising approach for the future of graph drawing. It also implies that there exists a mathematical function that efficiently projects any graph structure into a drawing, and that can be learned by Deep Learning models.

Future work leads include trying out other $(DNN)^2$ implementations, meaning other Deep Neural Network architectures, loss functions (*e.g.*, *stress*) and input node features (*e.g.*, *node2vec*, *DeepWalk*). Another interesting direction is to train and evaluate $(DNN)^2$ on other graph datasets or on specific graph families. Expanding the scale of the graphs size $(DNN)^2$ can handle, or apply it to other specific graph drawing applications are also promising leads for future work.

References

1. Hong, S.-H., Nishizeki, T., Quan, W. (eds.): GD 2007. LNCS, vol. 4875. Springer, Heidelberg (2008). https://doi.org/10.1007/978-3-540-77537-9
2. Brandes, U., Pich, C.: Eigensolver methods for progressive multidimensional scaling of large data. In: Kaufmann, M., Wagner, D. (eds.) GD 2006. LNCS, vol. 4372, pp. 42–53. Springer, Heidelberg (2007). https://doi.org/10.1007/978-3-540-70904-6_6
3. Cohen-Steiner, D., Kong, W., Sohler, C., Valiant, G.: Approximating the spectrum of a graph. In: Proceedings of the 24th ACM SIGKDD International Conference on Knowledge Discovery & Data Mining, pp. 1263–1271 (2018)
4. Conover, W.J., Iman, R.L.: On multiple-comparisons procedures. Tech. rep., Technical report, Los Alamos Scientific Laboratory (1979)
5. Defferrard, M., Bresson, X., Vandergheynst, P.: Convolutional neural networks on graphs with fast localized spectral filtering. arXiv preprint arXiv:1606.09375 (2016)
6. Deng, J., Dong, W., Socher, R., Li, L.-J., Li, K., Fei-Fei, L.: Imagenet: a large-scale hierarchical image database. In: 2009 IEEE Conference on Computer Vision and Pattern Recognition, pp. 248–255. IEEE (2009)
7. Espadoto, M., Hirata, N.S.T., Telea, A.C.: Deep learning multidimensional projections. Inf. Vis. **19**(3), 247–269 (2020)
8. Frick, A., Ludwig, A., Mehldau, H.: A fast adaptive layout algorithm for undirected graphs (extended abstract and system demonstration). In: Tamassia, R., Tollis, I.G. (eds.) GD 1994. LNCS, vol. 894, pp. 388–403. Springer, Heidelberg (1995). https://doi.org/10.1007/3-540-58950-3_393
9. Giovannangeli, L., Bourqui, R., Giot, R., Auber, D.: Toward automatic comparison of visualization techniques: application to graph visualization. Vis. Inform. **4**(2), 86–98 (2020)
10. Grover, A., Leskovec, J.: node2vec: scalable feature learning for networks. In: Proceedings of the 22nd ACM SIGKDD International Conference on Knowledge Discovery and Data Mining, pp. 855–864 (2016)
11. Haleem, H., Wang, Y., Puri, A., Wadhwa, S., Qu, H.: Evaluating the readability of force directed graph layouts: a deep learning approach. IEEE Comput. Graph. Appl. **39**(4), 40–53 (2019)
12. Hammond, D.K., Vandergheynst, P., Gribonval, R.: Wavelets on graphs via spectral graph theory. Appl. Comput. Harmon. Anal. **30**(2), 129–150 (2011)
13. Hinton, G., Roweis, S.T.: Stochastic neighbor embedding. In: NIPS, vol. 15, pp. 833–840. Citeseer (2002)
14. He, K., Zhang, X., Ren, S., Sun, J.: Deep residual learning for image recognition. In: The IEEE Conference on Computer Vision and Pattern Recognition (CVPR), pp. 770–778 (June 2016)
15. Kamada, T., Kawai, S., et al.: An algorithm for drawing general undirected graphs. Inf. Process. Lett. **31**(1), 7–15 (1989)
16. Kipf, T.N., Welling, M.: Semi-supervised classification with graph convolutional networks. arXiv preprint arXiv:1609.02907 (2016)
17. Kruiger, J.F., Rauber, P.E., Martins, R.M., Kerren, A., Kobourov, S., Telea, A.C.: Graph layouts by t-SNE. In: Computer Graphics Forum, vol. 36, pp. 283–294. Wiley Online Library (2017)
18. Kruskal, W.H., Wallis, W.A.: Use of ranks in one-criterion variance analysis. J. Am. Stat. Assoc. **47**(260), 583–621 (1952)
19. Kwon, O.-H., Crnovrsanin, T., Ma, K.-L.: What would a graph look like in this layout? A machine learning approach to large graph visualization. IEEE Trans. Vis. Comput. Graph. **24**(1), 478–488 (2017)

20. Kwon, O.-H., Ma, K.-L.: A deep generative model for graph layout. IEEE Trans. Vis. Comput. Graph. **26**(1), 665–675 (2019)
21. Leow, Y.Y., Laurent, T., Bresson, X.: GraphTSNE: a visualization technique for graph-structured data. arXiv preprint arXiv:1904.06915 (2019)
22. Perozzi, B., Al-Rfou, R., Skiena, S.: Deepwalk: online learning of social representations. In: Proceedings of the 20th ACM SIGKDD International Conference on Knowledge Discovery and Data Mining, pp. 701–710 (2014)
23. Purchase, H.: Which aesthetic has the greatest effect on human understanding? In: DiBattista, G. (ed.) GD 1997. LNCS, vol. 1353, pp. 248–261. Springer, Heidelberg (1997). https://doi.org/10.1007/3-540-63938-1_67
24. Purchase, H.C.: Metrics for graph drawing aesthetics. J. Vis. Lang. Comput. **13**(5), 501–516 (2002)
25. Purchase, H.C.: Experimental Human-computer Interaction: A Practical Guide with Visual Examples. Cambridge University Press, Cambridge (2012)
26. Purchase, H.C., Cohen, R.F., James, M.: Validating graph drawing aesthetics. In: Brandenburg, F.J. (ed.) GD 1995. LNCS, vol. 1027, pp. 435–446. Springer, Heidelberg (1996). https://doi.org/10.1007/BFb0021827
27. Tang, J., Qu, M., Wang, M., Zhang, M., Yan, J., Mei, Q.: Line: large-scale information network embedding. In: Proceedings of the 24th International Conference on World Wide Web, pp. 1067–1077 (2015)
28. Van der Maaten, L., Hinton, G.: Visualizing data using t-SNE. J. Mach. Learn. Res. **9**(11), 2579–2605 (2008)
29. Van Dongen, S.M.: Graph clustering by flow simulation. Ph.D. thesis (2000)
30. Wang, Q., Chen, Z., Wang, Y., Qu, H.: Applying machine learning advances to data visualization: a survey on ML4VIS. arXiv preprint arXiv:2012.00467 (2020)
31. Wang, Y., Jin, Z., Wang, Q., Cui, W., Ma, T., Qu, H.: DeepDrawing: a deep learning approach to graph drawing. IEEE Trans. Vis. Comput. Graph. **26**(1), 676–686 (2019)
32. Wang, Y., et al.: Ambiguityvis: visualization of ambiguity in graph layouts. IEEE Trans. Vis. Comput. Graph. **22**(1), 359–368 (2015)
33. Ware, C., Purchase, H., Colpoys, L., McGill, M.: Cognitive measurements of graph aesthetics. Inf. Vis. **1**(2), 103–110 (2002)
34. Wu, A., et al.: Survey on artificial intelligence approaches for visualization data. arXiv preprint arXiv:2102.01330 (2021)
35. Zheng, J.X., Pawar, S., Goodman, D.F.M.: Graph drawing by stochastic gradient descent. IEEE Trans. Vis. Comput. Graph. **25**(9), 2738–2748 (2018)

Implementation of Sprouts: A Graph Drawing Game

Tomáš Čížek$^{(\boxtimes)}$ and Martin Balko

Department of Applied Mathematics, Faculty of Mathematics and Physics,
Charles University, Prague, Czech Republic
{cizek,balko}@kam.mff.cuni.cz

Abstract. Sprouts is a two-player pencil-and-paper game invented by
John Conway and Michael Paterson in 1967. In the game, the players
take turns in joining dots by curves according to simple rules, until one
player cannot make a move. The game of Sprouts is very popular and
simple-looking, so it may come as a surprise that there are essentially no
AI Sprouts players available. This lack of computer opponents is caused
by the fact that the game hides a surprisingly high combinatorial complexity and implementing it involves fascinating programming challenges.
    We overcome all the implementation barriers and create the first user-
friendly Sprouts application with a strong artificial intelligence after
more than 50 years of the existence of the game. In particular, we
combine results from the theory of nimbers with new methods based
on Delaunay triangulations and crossing-preserving force-directed algorithms to develop an AI Sprouts player which plays a perfect game
on up to 11 spots.

Keywords: Sprouts · Combinatorial game · Graph drawing · Nimbers

1 Introduction

Sprouts is a 2-player combinatorial paper-and-pencil game with very simple rules.
The game starts with n initial spots and the players alternate in connecting the
spots by curves and adding a new spot on each newly drawn curve. No curve
can cross or touch another curve or itself and each spot can be incident to at
most three curves. The first player who cannot make a move loses the game; see
an example of a Sprouts game with 2 spots in Fig. 1.
    Sprouts was invented by John Conway and Mike Paterson at the University of
Cambridge in 1967 with an intention to create a game that is simple to play and
yet hard to analyze [5,19]. It was later popularized by Gardner [13] in a *Scientific*

T. Čížek and M. Balko were supported by the grant no. 21/32817S of the Czech Science Foundation (GAČR). T. Čížek was supported by the grant SVV-2020–260 578. M.
Balko acknowledges the support by the Center for Foundations of Modern Computer Science (Charles University project UNCE/SCI/004). This article is part of a project that
has received funding from the European Research Council (ERC) under the European
Union's Horizon 2020 research and innovation programme (grant agreement No 810115).

© Springer Nature Switzerland AG 2021
H. C. Purchase and I. Rutter (Eds.): GD 2021, LNCS 12868, pp. 391–405, 2021.
https://doi.org/10.1007/978-3-030-92931-2_28

Fig. 1. An example of a 2-spot game. The first player loses after 4 moves.

American article. In 1969, Anthony [3] mentioned Sprouts in his science-fiction novel *Macroscope* and, few years later, Pritchard [17] listed Sprouts in his compilation of world's best games for two players. Eventually, Sprouts became very popular with dozens of publications about the game and with the *World Game Of Sprouts Association*, which regularly held Sprouts championships. Yet, even after more than half a century, there are essentially no implementations of this game with a computer opponent [8]. This is very surprising at first, given how simple-looking and well-known the game of Sprouts is. However, the combinatorial complexity of Sprouts is very high and implementing the game thus involves various programming challenges.

We identify three main implementation barriers that cause many difficulties to potential developers. The first problem is to handle free-form input drawings. The drawn positions tend to degenerate and become confusing throughout the game and thus one has to come up with a method to maintain them stable and clear. Second, the number of possible moves can be exponential with respect to the number of initial spots, which makes creating a solid computer opponent very difficult. Finally, it is highly nontrivial to synchronize the free-form inputs of a human player with the game representation used by a computer.

In this paper, we overcome all these challenges and create the first user-friendly Sprouts implementation with a strong artificial intelligence. To do so, we apply techniques from the theory of *nimbers* used by the state-of-the-art Sprouts solver by Lemoine and Viennot [16] and combine them with our own *spindle method* for performing computer moves using Delaunay triangulations and crossing-preserving force-directed algorithms. Our program *Sprouts: A Drawing Game* supports games on up to 20 spots and contains an AI player that plays perfectly on n-spot positions with $n \leq 11$, surpassing all existing Sprouts implementations. The first version of our program is available at [9].

2   Related Work

In Table 1, we list an overview of all implementations of Sprouts that we are aware of. There are several Sprouts applications that allow to play only against human players. This includes *SproutsPlus* [2], *Sprouts Game* [11], and the *University of Utah Sprouts Applet* [1]. The application *3Graph* by Stefan Reiss [18] is the only Sprouts implementation with AI players. It plays a perfect game on n-spot positions with $n \leq 8$ and supports all features listed in Table 1 except of the remote game. Although 3Graph is, in our opinion, currently the best Sprouts implementation, there are some places for improvement. It is not very user-friendly, supports games with only at most 8 spots, and often crashes due to

Table 1. An overview of the existing Sprouts implementations with the following info: current availability of the program (CA), the support of the free-form inputs (FI), crossing detection (CD), position maintaining (PM), computer opponent (CO), remote game (RG), and the target platform (TP). Our application *Sprouts: A Drawing Game* supports all the listed features.

	CA	FI	CD	PM	CO	RG	TP
Sprouts - A Game of Maths!	✗	?	?	?	✓	?	iOS
SproutsPlus	✓	✗	✗	✗	✗	✓	iOS
Sprouts Game	✓	✓	✗	✗	✗	✗	iOS
UoU Sprouts Applet	✓	✓	✓	✗	✗	✗	Applet
3Graph	✓	✓	✓	✓	✓	✗	Windows

various internal errors. Browne [8] also mentions *Sprouts - A Game of Maths!* [15] as a Sprouts application with AI players, but he states that it did not work on any tested device and it is currently unavailable.

Besides these programs, there are several papers analyzing the implementation challenges of Sprouts [7,8,16] and even some intentions to create Sprouts applications with AI [1,8,20]. For example, Browne [8] mentions creating a complete Sprouts-playing app and investigating AI methods for playing the game at arbitrary sizes. However, no such result has been published yet.

Although almost no AI players are available, there are some computer Sprouts solvers. The outcomes of n-spot positions for $n \leq 7$ were determined by hand [12, 14]. Applegate, Jacobson, and Sleator [4] wrote the first computer analysis of Sprouts and successfully determined the winning player of n-spot games for each $n \leq 11$. They also introduced the famous *Sprouts conjecture* which states that each n-spot game is winning for the first player if and only if n is equal to 3, 4 or 5 modulo 6. Lemoine and Viennot [16] created the interactive Sprouts position editor *GLOP* that was used to determine all outcomes up to 44 spots and even some outcomes up to 53 spots. These are the strongest results to date and each of the computed outcomes agrees with the Sprouts conjecture, which, however, still remains open.

3   Preliminaries

We now introduce some notation and show basic properties of Sprouts. Let $\mathcal{P}$ be a class of plane graphs with maximum degree at most 3, obtained from a finite set of isolated vertices by a sequence of moves that obey the rules of Sprouts. The plane graphs from $\mathcal{P}$ are called *positions*. Each vertex of a position corresponds to one of the initial spots or to one of the spots added along the drawn curves. Edges of a position are formed by the portions of the curves between two spots; see part (a) of Fig. 2. For $n \in \mathbb{N}$, the *n-spot position* is a position consisting of n isolated vertices. The number of *lives* of a vertex v in a position P is $l(v) = 3 - \deg_P(v)$. A vertex v is *alive* if $l(v) > 0$ and *dead* otherwise. The faces

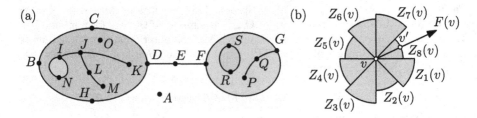

Fig. 2. (a) A Sprouts position with five regions. (b) A zone $Z(v)$ with a new location v' of the vertex v moved in the direction $F(v)$.

of P are called *regions*. The outer face of P is called the *outer region* of P and all other faces are called *inner regions* of P. For a region R of P, we let P_R be the plane subgraph of P induced by vertices that are incident to R. Every connected component of P_R is a *boundary* of R. For an inner region R, the unique boundary incident to the outer region of P_R is the *border boundary* of R. The other boundaries are called *inner boundaries*, including all boundaries of the outer region. A region R is *alive* if $\sum_{v \in R} l(v) > 1$ and *dead* otherwise.

It is quite easy to show that each game on the n-spot ends after at most $3n-1$ moves and lasts at least $2n$ moves [5]. It follows that Sprouts is a finite game. In fact, it is an *impartial game* [10] and thus there exists a winning strategy for one of the players in every position.

4   Graphical Representation

We assume that the edges of a Sprouts position P are drawn as piece-wise linear arcs and that P is contained in $[0, 1]^2$, which represents the playing area. Vertices of the *graphical representation* $gr(P)$ are the endpoints of the line segments forming the piece-wise linear arcs representing the edges of P or Sprouts spots that stand alone somewhere in a region of P. The vertices in the latter case are called *singletons*. The *game vertices* correspond to the real Sprouts spots. They are either singletons or the endpoints of the edges of P.

In $gr(P)$, we represent an edge e of P by a sequence $gr(e)$ of vertices of e starting and ending with the game vertices that are the endpoints of e. The ordering of $gr(e)$ is determined by the order in which e was drawn. The line segment connecting two consecutive vertices of $gr(e)$ is called a *small edge*. A boundary β of P is represented by a circular list of the sequences representing the edges of β. It is important for the move insertion that all inner boundaries in a region of P are oriented in the same way while the orientation of the border boundary is opposite. A region R is represented by a set of the lists representing the boundaries of P organized into a tree-like hierarchical structure, which is used to identify the region that contains a given point.

The graphical representation is used for correct insertion of the moves, crossing detection, and for redrawing of positions. Due to space limitations, we describe only the redrawing algorithm in detail.

4.1   Redrawing Algorithm

To keep the freely-drawn positions stable and clear, we implemented a modification of the redrawing algorithm *PrEd* by Bertauls [6] and its improved version *ImPrEd* by Simonetto et al. [21]. PrEd and ImPrEd are iterative force-directed algorithms that improve a given graph drawing while preserving its edge-crossing properties. At each iteration, we compute a force $F(v) \in \mathbb{R}^2$ for each vertex v of a position P and then move v in the direction $F(v)$. The amplitude of each move is restricted so that the edge-crossing properties are preserved.

We let V be the vertex set of P, E be the set of edges of P, and E_s be the set of small edges of P. We define the set S of four additional static edges around the game board $[0,1]^2$, which prevent vertices to move outside of the playing area.

We use three different forces between vertices and small edges: the attraction force F^a between vertices that are connected by a small edge, the repulsion force F^r between pairs of vertices, and the repulsion force F^e between vertices and small edges. To achieve better results, we count the force F^r only between vertices that do not lie on the same edge of P unless they are both game vertices. We use three different constants $\beta_{u,v}$, γ, and δ that customize the balance of these forces. The constant γ is the optimal distance of edges and δ is equal to the optimal length l_{opt} of a small edge. The value of $\beta_{u,v}$, which depends on two vertices u and v, is mainly used to strengthen the repulsion force between u and v if u and v are adjacent game vertices. The precise values of the attraction force $F^a \colon \binom{V}{2} \to \mathbb{R}^2$ and the repulsion force $F^r \colon \binom{V}{2} \to \mathbb{R}^2$ are

$$F^a(u,v) = \frac{\|u-v\|}{\delta}(u-v) \quad \text{and} \quad F^r(u,v) = \left(\frac{\beta_{u,v}}{\|u-v\|}\right)^2 (v-u),$$

where $F^r(u,v) = (0,0)$ if $\|u-v\| \geq \beta_{u,v}$. The force $F^e \colon V \times E_s \to \mathbb{R}^2$ is given by

$$F^e(v,ab) = \begin{cases} \frac{(\gamma - \|v - v_{ab}\|)^2}{\|v - v_{ab}\|}(v - v_{ab}) & v_{ab} \in ab, \|v - v_{ab}\| < \gamma \\ (0,0) & \text{otherwise,} \end{cases}$$

where v is a vertex disjoint from the edge ab and the point v_{ab} is the projection of the vertex v onto the line determined by the small edge ab. The overall force $F \colon V \to \mathbb{R}^2$ is computed for each vertex v as follows

$$F(v) = \sum_{\substack{u \in V \\ uv \in E_s}} F^a(u,v) + \sum_{\substack{u \in V \setminus \{v\} \\ \nexists e \in E:\, uv \in e}} F^r(u,v) + \sum_{\substack{ab \in E_s \cup S \\ v \notin ab}} F^e(v,ab) - \sum_{\substack{u,w \in V \\ vw \in E_s \\ u \notin vw}} F^e(u,vw).$$

After each force is computed, the vertices are moved in the direction $F(v)$. To preserve edge-crossings, we restrict the amplitude of these movements in the same way as in PrEd [6]. For each vertex v, we define a *zone* $Z(v)$ as a union of eight circular sectors $Z_1(v), \ldots, Z_8(v)$ with radii $r_1(v), \ldots, r_8(v)$, respectively; see part (b) of Fig. 2. At the end of each iteration, we compute the zone of each vertex v, find a sector $Z_i(v)$ that intersects the computed force $F(v)$, and then we move v in the direction $F(v)$ by $\min\{\|F(v)\|, r_i(v)\}$. We set the radii

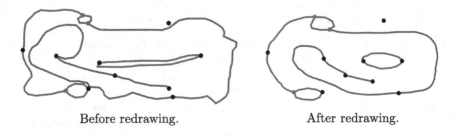

Before redrawing.   After redrawing.

Fig. 3. Redrawing a Sprouts position with 6 initial spots using 30 iterations.

$r_1(v), \ldots, r_8(v)$ so that the edge-crossings are preserved. To compute the radii of a zone $Z(v)$, we only consider small edges ab of P that are disjoint from v. Based on each such a small edge ab, we restrict the appropriate radii of the vertices v, a, and b. The initial values of the radii are set to a maximum size M_{max} of a movement within a single iteration. Updating the radii is then done as in [6], which implies that the same proof of the correctness still applies here.

To avoid having too short or too long edges, we merge too short small edges together after each iteration. If there are two adjacent small edges uv and vw with the distance $\|u - w\|$ below a certain value l_{mer}, then the small edges uv and vw are replaced by a new small edge uw. To prevent crossings, we merge uv and vw only if there is no vertex of P in the triangle uvw. We also subdivide the small edges of e whose length is greater than a certain value l_{sub}.

To speed up the computation, we update zones of vertices only for small edges that lie in the same region. Moreover, we use a graphical data structure *quadtree* as proposed by Simonetto et al. [21]. The use of quadtrees leads to a sublinear time of searching for vertices that are close to a given point. Since the amplitude of the movements of vertices is bounded by the finite value M_{max}, we update each zone $Z(v)$ using only small edges s whose distance from v is less than $2M_{max}$ using quadtrees. We also compute forces only between elements that are close enough. More precisely, we compute the repulsion force $F^r(u, v)$ between vertices u and v only if their distance is less than $\beta_{u,v}$. Similarly, we compute the edge repulsion force $F^e(v, ab)$ only if the distance between the vertex v and the small edge ab is less than γ. To determine close pairs of two vertices and of a vertex and a small edge, we implemented quadtrees for vertices and for edges.

Our redrawing algorithm is suitable for real-time computations in games with up to 20 spots; see Fig. 3. We use 30 iterations with a time limit after which the redrawing is terminated even if not all the iterations have finished. We also draw all curves with *Bezier splines* to make the moves look smoother.

5   Computer Opponent

To implement a strong computer opponent, we apply methods used in the currently best Sprouts solver *GLOP* [16]. First, we encode each Sprouts position P in a compact way using a *string representation* $sr(P)$, which was considered by many authors [4,8,16]. We describe the string representation and the methods very briefly, as we use them in the same way as Lemoine and Viennot [16].

The string representation $sr(P)$ of a position P is a string of game vertices, which are denoted by capital letters. Each boundary β forms a block of consecutive letters in $sr(P)$ in the order we meet the corresponding game vertices when traversing β. The boundaries are separated by dots in $sr(P)$. The regions are represented by consecutive blocks of boundaries separated by | in $sr(P)$. The position P is also split into *lands* which are mutually independent parts of P. The lands are separated by + in $sr(P)$. The strings of inner boundaries within the same region have the same orientation while the orientation of the string of the border boundary is opposite. For example, the position P from Fig. 4 can be represented by the string A.BCDEFGFEDH|DCBH.IJKJLMLJIN.O|IN+PQGFGQ.RS|RS.

5.1   String Simplification

A single Sprouts position can be represented by many different strings which leads to repetitive computations. Therefore, we simplify the string representation by applying two methods called the *string reduction* and the *string canonization*.

String Reduction. The string reduction simplifies strings using five steps. After the reduction, the string A.BCDEFGFEDH|DCBH.IJKJLMLJIN.O|IN|PQGFGQ.RS|RS of the position from part (a) of Fig. 2, which is not partitioned into lands, becomes O.AB2C|BAC.1a1a2.0+12.AB|AB; see Fig. 4.

1. **Delete dead parts**: We delete all dead vertices, then all boundaries that newly become empty, and then all dead regions and all empty lands.
2. **Apply generic names**: We rename each singleton to 0 and each vertex with two lives to 1. After merging vertices that occur twice in a row, we rename each letter that occurs only once in the string to 2.
3. **Split lands**: We split the position into lands.
4. **Rename letters**: We rename vertices contained in a single boundary starting from a in the order they occur in the string. Then we rename all other letter vertices within a single land starting from A in the order they occur.
5. **Merge boundaries**: For every region R with $\sum_{v \in R} l(v) \leq 3$, we merge all boundaries of R. This considerably reduces the number of strings.

String Canonization. Even after the string $sr(P)$ is reduced, there can be many reduced strings representing P since they might differ only by reordering of the boundaries, regions, lands and by relabeling of the vertices. A *canonization* selects the lexicographically minimal string representing P and thus makes the representation unique. However, the canonization takes too much running time. We thus perform only the *pseudocanonization* used by Lemoine and Viennot [16]. This does not make the string unique, but it considerably reduces the number of strings. For example, the string O.AB2C|BAC.1a1a2.0+12.AB|AB then becomes O.12a1a.ABC|O.2ABC+12.AB|AB, which we would get in the full canonization.

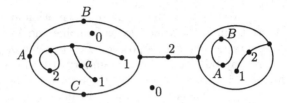

Fig. 4. The position from part (a) of Fig. 2 after the string reduction, which produces the string 0.AB2C|BAC.1a1a2.0+12.AB|AB.

Algorithm 1: Computing the nimber of a position P

Function *ComputeNimber(P)*

  $n \leftarrow 0$;
  while *True* **do**
    if *ComputeOutcome(P + n) is Loss* **then**
      return n;
    $n \leftarrow n + 1$;

5.2 Computation of Outcomes

Since Sprouts is an impartial game, we can determine the outcome of each position P. If the first player has a winning strategy, then P is *winning* and its outcome is *Win*. Otherwise, P is *losing* with the outcome *Loss*. To determine the outcome of P, we apply the theory of nimbers [5]. The *nimber* $|P|$ of a position P is the smallest non-negative integer that is not a nimber of a child of P. It follows that P is losing if and only if $|P| = 0$. The nimbers speed up the computation as positions tend to consist of several lands and the outcome of P can be obtained by merging the nimbers of the lands of P according to the following result.

Theorem 1 ([5,16]). *The nimber of a Sprouts position P consisting of lands $L_1, \ldots, L_n$ is $|P| = |L_1| \oplus \cdots \oplus |L_n|$, where $\oplus$ denotes the bitwise exclusive or.*

Instead of computing the outcome of P, we compute the outcome of a *couple* $(P+n)$ which consists of a parallel game of Sprouts on P and the game of Nim [5] on a single heap of n objects. This corresponds to computing nimbers of Sprouts positions which allows us to apply Theorem 1 and split positions into lands on which we proceed independently. We apply Algorithm 1 to efficiently compute the nimber of P using that the outcome of $(P+n)$ is Loss if and only if $|P| = n$.

We apply the Alpha-beta pruning algorithm to compute the outcome of $(P + n)$; see Algorithm 2. The set of children of $(P + n)$ is equal to the union of $\{ (C,n) \colon C$ is child of $P \}$ and $\{ (P,m) \colon m < n \}$. Note that the outcome of P is the same as the outcome of $(P + 0)$. We only store losing couples, which means that we only store positions whose nimber is known. Also, we store only single lands in our database as we are computing the lands separately.

Algorithm 2: Computing the outcome of a couple $(P + n)$

Function *ComputeOutcome(P + n)*

merge nimbers of lands whose nimbers are stored in the database with the nimber part of $(P + n)$ using the bitwise exclusive or $\oplus$;

compute the unknown nimbers of all lands remaining in P except of one land P' and merge them with the nimber part of $(P + n)$;

$(P' + n') \leftarrow$ the updated couple $(P + n)$ after the previous steps;

foreach *child C of $(P' + n')$* **do**

    outcome $\leftarrow$ *ComputeOutcome(C)*;

    if *outcome is Loss* **then**

        return *Win*;

store $(P' + n')$ into the database;

return *Loss*;

5.3 Creating the Computer Opponent

The computer opponent tries to compute the outcome of a position P. If it is Win, then he plays a move that leads to a losing child. Otherwise the outcome is Loss and all children are winning. The AI player can then select an arbitrary move. To make the exploration of the game tree faster, we use known techniques such as a suitable *children priorities* [16] and *boundary matching* on singletons [8].

We also implemented new features to improve the AI player. To make sure that the perfect AI never takes too long in finding a best move, we support *databases of pre-trained positions*. Unlike Sprouts solvers, the AI player has to consider all moves in the human player's turns so it is necessary to explore the game tree in a much larger width. This is a problem, as the game trees are very large. For example, already the tree of the 6-spot position contains 393103 strings. We solve this by restricting the set of possible moves of the AI player so that we do not have to explore the whole tree. The computer also selects children that are simpler to analyze in wining positions and children with proportionally most losing children in losing positions to increase a chance of opponent's mistake.

Lemoine and Viennot [16] also suggested to implement *distributed computing* for determining the outcomes. We implemented this improvement with threads using a global database that is equipped by synchronization primitives to prevent deadlocks. The outcome computation can be much faster with more threads.

6 Drawing a Computer Move

Here, we synchronize the graphical representation $gr(P)$ with the string representation $sr(P)$ of a position P so that we can draw a computer move found with $sr(P)$ into $gr(P)$. This is one of the most difficult steps we had to deal with and as far as we know, it is not fully described in the literature. Browne [8] sketched out the idea of using *Delaunay triangulations* and *Voronoi diagrams*.

Although his solution works for the n-spot positions, there are several missing parts for more complicated positions. So we apply our own new *spindle method*.

We also use Delaunay triangulations for computer's drawing. The *constrained conforming Delaunay triangulation* (CCDT) of a region of P is a constrained Delaunay triangulation using *Steiner points* to meet given constraints on the minimum angle and the maximum area of the triangles. We use the CCDT to triangulate a region whose edges are all constrained; see Fig. 5. Let p_1 and p_2 be two non-Steiner points of a CCDT $\mathcal{T}$. We define a plane graph $G_{\mathcal{T}} = G_{\mathcal{T}}(p_1, p_2)$ by letting the vertices of $G_{\mathcal{T}}$ be the points p_1, p_2 and the midpoints of all non-constrained edges of $\mathcal{T}$. Two vertices are connected by an edge in $G_{\mathcal{T}}$ if they lie in the same triangle of $\mathcal{T}$ but not on the same edge of $\mathcal{T}$. If p_1 and p_2 lie in the same triangle T, then we connect them through the center of gravity of T in $G_{\mathcal{T}}$.

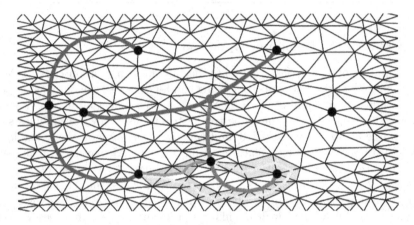

Fig. 5. A region triangulated by a CCDT $\mathcal{T}$ and of two paths (surrounded by shaded triangles) in $G_{\mathcal{T}}$ that connect different vertex occurrences of the same vertices.

We distinguish two types of moves depending on whether they connect vertices from different boundaries or from the same boundary of P. Consider two different boundaries $A_1 \cdots A_i \cdots A_m$ and $B_1 \cdots B_j \cdots B_n$ of P. The *double-boundary move* that connects two *vertex occurrences* A_i and B_j creates a new boundary $A_1 \cdots A_i Z B_j \cdots B_n B_1 \cdots B_j Z A_i \cdots A_m$ where Z is the newly added vertex. Let $A_1 \cdots A_i \cdots A_j \cdots A_n$ be a boundary in a region R with boundaries partitioned into sets $\mathcal{B}_{major}$, $\mathcal{B}_{minor}$, and $\{\beta\}$. The *single-boundary move* that connects (not necessarily different) vertex occurrences A_i and A_j and separates the boundaries $\mathcal{B}_{major}$ from the boundaries of $\mathcal{B}_{minor}$ splits R into the *major region* $A_i \cdots A_j Z.\mathcal{B}_{major}$ and into the *minor region* $A_1 \cdots A_i Z A_j \cdots A_n.\mathcal{B}_{minor}$.

Drawing a Double-Boundary Move. To draw a double-boundary move m between two vertex occurrences A_i and B_j in a region R, we construct a triangulation $\mathcal{T}$ of R and we let m be the shortest path in $G_{\mathcal{T}}(A_i, B_j)$ between A_i and B_j.

Drawing a Single-Boundary Move. Consider a single-boundary move m connecting vertex occurrences B_i and B_j with $i \leq j$ on a boundary β of a region R with boundaries $\mathcal{B}$ that splits the boundaries $\mathcal{B} \setminus \{\beta\}$ into a major partition $\mathcal{B}_{major}$ and a minor partition $\mathcal{B}_{minor}$. Drawing of m is much more complicated since we have to correctly split $\mathcal{B}$ into $\mathcal{B}_{major}$ and $\mathcal{B}_{minor}$. The first step is to connect all the inner boundaries from $\mathcal{B} \setminus \{\beta\}$ by a curve called *spindle* that starts and ends in the border boundary (or the border of the playing area if R is the outer region); see Fig. 6. Then we intertwine m with the spindle so that the partitions $\mathcal{B}_{major}$ and $\mathcal{B}_{minor}$ are on the correct sides of m; see Fig. 7. Intertwining m also uses triangulations and requires a lot of technical steps that are sketched below.

Setting up the Spindle. The spindle starts at an arbitrary vertex occurrence of the border boundary and leads to the closest vertex of an inner boundary from $\mathcal{B} \setminus \{\beta\}$. Then it continues from a vertex of the last visited boundary to the closest vertex of a non-visited inner boundary from $\mathcal{B} \setminus \{\beta\}$ until we visit all the inner boundaries from $\mathcal{B} \setminus \{\beta\}$. We end the spindle by connecting it to the closest non-visited vertex occurrence of the border boundary; see Fig. 6. The spindle divides the border polygon of R into the *primary polygon* P_{prim}, which is the polygon that contains β or the occurrences B_i and B_j if β is the border boundary, and the *secondary polygon* P_{sec}. The orientation of the spindle is opposite to the orientation induced by the counterclockwise orientation of P_{prim}.

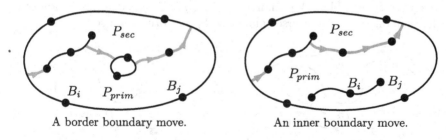

A border boundary move.   An inner boundary move.

Fig. 6. Spindles (heavier orange curves) that split the region into P_{prim} and P_{sec}. (Color figure online)

Intertwining the Spindle. Let $\mathcal{C}$ be a set of some of the inner boundaries visited by the spindle s. We intertwine the move m with s using an *enfolding* of $\mathcal{C}$; see Fig. 7. To enfold $\mathcal{C}$, we first draw a curve in P_{prim} from $v_0 = B_i$ to a vertex v_1 of the first segment of s that precedes a boundary from $\mathcal{C}$ with respect to the orientation of s. In P_{sec}, we then connect v_1 with a vertex v_2 of the first segment of s that precedes a boundary not in $\mathcal{C}$. We continue connecting v_i with v_{i+1} like this alternatingly in P_{prim} and P_{sec} until we get at the end of s. At the end, if we should continue drawing in P_{sec}, we just draw a curve to the last segment of s to get back to P_{prim}. Finally, we draw a curve in P_{prim} from the last connected vertex to B_j. If $\mathcal{C}$ is empty, we connect B_i with the first vertex of the last segment of s in P_{prim}, then we go to the last vertex of the last segment of s in P_{sec}, and then we return to B_j in P_{prim}. In a *reversed enfolding* of C, we intertwine s in the opposite direction.

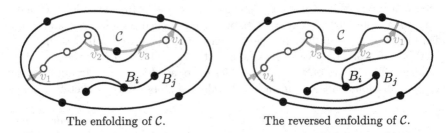

The enfolding of C.   The reversed enfolding of C.

Fig. 7. The enfoldings of the boundaries from C (green curves with empty discs). (Color figure online)

Choosing the Right Enfolding. We have four options how to enfold the partitions. We can choose C as the set of inner boundaries from B_{major} or from B_{minor}. We can also apply either the enfolding or the reversed enfolding. However, since the enfolded boundaries C always lie in the major region of m whereas the reversely enfolded boundaries C always lie in the minor region and since the border boundary cannot be enfolded nor reversely enfolded, we are left with a single option for enfolding. We enfold $C = B_{major}$ if B_{major} does not contain the border boundary and we reverse enfold $C = B_{minor}$ otherwise; see Fig. 8.

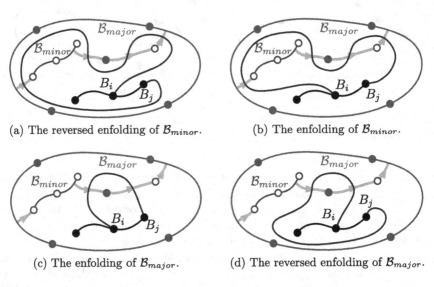

(a) The reversed enfolding of B_{minor}.   (b) The enfolding of B_{minor}.

(c) The enfolding of B_{major}.   (d) The reversed enfolding of B_{major}.

Fig. 8. The (a) correctly and (b–d) incorrectly chosen enfoldings for drawing a move (blue) from B_i to B_j with the partitions B_{major} (full red) and B_{minor} (empty green). (Color figure online)

Optimizing Moves. We use various techniques to make the moves nicer, as they should resemble moves drawn by a human player. For example, we take shorter equivalent moves if the border boundary is dead, we enfold only sets of close singletons as singletons are interchangeable, and we handle empty single-boundary moves separately. Finally, we note that some technical steps in the analysis of intertwining (for example the case $B_i = B_j$) are omitted in this extended abstract.

7   Conclusions and Future Work

We applied all the techniques described above to implement the Windows application *Sprouts: A Drawing Game* [9] that allows to play against a computer opponent; see Fig. 9 for a screenshot. Our program supports various forms of the game. We have a *Campaign mode* consisting of 150 positions derived from the n-spot positions with $n \leq 11$ where one can play against a perfectly playing computer opponent or against three other levels of AI. In a *Quick play mode*, it is possible to play against various computer opponents or against a human opponent on the n-spot positions with $n \leq 20$ or on custom positions made in the *Custom maps mode*. The program is not restricted to only local games but it also supports a *Remote game mode* where the players can connect to a Sprouts server and play against each other remotely.

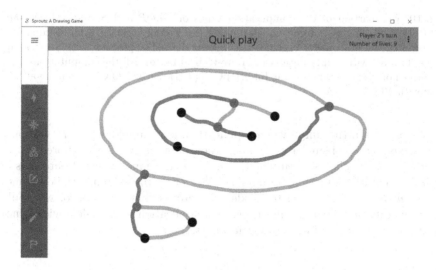

Fig. 9. A screenshot from our application *Sprouts: A Drawing Game.*

All the computed outputs agree with the results of *GLOP*. In particular, we obtained the same sizes of the game trees of the n-spot positions with $n \leq 6$. Moreover, the computation of outcomes with our program is slightly faster than with *GLOP*, even when using a single thread; see Fig. 10. With four threads, the computation can speed up significantly. We encode positions using the same notation as *GLOP* so its databases can be used in our program as well.

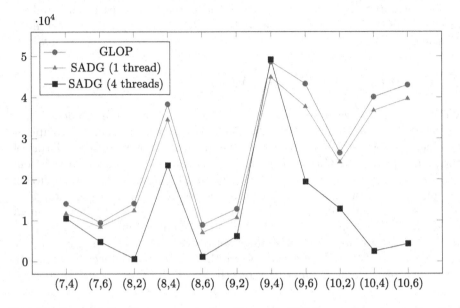

Fig. 10. Comparison of the computation times of *GLOP* and *Sprouts: A Drawing Game* (SADG). Every column (i, j) contains the average times (in milliseconds) of determining the outcome of 3 games starting on the i-spot position with j randomly selected moves with empty databases of pre-trained moves. All the computations were performed on a computer with the Intel(R) Core(TM) i7-6700HQ CPU running at 2.60 GHz with 16 GB of RAM.

We plan to further improve our application, for example, we would like opti-mize the enfolding of singletons to make the computer moves even more natural. Another possible plan is to employ the distributed computations of outcomes in an implementation of a new Sprouts solver in order to determine the outcomes of new n-spot positions and to tackle the Sprouts conjecture. Also, we would like to modify our program and develop an application for mobile devices since the game of Sprouts is ideal for touchscreens.

References

1. Alfeld, P.: The Game of Sprouts (1999). https://www.math.utah.edu/~alfeld/Sprouts/index.html. Accessed 15 Mar 2021
2. Altmann, C.E.: Sproutsplus (2019). apps.apple.com/au/app/sproutsplus/id1490125456#?platform=iphone. Accessed 15 May 2021
3. Anthony, P.: Macroscope. Avon, New York (1969)

4. Applegate, D., Jacobson, G., Sleator, D.: Computer analysis of Sprouts. Carnegie Mellon University Computer Science technical report CMU-CS (1991)
5. Berlekamp, E.R., Conway, J.H., Guy, R.K.: Winning ways for your mathematical plays, vol. 3. MA, second edn, A K Peters Ltd, Natick (2003)
6. Bertault, F.: A force-directed algorithm that preserves edge crossing properties. In: Kratochvyal, J. (eds.) Graph Drawing. GD 1999. LNCS, vol. 1731. Springer, Heidelberg (1999). https://doi.org/10.1007/3-540-46648-7_36
7. Brown, W., Baird, L.: A graph drawing algorithm for the game of Sprouts. In: Proceedings of the 2008 International Conference on Computer Graphics and Virtual Reality, pp. 117–121, CGVR 2008, 2008, Las Vegas, Nevada, USA. CSREA Press (2008)
8. Browne, C.: Algorithms for interactive Sprouts. Theoret. Comput. Sci. **644**, 29–42 (2016)
9. Čížek, T.: Sprouts: A Drawing Game (2021). https://kam.mff.cuni.cz/~cizek/ Sprouts/. Accessed 9 Jun 2021
10. Conway, J.H.: On Numbers and Games. Academic Press, London (1976)
11. Egri, G.: Sproutsgame (2019). apps.apple.com/uy/app/sprouts-game/id1479115453. Accessed 15 May 2021
12. Focardi, R., Luccio, F.L.: A modular approach to Sprouts. Discr. Appl. Math. **144**(3), 303–319 (2004)
13. Gardner, M.: Mathematical games: of sprouts and Brussels sprouts; games with a topological flavour. Sci. Amer. **217**, 112–115 (1967)
14. Gardner, M.: Mathematical Carnival. Penguin Books, New York (1990)
15. Gehrig, L.D.: Sprouts - A Game of Maths! (2011). https://appadvice.com/game /app/spouts-a-game-of-maths/426618463?fbclid=IwAR19mFHqisSybUHpvEpK4s 2ZJ2QO74sR1urst7Up6UtebT_dnerpyES8mT0. Accessed 15 May 2021
16. Lemoine, J., Viennot, S.: Computer analysis of Sprouts with nimbers. In: Games of no chance 4, Mathematical Science Research Institution Publication, vol. 63, pp. 161–181. Cambridge University Press, New York (2015)
17. Pritchard, D.: Brain Games: The World's Best Games for Two. Penguin Books, New York (1982)
18. Reiss, S.: 3Graph (2009). www.reisz.de/3graph.htm. Alternative link: https://www. heise.de/download/%20product/3graph-64693
19. Roberts, S.: Genius At Play: The Curious Mind of John Horton Conway. Bloomsbury Press, New York (2015)
20. Rocchini, C.: The (Computer Version of) Game of Sprouts (2006). game-of-sprouts.sourceforge.net/ Accessed 22 May 2021
21. Simonetto, P., Archambault, D., Auber, D., Bourqui, R.: ImPrEd: an improved force-directed algorithm that prevents nodes from crossing edges. Comput. Graph. Forum **30**(3), 1071–1080 (2011)

Graph Drawing Contest

Graph Drawing Contest Report

Philipp Kindermann[1]([✉]), Tamara Mchedlidze[2], and Wouter Meulemans[3]

[1] Universität Trier, Trier, Germany
kindermann@uni-trier.de
[2] Universiteit Utrecht, Utrecht, The Netherlands
t.mtsentlintze@uu.nl
[3] Eindhoven University of Technology, Eindhoven, The Netherlands
w.meulemans@tue.nl

Abstract. This report describes the 28th Annual Graph Drawing Contest, held in conjunction with the 29th International Symposium on Graph Drawing and Network Visualization (GD'21) in Tübingen, Germany. Due to the global COVID-19 pandemic, the conference and thus also the contest was held in a hybrid format, with both on-site and online participants. The mission of the Graph Drawing Contest is to monitor and challenge the current state of the art in graph-drawing technology.

1 Introduction

Following the tradition of the past years, the Graph Drawing Contest was divided into two parts: the *creative topics* and the *live challenge*.

Creative topics were comprised by two data sets. The first data set modeled *Movie Remakes* by different directors. The second data set shows a logical reconstruction of a scientific debate among 19th century geologists, namely the Great Devonian Controversy, as an *Argumentation Network*. The data sets were published about a year in advance, and contestants submitted their visualizations before the conference started.

The live challenge took place during the conference in a format similar to a typical programming contest. Teams were presented with a collection of *challenge graphs* and had one hour to submit their highest scoring drawings. This year's topic was to minimize edge-length ratio in a planar polyline drawing graph with vertex locations restricted to a grid and a maximum number of bends per edge allowed.

Overall, we received 26 submissions: 7 submissions for the creative topics and 19 submissions for the live challenge.

2 Creative Topics

The general goal of the creative topics was to model each data set as a graph and visualize it with complete artistic freedom, and with the aim of communicating

H. C. Purchase and I. Rutter (Eds.): GD 2021, LNCS 12868, pp. 409–417, 2021.
https://doi.org/10.1007/978-3-030-92931-2_29

as much information as possible from the provided data in the most readable and clear way.

We received 7 submissions for the first topic, and 0 for the second. Submissions were evaluated according to four criteria:

(i) Readability and clarity of the visualization,
(ii) aesthetic quality,
(iii) novelty of the visualization concept, and
(iv) design quality.

We noticed overall that it is a complex combination of several aspects that make a submission stand out. These aspects include but are not limited to the understanding of the structure of the data, investigation of the additional data sources, applying intuitive and powerful data visual metaphors, careful design choices, combining automatically created visualizations with post-processing by hand, as well as keeping the visualization, especially the text labels, readable. All submissions were printed on large poster boards and presented at the Graph Drawing Symposium. We also made all the submissions available on the contest website in the form of a virtual poster exhibition. During the conference, we presented these submissions and announced the winners. For a complete list of submissions, refer to http://www.graphdrawing.org/gdcontest/contest2021/results.html.

2.1   Movie Remakes

A movie remake is a production of a film that is based upon an earlier production. A remake tells the same story as the original but uses a different cast and may alter the theme or target audience. See Tang et al. [1] for related work on this topic.

For this topic, the task was to visualize a graph of movie remakes by different directors. The data contains a list of directors, and pairs of movies: the original and the remake (both with title, year, and directors). The data has been crawled from Wikipedia and consists of 91 directors and 102 pairs of movies. The participants were free to decide which parts of the data to visualize and how to visualize it.

Shared 2nd Place: Najla Amira Ochoa Leonor and Daniela Martinez Duarte (TU Wien). The authors used Louvain's method for community detection. Within each community, they identified film directors related to six or more directors as the most influential movie directors. Using these directors as the backbone for the visualization, they extracted a Steiner tree from the cinematographic graph. Then they used a spring layout for the visualization, where each influential director is shown in a radial drawing resembling a film reel. The authors also added context information for the data set, addressing common questions a reader could ask of the data set, such as the largest time span between a movie and its remake or the proportion of female directors. This was probably the prettiest submission and the committee appreciated the aesthetics and the design quality of the visualization.

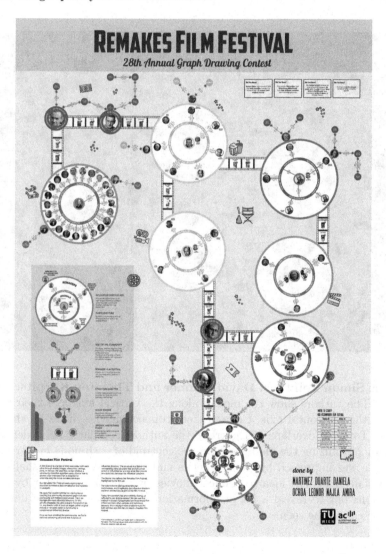

Shared 2nd Place: Michael Häglsperger, Sven Teufel, Rinor Kelmendi, and Henry Förster (Universität Tübingen). The authors arranged the original movies together with their remakes on a timeline where movies of the same director are grouped together. The dependency of movies is illustrated by edges pointing from the original to their remakes. The visualization mimics a video editor program, for this the movies of one director are arranged on the same height, so the grouped nodes look similar to tracks in an editing software. The authors also created an interactive version of the graph where it is possible to highlight dependencies by clicking on a movie, remake, director, or year. The interactive graph can be accessed at https://algo.inf.uni-tuebingen. de/movie-remakes. The committee liked the approach and the design quality of the submission, which made it easy to explore the data and explore interesting structures.

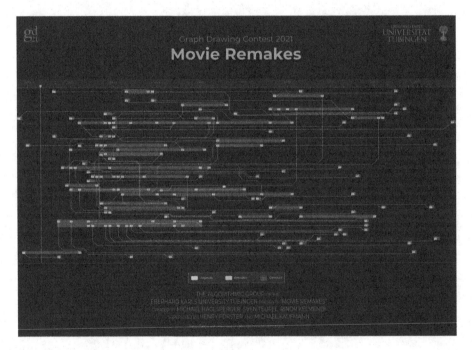

Winner: Simon Pointner, David Ammer and Thorsten Korpitsch (TU Wien). The authors show the movies along the vertical axis and their release dates on the horizontal axis. Colored trees connect movies made by the same director. To avoid overlaps of the trees, the authors computed an order of the movies by optimizing the distance from the centroid of each node in a tree. The committee was impressed by the aesthetics and the clarity of the visualization. We especially liked the idea to connect movies by the same director by a tree, which turns out to reveal a lot of information and structure in the data that otherwise cannot be seen.

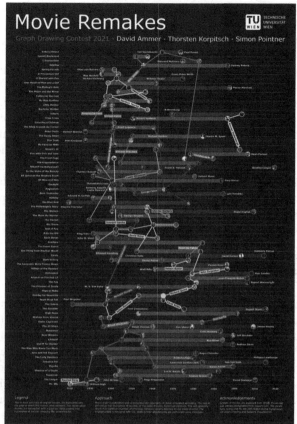

> **❝** Our main focus point was putting the movies as the central piece of information and therefore putting them on the horizontal axis while connecting the movies created by the same director mostly on the vertical axis. To minimize the number of crossings and the area for each director tree we optimized the order using a metaheuristic approach, in detail simulated annealing. The cost is defined as the divergence of movies, of the same director, from their centroid. **❞**
> *Simon Pointner*

2.2 Argumentation Network

The network shows a logical reconstruction of a scientific debate among 19th century geologists, namely the Great Devonian Controversy. The network contains 335 vertices which are of two types: statements and arguments. Each argument has one or more sentences as premises and one sentence as a conclusion. Vertices are grouped into 12 thematic clusters. The 1016 edges connect statements to statements and statements to arguments. Each edge has an associated type that describes the logical relation between vertices: `contrary`, `contradictory`, or `entails`.

Unfortunately, we have received no submission for this topic. During the conference, we conducted a brief survey to find the reasons why nobody submitted a drawing for this graph. The main reason seemed to be that the topic of the argumentation, the Great Devonian Controversy, is largely unknown and that it is difficult to find additional information on it. Most submissions have been done by students (supervised by more experienced researchers), when given the choice they all found the Movie Remakes graph to be more interesting and chose to work on that one instead. Overall, it was not the graph itself, but the topic that was not appealing enough for the participants. Based on the results of the survey, we decided not to pose the same graph again and will instead hand out two new topics for the Graph Drawing Contest 2022.

3   Live Challenge

The live challenge took place during the conference and lasted exactly one hour. During this hour, local participants of the conference could take part in the manual category (in which they could attempt to draw the graphs using a supplied tool: http://graphdrawing.org/gdcontest/tool/), or in the automatic category (in which they could use their own software to draw the graphs). Because of the global COVID-19 pandemic, we allowed everybody in both categories to participate remotely. To coordinate the contest, give a brief introduction, answering questions, and giving participants the possibility to form teams, we were kindly provided with both a room in the conference building, and a dedicated room in the gather.town of the conference.

The challenge focused on minimizing the planar polyline edge-length ratio on a fixed grid. The *planar edge-length ratio* of a straight-line drawing is defined as the ratio between the length of longest edge and the length of the shortest edge. There has been recent attention to this topic with several publications. The *planar polyline edge-length ratio* is a generalization of the planar edge-length ratio where edges do not have to be straight-line segments, but can be polylines with a maximum number of bends per edge defined by the input.

The input graphs were planar undirected graphs. For the manual category, each graph came already with a planar drawing.

The results were judged solely with respect to the edge-length ratio; other aesthetic criteria were not taken into account. This allows an objective way to evaluate each drawing.

3.1   The Graphs

In the manual category, participants were presented with six graphs. These were arranged from small to large and chosen to contain different types of graph structures. In the automatic category, participants had to draw the same six graphs as in the manual category, and in addition another seven larger graphs. Again, the graphs were constructed to have different structure.

For illustration, we include the fourth graph, which was given with a drawing where every edge has length 1, except one long diagonal edge with length $10\sqrt{2} \approx$ 14.14, in its initial state with vertices moved around randomly, the best manual solution we received (by team *New keyboard, who dis?*), and the best automatic solution we received (by team *The WorstLayoutProducers*).

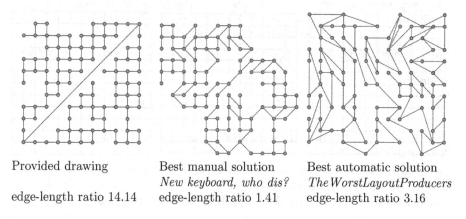

Provided drawing	Best manual solution	Best automatic solution
	New keyboard, who dis?	*The WorstLayoutProducers*
edge-length ratio 14.14	edge-length ratio 1.41	edge-length ratio 3.16

For the complete set of graphs and submissions, refer to the contest website at http://www.graphdrawing.org/gdcontest/contest2021/results.html. The graphs are still available for exploration and solving Graph Drawing Contest Submission System: https://graphdrawingcontest.appspot.com.

Similarly to the past years, the committee observed that manual (human) drawings of graphs often display a deeper understanding of the underlying graph structure than automatic and therefore gain in readability. The committee was also impressed by the fact that for all of the six small graphs the manual drawings were better than the automatic drawings. For the larger graphs, it turned out that the allowed grid sizes of the input were too restrictive and it was hard to get any planar drawing at all. Only for two of the seven large graphs, a feasible solution has been submitted.

3.2   Results: Manual Category

Below we present the full list of scores for all teams. The numbers listed are the edge-length ratios of the drawings; the horizontal bars visualize the corresponding scores.

graph	1	2	3	4	5	6
Mina et al.	1.11	2.23	5.06	2.82	6.72	6.0
Zinklos	1.0	1.67	1.27	1.41	1.86	1.0 **1.**
funnygame	1.22	2.3	2.77	2.23	4.25	2.51
Team Golden Ratio	1.19	1.58	3.28	2.23	5.71	3.55
WinterIsComing	1.16	1.21	1.38	2.23	3.75	4.5
Bako	1.09	1.25	2.76	2.23	5.06	2.63
Team perpendicular table	1.14	1.21	1.32	1.41	2.06	3.6 **2.**
Quickdraw	1.17	1.3	1.82	2.0	3.39	2.0
Q	1.0	2.64	1.18	2.23	2.54	2.4
unbowed, unbent, uncrossed	1.0	1.37	1.71	2.82	6.0	2.0
New keyboard, who dis?	1.11	1.17	1.75	1.41	2.68	1.94 **3.**
Cobbie	1.37	2.28	2.84	2.82	3.41	6.65
PG-team	1.2	1.49	2.78	2.82	6.4	5.0
Mihir Neve	1.13	1.19				
Fabri	1.41	2.23	6.47	2.23	7.07	4.24

Third place: **New keyboard, who dis?**, consisting of Soeren Nickel, Anaïs Villedieu, and Jules Wulms.
Second place: **Team perpendicular table**, consisting of Fouli Argyriou, Henry Förster, and Martin Gronemann
Winner: **Zinklos**, consisting of Jonathan Klawitter and Felix Klesen.

> 66 After the shocking event that gave our team its name, namely, a person with the name Zink leaving our team just minutes before the event started, we picked ourselves up again and got the adrenalin flowing. We then attacked each instance with the following steps. First, we checked if there is some structure in the graph that we can use such as a grid structure or other regularities. Second, since a good planar embedding is important for a good score, it was often worthwhile to search for a suitable one. In particular for the last instance, we made a timewise heavy investment to change the whole embedding such that a perfect edge length ratio could be achieved. Third, when the longest edges could not be shorted anymore, we iteratively extended the shortest edge by adding bends and drawing them zigzagily. 99
> *Jonathan Klawitter*

3.3   Results: Automatic Category

In the following we present the full list of scores for all teams that participated in the automatic category. The numbers listed are the edge-length ratios of the drawings; the horizontal bars visualize the corresponding scores.

graph	1	2	3	4	5	6	7	8	9	10	11	12	13
IQOutOfBoundsException	12.8	18.43	10.54				706				139.3		
Graphiti	1.52	2.49	2.47	10.63		2.86							3.
SPEIX	1.41	2.16	6.59	6.4	4.22	11.5							2.
TheWorstLayoutProducers	1.1	1.3	1.96	3.16	10.0	1.66		83.42				48.83	1.

Third place: **Graphiti**, consisting of Lukas Schmitt and David Rumpf.
Second place: **SPEIX**, consisting of Haolin Pan, Yiming Qin, and Kunhao Zheng.
Winner: **TheWorstLayoutProducers**, consisting of Moritz Greiner, Axel Kuckuk, Michael Bekos, and Maximilian Pfister.

> **"** While working on this task, we quickly realized that the drawings that optimize the given quality metric are far from nice. Hence, our team WorstLayoutProducers focused on producing the worst such layouts, which was enough to give us the first place. To be more precise, as a first step we compute a valid drawing of the (abstract) graph, and afterwards we used an iterative scheme that tries to improve the current solution by slightly modifying the current drawing without changing its embedding, as this turned out to be more efficient for larger graphs. **"**
> *Maximilian Pfister*

Acknowledgments. The contest committee would like to thank the organizing and program committee of the conference; the organizers who provided us with a room with hardware for the live challenge, monetary prizes, and beer glasses for the winners; the generous sponsors of the symposium; and all the contestants for their participation. Special thanks goes to Maarten Löffler for providing the data for the Movie Remakes graph. Further details including all submitted drawings and challenge graphs can be found at the contest website: http://www.graphdrawing.org/gdcontest/contest2021/results.html.

Reference

1. Tang, Y., Yu, J., Li, C., Fan, J.: Visual analysis of multimodal movie network data based on the double-layered view. Int. J. Distributed Sens. Netw. 11, 906316:1–906316:16 (2015)

Poster Abstracts

Graphomone: A Dynamic Network Player and Creation Tool

Eleni Katsanou[1]([✉]) [ID], Tamara Mchedlidze[2] [ID], Chrysanthi Raftopoulou[1] [ID], and Antonios Symvonis[1] [ID]

[1] Department of Mathematics, National Technical University of Athens, Athens, Greece
e.katsanou@outlook.com, {crisraft,symvonis}@math.ntua.gr
[2] Utrecht University, Utrecht, Netherlands
t.mtsentlintze@uu.nl

Abstract. We introduce the software tool Graphomone, designed to create and display dynamic networks in the form of node-link diagrams. With Graphomone, we aim to fill the existing gap in the variety of tools for network visualization, that is, to provide an easy and intuitive interface to create dynamic networks, to model dynamic attributes, and, finally to observe the changes in the network over time enhanced by the use of a change-warning colored halo.

Keywords: Dynamic network · Dynamic player · Dynamic editor

Introduction. Dynamic networks appear as models in diverse applications ranging from humanitarian [3] to industrial [10]. The information visualization community has developed various visualization methods for dynamic networks that contribute to their understanding and communication. Among them are animation (time-to-time mapping), applied to both node-link diagrams or matrix visualization, and timeline (time-to-space mapping), where node-link diagrams may appear juxtaposed [9]. While many scientific papers and software packages, e.g. [2, 4, 7, 8, 15] concentrate on the development of sophisticated visualization methods and user interfaces to present and analyse dynamic networks, such simple tasks as viewing a dynamic network and its creation are not easily accessible with the freely available tools. In this poster, we present Graphomone, an easy to use dynamic graph player and creator.

Dynamic Network Model and Its Visualization. At this state of the implementation, the network model of Graphomone is rather simple, consisting of nodes and edges, where both can have a small number of categorical and spatial attributes that are easy to be presented in a node-link diagram. Common visual attributes for nodes and edges are the visibility state (specifying whether the node/edge is visible), color and label. Additionally, nodes can change their shape (ellipse or rectangle), position and icon; and edges can change their width, style (solid, dashed or dotted) and arrow visibility (specifying whether the edge

H. C. Purchase and I. Rutter (Eds.): GD 2021, LNCS 12868, pp. 421–424, 2021.
https://doi.org/10.1007/978-3-030-92931-2

is directed). The dynamic nature of the network model in Graphomone can be represented in two different ways, using integers time stamps or dates. Both nodes and edges of the dynamic network model are "alive" for a specific interval. Note that, by using the visibility feature mentioned above nodes/edges can appear or disappear during their lifespan. All other attributes can change their values only within the object's lifespan, regardless of whether it is visible or not.

File Format. To describe the aforementioned dynamic network model and to allow network export and import actions into Graphomone, we introduce a new file format, GRDYN. The file is written in XML and the information is divided into four parts: the time format of the dynamic model (integers or dates), a declaration of the user-defined attributes, node information and edge information. For each node, it is specified when it is added and when it is deleted, when it changes its visibility state and when its visual attributes change values. The information regarding an edge is presented in a similar way, with the addition that its source and target nodes are specified together with info regarding its arrow visibility. The main reason for introducing a new file format and not using the already well-known GEXF (supported by Gephi) is that GEXF does not support dynamic changes of the visual attributes of nodes and edges.

Software Design. Graphomone offers two basic functionalities: dynamic network player and creation. The player displays the dynamic network transitions from one timestamp to the next; the changes that may take place are the addition/deletion of nodes/edges as well as alterations of their visual attributes. Graphomone supports both abrupt and smooth change of the visual attributes. In smooth mode, the user can choose the visual attributes that are changed smoothly. In particular, addition/deletion of elements is performed with a fade-in/out effect, relocation of nodes is performed with a smooth motion. Inspired by the tool Graph-Diaries [7], the changing object is also highlighted with a halo, having configurable color. Graphomone allows for an interactive creation of dynamic networks: the user can create a new dynamic network, insert nodes and edges into it, and edit their lifespan, visibility periods and visual attributes. Note that, while GraphDiaries [7] is a sophisticated tool for dynamic network visualization and analysis, it does not support any network creation functionality.

Use Case. Graphomone can be used for creation, editing and visualization of dynamic networks and it aims to make this process intuitive and user-friendly. As a use case we have considered the field of literature, where character networks are dynamic, describing the various interactions among the players over time. Using Graphomone, we have re-modeled the dynamic network of Hrafnkels Saga that appeared as one of the creative topics of Graph Drawing Challenge in 2020 [1].

Future Work. From the Hrafnkels Saga case, we concluded that in order to allow for an easy and natural applicability to literary dynamic networks, a slightly more complex network model is necessary which supports node classes and their dynamic changes. We plan to extend the design of the file format and the software support for more complex models of multilayer networks, by carefully choosing the right level of model complexity [12] necessary for the targeted applications. Having targeted a certain application domain, we will evaluate Graphomone with domain experts and improve it based on their input. A special effort will be devoted to making the software tool freely accessible and compatible with other network visualization tools such as e.g. Gephi by the mean of import an export functionalities that will allow an easy conversion among the file formats.

References

1. Graph drawing contest, creative topics. http://mozart.diei.unipg.it/gdcontest/contest2020/topics.html
2. The keylines toolkit. The Javascript toolkit for graph visualization. https://cambridge-intelligence.com/keylines/
3. Mapping the republic of letters, Voltaire's correspondence network. http://republicofletters.stanford.edu/publications/voltaire/
4. Palladio. Visualize complex historical data with ease. https://hdlab.stanford.edu/palladio/
5. Ahn, J.W., Plaisant, C., Shneiderman, B.: A task taxonomy for network evolution analysis. IEEE Trans. Visual. Comput. Graphics **20**, 365–376 (2014). https://doi.org/10.1109/TVCG.2013.238
6. Andrienko, N., Andrienko, G.: Exploratory Analysis of Spatial and Temporal Data: A Systematic Approach (2006). https://doi.org/10.1007/3-540-31190-4
7. Bach, B., Pietriga, E., Fekete, J.D.: Graphdiaries: animated transitions and temporal navigation for dynamic networks. IEEE Trans. Visual. Comput. Graphics **20**(5), 740–754 (2014). https://doi.org/10.1109/TVCG.2013.254. Institute of Electrical and Electronics Engineers
8. Bastian, M., Heymann, S., Jacomy, M.: Gephi: an open source software for exploring and manipulating networks. In: Proceedings of the International AAAI Conference on Web and Social Media, vol. 3, no. 1, 361–362 (2009). https://doi.org/10.13140/2.1.1341.1520
9. Beck, F., Burch, M., Diehl, S., Weiskopf, D.: A taxonomy and survey of dynamic graph visualization. Compu. Graphics Forum **36**, 133–159 (2016). https://doi.org/10.1111/cgf.12791
10. Cai, Q., Alam, S., Duong, V.: A spatial-temporal network perspective for the propagation dynamics of air traffic delays. Engineering **7**(4), 452–464 (2021). https://doi.org/10.1016/j.eng.2020.05.027
11. Kerracher, N., Kennedy, J., Chalmers, K.: A task taxonomy for temporal graph visualisation. IEEE Trans. Visual. Comput. Graphics **21**(10), 1160–1172 (2015). https://doi.org/10.1109/TVCG.2015.2424889
12. Kivelä, M., Arenas, A., Barthelemy, M., Gleeson, J.P., Moreno, Y., Porter, M.A.: Multilayer networks. J. Complex Netw. **2**(3), 203–271 (2014). https://doi.org/10.1093/comnet/cnu016
13. Lee, A., Archambault, D., Nacenta, M.: Dynamic network plaid: a tool for the analysis of dynamic networks, pp. 1–14. Association for Computing Machinery, New York (2019). https://doi.org/10.1145/3290605.3300360

14. Lee, B., Plaisant, C., Parr, C.S., Fekete, J.D., Henry, N.: Task taxonomy for graph visualization. In: Proceedings of the 2006 AVI Workshop on BEyond Time and Errors: Novel Evaluation Methods for Information Visualization, pp. 1–5. BELIV 2006. Association for Computing Machinery, New York (2006). https://doi.org/10.1145/1168149.1168168
15. Linhares, C.D.G., Travençolo, B.A.N., de Souza Paiva, J.G., Rocha, L.E.C.: Dynetvis: a system for visualization of dynamic networks. In: Seffah, A., Penzenstadler, B., Alves, C., Peng, X. (eds.) Proceedings of the Symposium on Applied Computing, SAC 2017, Marrakech, Morocco, 3–7 April 2017, pp. 187–194. ACM (2017). https://doi.org/10.1145/3019612.3019686

One-Way k-Crossable Graphs

Moritz Jagnow, Henry Förster$^{(\boxtimes)}$, and Michael Kaufmann

Wilhelm-Schickard-Institut für Informatik, Eberhard-Karls-Universität Tübingen,
Tübingen, Germany
moritz.jagnow@student.informatik.uni-tuebingen.de,
{foersth,mk}@informatik.uni-tuebingen.de

Motivation. Directed graphs appear in a plethora of practical applications. While real-world graphs often are not planar, the literature on beyond planarity so far lacks results on directed graphs [3] aside from one on upward RAC drawings [1]. Here we study the following version of the problem, which is inspired by fanplanar drawings, where each edge e can be intersected only by a fan of edges emerging from a common endpoint from either the left or the right side of e.

Let e be an edge in a directed graph $G = (V, E)$ and let Γ be a drawing of G. By the direction of e, the left and right side of e in Γ is well-defined. In Γ, each edge e' crossing e can be seen as entering from the left of e and leaving to the right of e or vice versa. Edge e is called *one-way crossed* if all edges e' that cross e enter from the same side. We call Γ *one-way k-crossed* if all edges of G are one-way crossed in Γ and each edge is involved in at most k crossings. Finally, we call G *one-way k-crossable* if it admits a one-way k-crossed drawing.

Results. We establish several density bounds for one-way k-crossable graphs:

k	Density upper bound	Density lower bound
1	$4(n-2)$	$4(n-2)$
2	$\frac{13}{3}(n-2)$	$\frac{13}{3}(n-2)$
3	$5(n-2)$	$5(n-2) - 2$
≥ 7	$6(n-2)$	$6(n-2) - 6$

We first prove a general density upper bound:

Theorem 1. *Let $G = (V, E)$ be one-way k-crossable. Then, G is a bi-planar graph, i.e., its thickness is 2. Hence, G has $m \leq 6n - 12$ edges.*

Proof. We color the edges of G using two colors, red and blue, so that no two edges of the same color cross each other, i.e. each color induces a planar subgraph of G: All edges e in G that are crossed by edges coming from the left hand side of e are colored blue. The remaining edges are then colored red; see Fig. 1. □

In fact, this upper bound can be reached for $k = 7$ up to a constant value:

Theorem 2. *There are infinitely many n-vertex one-way 7-crossable graphs with $m = 6n - 18$ edges.*

© Springer Nature Switzerland AG 2021
H. C. Purchase and I. Rutter (Eds.): GD 2021, LNCS 12868, pp. 425–427, 2021.
https://doi.org/10.1007/978-3-030-92931-2

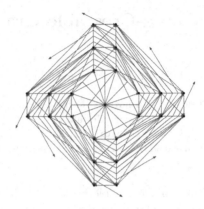

Fig. 1. Proof of Theorems 1 and 2 (Color figure online)

Proof. Figure 1 shows a one-way 7-crossable graph G that has $6n - 18$ edges. The red graph consists of nested rings of 8 triangulated quadrangles. Each quadrangle shares an edge with the corresponding quadrangle of the next ring. Blue edges are routed within the rings and connect vertices of two consecutive rings. □

Now that we established an almost-tight upper bound, a natural question is, whether there are smaller values of k so that there is a one-way k-crossable graph with $6n - O(1)$ edges. We address this question in the following. First, it is easy to see that for $k = 1$ the maximum density of one-way k-crossable graphs is restricted by the maximum density 1-planar graphs [4] as any orientation of a 1-planar graph is one-way crossed:

Theorem 3. *Let G be one-way 1-crossable. Then, G has at most $4n - 8$ edges which is a tight bound.*

Next, we provide a tight density bound for one-way 2-crossed graphs:

Theorem 4. *Let G be one-way 2-crossable. Then, G has at most $\frac{13}{3}(n-2)$ edges which is a tight bound.*

Proof. We claim that every blue edge of G can be assumed to have its endvertices in red faces of size 3 from which the maximum density of one-way 2-crossable can be directly derived since each triangular face of the red subgraph can contain the endvertex of at most two blue edges. Assume that a blue edge (u, v) crosses a red edge (p, q) as shown in Fig. 2. Clearly one of (p, u), (p, v), (u, q), (v, q) must exist in G. Let that edge be (p, u). Additionally another planar red edge can be added when removing at most 1 blue edge as shown in Fig. 2.

A corresponding lower bound is obtained from the optimal 2-planar graphs as characterized in [2] by removing one edge from each outer-2-planar K_5. □

Finally, it is possible to prove along the same lines the following:

Theorem 5. *Let G be one-way 3-crossable. Then, G has at most $5(n-2)$ edges. Also, there are infinitely many one-way 3-crossable graphs with $5n - 12$ edges.*

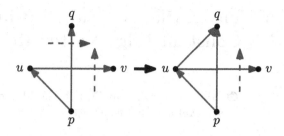

Fig. 2. Proof of Theorem 4 (Color figure online)

Conclusions. We initiated the study of one-way k-crossable graphs and investigate Turán type questions related to this new graph family. While it remains open to find the minimum k for which there are one-way k-crossable graphs with $6n - O(1)$ edges, our results imply that the correct k is between 3 and 7. Moreover, the recognition of one-way k-crossable graphs may be of interest.

References

1. Angelini, P., et al.: On the perspectives opened by right angle crossing drawings. J. Graph Algorithms Appl. **15**(1), 53–78 (2011). https://doi.org/10.7155/jgaa.00217
2. Bekos, M.A., Kaufmann, M., Raftopoulou, C.N.: On optimal 2- and 3-planar graphs. In: Aronov, B., Katz, M.J. (eds.) 33rd International Symposium on Computational Geometry, SoCG 2017. LIPIcs, vol. 77, pp. 16:1–16:16 (2017). https://doi.org/10.4230/LIPIcs.SoCG.2017.16
3. Didimo, W., Liotta, G., Montecchiani, F.: A survey on graph drawing beyond planarity. ACM Comput. Surv. **52**(1), 41–437 (2019). https://doi.org/10.1145/3301281
4. Pach, J., Tóth, G.: Graphs drawn with few crossings per edge. Combinatorica **17**(3), 427–439 (1997). https://doi.org/10.1007/BF01215922

Compact Convex Planar Grid Drawings of Graphs with Constant Edge-Vertex Resolution

Michael A. Bekos[1]([✉])[iD], Martin Gronemann[2][iD], Fabrizio Montecchiani[3][iD],
and Antonios Symvonis[4][iD]

[1] Department of Computer Science, University of Tübingen, Tübingen, Germany
bekos@informatik.uni-tuebingen.de
[2] Theoretical Computer Science, Osnabrück University, Osnabrück, Germany
martin.gronemann@uni-osnabrueck.de
[3] Department of Engineering, University of Perugia, Perugia, Italy
fabrizio.montecchiani@unipg.it
[4] School of Applied Mathematical and Physical Sciences, NTUA, Athens, Greece
symvonis@math.ntua.gr

1 Introduction

In this poster, we study planar straight-line grid drawings with constant *edge-vertex resolution*, that is, the minimum distance between a vertex and any point of a non-incident edge is at least a predefined constant independent of the number of vertices. The motivation to study such drawings stems from practical applications, in which each vertex is represented by a unit disk (a disk whose diameter is 1) and edge-vertex intersections should be avoided, that is, the edge-vertex resolution should be at least $\frac{1}{2}$. Early results date back to 1996, when Chrobak, Goodrich and Tamassia [2] claimed that every 3-connected planar graph admits a *convex* planar straight-line grid drawing on a grid of size $(3n - 7) \times (3n - 7)/2$ with edge-vertex resolution at least $\frac{1}{2}$. However, the details of the algorithm (and of its proof) supporting this claim never appeared in the literature. In this regard, very recently, Bekos et al. [1] proved (among other results) that every 3-connected planar graph admits a planar straight-line grid drawing on a grid of size $(3n - 7) \times (3n - 7)/2$ with edge-vertex resolution at least $\frac{1}{2}$. However, the obtained drawing is not necessarily convex.

We improve both results mentioned above by providing a linear-time algorithm to compute planar straight-line grid drawings with edge-vertex resolution at least $\frac{1}{2}$ that are convex and that fit on a grid of size $(n - 2 + a) \times (n - 2 + a)$, where a denotes the minimum of the number of faces f of the input graph and $n - 3$, that is, $a = \min\{f, n - 3\}$. In particular, if the input graph is maximal planar (that is, $f = 2n - 4$), our technique yields drawings of area $(2n - 5) \times (2n - 5)$. On the other hand, if the input graph is 3-connected cubic planar (that is, $f = n/2 + 2$), then our technique yields drawings of area $3n/2 \times 3n/2$. Our result is summarized in the next theorem, whose proof is sketched in the following section.

© Springer Nature Switzerland AG 2021
H. C. Purchase and I. Rutter (Eds.): GD 2021, LNCS 12868, pp. 428–430, 2021.
https://doi.org/10.1007/978-3-030-92931-2

Theorem 1. *Every 3-connected plane graph with n vertices and f faces admits a convex planar straight-line drawing with edge-vertex resolution at least $\frac{1}{2}$ on a grid of size $(n-2+a) \times (n-2+a)$, where $a = \min\{f, n-3\}$.*

2 Sketch of the Algorithm

To prove Theorem 1, we build on a well-known algorithm by Chrobak and Kant [3], which produces a convex planar straight-line drawing of a 3-connected plane graphs G with n vertices on a grid of size $(n-2) \times (n-2)$. Let $\pi = (P_0, \ldots, P_m)$ be a canonical ordering [5] of G, such that $P_0 = \{v_1, v_2\}$, $P_m = \{v_n\}$, and edges (v_1, v_2) and (v_1, v_n) exist and belong to the outer face of G. For $k = 0, \ldots, m$, let G_k be the subgraph induced by $\bigcup_{i=0}^{k} P_i$ and denote by C_k the *contour* of G_k defined as follows. If $k = 0$, then C_0 is the edge (v_1, v_2), while if $k > 0$, then C_k is the path from v_1 to v_2 obtained from the cycle delimiting the outer face of G_k after the removal of edge (v_1, v_2). Our algorithm also makes use of a Schnyder-like [4, 6] 4-coloring of the edges of G based on π. Namely, G_0 consists of a single edge (v_1, v_2), which is assigned the black color. Assuming that a 4-coloring has been constructed for G_{k-1}, we extend it for G_k as follows. We first color the edges of G_k that do not belong to G_{k-1} and that are on the contour C_k. We color the first such edge encountered in a traversal of C_k from v_1 to v_2 as blue, the last one as green and all remaining ones (i.e., those having both endpoints in P_k) as black. Similar to the Schnyder coloring of maximal planar graphs, we assign the color red to the remaining edges of G_k that do not belong to G_{k-1} (i.e., those are incident to P_k and are not part of C_k). Note that the latter case only arises if P_k is a singleton (i.e., it consists of a single vertex).

For $k = 1, \ldots, m$, assume that a planar internally convex grid drawing Γ_{k-1} of G_{k-1} with edge-vertex resolution at least $\frac{1}{2}$ has been computed such that the following condition, called *contour condition*, holds: the edges of C_{k-1} are drawn as straight-line segments with slopes 0, -1 and in $[+1, \infty)$; in particular, the slope of each blue edge of C_{k-1} is at least 1, the slope of each black edge of C_{k-1} is 0, while the slope of each green edge of C_{k-1} is -1. Moreover, each vertex v in G_{k-1} has been associated with a so-called *shift-set* $S(v)$.

Our algorithm works as follows. Initially, vertices v_1 and v_2 of P_0 are placed at points $(0,0)$ and $(1,0)$, respectively (as in the original algorithm). For placing P_k, with $k = 1, \ldots, m$, our algorithm first shifts the vertices of Γ_{k-1} appropriately to guarantee that the obtained drawing Γ_k has edge-vertex resolution at least $\frac{1}{2}$. If P_k contains more than one vertex, one can prove that this is guaranteed already by the construction proposed in the original algorithm. The difficulties arise in the case where P_k is a singleton, where additional shifts are required. In particular, we can prove that, for each red edge incident to the singleton vertex, shifting one more unit the shift set $S(v)$ of a suitably defined *critical* vertex v in Γ_{k-1} is enough to achieve the desired edge-vertex resolution.

The proof that the resulting drawing Γ_k is internally convex and preserves the contour condition easily follows by construction. On the other hand, proving that Γ_k has edge-vertex resolution at least $\frac{1}{2}$ is non-trivial and requires a careful analysis of the geometric configuration of each face modified by the additional shifts. Finally, the proof that the final width (and hence height) is $n - 2 + a$ is based on a charging scheme aimed at showing that the total number of additional shifts is bounded by the minimum between the number of faces and the number of red edges of G.

References

1. Bekos, M.A., Gronemann, M., Montecchiani, F., Pálvölgyi, D., Symvonis, A., Theocharous, L.: Grid drawings of graphs with constant edge-vertex resolution. Comput. Geom. **98**, 101789 (2021). https://doi.org/10.1016/j.comgeo.2021.101789
2. Chrobak, M., Goodrich, M.T., Tamassia, R.: Convex drawings of graphs in two and three dimensions (preliminary version). In: Whitesides, S. (ed.) SoCG. pp. 319–328. ACM (1996). https://doi.org/10.1145/237218.237401
3. Chrobak, M., Kant, G.: Convex grid drawings of 3-connected planar graphs. Int. J. Comput. Geom. Appl. **7**(3), 211–223 (1997). https://doi.org/10.1142/S0218195997000144
4. Felsner, S.: Geometric Graphs and Arrangements. Advanced Lectures in Mathematics. Vieweg+Teubner Verlag (2004). https://doi.org/10.1007/978-3-322-80303-0
5. Kant, G.: Drawing planar graphs using the canonical ordering. Algorithmica **16**(1), 4–32 (1996). https://doi.org/10.1007/BF02086606
6. Schnyder, W.: Embedding planar graphs on the grid. In: SODA, pp. 138–148. SIAM (1990)

What Is R? A Graph Drawer's Perspective

Max Sondag[1]($\boxtimes$) , Cagatay Turkay[2] , Sibylle Mohr[3] , Louise Matthews[3] ,
Kai Xu[4] , and Daniel Archambault[1]

[1] Swansea University, Swansea, UK
{m.f.m.sondag,d.w.archambault}@swansea.ac.uk
[2] Warwick University, Warwick, UK
Cagatay.Turkay@warwick.ac.uk
[3] University of Glasgow, Glasgow, UK
{Sibylle.Mohr,Louise.Matthews}@glasgow.ac.uk
[4] Middlesex University, Middlesex, UK
K.Xu@mdx.ac.uk

1 Introduction

In 2020, the worldwide pandemic of COVID-19 had a profound impact on society. One of the most important metrics that is being used to investigate the effectiveness of various interventions to control the spread of the pandemic is R, the effective reproduction number. It is talked about extensively in the news, and a wide-ranging array of different interventions are put in place by governments with the aim of getting R below 1 to curb the spread of the disease. In this abstract, we will investigate what R means from a graph drawer's perspective and aim to open up interesting and relevant research avenues.

2 Defining R

To define R, we start with a temporal network [2] G of the contacts between people, for example all contacts within a city or country. Each node u represents a person, and a temporal edge $e_t = (u, v)$ represents that person u was in contact with person v at discrete time t. We overlay the disease we are interested in on G. Some nodes will be index cases, the initial cases where the disease emanates from. These nodes are exposed to the disease from outside the network (i.e. a pangolin or international travel). A person v who has become infectious at time t, has a chance to expose their neighbors over edges with a time greater than t until v is no longer infectious. After an incubation time, these neighbors become infectious in turn, and can propagate the disease further. This propagation through the network creates an infection map: A set of rooted directed trees $\mathcal{T}$, where each node v represents a person and has a value $e(v)$ which indicates when this node was exposed to the disease. Using this infection map, we can define R.

This work was supported by UKRI EPSRC grant EP/V033670/1 and EP/V054236/1.

H. C. Purchase and I. Rutter (Eds.): GD 2021, LNCS 12868, pp. 431–433, 2021.
https://doi.org/10.1007/978-3-030-92931-2

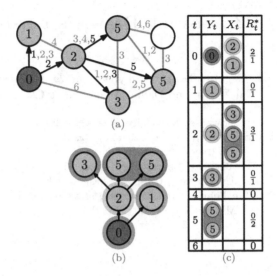

Fig. 1. Example calculation of R_t. (a) Contact graph with highlighted paths of infection. Values within nodes show exposure time $e(v)$. (b) The resulting infection map is a single infection tree. Nodes with the same exposure time have the same color. (c) R_t calculation table. R_t is the average over the 7 values of R_t^*: $(2+0+3+0+0+0+0)/7 \approx 0.71$

There are two different R's that are of interest to epidemiologists. R_0 (known as, "R naught" or "basic reproduction number"), which indicates the capacity for the disease to spread when the entire population is susceptible and no interventions are in place [1], and the R commonly used by the media: R_t ("R" or "effective reproduction number"), which takes interventions and immunity into account [3, 4]. The t in R_t is the time point of interest, often a day or a week, but is ultimately dependent on the characteristics of the disease and available data. In graph-theoretical terms, we can define R_t as follows. Let Y_t denote the nodes with $e(v) = t$. Let X_t be the set of children of Y_t in T. The R_t value for a single point in time is then: $R_t^* = \frac{|X_t|}{|Y_T|}$. Due to pragmatic issues such as reporting issues and weekly updates for policymakers, R_t is often averaged over a period of time (usually 7 days for COVID) to give the final R_t value. This calculation of R_t is shown in Fig. 1.

3   Open Problems in Real and Simulated Data

We briefly examine various (open) problems that are encountered when using this graph-based data. We first consider problems in the temporal network. In real data, both edges and nodes are missing from the graph. The time of contact could be incorrect, and is at best an approximation (often at day accuracy) even for simulated data. Continuing with problems in determining the spread of the disease over the network, the directionality of the edges is generally unknown. It is also not typically known for certain whether one nodes infected another, or if there was an outside influence. The time of exposure is often an estimation, as testing requires time, and the results might even be incorrect. For both real and simulated data, these data characteristics lead to the graph structure of the infection map being a large forest of small trees, with a few larger trees. While not all of these problems and characteristics are unique to epidemic graphs, we believe that there are nevertheless a number of interesting research questions present in this setting, and research from the Graph Drawing community could help in assisting with the current pandemic and mitigating future pandemics.

References

1. Anderson, R.M., May, R.M.: Infectious Diseases of Humans: Dynamics and Control. Oxford University Press, Oxford (1992)
2. Holme, P., Saramäki, J.: Temporal networks. Phys. Rep. **519**(3), 97–125 (2012)
3. Nishiura, H., Chowell, G.: The effective reproduction number as a prelude to statistical estimation of time-dependent epidemic trends. In: Mathematical and Statistical Estimation Approaches in Epidemiology, pp. 103–121. Springer, Dordrecht (2009). https://doi.org/10.1007/978-90-481-2313-1_5
4. Wallinga, J., Lipsitch, M.: How generation intervals shape the relationship between growth rates and reproductive numbers. Proc. R. Soc. B Biol. Sci. **274**(1609), 599–604 (2007)

Animating Disease Spread with Location Type

Kasper Krawczyk$^{(\boxtimes)}$ and Daniel Archambault

Swansea University, Swansea, UK
kasper.krawczyk@gmail.com, d.w.archambault@swansea.ac.uk

1 Introduction

One of the applications of graph drawing is visualising disease spread in a population. Specifically, event-based graph animation can help visualise contact tracing
model simulations. Building on the DynNoSlice algorithm [3, 4], we propose a
way of incorporating two types of relationships useful in epidemiology into a
dynamic graph animation: the vector-infected relationship as well as the 'location of infection' relationship. We focus on a COVID contact tracing model [2],
which provides information on an infected person's infection status, who they
were in contact with and the place or setting of infection. These characteristics
are accounted for in the presented algorithm, and encoded, respectively, with
colour (including saturation), insertion of edges (denoting either the location,
or the infection relationship), and grouping of nodes with using the 'cluster'
elements. Graphs are drawn synchronously on a $2D$ plane, and in a space-time
cube, with planes as time slices arranged perpendicularly to the time axis T.
Variations of the algorithm use graph forces in either $2D$ or $2D + T$, component
filtering, and node highlights to accentuate various aspects of the model.

2 Clusters and Cluster Forces

When drawing the temporal network in the space-time cube, we propose an
additional *cluster* graph element. It facilitates visual encoding of commonalities
between nodes. Related nodes become *members* of a cluster, attracted towards its
pole, represented as a location node in a $2D$ drawing (Fig. 1). Non-cluster-specific
forces in the algorithm follow the DynNoSlice algorithm [3, 4] and are in $2D + T$.
The presented algorithm introduces a number of new forces to position clusters
in the drawing, and which also are functions of regular nodes' relationships
to cluster pole nodes, other cluster member nodes and non-member nodes. A
member node, also marked out with an edge inserted between it and its pole, is
affected by the following forces:

1. A node-to-node attraction force is applied between the pole node and the
   cluster's member nodes.
2. A circumference repulsion force, itself a node-to-edge force, repelling nodes
   from an abstract polygon superimposed on the circumference, modelled after
   the PrEd algorithm [1].

© Springer Nature Switzerland AG 2021
H. C. Purchase and I. Rutter (Eds.): GD 2021, LNCS 12868, pp. 434–436, 2021.
https://doi.org/10.1007/978-3-030-92931-2

Infection status	Colour encoding
Exposed	
Asymptomatic	
Presymptomatic	
Symptomatic	
Severely symptomatic	
Recovered	
Dead	

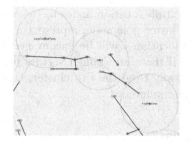

Fig. 1. Visualisation of graph topology and location of infection. Colour legend is on the left. Colours from green (exposed), through yellow (asymptomatic infection), orange (symptomatic infection) and red (severely symptomatic) encode various stages of infection progression. Blue encodes recovery, black - deceased persons. Three locations are shown in the drawing on the right (circles), with nodes in various stages of infection. The fully coloured version of the figure is available online. (Color figure online)

$$F(v, (a, b)) \longleftarrow \frac{(\gamma - \|p_v - v_e\|)^2}{\|p_v - v_e\|}(p_v - v_e)$$

$$v \in (a, b),\ a \neq v, b \neq v,\ \|p_v - v_e\| < \gamma$$

where p_x is the position of node x, γ is the desired distance from a node to an edge, and v_e is the position of node v on a line which is defined by the edge (a, b).

3. A node-to-pole repulsion force is applied to every non-member node (all nodes that are unassigned to a cluster).

$$F_u^r(v, u) \longleftarrow \left(\frac{\gamma}{\|p_v - v_e\|}\right)^8 \widehat{vu}$$

γ is the desired distance between two nodes, p_u and p_v are the positions of the member node and the pole node respectively, and $\widehat{vu}$ is the unit vector pointing in the direction from p_u to p_v. The exponent value of 8 was determined through experimentation on our data sets.

Variations of the algorithm allow for the pole nodes to be pinned to one location in *2D*, which then affects the space-time cube synchronised view. Then the algorithm generates forces that are different in *2D* and *2D + T*.

3 Functionalities

The implementation offers a range of options to highlight, filter or compare data set features from the perspective of location. Each of these options creates a dynamic graph animation corresponding to its space-time cube drawing. One of the algorithm variations is implemented as *Continuous with Multiple Locations Attraction*.

Given multiple location nodes n_i (where n stands for a cluster's index), each drawn as a cluster pole node, represent a given set of locations. All nodes associated with location i will be encouraged to stay within the i'th cluster's circumference. If there are enough vectors pulling node n_i in the attracted node's component away from the i'th cluster, there is a chance that n may be pulled out of the circumference.

References

1. Bertault, F.: A force-directed algorithm that preserves edge-crossing properties. Inf. Process. Lett. **74**(1–2), 7–13 (2000)
2. Mohr, S., et al.: Contact Tracing Model. https://github.com/ScottishCovid Response/Contact-Tracing-Model. Accessed 18 Sept 2020
3. Simonetto, P., Archambault, D., Kobourov, S.: Event-based dynamic graph visualisation. IEEE Trans. Visual. Comput. Graphics **26**(7), 2373–2386 (2020)
4. Simonetto, P., Archambault, D., Kobourov, S.: Drawing dynamic graphs without timeslices. In: Frati, F., Ma, K.-L. (eds.) GD 2017. LNCS, vol. 10692, pp. 394–409. Springer, Cham (2018). https://doi.org/10.1007/978-3-319-73915-1_31

Bitonicity and Upwardness

Splits, Bends and Their Relation to Area Requirements

Patrizio Angelini[1] , Michael A. Bekos[2] , Henry Förster[2(✉)] ,
and Martin Gronemann[3]

[1] Department of Mathematics and Applied Sciences,
John Cabot University, Rome, Italy
pangelini@johncabot.edu
[2] Department of Computer Science, University of Tübingen, Tübingen, Germany
{bekos,foersth}@informatik.uni-tuebingen.de
[3] Theoretical Computer Science, Osnabrück University, Osnabrück, Germany
martin.gronemann@uni-osnabrueck.de

In many practical applications, graphs are directed such that the direction of edges itself is an essential information. To convey this information to the reader, *upward* drawings are used. Namely, a planar drawing of a directed acyclic graph is *upward* if every edge (u, v) is drawn as a y-monotone curve, such that u lies below v; a graph is *upward planar* if it admits an upward planar drawing.

Here, we focus on *st*-planar graphs, i.e., directed acyclic planar graphs with a single source s and a single sink t. Since not all *st*-planar graphs admit upward planar straight-line drawings in polynomial area [3], one seeks for corresponding poly-line drawings with as few bends as possible. Gronemann [6] observed that every *st*-planar graph that admits a so-called bitonic *st*-ordering admits an upward planar straight-line drawing in quadratic area. Intuitively, a *bitonic st*-ordering forms a special type of *st*-ordering that takes the underlying embedding into account. More specifically, the successors of every vertex form an increasing and then decreasing subsequence (bitonic subsequence) in the *st*-ordering w.r.t. the embedding. While not every *st*-planar graph admits such an ordering, one may split certain edges to obtain one. Gronemann [6] proves that $n - 3$ splits are sufficient. In particular, splits are imposed by *forbidden configurations* [5].

It follows that every n-vertex *st*-planar graph admits an upward planar drawing in $O(n^2)$ area with at most one bend per edge and at most $n - 3$ bends in total [6]. If the maximum degree is 3, then the corresponding drawing is bendless [2]. Next, we prove that the bound on the total number of bends can be reduced from $n - 3$ to $\frac{n}{2}$ when the maximum degree is 5 using a technique of [7].

Theorem 1. *Let $G = (V, E)$ be an st-planar graph with n vertices and maximum degree 5. There is a set of edges $E' \subseteq E$ with $|E'| \leq \frac{n}{2}$ so that the graph G' obtained by splitting each edge in E' or the reversed graph $\tilde{G}'$ of G' is bitonic.*

Proof (sketch). Since a vertex $v \neq s$ has outdegree at most 4, v is a source of at most three internal faces. One belongs to all forbidden configurations with

© Springer Nature Switzerland AG 2021
H. C. Purchase and I. Rutter (Eds.): GD 2021, LNCS 12868, pp. 437–439, 2021.
https://doi.org/10.1007/978-3-030-92931-2

source v and its transitive edge can be split. Since an outdegree of at least 3 is required for v to be the source of a forbidden configuration in G, it cannot be a source of a forbidden configuration in the reversed graph $\tilde{G}$ of G. Hence, each vertex contributes at most one split in either G or $\tilde{G}$.   □

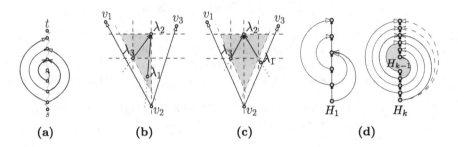

(a)   (b)   (c)   (d)

Fig. 1. (a) A 6-coil. (b–d) Illustration for the area requirement proof. In (a–c) V-shapes have empty triangular arrow heads while Λ-shapes have pointed arrow heads. In (d), different coils have different arrow heads.

When constructing upward planar drawings with the approach by Grone-mann [6], $n - 5$ bends may be required even for max. degree-6 graphs, while $\frac{n}{2} - 2$ bends may be required for max. degree 4 graphs. These limitations are tailored to the adopted approach, e.g., there are st-planar graphs that need a linear number of splits but admit bend-less upward planar drawings in linear area [6].

To investigate if these limitations are caused by the specific approach or are already imposed by the nature of upward planarity, we study lower bounds on the total number of bends under the polynomial-area requirement, independently of the required number of splits and of the allowed number of bends per edge. Our findings imply that the upper bounds on the number of bends obtained by the approach by Gronemann are worst-case almost tight, even if several bends per edge are allowed. A key ingredient is the notion of a k-coil, which is a path of k edges that spirals around an st-path; see Fig. 1a. A coil consists of alternating V- and Λ-shapes; red and blue in Fig. 1a, resp. A k-coil requires $\Omega(2^k)$ area in any upward planar straight-line drawing [4]. We extend this result as follows:

Lemma 1. *Let G be an st-plane graph containing a k-coil $\xi = \langle v_0, v_1, \ldots, v_k \rangle$. In any polyline upward planar drawing of G, ξ is drawn in $\omega(poly(k)) = \omega(2^{\log k})$ area unless $k/2 - O(\log k)$ edges of ξ have at least one bend.*

Proof (sketch). A V- or Λ-shape is *valid*, if it has no bend and the next bendless shape is of different type. We observe that a sequence of b bent edges along ξ invalidates at most $2b$ shapes, namely, the b edges are part of $b+1$ shapes and if b is even, there are two consecutive bendless shapes of the same type. Assume to the contrary that less than $k/2 - O(\log k)$ edges along ξ are bent. Then, there is an alternating sequence of $c = \omega(\log k)$ valid shapes. A valid V-shape will nest the next valid Λ-shape. We can prove via shearing that the V-shape covers at least 4 times the area that the Λ-shape covers; see Fig. 1b and 1c. As a result, ξ requires $\omega(4^c) = \omega(\text{poly}(k))$ area; a contradiction. □

Lemma 1 applied to the graph family in Fig. 1d and the lower bound construction in [1] yields the following:

Theorem 2. *There are infinitely many graphs that require $n - O(\log n)$ bent edges in any upward planar drawing of polynomial area. If the maximum degree is 4, the required number of bent edges is $n/2 - O(\log n)$.*

References

1. Angelini, P., Bekos, M.A., Förster, H., Gronemann, M.: Bitonic st-orderings for upward planar graphs: the variable embedding setting. In: Adler, I., Müller, H. (eds.) WG 2020. LNCS, vol. 12301, pp. 339–351. Springer, Cham (2020). https://doi.org/10.1007/978-3-030-60440-0_27
2. Bekos, M.A., Di Giacomo, E., Didimo, W., Liotta, G., Montecchiani, F.: Universal slope sets for upward planar drawings. In: Biedl, T., Kerren, A. (eds.) GD 2018. LNCS, vol. 11282, pp. 77–91. Springer, Cham (2018). https://doi.org/10.1007/978-3-030-04414-5_6
3. Battista, G.D., Tamassia, R., Tollis, I.G.: Area requirement and symmetry display of planar upward drawings. Discrete Comput. Geom. **7**(4), 381–401 (1992). https://doi.org/10.1007/BF02187850
4. Frati, F.: On minimum area planar upward drawings of directed trees and other families of directed acyclic graphs. Int. J. Comput. Geom. Appl. **18**(3), 251–271 (2008). https://doi.org/10.1142/S021819590800260X
5. Gronemann, M.: Bitonic st-orderings of biconnected planar graphs. In: Duncan, C., Symvonis, A. (eds.) GD 2014. LNCS, vol. 8871, pp. 162–173. Springer, Heidelberg (2014). https://doi.org/10.1007/978-3-662-45803-7_14
6. Gronemann, M.: Bitonic st-orderings for upward planar graphs. In: Hu, Y., Nöllenburg, M. (eds.) GD 2016. LNCS, vol. 9801, pp. 222–235. Springer, Cham (2016). https://doi.org/10.1007/978-3-319-50106-2_18
7. Rettner, C.: Flipped Bitonic st-Orderings of Upward Plane Graphs. Bachelor's thesis, University of Würzburg (2019)

Mixed Metro Maps with User-Specified Motifs

Tobias Batik[1]([📧]) [ID], Soeren Nickel[1] [ID], Martin Nöllenburg[1] [ID],
Yu-Shuen Wang[2] [ID], and Hsiang-Yun Wu[1] [ID]

[1] TU Wien, Vienna, Austria
e11701221@student.tuwien.ac.at,
{soeren.nickel,noellenburg}@ac.tuwien.ac.at, hsiang.yun.wu@acm.org
[2] National Chiao Tung University, Hsinchu, Taiwan
yushuen@cs.nctu.edu.tw

Abstract. In this poster, we propose an approach to generalize mixed metro map layouts with user-defined shapes for route-finding and advertisement purposes. In a mixed layout, specific lines are arranged in an iconic shape, and the remaining are in octilinear styles. The shape is expected to be recognizable, while the layout still fulfilling the classical octilinear design criteria for metro maps. The approach is in three steps, where we first search for the best fitting edge segment that approximates the guide shape and utilize least squares optimization to synthesize the layout automatically.

Keywords: Metro maps · Least squares optimization ·
Shape-preserving Dijkstra algorithm · Fréchet distance

1 Introduction

Metro maps are schematic representations of a metro network in order to facilitate effective route findings. Creating such maps manually is a challenging task, and thus several automatic approaches have been developed in the last two decades [10]. Most approaches aim for a single style, e.g., octolinear layouts (based on a set of slopes of multiples of 45°) [7, 8] or curvilinear layouts [3] (lines represented as smooth continuous curves). However, handcrafted metro maps often employ more than one style in a single map, to emphasize specific lines in the system, where some parts are arranged in an iconic shape and some in an octolinear style. The official Moscow metro map with its characteristic circles is a typical example [6]. Moreover, such maps could be used for advertising purposes or special events [11].

In this poster, we present an approach to generalize a mixed layout problem with a user-defined *guide shape*, represented as a polyline, as input. To guarantee the visual quality, the guide shape should be recognizable in the final layout, while the layout still fulfills the classical octolinear design criteria [7].

We acknowledge funding by the Austrian Science Fund (FWF) under grant P 31119.

H. C. Purchase and I. Rutter (Eds.): GD 2021, LNCS 12868, pp. 440–443, 2021.
https://doi.org/10.1007/978-3-030-92931-2

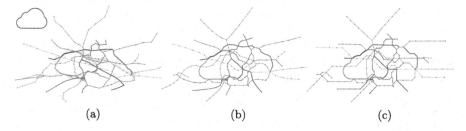

Fig. 1. Paris metro map, consisting of 236 stations and 268 edges. The layout is created with a cloud-shaped guide. (a) Geographically accurate layout with a potential path P highlighted. The input guide shape is shown on the top left. (b) The smooth layout and (c) the mixed layout with smooth edges highlighted.

2   Method

In our approach, we take a metro graph $G = (V, E)$ with geographic coordinates and a user-defined (open or closed) polyline L as input. This line $L = [l_0, l_1, \ldots, l_n]$ with line segments (l_i, l_{i+1}) serves as the user-specified guide shape for our mixed layout. Our approach is in three steps. First, the approach searches for the path that is the best representative of the user-defined guide shape. Then, we deform the metro graph to a smooth and finally to a mixed layout following the aforementioned design criteria.

To find a good alignment, we adapt the shape-preserving Dijkstra algorithm [5], which allows us to iteratively find potential paths $P = (v_i, v_j, \ldots)$ in G that are similar to the input guide shape L (Fig. 1(a)). In contrast to the original approach [4, 5], we quantify the difference between the path P and L using the Direction-Based Fréchet Distance [1]. After finding a good path P in the metro graph, we scale the guide shape and align the bounding boxes of the found path and the input guide shape for better deformation.

Our deformation is inspired by the approach of Wang et al. [9]. We first create a smooth layout (Fig. 1(b)), and then the mixed layout (Fig. 1(c)). Here, we optimize four constraints $\Omega = w_l\Omega_l + w_a\Omega_a + w_p\Omega_p + w_c\Omega_c$. Ω_l enforces uniform edge lengths, Ω_a maximizes angular resolutions, and Ω_p minimizes the distance of the position of a metro station $v \in V$ to its geographical location. In addition, Ω_c forces the layout to approximate the guide shape. This constraint pushes a station $v_i \in V$ in the direction of the guide shape L, if v_i is a close candidate that should be selected for shape approximation. More precisely, v_i will be moved toward the direction of point p_i, which is a projected point of v_i on the polyline L after bounding box alignment. To create a clear layout, we approximate each small segment of the guide shape with one single metro edge and, therefore, apply Ω_c only to a subset of the vertices. Let v_i' be a copy of v_i, which has been rotated around p_i by $180°$. To determine if the constraint Ω_c should be applied to the vertex $v_i \in V$, we check if any edge $e \in E$ is located between v_i and its projection p_i or between v_i' and p_i. If no such edge is found, the constraint Ω_c is added for this vertex.

442   T. Batik et al.

To create the mixed layout, we differentiate between two types of edges. Smooth edges (highlighted in Fig. 1(c)) approximate the shape and should align to a section of the guide shape L. The remaining edges are octolinear edges, which have to be drawn with an octolinear slope. We determine if an edge $(v_i, v_j) \in E$ is a smooth edge or not by checking if another edge is located between the vertices v_i and v_j and the guide shape L, as we do in the smoothing step. Similar to the smoothing step, we again create the mixed layout using the least squares optimization [9].

3 Results

Our approach creates visually pleasing and interesting results. The tested shapes are embedded well as shown in Fig. 1(c). Note that in our results, the guide shape is not approximated by a single connected line but multiple line segments in the metro system. Following the Gestalt Laws (Law of Closure, Law of Prägnanz) [2], we confirm that even paths with multiple discontinuities (resulting from the topology of the metro system) can be perceived and recognized by humans. In our experiments, the relative positions of vertices in the metro graph are not changed and no stations are located too close to another edge or station - making the final layout not only visually interesting but also usable for navigation purposes.

References

1. de Berg, M., Cook, A.F.: Go with the flow: the direction-based Fréchet distance of polygonal curves. In: Marchetti-Spaccamela, A., Segal, M. (eds.) TAPAS 2011. LNCS, vol. 6595, pp. 81–91. Springer, Heidelberg (2011). https://doi.org/10.1007/978-3-642-19754-3_10
2. Ellis, W.D.: A Source Book of Gestalt Psychology. Routledge, Abingdon (2013). https://doi.org/10.1037/11496-000
3. Fink, M., Haverkort, H., Nöllenburg, M., Roberts, M., Schuhmann, J., Wolff, A.: Drawing metro maps using Bézier curves. In: Didimo, W., Patrignani, M. (eds.) GD 2012. LNCS, vol. 7704, pp. 463–474. Springer, Heidelberg (2013). https://doi.org/10.1007/978-3-642-36763-2_41
4. Funke, S., Schirrmeister, R., Skilevic, S., Storandt, S.: Compass-based navigation in street networks. In: Gensel, J., Tomko, M. (eds.) W2GIS 2015. LNCS, vol. 9080, pp. 71–88. Springer, Cham (2015). https://doi.org/10.1007/978-3-319-18251-3_5
5. Funke, S., Storandt, S.: Path shapes: an alternative method for map matching and fully autonomous self-localization. In: Advances in Geographic Information Systems (SIGSPATIAL 2011), pp. 319–328. ACM (2011). https://doi.org/10.1145/2093973.2094016
6. Moskovsky Metropoliten: Moscow metro map and central circle (2019). https://mosmetro.ru/mcc/moscow-central-circle/. Accessed 08 Mar 2021
7. Nöllenburg, M., Wolff, A.: Drawing and labeling high-quality metro maps by mixed-integer programming. IEEE Trans. Vis. Comput. Graph. **17**(5), 626–641 (2010). https://doi.org/10.1109/TVCG.2010.81
8. Onda, M., Moriguchi, M., Imai, K.: Automatic drawing for Tokyo metro map. In: European Workshop on Computational Geometry (EuroCG 2018), pp. 62:1–62:6 (2018)

9. Wang, Y.S., Chi, M.T.: Focus+context metro maps. IEEE Trans. Vis. Comput. Graph. **17**(12), 2528–2535 (2011). https://doi.org/10.1109/TVCG.2011.205
10. Wu, H.Y., Niedermann, B., Takahashi, S., Roberts, M.J., Nöllenburg, M.: A survey on transit map layout from design, machine, and human perspectives. In: Computer Graphics Forum (Special Issue of EuroVis 2020), vol. 39, No. 3, June 2020. https://doi.org/10.1111/cgf.14030
11. ZERO PER ZERO: City railway system (2021). https://www.zeroperzero.com/city-railway-system. Accessed 25 Aug 2021

grApp

Ivelin Nikolaev Ivanov[1] and Jens M. Schmidt[2(✉)]

[1] TU Ilmenau, Ilmenau, Germany
[2] Institute for Algorithms and Complexity, Hamburg University of Technology,
Hamburg, Germany
jens.m.schmidt@tuhh.de

Abstract. We propose the software *grApp*, which is a web-based open-source editor for graphs. Its main feature is a smoothly animated visualization, which allows users to modify and analyze the graphs interactively.

Keywords: Smooth interactive graph visualization · Open-source · Graph6 · d3.js · NetworkX

1 Why Do We Need grApp?

When it comes to displaying graphs, there is an abundance of graph visualization tools available today, from which users may choose. E.g., there are commercial graph editors like *Microsoft Visio* and *yEd*, some of which are freely available but not open-source, there are graph drawing libraries like *NetworkX* [2] and *OGDF* [1], which are open-source but do not provide build-in interactive graph visualization, and there are the widely known web-based data libraries *d3.js* (https://d3js.org/) and *vis.js* (https://visjs.org/), which provide fast and smooth animation of data, but usually lack graph-specific functionalities standard graph editors have.

We therefore aim to combine the features of *d3.js* and *NetworkX* in order to provide an open-source and web-based graph editor that allows smoothly animated and interactive graph visualization.

The official websites of grApp are https://www3.tuhh.de/e11/schmidt/grApp/ and https://github.com/iviv62/grapp.

2 Features

In a nutshell, grApp supports the following features:

- Smooth and interactive graph visualization and modification in 2D and 3D
- Fully web-based and thus platform-independent
- Open-source
- Interface for user-created executables (interacting with the graph via .gml)

H. C. Purchase and I. Rutter (Eds.): GD 2021, LNCS 12868, pp. 444–446, 2021.
https://doi.org/10.1007/978-3-030-92931-2

– Import, export and conversion of 10 different graph formats
– Graph6 webinterface

In more detail, we build upon the force-based layout implementation used in *d3.js*, upon graph import and layout algorithms used in *NetworkX* and many other open-source third-party tools like e.g. *Django*. In order to display the graph, the layouting process of *d3.js* is performed continuously. This allows for a smooth animation of any interaction with the graph (like moving around vertices or deleting edges) by which we mean that there are many frames per second, for which the difference between the object coordinates in these frames is very small. The user may also select and define clusters of vertices with different specifications regarding the forces. This way vertices may be specified as having a fixed position, which means no forces will be applied (they are "pinned"), while others repel or attract each other.

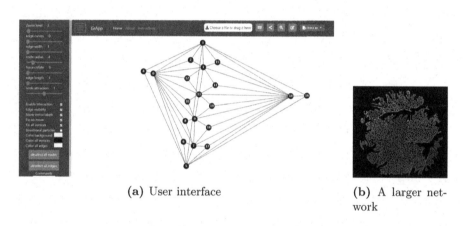

(a) User interface

(b) A larger network

Fig. 1. The graphical user interface.

grApp supports user-made binaries that may interact with the displayed graph and its current layout; here, the interface to the graph is provided by exchanging a .gml-specification of the graph, which the binary has to read in and output again (after possibly modifying it). The user may also manually modify all details of the current graph or its current visualization using the build-in JSON Editor. In addition, it is possible to share the abstract graph that is currently displayed as a web-link by using the share button on the toolbar (see Fig. 1a); then a web-link with the graph encoded in graph6-format will be generated and is available for anyone that has this link as long as the grApp server is online.

The displayed graph can be modified (creating and deleting vertices and edges and their visualization properties) and interacted with via mouse and key commands. A detailed manual of these commands is available on the grApp webpage on the *Instructions* page.

References

1. Chimani, M., Gutwenger, C., Jünger, M., Klau, G.W., Klein, K., Mutzel, P.: The open graph drawing framework (OGDF). In: Tamassia, R. (ed.) Handbook of Graph Drawing and Visualization, Chap. 17. CRC Press (2014). http://www.ogdf.net
2. Hagberg, A.A., Schult, D.A., Swart, P.J.: Exploring network structure, dynamics, and function using NetworkX. In: Varoquaux, G., Vaught, T., Millman, J. (eds.) Proceedings of the 7th Python in Science Conference, pp. 11–15 (2008)

Crossing Numbers of Beyond-Planar Graphs Revisited

Nathan van Beusekom⬦, Irene Parada⬦, and Bettina Speckmann$^{(\boxtimes)}$⬦

TU Eindhoven, Eindhoven, The Netherlands
{n.a.c.v.beusekom,i.m.de.parada.munoz,b.speckmann}@tue.nl

Graph drawing beyond planarity focuses on drawings of high visual quality for non-planar graphs which are characterized by certain forbidden edge configurations. A natural criterion for the quality of a drawing is the number of edge crossings. While empirical studies suggest that the number of crossings is not the only factor that influences human understanding of a drawing, it is nevertheless one of the most relevant aesthetic indicators [4–7]. However, it is NP-hard to compute the minimum number of crossings of a graph G across all possible drawings of G [3]. This minimum number of crossings is also referred to as the *crossing number* of G, denoted by $\mathrm{cr}(G)$.

In recent years there has been particular interest in drawings of *beyond-planar* graphs, which are characterized by certain forbidden crossing configurations of the edges; see the recent survey by Didimo *et al.* [2]. The question then arises whether beyond-planar drawings have a significantly larger crossing number than unrestricted drawings. The *restricted crossing number* $\mathrm{cr}_{\mathcal{F}}(G)$ of a graph G is the minimum number of crossings required to draw G such that the drawing satisfies the restrictions imposed by the beyond-planar family $\mathcal{F}$. For a beyond-planar family $\mathcal{F}$, the *crossing ratio* $\varrho_{\mathcal{F}}$ is defined as $\varrho_{\mathcal{F}} = \sup_{G \in \mathcal{F}} \mathrm{cr}_{\mathcal{F}}(G)/\mathrm{cr}(G)$, that is, the supremum over all graphs in $\mathcal{F}$ of the ratio between the restricted crossing number and the (unrestricted) crossing number.

Results. Chimani *et al.* [1] gave bounds on the crossing ratio for 1-planar, quasi-planar, and fan-planar graphs. Their 1-planarity bound also applies to k-planarity when allowing parallel edges. We show that parallel edges are not needed when extending their proof to k-planarity. To do so, we introduce the concept of k-*planar compound edges*, which exhibit essentially the same behavior as k parallel edges. We further show how to extend the proof constructions of Chimani *et al.* for quasi-planar and fan-planar graphs to two additional classes of beyond-planar graphs. These constructions use the concept of ℓ-compound edges; we also show how to use such ℓ-compound edges to prove lower bounds for two further families of beyond-planar graphs. Finally, we introduce the concept of ℓ-*bundles* which allow us to prove tight bounds on the crossing ratio for families of graphs where these bounds are quadratic in the number n of vertices. All bounds are summarized in Table 1. Bounds that are tight for a fixed k are indicated in boldface. For our upper bounds we assume the drawings to be *simple*, that is, two edges share at most one point. All bounds also apply to straight-line drawings.

Full version: https://arxiv.org/abs/2105.12452

© Springer Nature Switzerland AG 2021
H. C. Purchase and I. Rutter (Eds.): GD 2021, LNCS 12868, pp. 447–449, 2021.
https://doi.org/10.1007/978-3-030-92931-2

Table 1. Bounds for the crossing ratio.

Family	Forbidden configurations		Lower	Upper
k-planar	An edge crossed more than k times	$k = 2$	$\Omega(n/k)$	$O(k\sqrt{k}n)$
k-quasi-planar	k pairwise crossing edges	$k = 3$	$\Omega(n/k^3)$ [1]	$f(k)n^2\log^2 n$ [1]
Fan-planar	Two independent edges crossing a third or two adjacent edges crossing another edge from different "sides"		$\Omega(n)$ [1]	$O(n^2)$ [1]
(k, l)-grid-free	Set of k edges such that each edge crosses each edge from a set of l edges.	$k, l = 2$	$\Omega\left(\dfrac{n}{kl(k+l)}\right)$	$g(k,l)n^2$
k-gap-planar	More than k crossings mapped to an edge in an optimal mapping	$k = 1$	$\Omega(n/k^3)$	$O(k\sqrt{k}n)$
Skewness-k	Set of crossings not covered by at most k edges	$k = 1$	$\Omega(n/k)$	$O(kn + k^2)$
k-apex	Set of crossings not covered by at most k vertices	$k = 1$	$\Omega(n/k)$	$O(k^2n^2 + k^4)$
Planarly connected	Two crossing edges that do not have two of their endpoint connected by a crossing-free edge		$\Omega(n^2)$	$O(n^2)$
k-fan-crossing-free	An edge that crosses k adjacent edges	$k = 2$	$\Omega(n^2/k^3)$	$O(k^2n^2)$
Straight-line RAC	Two edges crossing at an angle $< \frac{\pi}{2}$		$\Omega(n^2)$	$O(n^2)$

References

1. Chimani, M., Kindermann, P., Montecchiani, F., Valtr, P.: Crossing numbers of beyond-planar graphs. In: Archambault, D., Tóth, C.D. (eds.) GD 2019. LNCS, vol. 11904, pp. 78–86. Springer, Cham (2019). https://doi.org/10.1007/978-3-030-35802-0_6
2. Didimo, W., Liotta, G., Montecchiani, F.: A survey on graph drawing beyond planarity. ACM Comput. Surv. **52**(1), 1–37 (2019). https://doi.org/10.1145/3301281
3. Garey, M., Johnson, D.S.: Crossing number is NP-complete. SIAM J. Algebraic Discrete Methods **4**(3), 312–316 (1983). https://doi.org/10.1137/0604033
4. Kobourov, S.G., Pupyrev, S., Saket, B.: Are crossings important for drawing large graphs? In: Duncan, C., Symvonis, A. (eds.) GD 2014. LNCS, vol. 8871, pp. 234–245. Springer, Heidelberg (2014). https://doi.org/10.1007/978-3-662-45803-7_20
5. Purchase, H.: Which aesthetic has the greatest effect on human understanding? In: DiBattista, G. (ed.) GD 1997. LNCS, vol. 1353, pp. 248–261. Springer, Heidelberg (1997). https://doi.org/10.1007/3-540-63938-1_67
6. Purchase, H.C., Cohen, R.F., James, M.I.: An experimental study of the basis for graph drawing algorithms. J. Exp. Algorithmics **2**, 1–17 (1997). https://doi.org/10.1145/264216.264222
7. Ware, C., Purchase, H., Colpoys, L., McGill, M.: Cognitive measurements of graph aesthetics. Inf. Visual. **1**, 103–110 (2002). https://doi.org/10.1057/palgrave.ivs.9500013

Compact 3D Grid Drawings of Trees

Irene Parada[ORCID], Kees Voorintholt[✉], and Zeyu Zhang

Eindhoven University of Technology, Eindhoven, The Netherlands
i.m.de.parada.munoz@tue.nl, {k.d.voorintholt,z.zhang3}@student.tue.nl

Abstract. We show that perfect binary trees with $n - 1$ vertices can be optimally embedded in 3D, that is, they admit a straight-line drawing on a $\sqrt[3]{n}$ by $\sqrt[3]{n}$ by $\sqrt[3]{n}$ grid without intersecting edges. To show it, we adapt a recursive approach used in 2D by Akitaya *et al.* [GD'18] to construct compact embeddings of perfect binary trees on a square grid.

Recently, Dujmović *et al.* [3] showed that planar graphs admit a 3D straight-line grid drawing in linear volume. This was previously shown for planar graphs of bounded degree [2]. The bound in particular applies to perfect binary trees. In this work we show that this class of graphs in fact admits a *compact* embedding in 3D with optimal aspect ratio. In a compact embedding all points of a given grid except for maybe one are used. We follow a similar strategy to that in Akitaya *et al.* [1]. The authors recursively construct two types of compact embeddings of perfect binary trees in 2D.

Our construction in 3D uses three recursive blocks F_k, G_k, and H_k. Every block is a compact embedding of a perfect binary tree T_k of height k, where on one grid point the root of T_k is located and one grid point is empty and does not correspond to any vertex of T_k. The critical difference between the blocks is the placement of the empty node with respect to the root node. The coordinates of these nodes in the three blocks can be found in Table 1.

Table 1. Root node and free node placement for F, G, and H.

F_k	**Root node:** $(2 \cdot 2^{(k-2)/3} - 1, 2^{(k-2)/3} - 1, 2^{(k-2)/3} - 1)$ Empty node: $(2 \cdot 2^{(k-2)/3} - 1, 2^{(k-2)/3}, 2^{(k-2)/3})$	
G_k	**Root node:** $(2 \cdot 2^{(k-2)/3} - 1, 2^{(k-2)/3} - 1, 2^{(k-2)/3} - 1)$ Empty node: $(0, 0, 2 \cdot 2^{(k-2)/3} - 1)$	
H_k	**Root node:** $(2 \cdot 2^{(k-2)/3} - 1, 2^{(k-2)/3} - 1, 2^{(k-2)/3} - 1)$ Empty node: $(2 \cdot 2^{(k-2)/3} - 1, 2 \cdot 2^{(k-2)/3} - 1, 2 \cdot 2^{(k-2)/3} - 1)$	

The trivial blocks correspond to drawings of a perfect binary tree of height 2 in a $2 \times 2 \times 2$ grid. Each non-trivial block can be recursively constructed from eight

© Springer Nature Switzerland AG 2021
H. C. Purchase and I. Rutter (Eds.): GD 2021, LNCS 12868, pp. 450–452, 2021.
https://doi.org/10.1007/978-3-030-92931-2

smaller blocks, which are then connected by non-intersecting edges to complete the drawing of the perfect binary tree. More precisely, to draw a perfect binary tree T_k with height k, the eight subtrees of height $k - 3$ are recursively drawn using sub-blocks and the top three levels of T_k carefully connect the sub-blocks.

We next present the recursive definition of the three types of blocks.

The eight sub-blocks are listed from bottom to top ($-z$ to $+z$), from front to back ($-y$ to $+y$), and from left to right ($-x$ to $+x$).

Block F_k. (see Fig. 1 (left)).
Sub-blocks: F_{k-3}, G_{k-3} (rotated π around the z-axis), F_{k-3}, H_{k-3} (rotated π around the z-axis), F_{k-3}, H_{k-3} (rotated π around the z-axis and rotated π around the x-axis), F_{k-3}, and G_{k-3} (rotated π around the z-axis and rotated π around the x-axis).

Block G_k. (see Fig. 1 (middle)).
Sub-blocks: F_{k-3}, G_{k-3} (rotated π around the z-axis), F_{k-3}, H_{k-3} (rotated π around the z-axis), G_{k-3}, F_{k-3} (rotated π around the z-axis), H_{k-3} (rotated π around the x-axis), and F_{k-3} (rotated π around the z-axis).

Block H_k. (see Fig. 1 (right)).
Sub-blocks: F_{k-3}, G_{k-3} (rotated π around the z-axis), F_{k-3}, H_{k-3} (rotated π around the z-axis), F_{k-3}, H_{k-3} (rotated π around the z-axis and rotated π around the x-axis), F_{k-3}, and G_{k-3} (rotated π around the z-axis).

The connections (top three levels of the tree) are all similar in all blocks: the third level connects the root nodes of the sub-blocks with the empty node of the same or an adjacent F sub-block; the second level connects these empty nodes with the two empty nodes of the two H sub-blocks on a different yz plane. Finally, these two nodes are connected to the empty node of a G sub-block.

All blocks define straight-line drawings without crossings. To see it is enough to focus on the edges that are not recursively defined as part of a sub-block. The projection of these edges on the yz plane defines a crossing-free drawing. Moreover, these edges either connect the root and empty nodes of an F sub-block or they connect the two sides of one of the three planes that split the sub-blocks. From these observations it follows that each of the blocks F_k, G_k, and H_k defines a crossing-free compact drawing of the perfect binary tree of height k.

Fig. 1. Recursive definitions of the blocks F (left), G (middle), and H (right). (For extra clarity, the online colored version of the paper depicts the different types of sub-blocks with different colors.) (Color figure online)

Theorem 1. *The perfect binary tree with height* $k = 3x - 1$ *for* $x \in \mathbb{N}$ *with* $n - 1 = 2^{k+1} - 1$ *vertices has a compact embedding on the* $\sqrt[3]{n} \times \sqrt[3]{n} \times \sqrt[3]{n}$ *grid.*

References

1. Akitaya, H.A., Löffler, M., Parada, I.: How to fit a tree in a box. In: Biedl, T., Kerren, A. (eds.) GD 2018. LNCS, vol. 11282, pp. 361–367. Springer, Cham (2018). https://doi.org/10.1007/978-3-030-04414-5_26
2. Bekos, M.A., et al.: Planar graphs of bounded degree have bounded queue number. SIAM J. Comput. **48**(5), 1487–1502 (2019). https://doi.org/10.1137/19M125340X
3. Dujmović, V., Joret, G., Micek, P., Morin, P., Ueckerdt, T., Wood, D.R.: Planar graphs have bounded queue-number. J. ACM **67**(4), 22:1-22:38 (2020). https://doi.org/10.1145/3385731

A New Heuristic for Rectilinear Crossing Minimization

François Doré[(✉)] and Enrico Formenti

Université Côte d'Azur, CNRS, I3S, Nice, France
{francois.dore,enrico.formenti}@univ-cotedazur.fr

Abstract. A new heuristics for rectilinear crossing minimization is proposed. It is based on the idea of iteratively repositioning nodes after a first initial graph drawing. The new position of a node is computed by casting *rays* from the node towards graph edges. Each ray receives a mark and the one with the best mark determines the new position.

The heuristic has interesting performances when compared to the best competitors which can be found in classical graph drawing libraries like Tulip [1].

Keywords: Rectilinear crossing minimization · Graph drawing · Computational geometry

Problem Statement. The problem of graph design has been studied for more than sixty years producing an impressive number of publications. Several criteria are commonly accepted as characterizing a *good* graph drawing, such as the angle resolution, the distribution of vertices in the plane and the number of edge crossings. In this work, we focus on the last one. Indeed, we search for a drawing which minimizes the number of crossings when the edges are drawn as straight lines. We call this problem the rectilinear crossing minimization problem.

Remark that the problem of determining the minimal crossing number of a generic graph is complete for the existential theory over the reals and hence its complexity (in the classical setting) is somewhere between NP and PSPACE [4]. This fact motivates the introduction of heuristics which try to find a good compromise between exact solutions and reasonable computational time. Our work follows this direction.

The New Heuristic. A rectilinear drawing (or embedding) D of a graph $G = \langle V, E \rangle$ is an injective mapping from V to $\mathbb{R}^2$. The algorithm starts with an initial rectilinear drawing (all vertices positioned on a circle for example) and then iteratively tries to change the vertex positions to decrease the number of crossings.

The (potentially) new position p of a vertex v is obtained by casting a number R of *rays* from $D(v)$ (the actual position of v). A ray segment r_i is a half-line defined by a position p_i and a direction d_i. These two values determine a new point p_{i+1}, which is the first intersection with an edge of G encountered by r_i, or with one of the four false edges which delimits the bounding box of the initial

© Springer Nature Switzerland AG 2021
H. C. Purchase and I. Rutter (Eds.): GD 2021, LNCS 12868, pp. 453–454, 2021.
https://doi.org/10.1007/978-3-030-92931-2

drawing, if it intersects none of the edges. Next, a suitable score function decides if r_i crosses this edge (and hence $d_{i+1} = d_i$) or reflects on it like if the edge acted like a mirror (in this case d_{i+1} is the incident angle of reflection). Repeating this routine provides a sequence of $p_0, \ldots, p_{n_r}$ where $p_0 = D(v)$, the p_i are obtained by applying the routine i times and $n_r - 1$ is the number of ray segments. Finally, p is the middle between p_{n_r-1} and p_{n_r}.

Each *ray* r determines a position $p(r)$. To update $D(v)$, we compare all the positions obtained on two criteria: first, the number of crossings induced by the movement of v to $p(r)$; second, the sum of the squared norms of the Hooke's law forces applied to the edges of v, when positioned on $p(r)$. If none of the evaluated positions gives better results then $D(v)$ is kept unchanged and we process another vertex. The algorithm keeps trying to move vertices until no ray offers a position which significantly improves the above criteria.

Complexity. To move one node v the algorithm runs in $O(kERn_r)$, where E is the number of edges and k is the degree of v. At each iteration, it runs through the nodes until one can be moved to a better location. This leads to a total complexity of $O(E^2Rn_r)$ per iteration.

Experiments. Our heuristic has been tested using the datasets provided on the site of the International Symposium on Graph Drawing. Namely, we used graphs from three classes: *NORTH, ROME* and *DAG*. For each class, two parametrizations of our algorithm were compared with the best rectilinear algorithm of OGDF implemented in Tulip: stress minimization. Regardless of the class, our algorithm presents significantly less crossings. Moreover, it finds more planar drawings if the given graph is planar, at least for the *NORTH* and *ROME* classes. More experiments are needed to better assess the performance of our heuristics especially for large and dense graphs. In particular, it would be intesting to compare them to similar approaches based on the same principle of vertex movement [2, 3].

References

1. Auber, D., et al.: Tulip 5. In: Alhajj, R., Rokne, J. (eds.) Encyclopedia of Social Network Analysis and Mining. Springer, New York (2017). https://doi.org/10.1007/978-1-4614-7163-9_315-1
2. Radermacher, M., Reichard, K., Rutter, I., Wagner, D.: Geometric heuristics for rectilinear crossing minimization. ACM J. Exp. Algorithm. **24**, 1–21 (2019). https://doi.org/10.1145/3325861
3. Radermacher, M., Rutter, I.: Geometric crossing-minimization - a scalable randomized approach. In: Bender, M.A., Svensson, O., Herman, G. (eds.) 27th Annual European Symposium on Algorithms (ESA 2019). LIPIcs, vol. 144, pp. 76:1–76:16. Schloss Dagstuhl-Leibniz-Zentrum fuer Informatik, Dagstuhl (2019). https://doi.org/10.4230/LIPIcs.ESA.2019.76
4. Schaefer, M.: Complexity of some geometric and topological problems. In: Eppstein, D., Gansner, E.R. (eds.) GD 2009. LNCS, vol. 5849, pp. 334–344. Springer, Heidelberg (2010). https://doi.org/10.1007/978-3-642-11805-0_32

Drawing Clusterings of Bipartite Graphs

Thibault Marette[1(✉)] and Stefan Neumann[2]

[1] Ecole Normale Supérieure de Lyon, Lyon, France
thibault.marette@ens-lyon.fr
[2] KTH Royal Institute of Technology, Stockholm, Sweden
neum@kth.se

Abstract. Finding clusters in bipartite graphs is a popular problem in many communities such as data mining or bioinformatics. Drawing these clusters is simple as long as the clusters are disjoint but when the clusters share vertices then the task becomes more complicated. We formalize the problem of drawing *a given clustering* of overlapping clusters in bipartite graphs and we formulate an objective function that captures how well the clustering is visualized. We also present algorithms that optimize our objective function and we evaluate our methods on real-world datasets.

In many data science tasks the goal is to understand the relationship of two different sets of entities. This problem is often modelled using bipartite graphs where the two sides of the bipartite graph correspond to the two sets of different entities. An edge indicates that two entities interact with each other (e.g., a customer has bought a product). Many (bi-)clustering algorithms try to find groups of related entities by finding dense subgraphs in the bipartite graph.

The development of algorithms for finding such clusterings has been an active research area for decades [2, 4, 6]. However, these algorithms typically return their solutions in a format that is not human-readable. Therefore, inspecting the results can be difficult and visualization tools can be helpful.

Computing such a visualization is simple as long as all clusters are *disjoint*, i.e., they do not share any vertices. However, when the clusters *overlap*, i.e., some vertices are contained in multiple clusters, the visualization becomes more difficult [8]. The main challenge is that, due to the overlap of the clusters, we might be forced to draw some clusters non-consecutively and the visualization needs to make a choice which clusters should be split up. Such overlapping clusters occur, for instance, in data mining [4, 5] and in bioinformatics [3].

In our work, we develop novel algorithms for drawing *a given clustering* of a bipartite graph even when the clusters overlap. This is in contrast to seriation algorithms [1, 9] which also allow us to find clusters in graphs but they cannot visualize a clustering that they obtain as input. Unlike many leaf-ordering approaches, we do not assume access to a (dis-)similarity function for the clusters or for the rows and columns of the input matrix; we elaborate on this below.

This work was supported by the ERC Advanced Grant REBOUND (834862) and the EC H2020 RIA project SoBigData++ (871042).

© Springer Nature Switzerland AG 2021
H. C. Purchase and I. Rutter (Eds.): GD 2021, LNCS 12868, pp. 455–457, 2021.
https://doi.org/10.1007/978-3-030-92931-2

Fig. 1. Two visualizations of the same dataset using different clusterings obtained from the PCV algorithm [6] (left) and the Basso algorithm [4] (right).

In our visualization we decide to draw the (bi-)adjacency matrix of the bipartite graph. Since the clusters represent dense subgraphs within the bipartite graph, each cluster induces a dense area of 1-entries in the adjacency matrix. In Fig. 1 we visualize the outputs of two different clustering algorithms that were run on the *same* dataset. The figure suggests that the left method finds smaller, denser clusters, whereas the right method finds larger, slightly sparser clusters. The left clustering in Fig. 1 also shows that some of the clusters have significant overlap (several columns appear in different row-clusters).

To obtain a good visualization of the clustering we need to reorder the rows and the columns of the biadjacency matrix based on the clustering. We propose an objective function that generalizes to the overlapping setting and incentivizes orderings in which each cluster is drawn as a consecutive rectangle of 1-entries. When this is not possible, our objective function tries to maximize the squared area of the different non-consecutive parts of the clusters. We show that our objective function is NP-hard to optimize by proving that the associated decision problem: "Can all clusters be drawn as a consecutive rectangle?" is NP-complete via a reduction from MAX-Hamiltonian Path.

We also provide heuristics to optimize our objective function. To improve efficiency, we run a *pre-processing step* that decomposes the overlapping clusters into disjoint *blocks*. The blocks form a partition of the rows and columns of the matrix and each cluster corresponds to the union of a set of blocks. To obtain our visualization, it suffices to *reorder the blocks*, as an ordering of the blocks implies an ordering of the rows and columns. The complexity of our ordering problem only depends on the number of blocks, which is small in practice.

In a *post-processing step* we aim to improve the *global* visualization; this is in contrast to our objective function which only *locally* ensures that each clusters is drawn well individually. To ensure that the local structure of the visualization is not altered, we guarantee that we do not decrease the objective function.

An interesting question for future work is whether leaf-ordering methods such as *dendosort* [7] can be used to order the blocks that we create in the pre-processing. The main conceptual difference is that when drawing dendrograms, one has access to a function that measures the (dis-)similarity of a pair of clusters. However, in our setting we do not have such a similarity function for the blocks and proposing such a function seems like an interesting direction.

References

1. Behrisch, M., Bach, B., Henry Riche, N., Schreck, T., Fekete, J.D.: Matrix reordering methods for table and network visualization. In: Computer Graphics Forum, vol. 35, pp. 693–716. Wiley Online Library (2016)
2. Hartigan, J.A.: Direct clustering of a data matrix. J. Am. Stat. Assoc. **67**(337), 123–129 (1972)
3. Madeira, S., Oliveira, A.: Biclustering algorithms for biological data analysis: a survey. IEEE/ACM Trans. Comput. Biol. Bioinform. **1**(1), 24–45 (2004)
4. Miettinen, P., Mielikäinen, T., Gionis, A., Das, G., Mannila, H.: The discrete basis problem. IEEE Trans. Knowl. Data Eng. **20**(10), 1348–1362 (2008)
5. Miettinen, P., Neumann, S.: Recent developments in Boolean matrix factorization. In: IJCAI, pp. 4922–4928 (2020)
6. Neumann, S.: Bipartite stochastic block models with tiny clusters. In: NeurIPS, vol. 31, pp. 3871–3881 (2018)
7. Sakai, R., Winand, R., Verbeiren, T., Vande Moere, A., Aerts, J.: Dendsort: modular leaf ordering methods for dendrogram representations in R. F1000Research **3**, 177 (2014)
8. Vehlow, C., Beck, F., Weiskopf, D.: Visualizing group structures in graphs: a survey. In: Computer Graphics Forum, vol. 36, pp. 201–225 (2017)
9. Xu, P., Cao, N., Qu, H., Stasko, J.: Interactive visual co-cluster analysis of bipartite graphs. In: PacificVis, pp. 32–39 (2016)

The Universe Beyond Planarity

Michael A. Bekos$^{(\boxtimes)}$, Paul Göhring, Michael Kaufmann, and Axel Kuckuk

Department of Computer Science, University of Tübingen, Tübingen, Germany
{bekos,mk,kuckuk}@informatik.uni-tuebingen.de,
paul.goehring@student.uni-tuebingen.de

Abstract. In this poster, we give a pictorial depiction of graph drawing beyond planarity, which illustrates several edge density results, inclusion-relationships and incomparabilities between different graph classes.

1 Introduction

Graph drawing beyond planarity is a relatively new research direction, which deals with non-planar graphs that locally avoid specific edge-crossing configurations or guarantee specific properties for the edge crossings. Its primary motivation stems from cognitive experiments [23] showing that the absence of specific kinds of edge-crossing configurations has a positive impact on the human understanding of a graph drawing.

In this context, different restrictions on the edge crossings naturally give rise to different classes of beyond-planar graphs. Some of the most studied classes include: (i) *k-planar graphs*, in which each edge is crossed at most $k \geq 1$ times [1, 10, 25–28], (ii) *k-quasiplanar graphs*, which disallow $k \geq 1$ pairwise crossing edges [2, 3, 8], (iii) *fan-planar* graphs, in which no edge is crossed by two independent edges or by two adjacent edges from different directions [7–9, 24], and (iv) *RAC graphs*, in which edge crossings only happen at right angles [15–17].

Two notable sub-classes of 1-planar graphs are the *IC-planar graphs* [11, 30], in which crossings are *independent* (i.e., no two crossed edges share an endpoint), and the *NIC-planar graphs* [29], where crossings are *nearly independent* (i.e., no two pairs of crossed edges share two endpoints). Additional subclasses are defined by placing extra restrictions on the obtained drawings, namely, that the vertices are required to lie either on two parallel lines (*2-layer* setting; see, e.g., [8, 9, 13, 14]) or on the outer face of the drawing (*outer* setting; see, e.g., [6, 7, 12, 20, 21]). For an extensive introduction refer to [22].

From the combinatorial point of view, the main question concerns the maximum number of edges for a graph in a certain class, which is in general linear in the number of vertices. Also, the study of possible inclusion-relationships between different graph classes has received attention [4, 17]. From the complexity side, recognizing whether a graph belongs to a beyond-planar graph class has often been proven to be NP-hard [5, 7, 9, 11, 19], while polynomial-time testing algorithms can be derived for the outer and for the 2-layer setting [7, 8, 20].

© Springer Nature Switzerland AG 2021
H. C. Purchase and I. Rutter (Eds.): GD 2021, LNCS 12868, pp. 458–461, 2021.
https://doi.org/10.1007/978-3-030-92931-2

2 Contribution

Our contribution is a pictorial depiction of several results from the literature on graph drawing beyond planarity, which we call *the universe beyond planarity*. In our poster, we use planet systems as a metaphor to stress different types of information. Our universe has three suns of different colors representing general (yellow), outer (blue) and 2-layer planarity (red). Around each sun there exist planets (arranged in orbits) corresponding to different classes of graphs beyond planarity. Each orbit stands for a certain density range, such that the closer an orbit is to the sun the sparser are the corresponding graph classes; planets along the same orbit have the same edge density up to additive constants. The size of each planet additionally denotes the density of the corresponding graph class. Comets denote graph classes whose density is nearly identical to the planets next to the comets.

Directed edges denote inclusion relationships between different graph classes, while an incomparability between a pair of graph classes is denoted by a dashed edge. An additional information that is included in the poster is the fact that several classes of graphs beyond planarity have corresponding *optimal* versions, which correspond to proper subclasses achieving the maximum edge density. This information is indicated by a ring surrounding the corresponding planet. Planets whose corresponding graph classes admit an efficient recognition algorithm are shown illuminated. The poster is available at: algo.inf.uni-tuebingen.de/gd2021/posters/beyond-planarity.pdf.

References

1. Ackerman, E.: On topological graphs with at most four crossings per edge. CoRR abs/1509.01932 (2015)
2. Ackerman, E., Tardos, G.: On the maximum number of edges in quasi-planar graphs. **114**(3), 563–571 (2007). https://doi.org/10.1016/j.jcta.2006.08.002
3. Agarwal, P.K., Aronov, B., Pach, J., Pollack, R., Sharir, M.: Quasi-planar graphs have a linear number of edges. Combinatorica **17**(1), 1–9 (1997). https://doi.org/10.1007/BF01196127
4. Angelini, P., et al.: Simple k-planar graphs are simple (k+1)-quasiplanar **142**, 1–35 (2020). https://doi.org/10.1016/j.jctb.2019.08.006
5. Argyriou, E.N., Bekos, M.A., Symvonis, A.: The straight-line RAC drawing problem is NP-hard **16**(2), 569–597 (2012). https://doi.org/10.7155/jgaa.00274
6. Auer, C., et al.: Outer 1-planar graphs. Algorithmica **74**(4), 1293–1320 (2016). https://doi.org/10.1007/s00453-015-0002-1
7. Bekos, M.A., Cornelsen, S., Grilli, L., Hong, S.-H., Kaufmann, M.: On the recognition of fan-planar and maximal outer-fan-planar graphs. Algorithmica **79**(2), 401–427 (2017). https://doi.org/10.1007/s00453-016-0200-5
8. Binucci, C., et al.: Algorithms and characterizations for 2-layer fan-planarity: from caterpillar to stegosaurus **21**(1), 81–102 (2017). https://doi.org/10.7155/jgaa.00398
9. Binucci, C., et al.: Fan-planarity: properties and complexity. Theor. Comput. Sci. **589**, 76–86 (2015). https://doi.org/10.1016/j.tcs.2015.04.020

10. Bodendiek, R., Schumacher, H., Wagner, K.: Über 1-optimale graphen. Math. Nachrichten **117**(1), 323–339 (1984). https://doi.org/10.1002/mana.3211170125

11. Brandenburg, F.J., Didimo, W., Evans, W., Kindermann, P., Liotta, G., Montecchiani, F.: Recognizing and drawing IC-planar graphs. Theor. Comp. Sci. **636**, 1–16 (2016). https://doi.org/10.1016/j.tcs.2016.04.026

12. Dehkordi, H.R., Eades, P., Hong, S., Nguyen, Q.H.: Circular right-angle crossing drawings in linear time. Theor. Comput. Sci. **639**, 26–41 (2016). https://doi.org/10.1016/j.tcs.2016.05.017

13. Di Giacomo, E., Didimo, W., Eades, P., Liotta, G.: 2-layer right angle crossing drawings. Algorithmica **68**(4), 954–997 (2014). https://doi.org/10.1007/s00453-012-9706-7

14. Didimo, W., Eades, P., Liotta, G.: A characterization of complete bipartite RAC graphs. Inf. Process. Lett. **110**(16), 687–691 (2010). https://doi.org/10.1016/j.ipl.2010.05.023

15. Didimo, W., Eades, P., Liotta, G.: Drawing graphs with right angle crossings. Theor. Comput. Sci. **412**(39), 5156–5166 (2011). https://doi.org/10.1016/j.tcs.2011.05.025

16. Didimo, W., Liotta, G.: The crossing-angle resolution in graph drawing. In: Pach, J. (ed.) Thirty Essays on Geometric Graph Theory, pp. 167–184. Springer, New York (2013). https://doi.org/10.1007/978-1-4614-0110-0_10

17. Eades, P., Liotta, G.: Right angle crossing graphs and 1-planarity **161**(7–8), 961–969 (2013). https://doi.org/10.1016/j.dam.2012.11.019

18. Fox, J., Pach, J., Suk, A.: The number of edges in k-quasi-planar graphs **27**(1), 550–561 (2013). https://doi.org/10.1137/110858586

19. Grigoriev, A., Bodlaender, H.L.: Algorithms for graphs embeddable with few crossings per edge. Algorithmica **49**(1), 1–11 (2007). https://doi.org/10.1007/s00453-007-0010-x

20. Hong, S.-H., Eades, P., Katoh, N., Liotta, G., Schweitzer, P., Suzuki, Y.: A linear-time algorithm for testing outer-1-planarity. Algorithmica **72**(4), 1033–1054 (2015). https://doi.org/10.1007/s00453-014-9890-8

21. Hong, S.-H., Nagamochi, H.: Testing full outer-2-planarity in linear time. In: Mayr, E.W. (ed.) WG 2015. LNCS, vol. 9224, pp. 406–421. Springer, Heidelberg (2016). https://doi.org/10.1007/978-3-662-53174-7_29

22. Hong, S.-H., Tokuyama, T. (eds.): Beyond-Planar Graphs: Communications of NII Shonan Meetings. Springer, Singapore (2020). https://doi.org/10.1007/978-981-15-6533-5

23. Huang, W., Hong, S., Eades, P.: Effects of crossing angles. In: PacificVis 2008, pp. 41–46. IEEE (2008). https://doi.org/10.1109/PACIFICVIS.2008.4475457

24. Kaufmann, M., Ueckerdt, T.: The density of fan-planar graphs. CoRR 1403.6184 (2014)

25. Kobourov, S.G., Liotta, G., Montecchiani, F.: An annotated bibliography on 1-planarity. Comput. Sci. Rev. **25**, 49–67 (2017). https://doi.org/10.1016/j.cosrev.2017.06.002

26. Pach, J., Radoicic, R., Tardos, G., Toth, G.: Improving the crossing lemma by finding more crossings in sparse graphs. Discrete Comput. Geom. **36**(4), 527–552 (2006). https://doi.org/10.1007/s00454-006-1264-9

27. Pach, J., Tóth, G.: Graphs drawn with few crossings per edge. Combinatorica **17**(3), 427–439 (1997). https://doi.org/10.1007/BF01215922

28. Ringel, G.: Ein Sechsfarbenproblem auf der Kugel. Abh. Math. Sem. Univ. Hamb. **29**, 107–117 (1965). https://doi.org/10.1007/BF02996313

29. Zhang, X.: Drawing complete multipartite graphs on the plane with restrictions on crossings. Acta Math. Sin. Engl. Ser. **30**(12), 2045–2053 (2014). https://doi.org/10.1007/s10114-014-3763-6
30. Zhang, X., Liu, G.: The structure of plane graphs with independent crossings and its applications to coloring problems. Cent. Eur. J. Math. **11**(2), 308–321 (2013). https://doi.org/10.2478/s11533-012-0094-7

Author Index

Printed in the United States
by Baker & Taylor Publisher Services